Advances in Experimental Medicine and Biology

Proteomics, Metabolomics, Interactomics and Systems Biology

Volume 1291

Sub-Series Editor

Daniel Martins-de-Souza
University of Campinas (UNICAMP)
Institute of Biology
Laboratory of Neuroproteomics
Campinas, Brazil

Advances in Experimental Medicine and Biology provides a platform for scientific contributions in the main disciplines of the biomedicine and the life sciences. This series publishes thematic volumes on contemporary research in the areas of microbiology, immunology, neurosciences, biochemistry, biomedical engineering, genetics, physiology, and cancer research. Covering emerging topics and techniques in basic and clinical science, it brings together clinicians and researchers from various fields.

Advances in Experimental Medicine and Biology has been publishing exceptional works in the field for over 40 years, and is indexed in SCOPUS, Medline (PubMed), Journal Citation Reports/Science Edition, Science Citation Index Expanded (SciSearch, Web of Science), EMBASE, BIOSIS, Reaxys, EMBiology, the Chemical Abstracts Service (CAS), and Pathway Studio.

2019 Impact Factor: 2.450 5 Year Impact Factor: 2.324

Content of this series is reviewed.

This series of volumes focuses on concepts, techniques and recent advances in the field of proteomics, interactomics, metabolomics and systems biology. Recent advances in various 'omics' technologies enable quantitative monitoring of myriad various biological molecules in a high-throughput manner, and allow determination of their variation between different biological states on a genomic scale. Now that the sequencing of various genomes, from prokaryotes to humans, has provided the list and linear sequence of proteins and RNA that build living organisms, defining the complete set of interactions that sustain life constitutes one of the key challenges of the postgenomic era. This series is intended to cover experimental approaches for defining protein-protein, protein-RNA, protein-DNA and protein-lipid interactions; as well as theoretical approaches dealing with data analysis, integration and modeling and ethical issues.

More information about this series at http://www.springer.com/series/15040

Paul C. Guest
Editor

Studies on Biomarkers and New Targets in Aging Research in Iran

Focus on Turmeric and Curcumin

Springer

Editor
Paul C. Guest
University of Campinas
Sao Paulo, Brazil

ISSN 0065-2598 ISSN 2214-8019 (electronic)
Advances in Experimental Medicine and Biology
ISSN 2730-6216 ISSN 2730-6224 (electronic)
Proteomics, Metabolomics, Interactomics and Systems Biology
ISBN 978-3-030-56152-9 ISBN 978-3-030-56153-6 (eBook)
https://doi.org/10.1007/978-3-030-56153-6

This Springer imprint is published by the registered company Springer Nature Switzerland AG
The registered company address is: Gewerbestrasse 11, 6330 Cham, Switzerland

Preface

The Middle East is known as the cradle of civilization. It was the crossroads of ancient empires and the birthplace of major world religions. Today it is the center of many world issues due to its economic, religious, and political importance. Although it has lagged behind many other regions of the world in medicinal research, this has increased dramatically in recent years with increasing numbers of relevant publications, and the country of Iran has spearheaded this progress. Much of the research has focused on increasing our understanding of the aging process and attempting to identify biomarkers and natural products to improve the human health span. This book provides a comprehensive overview of the research conducted in the Middle East on the health benefits of curcumin, a phytochemical derived from the famous spice turmeric. Hundreds of studies have now been published describing the health benefits of this spice. The importance of this research is exemplified by poor data regarding health and longevity as only 0.08% of the population in Iran consists of individuals over 90 years of age. This is approximately 10 times lower than the percentage of this same age group in the United Kingdom and the United States of America and almost 20 times lower than that in Japan.

This book presents a series of reviews and meta-studies describing research which has resulted in identification of potential new biomarkers and drug targets for age-related disorders. All of the studies have focused on the testing of curcumin and related products, which have already shown some promising leads in age-related conditions such as heart disease, diabetes, cognitive impairment, and cancer. The authors in this series come from different centers and cities of Iran, including Mashhad, Tehran, Isfahan, Ahvaz, Birjand, Quchan, and Yazd, and many of the chapters feature collaborations with other countries of the Middle East and throughout the world, including Brazil, Italy, Mexico, Oman, Poland, Switzerland, the United Kingdom, and the United States of America. This underscores the emergence of the Middle East into this arena of research.

The book will be of high interest to scientific and clinical researchers in the subject of aging and age-related disease, and to physicians and pharmaceutical company scientists since it gives insights into the latest strategies, biomarkers, and targets involved in the mechanism of action of curcumin to promote healthy aging. It will also provide important information on disease mechanisms related to age-related disorders as each chapter will be presented in the context of specific chronic diseases.

Sao Paulo, Brazil — Paul C. Guest

Contents

Research in the Middle East into the Health Benefits of Curcumin

1

Paul C. Guest and Amirhosein Sahebkar

Abstract

The Middle East is known as the cradle of civilization. It was the crossroads of ancient empires and the birthplace of major world religions. Today, it is the center of many world issues due to economic, religious, and political reasons. Although it has lagged behind many other regions of the world in medicinal research, this has increased dramatically in recent years with increasing numbers of relevant publications. Much of this research has focused on increasing our understanding of the aging process and attempting to identify biomarkers and natural products to improve the human health span. This review provides a brief overview of the research conducted in the Middle East on the health benefits of curcumin, a phytochemical derived from the famous spice turmeric. These efforts have been mainly spearheaded by Iran.

Keywords

Middle East · Iran · Turmeric · Curcumin · Research · Clinical trial · New drug targets

P. C. Guest
Laboratory of Neuroproteomics, Department of Biochemistry and Tissue Biology, Institute of Biology, University of Campinas (UNICAMP), Campinas, Brazil

A. Sahebkar (✉)
Biotechnology Research Center, Pharmaceutical Technology Institute, Mashhad University of Medical Sciences, Mashhad, Iran

Neurogenic Inflammation Research Center, Mashhad University of Medical Sciences, Mashhad, Iran

Halal Research Center of IRI, FDA, Tehran, Iran

Polish Mother's Memorial Hospital Research Institute (PMMHRI), Lodz, Poland
e-mail: sahebkara@mums.ac.ir

1.1 Introduction

The Middle East is known as the birthplace of civilizations and has always served as crossroads of world cultures and religions. The official countries recognized by the United Nations as being part of the Middle East are Bahrain, Cyprus, Egypt, Iran, Iraq, Israel, Jordan, Kuwait, Lebanon, Oman, Qatar, Saudi Arabia, Syria, Turkey, United Arab Emirates (UAE), and Yemen, although this can vary depending on the source and can include other countries from Africa and Asia (Fig. 1.1) (Table 1.1). The largest of the countries listed above in terms of area is Saudi Arabia at 2,149,690 km^2 and the smallest is Bahrain at 780 km^2. In terms of population, Egypt has the largest number at approximately 82,798,000 people and Cyprus has the smallest with 1,088,503 people [1].

P. C. Guest (ed.), *Studies on Biomarkers and New Targets in Aging Research in Iran*, Advances in Experimental Medicine and Biology 1291, https://doi.org/10.1007/978-3-030-56153-6_1

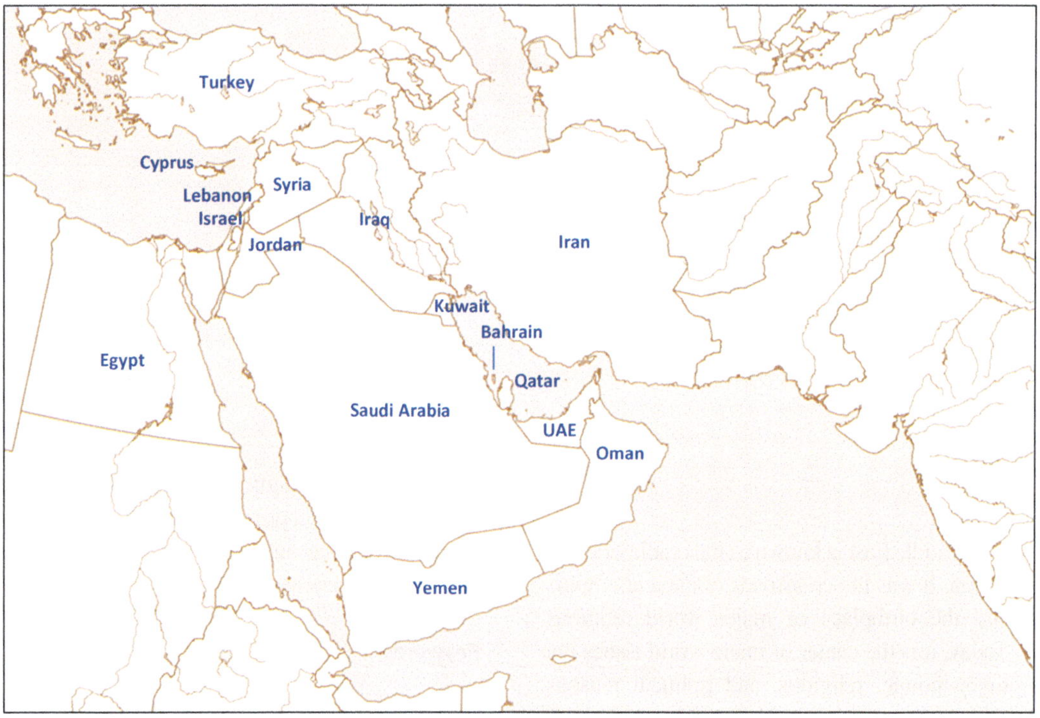

Fig. 1.1 The Middle East. UAE: United Arab Emirates

Table 1.1 Middle East countries in terms of area and population

Country	Area (km^2)	Population
Bahrain	780	1,701,582
Cyprus	9250	1,207,360
Egypt	1,010,407	102,334,403
Iran	1,648,195	83,992,953
Iraq	438,317	40,222,503
Israel	20,770	8,655,540
Jordan	92,300	10,203,139
Kuwait	17,820	4,270,563
Lebanon	10,452	6,825,441
Oman	212,460	5,106,622
Qatar	11,437	2,881,060
Saudi Arabia	2,149,690	34,813,867
Syria	185,180	17,500,657
Turkey	783,562	84,339,067
United Arab Emirates	82,880	9,890,400
Yemen	527,970	29,825,968

Consistent with the rest of the world, the proportion of aged people (≥65 years-old) has increased dramatically in recent years in most of these countries [1, 2]. Table 1.2 shows the population over 65 years old in each country in the Middle East in terms of total number and percentage relative to the total population of that country. This shows that Turkey has the greatest number of individuals over the age of 65 years at 7,574,551 and Qatar has the lowest at 48,663 per-

Table 1.2 Middle East countries in terms of population over 65 years old and change since the year 2000

Country	Population ≥ 65 years old	% Population ≥ 65 years old	2020:2000 ≥ 65 years old ratio
Bahrain	874,118	2.93	1.08
Cyprus	173,965	14.41	1.41
Egypt	5,456,144	5.33	1.08
Iran	5,513,595	6.56	1.50
Iraq	1,384,755	3.44	0.99
Israel	1,074,531	12.41	1.24
Jordan	403,405	4.00	1.29
Kuwait	129,642	3.03	1.92
Lebanon	515,256	7.54	1.32
Oman	128,229	2.51	1.03
Qatar	48,663	1.69	0.97
Saudi Arabia	1,217,949	3.50	1.17
Syria	853,060	4.87	1.46
Turkey	7,574,551	9.00	1.47
United Arab Emirates	125,051	1.26	1.16
Yemen	767,679	2.57	1.05

sons. However, Cyprus has the highest proportion of individuals in the "over 65 age" group at 14.41% of the total population and the UAE has the lowest at 1.26%. In addition, Kuwait has seen the largest increase in the proportion of the "over 65 age" group since the year 2000 at 1.92-fold (2020 compared with 2000), followed by Iran, Turkey, and Syria at 1.50-, 1.47-, and 1.46-fold, respectively (Table 1.2).

In line with this aging population trend, medical expenses are expected to rise due to the increased prevalence of age-related diseases and associated disabilities [3]. Thus, there are increased efforts by researchers and clinicians to come up with widely generalizable preventive and therapeutic approaches to reduce the burden of diseases such as type 2 diabetes mellitus, cardiovascular disorders, Alzheimer's disease, frailty, and cancer [4]. Complementary and alternative medicine has a long history of use in the Middle East countries, and there are rich anecdotal and traditional medicine reports on the widespread use and efficacy of herbal medicines for a variety of diseases in this region. During the recent decades, overwhelming evidence from well-designed pharmacological and clinical studies has corroborated the traditional uses of medicinal plants and their active ingredients (phytochemicals) in different pathologies [5].

Among the vast numbers of herbal medicines, turmeric (*Curcuma longa* L.) has received a considerable reputation in recent years given its well-known pharmacological and healing properties [5]. It has been used in traditional medicine in parts of Asia and the Middle East for more than 4000 years. There are reports of its use in the Vedic culture in India dating back to 2500 BC. Also, it appeared as part of Ayurvedic medicine around 500 BC and this is still practiced today. Inhalation of burning turmeric fumes has been claimed to relieve congestion, and the topical application of turmeric juice or paste was thought to help in the wound healing process. Just two centuries ago, Vogel and Pelletier reported the isolation of a "yellow-colored matter" from the rhizomes of turmeric which they called curcumin [6]. In 1910, Milobedzka and Lampe identified the chemical structure of curcumin as 1,6-heptadiene-3,5-dione-1,7-bis (4-hydroxy-3-methoxyphenyl)-(1E, 6E) (Fig. 1.2) [7]. It was not until 1949 that curcumin first appeared in the scientific literature in a significant way through the finding that it had bacteriostatic properties [8].

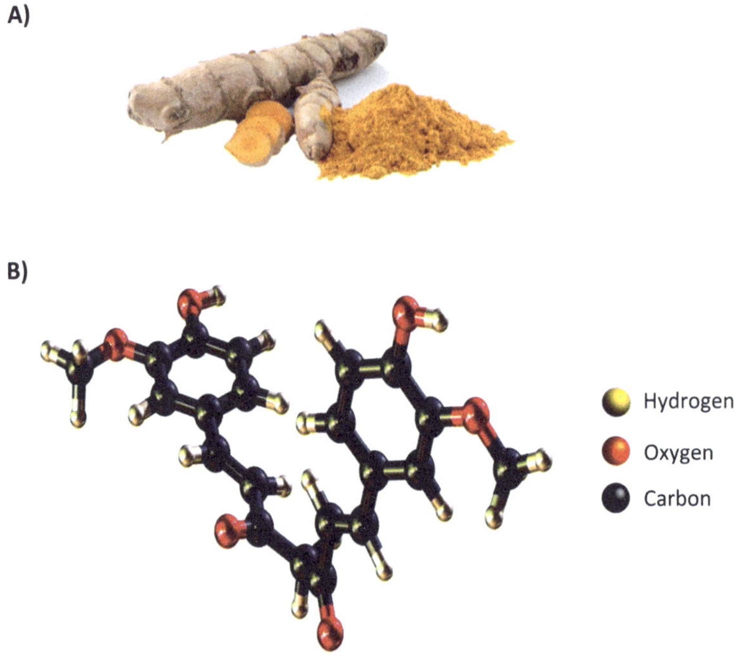

Fig. 1.2 (**a**) Turmeric rhizomes and powder. (**b**) Structure of curcumin

This review describes some of the emerging potential medicinal uses for curcumin and the role that the Middle East has played in these studies. In addition, it will discuss the current health issues in these countries as well as the research taking place there into the potential therapeutic uses of curcumin.

1.2 Current Uses

Turmeric is thought to have originated in India, the largest producer and consumer of this spice in the world (Fig. 1.3a). Almost 80% of the world's turmeric is produced in India, with the remaining 20% produced in China, Myanmar, Bangladesh, Nigeria, and others [9]. Economically, it is important due to its many uses and it is exported throughout the world. The turmeric plant is cultivated and propagated via its rootstalk in hot and humid climates at low to medium altitudes (up to 1200 m). The plant is harvested by digging up the rhizomes when the leaves start to turn yellow (approximately 8 months after plantation). The top importers of this spice are the United States of America (USA) (29%), followed by Iran (15%) and the UAE (10%), with Malaysia, Morocco, the United Kingdom (UK), Sri Lanka, Germany, Japan, and Saudi Arabia accounting for most of the remaining 46% of the imports (Fig. 1.3b) [10].

Turmeric is perhaps best known as a major ingredient of curries and is also known for its use as a dye, and for applications in the cosmetic and pharmaceutical industries [11]. Turmeric has attracted considerable interest over recent decades as it contains the curcuminoids, curcumin, demethoxycurcumin, and bisdemethoxycurcumin, which have bioactivity. Consumption of curcumin appears to be safe and confer beneficial effects against multiple chronic diseases [12, 13], including obesity, metabolic syndrome, diabetes, various cancers, fatty liver disease, major depression, Alzheimer's disease, and arthritis. Thus far, no precise conclusions can be made and thus additional large-scale, well-designed clinical studies are required which also take into account potential confounding factors specific for each disease/medical area. A major consideration has concerned the low aqueous solubility and oral bioavailability of curcumin [14]. With this in mind, several approaches have now been developed to increase curcumin's bioavailability, including the use of curcumin analogues [15], hyaluronic acid composite microneedles containing curcumin-

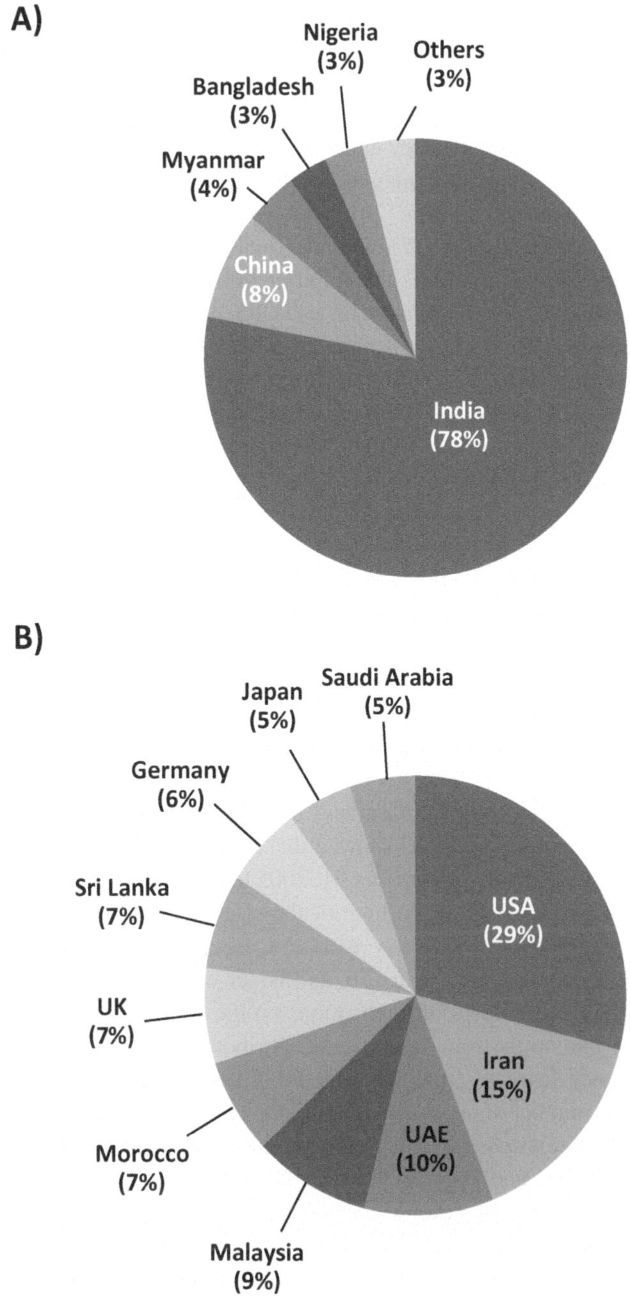

Fig. 1.3 Top (**a**) producers and (**b**) importers of turmeric in the world

loaded micelles [16], different types of bio-nanocarriers [17], and encapsulation in liposome nanocarriers [18].

A number of studies have now been conducted which aimed to determine the molecular mechanism of curcumin function in health and disease, as a means of finding molecular targets as well as the pharmacological profile and interactions of this phytochemical [19]. In tumor cells, curcumin regulates the cell cycle, reactive oxygen species (ROS) generation, cytosolic Ca^{2+} levels, nuclear factor kappa B (NF-κB) signaling, and cytokine responses. In turn, this triggers endoplasmic reticulum (ER) and mitochondrial stress, leading to reduction in cancer cell propagation or increased apoptosis and/or death. In addition, curcumin can cross the blood brain barrier, enabling the protection of neurons from oxidative stress and inflammatory damage. It has also been found that curcumin may alleviate diseases such as rheumatoid arthritis [20], cardiac hypertrophy [21], breast cancer [22], and non-small cell lung cancer (NSCLC) [23], via inhibition of the mammalian target of rapamycin (mTOR) pathway. The mTOR pathway was first described as a regulator of G1- to S-phase transition in eukaryotic cells [24] and is now well known as a promising target in cancer in multiple disease areas as well as in the aging process [25–28].

1.3 The Growth in Curcumin Research in the Middle East

A PubMed search using the terms "curcumin" and "Bahrain" or "Cyprus" or "Egypt" or "Iran" or "Iraq" or "Israel" or "Jordan" or "Kuwait" or "Lebanon" or "Oman" or "Qatar" or "Saudi Arabia" or "Syria" or "Turkey" or "United Arab Emirates" or "Yemen" showed that Iran had the most publications since Jan 01, 2005 to the present day (Feb 24 at the time of the search) with 795, followed by Turkey with 238, Egypt with 233, Saudi Arabia with 156, and Israel with 98. The large number of publications on the subject of curcumin from Iran can be attributed to a steady increase from the year 2011 (Fig. 1.4). The number of publications by Iran over this time is only exceeded by China, The United States of America, and India at 3263, 1989, and 1941, respectively. However, if this value is normalized according to population size, the publication rates in this field are highest for Iran (9.47 per million people) and lowest for India (1.41 per million people) (Table 1.3).

In Iran alone, the last 10 research articles have shown positive effects of curcumin in disease areas such as rheumatoid arthritis [29], wound healing [30], ulcerative colitis [31], non-alcoholic fatty liver disease [32–34], diabetes cell therapy [35], osteoarthritis [36], endodontic infection [37], and insulin resistance [38]. In addition, recent research has attempted to address the low bioavailability issue through the development of different curcumin formulations or delivery systems [39–48]. These have been tested mainly in disease or medical areas such as cancer [39, 43, 45–48], wound healing [40], malaria [41], drug delivery [42], and polycystic ovary syndrome [44]. Further research at both the preclinical and clinical levels is required in order to determine the most efficacious delivery systems for each specific disease area.

1.4 The Health Situation in the Middle East

As with most of the rest of the world, there should be a focus on improved nutrition due to the rising obesity epidemic in the Middle East. Strikingly, 8 countries from the Middle East (Kuwait, Jordan, Saudi Arabia, Qatar, Turkey, Lebanon, Egypt, and the UAE) have been ranked as having the highest adult obesity prevalence out of the top 20 most obese countries in the world (the top 10 positions are occupied by Pacific Island nations or territories: Nauru; Cook Islands; Palau; Marshall Islands; Tuvalu; Niue; Tonga; Samoa; Kiribati; and Micronesia) [49]. Therefore, research into natural dietary solutions such as supplementation with curcumoids is essential in the Middle East. In the sections below, the health needs of each of country are presented in terms the highest causes of death and disability combined, as well as the highest risk factors which may have contributed to these. All of the information is based on 2017 statistics from the Institute for Health Metrics and Evaluation (IHME), an independent global health research center at the University of Washington, Seattle, WA, USA [50]. The data regarding body mass index (BMI) was obtained from IndexMundi [51] and were all recorded in 2016.

1.4.1 Bahrain

In Bahrain, the highest cause of death and disability was diabetes, followed by ischemic heart disease, low back pain, headache disorders, drug use disorders, and depressive disorders. Not sur-

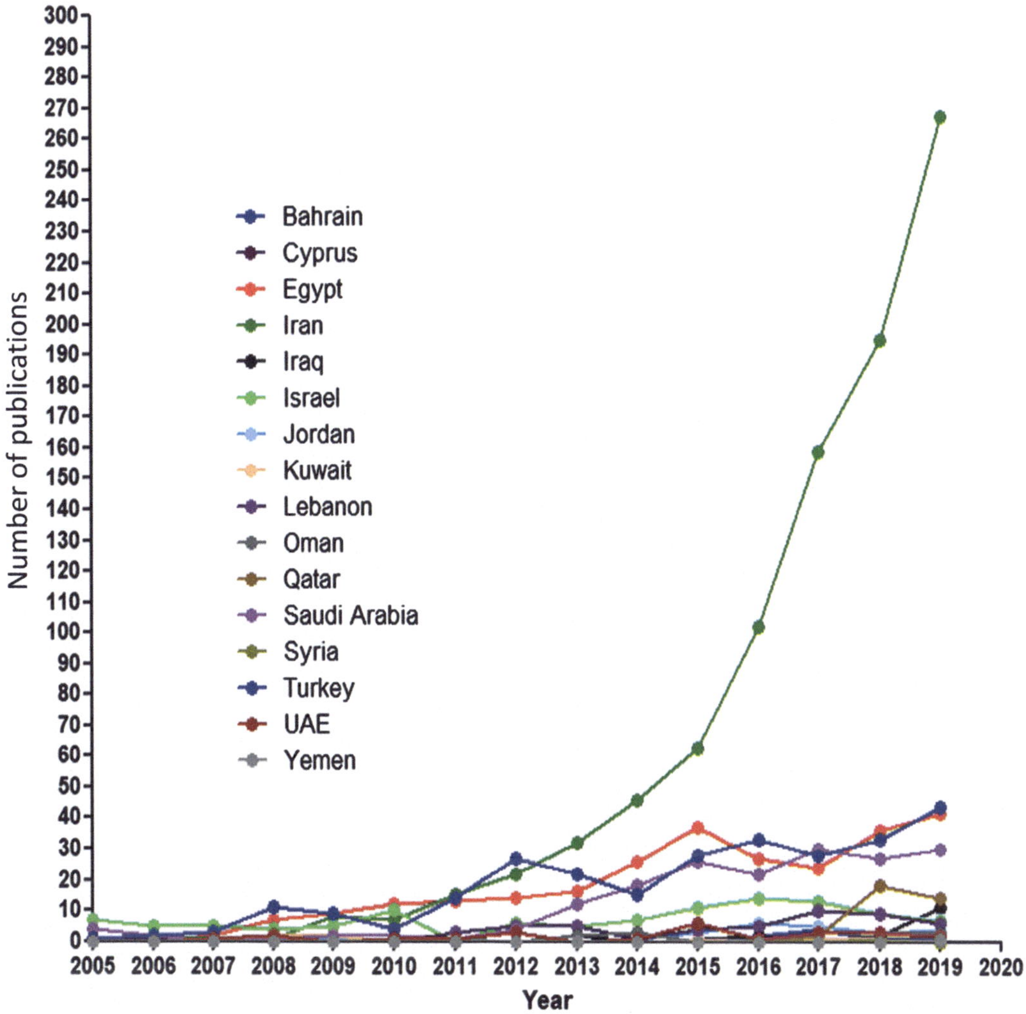

Fig. 1.4 Number of publications from Jan 1, 2005 to Dec 31, 2019 indexed on PubMed per country in the Middle East

Table 1.3 Number of publications regarding curcumin research per one million people from Jan 01, 2005 to Feb 24, 2020

Country	Total number of publications	Publications/million people
China	3263	2.27
India	1941	1.41
Iran	795	9.47
USA	1989	6.01

prisingly, the main risk factors driving this were high BMI, high fasting plasma glucose, dietary risks, high blood pressure (BP), and tobacco use. The data regarding BMI as having the highest risk association with the causes of death and disability is consistent with the obesity rate in 2016 amounting to 29.8% of the adult population.

1.4.2 Cyprus

In Cyprus, ischemic heart disease was the number one cause of death and disability combined, followed by low back pain, diabetes, headache disorders, stroke, and lung cancer. The main risk factors which drove these were tobacco use, high fasting plasma glucose, dietary risks, high BP, and BMI. In 2016 the adult obesity prevalence was 21.8%, consistent with the finding that high BMI was not the highest driver of deaths and disability. The statistic showing that tobacco use was the highest risk factor is likely to be linked to the fact that Cyprus has one of the highest tobacco consumption rates in Europe [52].

1.4.3 Egypt

Ischemic heart disease was the top-ranked cause of death and disability followed by road injuries, lower respiratory infection, stroke, neonatal disorders, and cirrhosis. This was linked with the top risk factor of deleterious diets, followed by high BP, high BMI, high fasting plasma glucose, and tobacco use. The main risks of poor diet, high BP and high BMI, are consistent with the observation that the adult obesity prevalence was 32% in 2016. This translated to Egypt being the 19th most obese nation in the world. To account for death and disability due to road accidents, Egypt reports approximately 0.42 deaths per 1000 people due to traffic accidents, and pedestrians account for approximately 20% of these fatalities [53].

1.4.4 Iran

The leading causes of death and disability in Iran were ischemic heart disease, neonatal disorders, road injuries, headache disorders, low back pain, and drug use disorders. In terms of risk factors, these were driven mostly by high BP, dietary risks, high fasting plasma glucose, high BMI, and tobacco use. The adult obesity prevalence in Iran was 25.8%. These factors are linked to the rising epidemic in cardiovascular diseases and may be related to socioeconomic and nutritional changes, combined with inadequate physical activity levels [54]. Thus, nutrition and exercise-based strategies to promote health and control these risk factors are essential to reduce the burden of this disease.

1.4.5 Iraq

The leading causes of death in Iraq were markedly different from the other countries. In descending order these were conflict and terror, neonatal disorders, congenital defects, ischemic heart disease, stroke, and headache disorders. These were linked to malnutrition, dietary risks, high BMI, BP, and fasting plasma glucose. The adult obesity prevalence was 30.4% which accounts for the metabolic-related risk factors. The cause of death or disability due to conflict and terror is most likely an outcome of the various wars that have plagued the region since 2003 [55].

1.4.6 Israel

In Israel, the main cause of death and disability was low back pain and headache disorders, followed by ischemic heart disease, diabetes, depressive disorders, and blindness and vision impairment. These appeared to be linked to the same causative factors as for most other countries in the Middle East with high fasting plasma glucose contributing the highest risk, followed by tobacco use, high BMI, high BP, and dietary risks. The adult obesity prevalence rate was 26.1% which was marginally lower than other Middle East countries apart from Iran and Yemen (see below). This may account for the fact that metabolic disorders were not the leading cause of death or disability.

1.4.7 Jordan

The main cause of death and disability in Jordan was neonatal disorders, followed by congenital defects, ischemic heart disease, headache disorders, low back pain, and diabetes. The main risk factors associated with these were malnutrition, high BMI, high fasting plasma glucose, dietary risks, and high BP. The obesity prevalence rate was relatively high in Jordan at 35.5%, leading to its ranking as the 13th most obese nation in the world.

These findings are most likely due to a shift from undernutrition to one of overnutrition which has led to premature morbidities and mortality in the population [56]. More recent studies have a prevalence of short stature, underweight, overweight, and obesity among Jordanian school children [57].

1.4.8 Kuwait

Ischemic heart disease was the highest cause of death and disability in Kuwait, and this was followed by low back pain, headache disorders, drug use disorders, neonatal disorders, and road injuries. In line with this, the highest risk factors were high BMI, dietary risks, tobacco use, high fasting plasma glucose, and BP. The finding of ischemic heart disease as the highest cause of disability and high BMI as the greatest risk factor is not surprising given that Kuwait has an adult obesity prevalence rate of 37.9%, which makes it the highest of the Middle East countries listed here (ranked 11th in the world).

1.4.9 Lebanon

The condition which caused the most death and disability in Lebanon was ischemic heart disease, and this was followed by drug use disorders, diabetes, neonatal disorders, headache disorders, and low back pain. The risk factors which drove this were high BMI, tobacco use, high fasting plasma glucose, dietary risks, and high BP. The obesity prevalence rate was high at 32% making it the 18th most obese nation out of the world ranking.

1.4.10 Oman

The highest cause of death and disability in Oman was road injuries. This was followed by ischemic heart disease, neonatal disorders, drug use disorders, headache disorders, and low back pain. However, the greatest risk factors associated with these causes of death and disability were mostly nutrition-based with high BMI at the top of the list, followed by dietary risks, high BP, high fasting plasma glucose, and malnutrition. The adult obesity prevalence rate was lower than most of the Middle East countries at 27%. The fact that road traffic injury was the highest cause of death and disability is consistent with a report from 2013 which showed that the incidence of road crashes in Middle East countries such as Oman has risen dramatically over recent years and is now a leading cause of death for younger adults [58].

1.4.11 Qatar

In Qatar, the greatest causes of death and disability were different than those seen in most other countries. The highest cause was drug use disorders, followed by road injuries, low back pain, headache disorders, diabetes, and depressive disorders. In line with this, the risk factors were also slightly different with high BMI having the highest risk, followed by drug use, high fasting plasma glucose, dietary risks, and occupational risks. The statistic on drug use disorders is most likely linked to the unfortunately high rate of drug and alcohol addiction in this country [59]. The finding that BMI was the highest risk factor is consistent with an adult obesity prevalence rate of 35.1% (equivalent to 15th position in the world obesity prevalence ranking).

1.4.12 Saudi Arabia

The highest causes of death and disability combined in Saudi Arabia were road injuries, ischemic heart disease, headache disorders, low back pain, drug use disorders, and neonatal disorders. Again, these were slightly different from most other Middle East countries and the traffic accident statistic is in line with reports that a car accident happens every minute on average in Saudi Arabia [60]. The greatest associated risks appeared to be linked with diet with high BMI, dietary risks, and high fasting plasma glucose, BP, and low density lipoprotein (LDL) levels. This was correlated with a relatively high adult obesity prevalence rate of 35.4%, which made Saudi Arabia the 14th most obese nation in the world.

1.4.13 Syria

The main causes of death and disability in Syria were similar to those in Iraq. The highest cause was conflict and terror, followed by ischemic heart disease, stroke, low back pain, headache disorders, and neonatal disorders. This is likely to be linked to the ongoing civil war in the country since 2011 [61]. As with most other countries, however, the main risk factors were related to nutrition with dietary risks occupying the top position, followed by high BP, BMI, LDL, and fasting plasma glucose. The obesity prevalence was 27.8% of adults.

1.4.14 Turkey

The highest cause of death and disability combined in turkey was ischemic heart disease, low back pain, neonatal disorders, chronic obstructive pulmonary disease (COPD), stroke, and diabetes. The most causative risk factors were unique to Turkey with tobacco use topping the list, followed by high BMI, high fasting plasma glucose, dietary risks, and air pollution. Although several measures against smoking have been taken over the last decade, smoking still remains a significant burden on health and the economy in Turkey [62]. The finding that the second highest cause of death and disability was high BMI is consistent with an adult obesity rate of 32.1% (17th in the world ranking).

1.4.15 UAE

The highest causes of death in the UAE appeared similar to that seen in Qatar with drug use disorders being the greatest cause, followed by road injuries, ischemic heart disease, low back pain, diabetes, and headache disorders. The main risk factors which drove these causes of death were also similar to those found for Qatar with high BMI, drug use, dietary risks, high fasting plasma glucose, and BP. The obesity rate for adults was 31.7%, which made Qatar the 20th most obese nation.

1.4.16 Yemen

The highest causes of death in Yemen were also slightly different from the other Middle East countries with neonatal disorders, conflict and terror, ischemic heart disease, diarrheal disease, dietary iron deficiency, and congenital defects. These were also driven by somewhat unique risk factors of malnutrition, dietary risks, water, sanitation, and hygiene, high BP, and tobacco use. The ongoing conflict in Yemen has caused the health system to disintegrate, which has left children in particular more vulnerable to malnutrition and disease [63]. Thus, this country has the lowest adult obesity prevalence rate in the Middle East at 17.1%.

1.5 Conclusions and Future Prospects

Recent clinical and experimental studies have shown that the ancient spice turmeric has many beneficial medicinal properties which are thought to be mediated, at least in part, by bioactive curcuminoids. The main properties reported include anti-inflammatory, antioxidant, anti-tumor, anti-aging, and antidiabetic, as well as cardioprotective, hepatoprotective, and neuroprotective activities. The beneficial effects are usually achieved through consumption as part of the diet or as a supplement with other medicines. Currently, one of the most pressing needs is to increase the bioavailability of curcumin to see if the pharmacological effects can be augmented. Thus far, considerable progress has been made toward this objective such as through the development of nano-composite curcumin formulations. Further research on the potential benefits of curcumin is essential considering the alarming rise in age-related diseases such as obesity in the Middle East and the rest of world. Finally, although compelling evidence from the clinical studies has recently emerged to support the efficacy and safety of curcumin supplementation, large-scale event-driven randomized controlled trials are yet to be conducted in order to verify the role of curcumin in the armamentarium of drugs for chronic age-related diseases.

References

1. https://www.populationpyramid.net/world/2020/
2. Hajjar RR, Atli T, Al-Mandhari Z, Oudrhiri M, Balducci L, Silbermann M (2013) Prevalence of aging population in the Middle East and its implications on cancer incidence and care. Ann Oncol 24(Suppl 7):vii11–vii24. https://doi.org/10.1093/annonc/mdt268
3. Cao B, Bray F, Ilbawi A, Soerjomataram I (2018) Effect on longevity of one-third reduction in premature mortality from non-communicable diseases by 2030: a global analysis of the Sustainable Development Goal health target. Lancet Glob Health 6(12):e1288–e1296. https://doi.org/10.1016/S2214-109X(18)30411-X
4. Ergin V, Bali EB, Hariry RE, Karasu C (2013) Natural products and the aging process. Horm Mol Biol Clin Investig 16(2):55–64
5. Prasad S, Aggarwal BB (2011) Turmeric, the golden spice: from traditional medicine to modern medicine. In: Benzie IFF, Wachtel-Galor S (eds) Herbal medicine: biomolecular and herbal medicine: biomolecular and clinical aspects, 2nd edn. CRC Press/Taylor & Francis, Boca Raton, FL, pp 263–288
6. Vogel H, Pelletier J (1815) Curcumin – biological and medicinal properties. J de Pharmacie I:289
7. Milobedzka J, Kostanecki S, Lampe V (1910) Zur Kenntnis des curcumins. Ber Deut Chem Ges 43:2163–2170
8. Schraufstatter E, Bernt H (1949) Antibacterial action of curcumin and related compounds. Nature 164(4167):456. https://doi.org/10.1038/164456a0
9. http://www.karvycommodities.com/downloads/karvySpecialReports/karvysSpecialReports_20080820_01.pdf
10. https://www.quora.com/Which-is-the-major-country-for-exporting-turmeric
11. Kunnumakkara AB, Bordoloi D, Padmavathi G, Monisha J, Kishor Roy N, Prasad S et al (2017) Curcumin, the golden nutraceutical: multitargeting for multiple chronic diseases. Br J Pharmacol 174(11):1325–1348
12. Mantzorou M, Pavlidou E, Vasios G, Tsagalioti E, Giaginis C (2018) Effects of curcumin consumption on human chronic diseases: a narrative review of the most recent clinical data. Phytother Res 32(6):957–975
13. Soleimani V, Sahebkar A, Hosseinzadeh H (2018) Turmeric (*Curcuma longa*) and its major constituent (curcumin) as nontoxic and safe substances: Review. Phytother Res 32(6):985–995. https://doi.org/10.1002/ptr.6054. Epub 2018 Feb 26. PMID: 29480523
14. Dei Cas M, Ghidoni R (2019) Dietary curcumin: correlation between bioavailability and health potential. Nutrients 11(9):E2147. https://doi.org/10.3390/nu11092147
15. Zhao S, Pi C, Ye Y, Zhao L, Wei Y (2019) Recent advances of analogues of curcumin for treatment of cancer. Eur J Med Chem 180:524–535
16. Cheng Z, Lin H, Wang Z, Yang X, Zhang M, Liu X et al (2020) Preparation and characterization of dissolving hyaluronic acid composite microneedles loaded micelles for delivery of curcumin. Drug Deliv Transl Res. https://doi.org/10.1007/s13346-020-00735-2
17. Moballegh Nasery M, Abadi B, Poormoghadam D, Zarrabi A, Keyhanvar P, Khanbabaei H et al (2020) Curcumin delivery mediated by bio-based nanoparticles: a review. Molecules 25(3):E689. https://doi.org/10.3390/molecules25030689
18. Swami Vetha BS, Oh PS, Kim SH, Jeong HJ (2020) Curcuminoids encapsulated liposome nanoparticles as a blue light emitting diode induced photodynamic therapeutic system for cancer treatment. J Photochem Photobiol B 205:111840. https://doi.org/10.1016/j.jphotobiol.2020.111840
19. Liczbiński P, Michałowicz J, Bukowska B (2020) Molecular mechanism of curcumin action in signaling pathways: review of the latest research. Phytother Res. https://doi.org/10.1002/ptr.6663
20. Dai Q, Zhou D, Xu L, Song X (2018) Curcumin alleviates rheumatoid arthritis-induced inflammation and synovial hyperplasia by targeting mTOR pathway in rats. Drug Des Devel Ther 12:4095–4105
21. Liu R, Zhang HB, Yang J, Wang JR, Liu JX, Li CL (2018) Curcumin alleviates isoproterenol-induced cardiac hypertrophy and fibrosis through inhibition of autophagy and activation of mTOR. Eur Rev Med Pharmacol Sci 22(21):7500–7508
22. Hu S, Xu Y, Meng L, Huang L, Sun H (2018) Curcumin inhibits proliferation and promotes apoptosis of breast cancer cells. Exp Ther Med 16(2):1266–1272
23. Wang N, Feng T, Liu X, Liu Q (2020) Curcumin inhibits migration and invasion of non-small cell lung cancer cells through up-regulation of miR-206 and suppression of PI3K/AKT/mTOR signaling pathway. Acta Pharma 70(3):399–409
24. Wiederrecht GJ, Sabers CJ, Brunn GJ, Martin MM, Dumont FJ, Abraham RT (1995) Mechanism of action of rapamycin: new insights into the regulation of G1-phase progression in eukaryotic cells. Prog Cell Cycle Res 1:53–71
25. Rosner M, Hanneder M, Siegel N, Valli A, Fuchs C, Hengstschläger M (2008) The mTOR pathway and its role in human genetic diseases. Mutat Res 659(3):284–292

26. O' Neill C (2013) PI3-kinase/Akt/mTOR signaling: impaired on/off switches in aging, cognitive decline and Alzheimer's disease. Exp Gerontol 48(7):647–653
27. Maiese K (2017) Moving to the rhythm with clock (circadian) genes, autophagy, mTOR, and SIRT1 in degenerative disease and cancer. Curr Neurovasc Res 14(3):299–304
28. Papadopoli D, Boulay K, Kazak L, Pollak M, Mallette F, Topisirovic I et al (2019) mTOR as a central regulator of lifespan and aging. F1000Res 8:F1000 Faculty Rev-998. https://doi.org/10.12688/f1000research.17196.1
29. Javadi M, Khadem Haghighian H, Goodarzy S, Abbasi M, Nassiri-Asl M (2019) Effect of curcumin nanomicelle on the clinical symptoms of patients with rheumatoid arthritis: a randomized, double-blind, controlled trial. Int J Rheum Dis 22(10):1857–1862
30. Mohammadi Z, Sharif Zak M, Majdi H, Mostafavi E, Barati M, Lotfimehr H et al (2019) The effect of chrysin-curcumin-loaded nanofibres on the wound-healing process in male rats. Artif Cells Nanomed Biotechnol 47(1):1642–1652
31. Sadeghi N, Mansoori A, Shayesteh A, Hashemi SJ (2019) The effect of curcumin supplementation on clinical outcomes and inflammatory markers in patients with ulcerative colitis. Phytother Res. https://doi.org/10.1002/ptr.6581
32. Mirhafez SR, Rezai A, Dehabeh M, Nobakht M Gh BF, Bidkhori M, Sahebkar A et al (2019) Efficacy of phytosomal curcumin among patients with non-alcoholic fatty liver disease. Int J Vitam Nutr Res; Dec:10:1–10:9
33. Saberi-Karimian M, Keshvari M, Ghayour-Mobarhan M, Salehizadeh L, Rahmani S, Behnam B et al (2020) Effects of curcuminoids on inflammatory status in patients with non-alcoholic fatty liver disease: A randomized controlled trial. Complement Ther Med 49:102322. https://doi.org/10.1016/j.ctim.2020.102322
34. Cicero AFG, Sahebkar A, Fogacci F, Bove M, Giovannini M, Borghi C (2020) Effects of phytosomal curcumin on anthropometric parameters, insulin resistance, cortisolemia and non-alcoholic fatty liver disease indices: a double-blind, placebo-controlled clinical trial. Eur J Nutr 59(2):477–483
35. Javidi MA, Kaeidi A, Mortazavi Farsani SS, Babashah S, Sadeghizadeh M (2019) Investigating curcumin potential for diabetes cell therapy, in vitro and in vivo study. Life Sci 239:116908. https://doi.org/10.1016/j.lfs.2019.116908
36. Hashemzadeh K, Davoudian N, Jaafari MR, Mirfeizi Z (2019) The effect of nanocurcumin on the improvement symptoms of knee osteoarthritis: a randomized clinical trial. Curr Rheumatol Rev. https://doi.org/10.2174/1874471013666191223152658
37. Ghorbanzadeh R, Assadian H, Chiniforush N, Parker S, Pourakbari B, Ehsani B et al (2019) Modulation of virulence in Enterococcus faecalis cells surviving antimicrobial photodynamic inactivation with reduced graphene oxide-curcumin; an ex vivo biofilm model. Photodiagnosis Photodyn Ther 30:101643. https://doi.org/10.1016/j.pdpdt.2019.101643
38. Dolati S, Namiranian K, Amerian R, Mansouri S, Arshadi S, Azarbayjani MA (2020) The effect of curcumin supplementation and aerobic training on anthropometric indices serum lipid profiles, C-reactive protein and insulin resistance in overweight women: a randomized, doubleblind, placebo-controlled trial. J Obes Metab Syndr. https://doi.org/10.7570/jomes19055
39. Karimi Alavijeh R, Akhbari K (2020) Biocompatible MIL-101(Fe) as a smart carrier with high loading potential and sustained release of curcumin. Inorg Chem. https://doi.org/10.1021/acs.inorgchem.9b02756
40. Esmaeili E, Eslami-Arshaghi T, Hosseinzadeh S, Elahirad E, Jamalpoor Z, Hatamie S et al (2020) The biomedical potential of cellulose acetate/polyurethane nanofibrous mats containing reduced graphene oxide/silver nanocomposites and curcumin: Antimicrobial performance and cutaneous wound healing. Int J Biol Macromol 152:418–427
41. Elmi T, Ardestani MS, Hajialiani F, Motevalian M, Mohamadi M, Sadeghi S et al (2020) Novel chloroquine loaded curcumin based anionic linear globular dendrimer G2: a metabolomics study on Plasmodium falciparum in vitro using 1HNMR spectroscopy. Parasitology 147(7):747–759. https://doi.org/10.1017/S0031182020000372
42. Omidi S, Pirhayati M, Kakanejadifard A (2020) Co-delivery of doxorubicin and curcumin by a pH-sensitive, injectable, and in situ hydrogel composed of chitosan, graphene, and cellulose nanowhisker. Carbohydr Polym 231:115745. https://doi.org/10.1016/j.carbpol.2019.115745
43. Firouzi Amoodizaj F, Baghaeifar S, Taheri E, Farhoudi Sefidan Jadid M, Safi M, Seyyed Sani N et al (2020) Enhanced anticancer potency of doxorubicin in combination with curcumin in gastric adenocarcinoma. J Biochem Mol Toxicol:e22486. https://doi.org/10.1002/jbt.22486
44. Abhari SMF, Khanbabaei R, Roodbari NH, Parivar K, Yaghmaei P (2020) Curcumin-loaded super-paramagnetic iron oxide nanoparticle affects on apoptotic factors expression and histological changes in a prepubertal mouse model of polycystic ovary syndrome-induced by dehydroepiandrosterone - A molecular and stereological study. Life Sci:117515. https://doi.org/10.1016/j.lfs.2020.117515
45. Varshosaz J, Jandaghian S, Mirian M, Sajjadi SE (2020) Co-delivery of rituximab targeted curcumin and imatinib nanostructured lipid carriers in non-Hodgkin lymphoma cells. J Liposome Res 6:1–15. https://doi.org/10.1080/08982104.2020.1720718
46. Karimi M, Gheybi F, Zamani P, Mashreghi M, Golmohammadzadeh S, Amiri Darban S et al (2020) Preparation and characterization of stable nanoliposomal formulations of curcumin with high loading efficacy: in vitro and in vivo anti-tumor study. Int J Pharm:119211. https://doi.org/10.1016/j.ijpharm.2020.119211
47. Alizadeh N, Malakzadeh S (2020) Antioxidant, antibacterial and anti-cancer activities of β-and γ-CDs/

curcumin loaded in chitosan nanoparticles. Int J Biol Macromol 147:778–791
48. Ghaffari M, Dehghan G, Baradaran B, Zarebkohan A, Mansoori B, Soleymani J et al (2020) Co-delivery of curcumin and Bcl-2 siRNA by PAMAM dendrimers for enhancement of the therapeutic efficacy in HeLa cancer cells. Colloids Surf B Biointerfaces 188:110762. https://doi.org/10.1016/j.colsurfb.2019.110762
49. https://www.indexmundi.com/g/r.aspx?v=2228
50. http://www.healthdata.org/
51. https://www.indexmundi.com/
52. Christophi CA, Pampaka D, Paisi M, Ioannou S, DiFranza JR (2016) Levels of physical dependence on tobacco among adolescent smokers in cyprus. Addict Behav 60:148–153
53. https://www.who.int/violence_injury_prevention/road_traffic/countrywork/egy/en/
54. Sarrafzadegan N, Mohammmadifard N (2019) Cardiovascular disease in iran in the last 40 years: prevalence, mortality, morbidity, challenges and strategies for cardiovascular prevention. Arch Iran Med 22(4):204–210
55. https://www.cfr.org/interactive/global-conflict-tracker/conflict/political-instability-iraq
56. Madanat HN, Troutman KP, Al-Madi B (2008) The nutrition transition in Jordan: the political, economic and food consumption contexts. Promot Educ 15(1):6–10
57. Zayed AA, Beano AM, Haddadin FI, Radwan SS, Allauzy SA, Alkhayyat MM et al (2016) Prevalence of short stature, underweight, overweight, and obesity among school children in Jordan. BMC Public Health 16(1):1040. https://doi.org/10.1186/s12889-016-3687-4
58. http://siteresources.worldbank.org/INTTOPGLOROASAF/Resources/TN-MENA-Road-Safety-Dahdah-and-Bose-revised-print.pdf
59. https://www.thecabinchiangmai.com/international-rehab/drug-rehab-qatar/
60. https://stepfeed.com/a-car-crash-happens-every-minute-in-saudi-arabia-3428
61. https://www.bbc.co.uk/newsround/16979186
62. http://www.xinhuanet.com/english/2019--01/08/c_137728824.htm
63. https://www.unicef.org/nutrition/yemen_94543.html

Turmeric and Curcumin: From Traditional to Modern Medicine

2

Maryam Akaberi, Amirhossein Sahebkar, and Seyed Ahmad Emami

Abstract

The rhizome of turmeric (*Curcuma longa* L.) has been used as an herbal medicine, coloring agent, spice, and food additive for thousands of years in different parts of the world particularly in Asian countries. It has been used for a range of diseases in many traditional medical schools, including Islamic traditional medicine, Chinese traditional medicine, and Ayurveda. It has been used mainly for digestive problems, as a cardio-, hepato-, and neuroprotective agent as well as in many inflammatory conditions such as arthritis and for enhancing immune system. Curcumin, a diarylheptanoid derivative found in turmeric, has anti-inflammatory, antioxidant, and anticancer properties; controls obesity and metabolic problems; and improves memory and mood disorders. Therapeutically, curcumin exhibits promising potential in preclinical and clinical studies and is currently in human trials for a variety of conditions, including metabolic syndrome, nonalcoholic fatty liver disease, rheumatoid arthritis, migraine, premenstrual syndrome, ulcerative colitis, knee osteoarthritis, polycystic ovarian syndrome, atherosclerosis, liver cirrhosis, amyotrophic lateral sclerosis, depression, psoriasis, and Alzheimer's disease. Among all beneficial activities reported for curcumin, the research toward the obesity and metabolic-preventing/suppressing aspects of curcumin is growing. These findings emphasize that most of the traditional applications of turmeric is due to the presence of its key constituent, curcumin. According to the traditional background of turmeric use and clinical values of curcumin, further preclinical studies for unstudied properties and clinical studies with larger sample sizes for confirmed activities are expected.

M. Akaberi
Department of Pharmacognosy, School of Pharmacy, Mashhad University of Medical Sciences, Mashhad, Iran

A. Sahebkar (✉)
Biotechnology Research Center, Pharmaceutical Technology Institute, Mashhad University of Medical Sciences, Mashhad, Iran

Neurogenic Inflammation Research Center, Mashhad University of Medical Sciences, Mashhad, Iran

Polish Mother's Memorial Hospital Research Institute (PMMHRI), Lodz, Poland

Halal Research Center of IRI, FDA, Tehran, Iran
e-mail: sahebkara@mums.ac.ir

S. A. Emami (✉)
Department of Traditional Pharmacy, School of Pharmacy, Mashhad University of Medical Sciences, Mashhad, Iran
e-mail: emamia@mums.ac.ir

P. C. Guest (ed.), *Studies on Biomarkers and New Targets in Aging Research in Iran*, Advances in Experimental Medicine and Biology 1291, https://doi.org/10.1007/978-3-030-56153-6_2

Keywords

Turmeric · *Curcuma longa* · Zingiberaceae · curcumin · traditional medicine · clinical trials · hepato-protective · cardioprotective · neuroprotective

2.1 Introduction

Turmeric, belonging to Zingiberaceae family, is a perennial plant growing up to one meter high with oblong or cylindrical rhizomes. The rhizomes, being externally brown, consists of an egg-shaped primary rhizome called the "tuber" and several branched secondary rhizomes called the "rhizome." The internal color of the rhizomes ranges from yellow to yellow-orange. This yellow color is due to the presence of pigments known as curcuminoids, which possess diverse pharmacological activities [1]. Chemically, curcuminoids known as diarylheptanoids consist of two aryl groups which are connected to each other via a chain with seven carbons. Curcumin (diferuloylmethane) is the most important bioactive curcuminoids. However, there exist other curcuminoids like desmethoxycurcumin and bisdesmethoxycurcumin in turmeric rhizome (Fig. 2.1) [2]. According to preliminary, preclinical, and clinical studies, the yellow pigment curcumin possesses a worthy pharmacology including anti-inflammatory [3, 4], immunomodulatory [5], antioxidant [6], hypolipidaemic [7], antimicrobial [8], anticarcinogenic [9], antitumor [10, 11, 12], radioprotective [13], neuroprotective [14], hepato-protective [15], nephroprotective [16], cardio-protective [17], and vasoprotective activities. Curcumin has been shown to interact with a wide range of biochemical pathways and influence a variety of molecular targets such as cytokines [18], transcription factors [19], kinases [20], growth factors [21], and microRNAs [22].

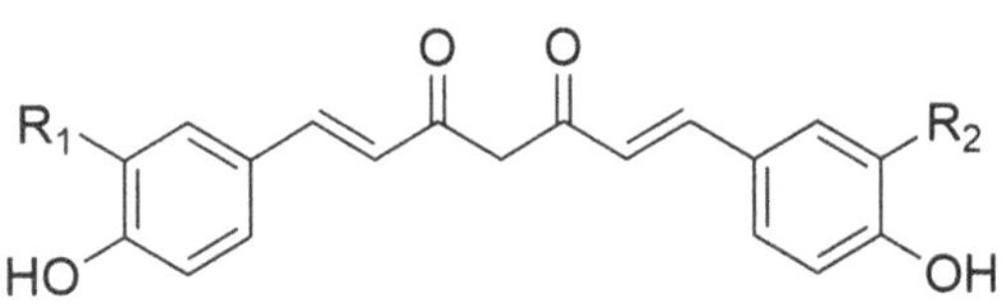

Curcuminoids:	R_1	R_2
Curcumin	OCH_3	OCH_3
Desmethoxycurcumin	OCH_3	H
Bisdesmethoxycurcumin	H	H

Fig. 2.1 The chemical structure of the main curcuminoids (the key constituents of turmeric)

Turmeric, also known as Indian saffron, has a long history of use as an herbal medicine, spice, and a coloring agent [23]. The first record of turmeric use dates back to 600 BC in an Assyrian herbal. It was also mentioned by famous Greek physician, Dioscorides, and many traditional scholars in Islamic traditional medicine (ITM) [24–37]. It has also a long-standing reputation in Chinese traditional medicine (TCM), Ayurveda, and different folk medicines around the world [38]. It has been traditionally used for the treatment of a range of diseases including skin disorders (topically) and poor digestion and liver function (internally) [38]. As the knowledge from traditional medicine is a guide for the scientists in the field of natural product–based drug discovery, we decided to investigate the medicinal applications of turmeric in different systems of traditional medicine and the pharmacological activities of curcumin in modern medicine to see the progress from ancient medicine to current clinical trials.

2.2 Traditional View

Turmeric has a long history of medical use in many cultures, particularly in Asia. Turmeric is originally a plant from India and South-East Asia, but nowadays it is cultivated in different regions of the world. However, India is still the largest producer of this valuable plant. Several traditional textbooks [24–37] and electronic scientific literature were searched to find the applications of turmeric in ancient times.

2.2.1 Turmeric in Islamic Traditional Medicine (ITM)

Turmeric, known as Zardchoobeh in Farsi (Persian language) and Hūrd, Kūrkūm, ʿOruqṢūfer, ʿOruqal-ṢabaghinKabir in Arabic, refers to *Curcumalonga* L. [Syn. *C. domestica* Valeton] in ITM textbooks. The latter name has made many of the ITM scientists a bit confused since *Chelidoniummajus* L. is called ʿOruqal-Ṣabaghin Ṣaghir. Some authors do not distinguish between these two names regardless of the suffixes Kabir or Ṣaghir. Even, some scientists believe that ʿOruqal-Ṣabaghin is the name of *Rubiatinctorum* L.

According to turmeric monographs in the most important textbooks of Islamic medicine, it is deduced that most authors consider it as warm and dry in the second degree and others in the third degree. Pharmacologically, different parts of turmeric plants have been prescribed by ITM physicians to manage various health problems particularly, inflammatory-related disorders. For example, the wood of turmeric has been used for the treatment of toothache. Its extract has been applied for eye problems such as cataract and corneal opacity as well as strengthening eyesight. It was administered topically to relieve the aches in the joints. Putting turmeric powder on wet and infectious wounds can dry them out.Topical administration of turmeric is useful for treating scabies, moist scabies, and herpes. In addition, this spice has been highly recommended for obstructive jaundice (caused by bile duct obstruction), cardiac problems like tachycardia, and nervous system–related problems such as epilepsy and paralysis [24–37].

2.2.2 Indian Traditional Medicine (Ayurveda)

Turmeric is commonly called "Haldi" in north India, and "Manjal" in the south [39]. In India, turmeric is used for a range of diseases similar to ITM. Indian people has used turmeric as a gastric tonic, as well as a blood purifier, digestive, anti-fever, and wound healing agent. It is also used for pregnancy nausea and skin and liver disease. It has been externally used to relieve conjunctivitis, skin infections, arthritis, hemorrhoids, and eczema. In addition, Indian women apply turmeric on their skin to reduce hair growth [39, 40].

2.2.3 Traditional Chinese Medicine (TCM)

In TCM, a range ofapplications have been reported for turmeric rhizome and tuber. Turmeric rhizomes, being a blood and Qi (vital energy) stimulant, have been used as an analgesic, painkiller, and wound healing agent. It has been prescribed for treating jaundice, chest and abdominal pain and swelling, frozen scapula, postpartum abdominal pain, and amenorrhea. Turmeric tuber, known as Yu Jin, is spicy and bitter and has been used in hot conditions like viral hepatitis since it is considered as a cooling agent. Ithas been used to relieve menstrual pain, heal traumatic injuries, treat enlarged liver cirrhosis, and treat mental problems like mania and epilepsy [41].

2.2.4 Traditional Thai Medicine (TTM)

In TTM, turmeric has been used in cases of gastrointestinal ulcers, anal hemorrhoids, vaginal hemorrhoids, skin diseases, ringworms, and insect bites and for the prevention of gonorrhea and common colds [23].

2.3 Clinical View

The biological and pharmacological effects of curcumin have been described by several in vitro and in vivo studies, showing the potential antioxidant, anti-inflammatory, cardio-protective, anti-microbial, nephro-protective, hepatoprotective, hypoglycaemic, immunomodulatory, and anti-rheumatic activities of this valuable compound [37]. Because of potent activities and rare serious side effects of curcumin observed in many preclinical studies, there have been a growing trend in clinical trials on curcumin in the last two decades. A search through scientific databases PubMed and Scopus using the words "curcumin" and "clinical trial" or "trial" yielded a total of 262 articles of which 175 are included in the present paper (Table 2.1). Other articles were systematic reviews, meta-analysis, or irrelevant topics. Similar to in vitro and animal studies, various human trials have shown anti-inflammatory, antioxidant, cardio-protective, antimicrobial properties (especially topically), and tissue protection effects including neuroprotective and hepato-protective activities for curcuminoids (Table 2.1).

Table 2.1 Clinical evidences for biological activities of curcumin

No.	Activity	Health problem	Study design	Participant/sample size	Dosage	Age (years)	Results	Ref.
1	Analgesic	Postoperative pain after laparoscopic gynecologic surgery	Pilot RCT	60 patients who were to undergo laparoscopic gynecologic surgery	Curcuminoid extract tablets (250 mg) four times a day on postoperative days 1–3 ($n = 30$)	33.1 ± 6.3	↓ Pain severity postoperatively ($p < 0.001$)	[88]
2	Anti-inflammatory	Rheumatoid arthritis (RA)	Double-blind RCT	65 patients with RA	Curcumin nanomicelle capsules (40 mg) 3 times a day for 12 weeks ($n = 30$)	–	Some positive changes in the DAS-28 (disease activity score of 28 joints), TJC (tender joint count), and SJC (swollen joint count), although not significantly	[45]
3	Anti-inflammatory	Migraine	RCT	80 episodic migraine patients	Combination of ω-3 fatty acids (2500 mg) + nano-curcumin (80 mg) over 2 months	–	↓ IL-6 mRNA ↓ hs-CRP serum ($p < 0.05$)	[47]
4	Anti-inflammatory	PMS	RCT	40 women with PMS	Combination of aerobic exercise (for 8 weeks, 3 times in a week) and curcumin capsules (100 mg of curcumin) daily from 7 days before the menstrual period up to 3 first days	18–35	↓ Physical symptoms ($p < 0.05$) ↓ Estrogen levels Improve behavioral and PMS mood symptoms	[89]
5	Anti-inflammatory	Takayasu arteritis (TA)	Double-blind RCT	246 patients with acute TA	Curcumin capsule (300 mg) curcumin daily for 4 weeks	19–52	↓ CRP ↓ TNF-α ↓ ESR (Erythrocyte sedimentation rate) ($p < 0.01$) ↓ BVAS (Birmingham vascular activity score) ($p < 0.05$)	[90]
6	Anti-inflammatory	Mastitisin breastfeeding women	Double-blind RCT	63 breastfeeding women with lactational mastitis	Curcumin topical cream, one pump wvery 8 h for 3 days ($n = 32$)	21–35	↓ Rate of moderate ($p = 0.019$) and mild ($p = 0.002$) mastitis ↓scores for tension ($p < 0.001$), erythema ($p < 0.001$), and pain ($p < 0.001$)	[50]

7	Anti-inflammatory	Radiation dermatitis (RT)	Double-blind RCT	30 breast cancer patients	Curcumin (6 g/d) orally three times per day throughout the course of RT	58.1	↓ Radiation dermatitis severity ($p = 0.008$) ↓ Moist desquamation rate ($p = 0.002$)	[51]
8	Anti-inflammatory	Postmenopausal syndrome	Triple-blind RCT	93 postmenopausal women	Curcumin (500 mg) per day for 8 weeks	51.7	↓ Hot flashes ($p = 0.001$)	[52]
9	Anti-inflammatory	Ulcerative colitis	Double-blind RCT	56 patientswith ulcerative colitis	Curcuminoids nanomicelles (80 mg, three times daily, orally) plus mesalamine (3 g/24 h, orally) for 4 weeks ($n = 28$)	38.2 ± 16.4	↓ Simple clinical colitis activity index	[53]
10	Anti-inflammatory	MeS	Double-blind parallel-group RCT	117 patients with MeS	Curcumin (1 g/day) for 8 weeks ($n = 59$)	44.8 ± 8.7	↓ TNF-α ($p < 0.001$) ↓ IL-6 ($p < 0.001$) ↓ TGF-β ($p < 0.001$) ↓ MCP-1 ($p < 0.001$)	[18]
11	Anti-inflammatory	CPC-SM	Double-blind RCT	89 male subjects	Curcuminoids (500 mg) for 4 weeks ($n = 45$)	51.0 ± 7.3	Improve spirometric parameters FEV1/FVC ($p = 0.002$) ↓ IL-6 ($p < 0.001$) ↓ IL-8 ($p = 0.035$) ↓ TNF-α ($p < 0.001$) ↓ TGF-β ($p < 0.001$) ↓ substance P ($p = 0.016$) ↓ hs-CRP ($p < 0.001$) ↓ CGRP (calcitonin gene-related peptide) ($p < 0.001$) ↓ MCP-1 ($p < 0.001$)	[54]
12	Anti-inflammatory	Obesity	Randomized crossover trial	30 obese individuals	Curcumin (1 g/day) for 4 weeks	18–65	↓ IL-1β ($P = 0.042$) ↓ IL-4 ($P = 0.008$) ↓ VEGF ($P = 0.01$)	[91]
13	Anti-inflammatory	KOA	Pilot double-blind parallel-group RCT	40 patients suffering from knee OA	Curcuminoids (1500 mg/day) for 6 weeks ($n = 19$)	57.3 ± 8.8	↓ WOMAC (Western Ontario and McMaster universities osteoarthritis index) ($p = 0.001$) ↓ VAS ($p < 0.001$) ↓ Lequesne's pain functional index ($p = 0.013$)	[57]

(continued)

Table 2.1 (continued)

No.	Activity	Health problem	Study design	Participant/sample size	Dosage	Age (years)	Results	Ref.
14	Anti-inflammatory	Solid tumors (ST)	Double-blind RCT	80 subjects with ST	Curcuminoids (180 mg/day) for 8 weeks ($n = 40$)	25–65	↑ QoL ($p < 0.001$) ↓ TNF-α ($p < 0.001$) ↓ TGFβ ($p < 0.001$) ↓ IL-6 ($p = 0.061$) ↓ Substance P ($p = 0.005$) ↓ hs-CRP ($p < 0.001$) ↓ CGRP ($p < 0.001$) ↓ MCP-1 ($p < 0.001$)	[92]
15	Anti-inflammatory	Chronic SM-induced pruritic skin lesions	Double-blind RCT	96 sulfur mustard (SM)-exposed patients	Curcumin (1 g/day) for 4 weeks ($n = 46$)	37–59	↓ IL-8 ($p < 0.001$) ↓ hs-CRP ($p < 0.001$) ↓ CGRP ($p < 0.001$) ↑ QoL	[93]
16	Anti-inflammatory	KOA	Double-blind RCT	44 patients	Diclofenac (75 mg/day) with curcumin (1000 mg/day) for 3 months	–	Decrease pain and improve function in daily living though not significantly	[58]
17	Anti-inflammatory	PMS	Double-blind RCT	70 patients with PMS	Curcumin capsules (100 mg) twice daily for 7 days before menstruation and for 3 days after menstruation for 3 successive cycles ($n = 35$)	25.2 ± 9.2	↓ Total severity of PMS score	[49]
18	Anti-inflammatory, Anti-oxidant	Obesity	Parallel RCT	60 overweight and obese adolescent girls	Curcumin (500 mg/day) + a slight weight loss diet for 10 weeks	13–18	↓ IL-6 ↓TAC (total antioxidant capacity) ↓ MDA ↓ hs-CRP ↓ IL-6	[94]
19	Anti-inflammatory, antioxidant	Infertility in men	Double-blind RCT	60 infertile men	80 mg curcumin nanomicelle daily for 10 weeks ($n = 30$)	20–45	↑ Total sperm count ($p < 0.001$) ↑ Sperm concentration ($p < 0.001$) ↑ Sperm motility ↑ TAC ($p < 0.001$) ↑ MDA ($p < 0.001$) ↑ CRP ↑ TNF-α ($p < 0.01$)	[95]

20	Antioxidant	Polycystic ovarian syndrome (PCOS)	RCT	72 patients with PCOS	Curcumin capsules (1500 mg) 3 times per day for 12 weeks	18–41	↑ Gene expression of PGC1α ($p = 0.011$) ↑ Activity of the Gpx ($p = 0.045$) ↑ Gene expression of SIRT1 ↑ SOD enzymenon-significantly	[61]
21	Antioxidant	β-Thalassemia major	Double-blind RCT	61 patients with β-thalassemia major	Curcumin capsules (500 mg) twice daily for 12 weeks	18–40	↓ Serum MDA ($p = 0.002$), ↓ Total ($p < 0.001$) and direct ($p < 0.001$) bilirubin ↑ TAC ($p = 0.005$)	[96]
22	Antioxidant	CPC-SM	RCT	89 subjects	Curcuminoids (1500 mg/day) in combination with piperine (15 mg/day) for a period of 4 weeks ($n = 45$)	51.0 ± 7.3	↑ Serum GSH, ↓ MDA, Improving COPD assessment test and St. George respiratory questionnaire ($p < 0.001$)	[55]
23	Antioxidant	Type 2 diabetes mellitus (T2DM)	Double-blind RCT	118 patients with T2DM	Curcumin (1000 mg/day) + piperine (10 mg/day) for 8 weeks	18–65	↑ TAC ($p < 0.001$) ↑ SOD activities ($p < 0.001$) ↓ Serum MDA levels ($p < 0.001$)	[97]
24	Antioxidant	KOA	Double-blind RCT	40 patients with KOA	Curcuminoid capsules (1500 mg/day) + piperine (15 mg/day) for 6 weeks ($n = 19$)	57.3 ± 8.8	↑ SOD ($p < 0.001$) ↑ GSH ($p = 0.064$) ↓ MDA ($p = 0.044$)	[98]
25	Antioxidant	CPC-SM	Double-blind RCT	89 patients with CPC-SM	Curcumin (500 mg/day) + piperine (15 mg/day) for 4 weeks ($n = 45$)	51.0 ± 7.3	↑ GSH ↓ MDA Improve COPD assessment test and St. George respiratory questionnaire ($p < 0.001$)	[55]
26	Antioxidant	ST	Double-blind RCT	80 subjects with ST	Curcuminoids (180 mg/day) for 8 weeks ($n = 40$)	25–65	↓ Serum thiobarbituric acid reactive species ($p < 0.001$) ↑ SOD and catalase ↑ GSH ($p < 0.001$) ↑ QoL ($p = 0.003$)	[99]

(continued)

Table 2.1 (continued)

No.	Activity	Health problem	Study design	Participant/sample size	Dosage	Age (years)	Results	Ref.
27	Antioxidant	SM-induced pruritus	Double-blind RCT	96 male Iranian veterans	Curcumin (1 g/day) for 4 weeks ($n = 46$)	37–59	↓ Substance P ($p < 0.001$) ↓ SOD ($p = 0.02$), ↓ Gpx ($p = 0.006$) and catalase ($p < 0.001$) ↓ Pruritus score ($p < 0.001$) ↓ VAS (visual analogue scale) score ($p < 0.001$) ↓ Overall ($p < 0.001$) and objective SCORAD (scoring atopic dermatitis) ($p = 0.009$) ↓ Dermatology life quality index ($p < 0.001$)	[59]
28	Anti-infective	Oral infections	RCT	27 adults	Photodynamic therapy with blue light and curcumin (30 mg/L)	20–35	↓ CFU (colony-forming units) ($p = 0.001$) ↓ Microbial reduction ($p < 0.05$)	[100]
29	Anti-infective	Chronic periodontitis	RCT	10 patients with two sites in the contralateral quadrants having probing pocket depths of ≥5 mm	Curcumin gel tropically+ SRP (scaling and root planning) for 4 weeks	–	↑ Periodontal parameters ↓ Plaque index ↓ Probing depth	[101]
30	Cardio-protective	Metabolic disorders in PCOS subjects	Double-blind RCT	60 women with PCOS	Curcumin (500 mg/day) for 12 weeks ($n = 30$)	18–40	↓ Weight and BMI ($p = 0.03$) ↓ FBG ($p = 0.002$) ↑ Serum insulin ($p = 0.02$) ↑ Insulin resistance ($p = 0.02$) ↑ Insulin sensitivity ($p = 0.02$) ↓ Total cholesterol ($p = 0.001$) ↓ LDL-cholesterol ($p = 0.001$) ↓ TC/HDL-C ($p < 0.001$) ↑ HDL-C levels ($p = 0.01$) ↑ Gene expression of PPAR-γ ($p = 0.03$) ↑ LDL receptor ($p < 0.001$)	[102]

31	Cardio-protective	PCOS-associated diabetes	Double-blind RCT	60 women with PCOS	1 g turmeric extract 95% (475 mg curcuminoids covering 70–80% curcumin, 15–20% demethoxycurcumin and 2.5–6.5% bisdemethoxycurcumin) daily for 6 weeks	18–40	↑ Serum insulin ($p = 0.020$) ↑ Insulin sensitivity ($p = 0.003$) ↓HOMA-IR ($p = 0.067$)	[103]
32	Cardio-protective	Atherosclerosis	Double-blind RCT	226 patients with T2DM	The extract (250 mg curcuminoids) Three capsules curcumin twice a day for 6 months ($n = 113$)	59.2 ± 1.1	↓ Pulse wave velocity ↑ Serum adiponectin ↓ Leptin	[104]
33	Cardio-protective	Hemodialysis-induced inflammation	Parallel, double-blind RCT	26 hemodialysis patients	Nano-curcumin soft gel three times/day (120 mg/d) over 12 weeks	18–80	↓ hs-CRP ($p < 0.001$) ↓ Serum adhesion molecule VCAM-1 ($p < 0.001$), ↓ TG, TC and LDL-C	[44]
34	Cardio-protective	MeS	Double-blind clinical trial	120 patients with MeS	Curcumin group (CG) (1 g/day), phospholipidated curcumin group (PCG) (1 g/day) for 6 weeks	18–65	↑ Serum Zn in both CG and PCG ($p < 0.001$) though higher in PCG in the ($p < 0.05$) ↑ Serum Zn/Cu level, more significant in PCG ($p < 0.001$)	[105]
35	Cardio-protective	Myocardial injury following elective percutaneous coronary intervention (PCI)	Pilot RCT	110 patients undergoing elective PCI	Single dose of 480 mg nanomicelle curcumin orally and the standard treatment ($n = 55$)	58.6 ± 9.1	↓ The raise of CK-MB (creatine kinase-MB)	[17]
36	Cardio-protective	Cardiovascular risk factors	RCT	60 overweight and obese female	95% turmeric extract tablet (500 mg) per day for 10 weeks	13–18	↓ BMI ($p = 0.019$) ↓ WC ($p = 0.008$) ↓ HC ($p = 0.030$) ↑ HDL ($p = 0.042$) ↓TG/HDL ($p = 0.021$)	[75]
37	Cardio-protective	Diabetes	Double-blind RCT	53 patients with T2DM	Curcumin capsule (1500 mg) 3 times a day for 10 weeks ($n = 25$)	40–70	↓ Mean weight ($p < 0.05$) ↓ BMI ($p < 0.05$) ↓ WC ($p < 0.05$) ↓ FBG ($p < 0.05$)	[106]

(continued)

Table 2.1 (continued)

No.	Activity	Health problem	Study design	Participant/sample size	Dosage	Age (years)	Results	Ref.
38	Cardio-protective	MeS	Double-blind, placebo-controlled clinical trial	80 overweight subjects with suboptimal values of FPG (fasting plasma glucose)	800 mg phytosomal curcumin (200 mg curcumin, 120 mg phosphatidylserine, 480 mg phosphatidylcholine and 8 mg piperine) twice/day for 8 weeks	18–70	↑ Fasting plasma insulin (FPI) ↑ HOMA index ↓ WC ↓ Blood pressure, TG ↓ Liver transaminases ↓ γ-GT ↓ Liver steatosis index ↓ Serum cortisol	[107]
39	Cardio-protective	β-Thalassemia major	Double-blind RCT	68 patients with β-thalassemia major	Curcumin capsules (500 mg/day) for 12 weeks	26.0 ± 6.9	↓ HOMA-IR ($p = 0.048$) ↓ TG ($p = 0.020$) ↓ TG/HDL-C ratio ($p = 0.024$) ↓ hs-CRP levels ($p = 0.022$)	[108]
40	Cardio-protective	Metabolic disorders in HIV/AIDS subjects	Double-blind, crossover, randomized clinical trial	20 patients with HIV/AIDS under antiretroviral therapy	Curcumin (1000 mg/day) for 30 days	45.5 (9.7)	↓ TG	[109]
41	Cardio-protective	Cardiovascular disease risk factors and arterial function	Double-blinded, placebo-controlled trial	22 obese men	Curcumin (500 mg/day) formulated with fenugreek soluble fiber for 2 weeks	18–35	Improve homocysteine and HDL concentrations	[74]
42	Cardio-protective	Diabetes	Double-blind RCT	44 patients with T2DM	Curcumin (1500 mg/day) for 10 weeks	40–70	↓ hs-CRP ($p < 0.05$) ↑ Adiponectin ($p < 0.05$)	[110]
43	Cardio-protective	Hypercholes-terolemia	Double-blinded, 2 × 2 factorial RCT	76 individuals with hypercholeste-rolemia	Phytosterols (2 g/day) + curcumin (200 mg/day) for 4 weeks	18–70	↑ Cholesterol-lowering effects of phytosterols	[111]
44	Cardio-protective	Hyperlipidemia	Double-blind RCT	118 patients with T2DM	Curcumin (1000 mg/day) + piperine (10 mg/day) for 12 weeks	18–65	↓ TC ($p = 0.023$), non-HDL-C ($p = 0.014$), ↓ lipoprotein(α) [Lp(α)] ($p = 0.001$) ↑ HDL-C ($p = 0.048$)	[112]

45	Cardio-protective	MeS	Pilot, double-blind RCT	127 patients with MeS	Curcumin (1000 mg/day) for 8 weeks ($n = 59$)	44.8 ± 8.7	↑ Adiponectin ($p < 0.001$) ↓ Leptin ($p < 0.001$)	[113]
46	Cardio-protective	Obesity	RCT	44 overweight subjects	800 mg *Curcuma longa* extract (95% curcumin) + 8 mg piperine twice a day for 1 month	39.1 ± 16.8	↑ Weight loss ↑ Body fat ↓ WC ↓ HC ↓ BMI ($p < 0.01$)	[67]
47	Cardio-protective	MeS, dislipidemia	RCT	100 patients with MeS	Curcuminoids (1000 mg/day) for 8 weeks ($n = 50$)	25–75	↓ LDL-C, TC, TG and Lp(α) ↑ HDL-C	[114]
48	Cardio-protective	Hyperlipidemia	Double-blind RCT	65 patients with MeS	Curcumin extract capsules (630 mg thrice daily) for 12 weeks ($n = 33$)	59.0 ± 10.1	↑ HDL-C ($p < 0.05$) ↓ LDL ($p < 0.05$)	[115]
49	Cardio-protective	Diabetes	Double-blind, placebo-controlled trial	100 overweight/obese patients with T2DM	Curcuminoids (300 mg/day) for 3 months ($n = 50$)	55.4 ± 6.4	↓ FBG ($p < 0.01$), ↓ HbA1c ($p = 0.031$), ↓ HOMA-IR ($p < 0.01$) ↓ Total free fatty acid ($p < 0.01$) ↓ TG ($p = 0.018$) ↑ Lipoprotein lipase ($p < 0.01$)	[72]
50	Cardio-protective	Hyperlipidemia	Double-blind RCT	30 obese individuals	Curcumin (1 g/day) for 30 days	15–65	↓ TG ($p = 0.009$)	[116]
51	Hepato-protective	NAFLD	Randomized controlled trial	102 patients with NAFLD (grades 1–3)	Phytosomal form of curcumin (1000 mg/day) for 8 weeks ($n = 50$)	45.0 ± 12.6	↓ BMI ($p = 0.003$) ↓ WC ($p = 0.024$) ↓ AST ($p < 0.001$) ↓ ALT ($p < 0.001$)	[117]
52	Hepato-protective	NAFLD	Double-blind RCT	80 patients with NAFLD	Curcumin formulation (500 mg/day equivalent to 70 mg curcumin) for 8 weeks ($n = 40$)	46.4 ± 11.6	↓ Liver fat content ↓ BMI ↓ TC, LDL-C, TG ↓ ALT, AST ↓ Glucose ↓ Glycated hemoglobin	[118]
53	Hepato-protective	NAFLD	RCT	87 patients with NAFLD (grades 1–3)	1000 mg/day for 8 weeks ($n = 50$)	45.0 ± 12.6	↓ TC ($p < 0.001$) ↓ LDL ($p < 0.001$) ↓ TG ($p < 0.001$) ↓ non-HDL-C ($p < 0.001$) ↓ Uric acid ($p < 0.001$)	[119]

(continued)

Table 2.1 (continued)

No.	Activity	Health problem	Study design	Participant/sample size	Dosage	Age (years)	Results	Ref.
54	Hepato-protective	Diabetes	Double-blind RCT	100 patients with T2DM	Curcumin (500 mg) and piperine (5 mg) daily for 3 months	18–65	↓ Glucose ($p = 0.048$), ↓ C-peptide ($p < 0.001$) ↓ HbA1c ($p < 0.001$) ↓ ALT ($p = 0.032$) ↓ AST ($p = 0.002$)	[120]
55	Hepato-protective	NAFLD	Double-blind RCT	84 overweight/obese patients with NAFLD	Curcumin capsules (40 mg) twice/day for 3 months ($n = 42$)	25–50	↑ HDL, ↑ Insulin sensitivity ↑ Nesfatin ↓ Fatty liver degree ↓ Liver transaminases ↓ WC, FBS, FBI, HbA1c, TG, TC, LDL, HOMA-IR, TNF-α, hs-CRP, and IL-6 ($p < 0.05$)	[70]
56	Hepato-protective	NAFLD	Double-blind RCT	65 eligible patients	Phospholipidated curcumin capsules (250 mg/day) equivalent to 50 mg/day pure curcumin) for 8 weeks	44.8 (11.14)	↑ HDL-C ($p = 0.01$) ↓ Leptin: adiponectin ratio ($p < 0.001$)	[121]
57	Hepato-protective	Iron deficiency	Double-blind RCT	68 β-thalassemia major patients	Curcumin capsules (500 mg) twice/day for 12 weeks	18–40	Alleviate iron burden and liver dysfunction by: ↓ NTBI (nontransferrin bound iron) ($p = 0.001$) ↓ ALT ($p = 0.018$) ↓ AST ($p = 0.009$)	[122]
58	Hepato-protective	NAFLD	Double blind RCT	80 healthy participants with moderately high BMI (24–30)	Tablets containing a combination of hot water extract of *Curcuma longa* and curcumin thrice a day for 12 weeks	51.7	↓ Serum liver enzyme levels possibly through the suppression of systemic inflammation	[43]
59	Hepato-protective	Liver cirrhosis	Double-blind RCT	70 patients with liver cirrhosis	Curcumin (1000 mg/day) for 12 weeks ($n = 35$)	20–70	Improve quality of life (QoL)	[123]

60	Hepato-protective	NAFLD	Double-blind RCT	55 patients with NAFLD	One capsule containing 500 mg curcuminoids (plus 5 mg piperineper day for 8 weeks	18–70	Improve the severity of NAFLD ($p = 0.002$) Improve serum concentrations of TNF-α ($p = 0.024$), MCP-1 ($p = 0.008$) and EGF ($p = 0.0001$)	[124]
61	Hepato-protective	Liver cirrhosis	Double-blind RCT	70 patients with liver cirrhosis	Curcumin (1000 mg/day) for 3 months ($n = 35$)	20–70	↓ Severity of cirrhosis ↓ MELD(i), MELD, MELD-Na, (model for end-stage liver disease) ↓ Child–Pugh scores ($p < 0.001$)	[125]
62	Hepato-protective	NAFLD	Double-blind RCT	58 NAFLD patients	Phospholipid curcumin (250 mg) per day for 8 weeks	25–65	↓ Some serum metabolites including: 3-methyl-2-oxovaleric acid, 3-hydroxyisobutyrate, kynurenine, succinate, citrate, α-ketoglutarate, methylamine, trimethylamine, hippurate, indoxyl sulfate, chenodeoxycholic acid, taurocholic acid, and lithocholic acid	[126]
63	Neuroprotective	Diabetic sensorimotor polyneuropathy	Parallel, double-blind RCT	80 diabetic patients	80 mg nano-curcumin supplementation for 8 weeks ($n = 40$)	30–60	↓ HbA1c ($p < 0.001$) ↓FBS ($p = 0.004$) ↓ Total score of neuropathy ($p < 0.001$) ↓ Total reflex score ($p = 0.04$) ↓ Temperature ($p = 0.01$)	[127]
64	Neuroprotective	Diabetic peripheral neuropathy (DPN)	Double-blind RCT	80 patients with T2DM and DPN	Nano-curcumin capsules (80 mg) daily for 8 weeks ($n = 40$)	30–60	↓ Mean score of depression (0.02) ↓ Mean score of anxiety ($p = 0.009$)	[128]
65	Neuroprotective	ALS	Double-blind RCT	54 eligible patients with ALS	80 mg nanocurcumin daily for 12 month ($n = 27$)	18–85	↑ Survival curves ($p = 0.036$)	[129]

(continued)

Table 2.1 (continued)

No.	Activity	Health problem	Study design	Participant/sample size	Dosage	Age (years)	Results	Ref.
66	Neuroprotective	Migraine	Double-blind, phase II RCT	100 men and women with episodic migraine	A combination of nano-curcumin (80 mg) + CoQ10 (300 mg)	32	↓ Frequency, severity, duration of migraine attacks and HDR (headache diary results) ($p < 0.001$) Better scores in migraine-specific questionnaires ($p < 0.001$)	[48]
67	Neuroprotective	Migraine	Placebo-controlled clinical trial	74 episodic migraine patients	ω-3 fatty acids and nano-curcumin for 2 months	–	↓ COX-2/iNOS mRNA and serum levels ↓ Frequency, severity, and duration of headaches ($p < 0.05$)	[46]
68	Neuroprotective	Alzheimer's disease	Double-blind, placebo-controlled trial	40 subjects	Theracurmin® containing 90 mg of curcumin twice daily for 18 months ($n = 21$)	51–84	↑ Memory and attention in non-demented adults ↓ Amyloid and tau accumulation in brain regions	[130]
69	Neuroprotective	ALS	Double-blind therapeutic trial	Patients with ALS	Curcumin (600 mg/day) for 6 months	–	↓ Disease progression ↓ Oxidative damage ↑ Aerobic metabolism	[131]
70	Neuroprotective	PMS	Double-blinded RCT	70 patients with PMS	Curcumin capsules (100 mg) twice daily for 7 days before menstruation and for 3 days after menstruation for 3 successive menstrual cycles ($n = 35$)	25.2 ± 9.2	↑ Serum brain-derived neurotrophic factor levels	[132]
71	Neuroprotective	Depression	Double-blind RCT	50 patients with major depressive disorder	Curcumin extract (500 mg, twice daily) for 8 weeks	18–65	↑ Urinary thromboxane B2 ($p < 0.05$) ↑ Substance P ($p < 0.001$)	[133]

72	Wound healing	Wound and pain associated with episiotomy	Double-blind RCT	120 healthy primiparous women with a vaginal delivery	Topical use of curcumin for 10 days after delivery	–	↓ Total scores of the REEDA ($p < 0.001$)	[134]
73	Wound healing	Aphthous stomatitis	Double blind RCT	57 patients	Curcumin gel (2% curcumin) for 2 weeks ($n = 28$)	18–65	↓ Pain intensity and size of aphthous ulcer	[135]
74	Wound healing	Oral submucous fibrosis (OSMF)	Parallel RCT	60 patients with OSMF	Curcumin tablet (300 mg) twice daily for 6 months ($n = 30$)	17–60	Improvement in mouth opening, burning sensation, tongue protrusion, and cheek flexibility	[136]

2.3.1 Anti-Inflammatory and Antioxidant Activities

The anti-inflammatory activity of curcumin, being reported for the first time in 1971, has been investigated in many preclinical and clinical trials so far [41, 42]. In many cases, curcumin exerts its beneficial activity via decreasing acute or chronic inflammatory responses (Table 2.1). Curcumin inhibits the production of pro-inflammatorymonocyte/macrophage-derived cytokines such as interleukin-8 (IL-8), monocyte inflammatory protein-1(MIP-1), monocyte chemotactic protein-1 (MCP-1), IL-1b, and tumor necrosis factor-α (TNF-α) in several in vitro studies. Also, in vivo studies demonstrated the beneficial effect of curcumin on adipose tissue through the inhibition of several pro–inflammatory mediators, such as MCP-1, IL-1β, TNFα, IL-6, and cyclooxygenase 2 (COX2) [6, 43].

The clinical trials also support the benefit of curcumin in the treatment of acute and chronic inflammatory problems including advanced chronic kidney disease with hemodialysis [19, 44], cardiovascular diseases, rheumatoid arthritis [45], migraine [46–48], premenstrual syndrome (PMS) [49], mastitis [50], radiation dermatitis [51], post-menopausal syndrome [52], ulcerative colitis [53], chronic pulmonary complications of sulfur mustard (CPC-SM) [54, 55], metabolic syndrome (MeS) [56], knee osteoarthritis (KOA) [57, 58], and chronic SM-induced pruritic skin lesions [59] via suppressing the production of cytokines including IL-1β, IL-4, IL-6, IL-8, and TNF-α [60].

The protective effect of curcumin against oxidative damage has been proven in several cell lines and animal models. Curcumin reduces lipid peroxidation through the normalization of antioxidant enzyme levels, such as superoxide dismutase, catalase, and glutathione peroxidase (Gpx). It is a free radical scavenger of hydroxyl (OH) and nitric oxide (NO) radicals, protects DNA against oxidative injury, and reduces reactive oxygen species (ROS). Moreover, it can activate other cellular antioxidants such as heme oxygenase-1, NADPH:quinone oxidoreductase-1 and glutathione, upregulated Nrf2-ARE-regulated pathways [6, 61–63].

A systematic review and meta-analysis of curcumin in the management of inflammatory and oxidative stress markers concluded that there was a significant tendencyin favor of taking curcumin-containing supplements through a significant reduction in IL-6, high sensitive c-reactive protein(hs-CRP), and malondialdehyde (MDA) levels [64]. Another meta-analysis found that consumption of curcumin for 4 weeks or longer may play an antioxidative role by reducing circulating MDA concentrations and increasing superoxide dismutase (SOD) activity and the effect is greater when combined with piperine [65].

2.3.2 Metabolic Disorders

Metabolic disorders have been one of the major health problems in recent decades in both developed and developing countries. Interestingly, many of the studied clinical trials have been focused on investigating the effects of curcumin on different components of metabolic problems [56, 63, 66–69]. Accordingly, most of these studies have shown the beneficial effects of curcumin on regulating glucose and lipid contents of people with MeS or subjects prone to it. In the following paragraphs, the effects of curcumin are discussed in two main categories, namely, its cardio- and hepato-protective activities.

2.3.2.1 Cardio- and Vaso-Protective

Curcumin is gaining growing interest in the scientific community for treating cardiovascular problems. Curcumin has cardio-protective activities through different mechanisms and multiple molecular targets mainly due to its exogenous and endogenous antioxidant capabilities (Table 2.1). In particular, curcumin is involved in the regulation of lipid and glucose metabolism. Therefore, curcumin exerts its cardio-protective activity either through decreasing blood glucose levels via increasing insulin sensitivity and secretion [70–72] and decreasing insulin resistance or by regulating lipid metabolism such as decreasing triglycerides (TG), low density lipoprotein (LDL) and increasing high density lipoprotein

(HDL) levels as well as decreasing obesity risk factors including body mass index (BMI), waist and hip circumferences [63, 73–75]. Curcumin regulates gene expression of PPAR-γ and LDL receptor, decreases serum CRP, adhesion molecule VCAM-1, creatine kinase-MB (CK-MB), leptin and cortisol, and increases serum adiponectin (Table 2.1).

Owing to the extensive clinical studies on curcumin, there are a large number of systematic reviews and meta-analysis from published works. Most of these analyses emphasized on the beneficial effects of curcumin in regulating blood glucose and lipid levels though there are some inconsistencies in the results. For instance, a meta-analysis, published in 2017, included seven double-blind, randomized, controlled trials ($n = 649$) of curcumin for dyslipidemia. Curcumin significantly reduced serum LDL-C and TG levels as compared to those in the control group though serum HDL-C levels were not obviously improved [76]. A systematic review of the efficacy of curcumin in patients with MeS concluded that curcumin appears to reduceTG, and total cholesterol (TC), as well as fasting glucose levels, homeostasis model of assessment-estimated insulin resistance (HOMA-IR), glycosylated hemoglobin (HbA1c) while no significant effects on HDL and LDL cholesterol levels [77]. A systematic review and meta-analysis of the role of curcumin as an adjunct therapy in patients with MeS found seven clinical trials published up to late 2018. Data analysis showed that curcumin can significantly improve FBG, TG, HDL-C, and diastolic blood pressure levels [56]. In another systematic review studying eleven clinical trials on curcumin given to patients with disglycemia found benefit in improving FBG and HbA1c compared with placebo but not HOMA-IR [71]. A systematic review and meta-analysis of seven clinical trials investigating the effects of curcumin glucose profile in prediabetic patients revealed that curcumin can significantly reduce HbA1c and FBG [78].

In addition to direct effects of curcumin on blood glucose and lipid levels, curcumin appears to increase adiponectin and decrease leptin content. A 2019 systematic review of randomized, double–blind clinical trials identified six studies and concluded that curcumin supplementation significantly increased adiponectin concentrations in comparison with placebo. Greater effects on adiponectin were observed in trials lasting ≤10 weeks [79]. Another systematic review also published in 2019 investigated the influence of curcumin intake on weight loss among patients with MeS and related disorders. Eighteen clinical studies were identified from up to 2018 involving a total of 1604 patients. Curcumin intake significantly reduced BMI, weight, waist circumference (WC), as well as leptin levels and increased adiponectin levels [66].

2.3.2.2 Hepato-Protective

Curcumin has demonstrated promising hepatoprotective activity through decreasing serum liver enzyme levels such as AST (aspartate aminotransferase) and ALT (alanine aminotransferase) [80], improving BMI, WC and hip circumference (HC) [81], decreasing liver fat content [69], and improving serum concentrations of TNF-α, MCP-1 and epidermal growth factor (EGF).

Among the reviewed clinical trials investigating the hepato-protective activities of curcumin, nonalcoholic fatty liver disease (NAFLD) is the most studied disorder. Unfortunately, NAFLD, a major cause of liver-related morbidity, is prevailing due to the rising epidemic of obesity. A systematic review and meta-analysis of curcumin versus placebo in the treatment of NAFLD concluded that curcumin supplementation significantly reduced BMI and WC [73]. Another meta-analysis found that curcumin supplementation in NAFLD patients makes a significant reduction in ALT, AST, serum TC, LDL, FBG, HOMA-IR, serum insulin, and WC [68]. Similarly, a meta-analysis that included 4 RCTs with a total of 229 NAFLD patients found that curcumin decreases LDL-C, TG, fasting blood sugar (FBS), HOMA-IR, weight, and AST levels compared with placebo [69].

2.3.3 Neuroprotective and Antidepressant

Neuroprotective potential of curcumin thought to be largely mediated by antioxidantmechanisms, anti-inflammatory responses such as inhibition of cytokine production [82], and reducing amyloid and tau accumulation in brain regions. Several clinical trials have examined the effects of curcumin supplementation on nervous system–related disorders like age-related cognitive decline, dementia, depression, and mood disorders. As it is shown in Table 2.1, curcumin supplementation has neuroprotective effects in a wide range of diseases including diabetic polyneuropathy, amyotrophic lateral sclerosis (ALS), migraine, Alzheimer's disease, depression, and neurological symptoms associated with PMS.

Among all neuroprotective activities reported for curcumin, the efficacy of curcumin in Alzheimer's disease and depression is receiving considerable research attention. In vitro and in vivo studies show that curcumin can significantly inhibit monoamine oxidase MAO-A and MAO-B activities in a dose-dependent manner, increase serotonin, noradrenaline, and dopamine levels markedly. A recent meta-analysis of six clinical trials with a total of 289 subjects showed that curcumin improved the memory and cognitive function in elderly people [83]. A meta-analysis evaluating the efficacy and safety of curcumin for the treatment of depression concluded a significant effect of curcumin in ameliorating depressive symptoms. Out of all six included trials, significant anti-anxiety effects were observed in 3 of the trials [84]. Another meta-analysis of six clinical trials, up to 2015, on curcumin in patients with major depression found that although there is a significant reduction of depression symptoms in all the patients, curcumin had the highest effect when given to middle-aged patients, for longer duration of administration, and at higher doses [85]. The anti-depressant and anti-anxiety activities of curcumin have also been confirmed by a recent meta-analysis of nine clinical trials [86].

2.3.4 Indications Supported by Both Traditional Medicine and Clinical Trials

From applications of turmeric in different systems of traditional medicine and the observed clinical efficacy and safety of curcumin in several trials, it can be concluded that there is a correlation between the data extracted from both sources. As indicated in the section "traditional view," turmeric is mostly used in inflammatory and oxidative conditions and has been prescribed as a heart, liver, and neurological tonic as well as digestive problems in ITM, TCM, and Ayurveda. Islamic and Chinese traditional medicines and Ayurveda as the ancient schools of medicine have a large contribution to the distribution of old medicine to the modern world. The prescriptions in these schools, mostly being based on phytotherapy, are full of clues to find lead compounds with great medicinal properties. For instance, in ITM like many other medical systems, herbs have provided the basis for solving health problems. In this medical system, considerable respect was paid to the qualities of individual herbs. Physicians were familiar with the nature of each remedy, its actions, energy pattern, indications, and duration of action. Moreover, they were expected to understand the specific relationships of the herbal medicine to the organs, its toxicity contraindications, and antidotes, types of preparation that can be made, its dosage, administration and even its natural habitat. There were many outstanding physicians who made their scientific knowledge about the medicinal herbs written as Materia Medica. Persians ar-Rhazi (Rhazes), al-Majusi (Haly Abbas), and Ibn-Sina (Avicenna) were among the well-known physician who did notable developments. They were not only well familiar with the theoretical aspects of phytotherapy but also in practice [41, 87]. Turmeric is one of the well-known spices and medicinal herbs with many monographs written about its characteristics and medicinal values. It is almost mentioned in all ITM medicinal textbooks [87]. Curcuminoids are the yellow pigments in turmeric. Curcumin is the most important bioactive curcuminoids. As discussed in the previous sections, there are thousands of articles investigating the pharmacological activities of curcumin. Interestingly, the trend of clinical trials is toward to the claimed traditional uses. In the case of cardiovascular, metabolic, gastrointestinal, and neurological disorders, curcumin shows promising activities. It has also shown beneficial activities in inflammatory conditions and as a chemopreventive agent. Thus, it can be concluded that most of the reported traditional properties of turmeric are because of the presence of curcumin.

Taken together, curcumin is a promising lead compound for therapeutically purposes. Further research projects including in vitro and in vivo studies for the undetermined potential activities of curcumin as well as clinical trials for the potent properties reported in preliminary studies are expected. In addition, for the activities investigated in many human trials with positive results, more comprehensive studies are necessary.

Acknowledgements The authors are grateful to Mashhad University of Medical Sciences (MUMS) for help and supports.

Conflict of Interest: None.

References

1. Moreira J, Saraiva L, Pinto MM, Cidade H (2020) Diarylpentanoids with antitumor activity: a critical review of structure-activity relationship studies. Eur J Med Chem 92:112177. https://doi.org/10.1016/j.ejmech.2020.112177
2. Hatamipour M, Ramezani M, Tabassi SAS, Johnston TP, Sahebkar A (2019) Demethoxycurcumin: a naturally occurring curcumin analogue for treating non-cancerous diseases. J Cell Physiol 234(11):19320–19330
3. Mollazadeh H, Cicero AFG, Blesso CN, Pirro M, Majeed M, Sahebkar A (2019) Immune modulation by curcumin: the role of interleukin-10. Crit Rev Food Sci Nutr 59(1):89–101
4. Hasanzadeh S, Read MI, Bland AR, Majeed M, Jamialahmadi T, Sahebkar A (2020) Curcumin: an inflammasome silencer. Pharmacol Res 159:104921. https://doi.org/10.1016/j.phrs.2020.104921. Epub 2020 May 25. PMID: 32464325
5. Abdollahi E, Momtazi AA, Johnston TP, Sahebkar A (2018) Therapeutic effects of curcumin in inflammatory and immunemediated diseases: a nature-made jack-of-all-trades? J Cell Physiol 233(2):830–848
6. Boroumand N, Samarghandian S, Hashemy SI (2018) Immunomodulatory, anti-inflammatory, and antioxidant effects of curcumin. J Herb Med Pharmacol 7(4):211–219
7. Panahi Y, Ahmadi Y, Teymouri M, Johnston TP, Sahebkar A (2017) Curcumin as a potential candidate for treating hyperlipidemia: A review of cellular and metabolic mechanisms. J Cell Physiol 233(1):141–152. https://doi.org/10.1002/jcp.25756. Epub 2017 Jun 6. PMID: 28012169
8. Naz RK, Lough ML, Barthelmess EK (2016) Curcumin: a novel non-steroidal contraceptive with antimicrobial properties. Front Biosci (Elite Ed) 8(1):113–128
9. Teymouri M, Pirro M, Johnston TP, Sahebkar A (2017) Curcumin as a multifaceted compound against human papilloma virus infection and cervical cancers: A review of chemistry, cellular, molecular, and preclinical features. BioFactors 43(3):331–346.
10. Mohajeri M, Bianconi V, Ávila-Rodriguez MF, Barreto GE, Jamialahmadi T, Pirro M, Sahebkar A (2020) Curcumin: a phytochemical modulator of estrogens and androgens in tumors of the reproductive system. Pharmacol Res 156:104765. https://doi.org/10.1016/j.phrs.2020.104765. Epub 2020 Mar 23. PMID: 32217147
11. Doello K, Ortiz R, Alvarez PJ, Melguizo C, Cabeza L, Prados J (2018) Latest in vitro and in vivo assay, clinical trials and patents in cancer treatment using curcumin: a literature review. Nutr Cancer 70(4):569–578
12. Iranshahi M, Sahebkar A, Hosseini ST, Takasaki M, Konoshima T, Tokuda H (2010) Cancer chemopreventive activity of diversin from Ferula diversivittata in vitro and in vivo. Phytomedicine 17(3-4):269–273. https://doi:10.1016/j.phymed.2009.05.020. Epub 2009 Jul3. PMID: 19577457
13. Kolivand S, Amini P, Saffar H, Rezapoor S, Motevaseli E, Najafi M et al (2019) Evaluating the radioprotective effect of curcumin on rat's heart tissues. Curr Radiopharm 12(1):23–28
14. Forouzanfar F, Read MI, Barreto GE, Sahebkar A (2020) Neuroprotective effects of curcumin through autophagy modulation. IUBMB Life 72(4):652–664
15. Panahi Y, Kianpour P, Mohtashami R, Jafari R, Simental-Mendía LE, Sahebkar A (2017) Efficacy and Safety of Phytosomal Curcumin in Non-Alcoholic Fatty Liver Disease: A Randomized Controlled Trial. Drug Res (Stuttg) 67(4):244–251. https://doi.org/10.1055/s-0043-100019. Epub 2017 Feb 3. PMID: 28158893.
16. Abd El-Kader M, Taha RI (2020) Comparative nephroprotective effects of curcumin and etoricoxib against cisplatin-induced acute kidney injury in rats. Acta Histochem 122(4):151534. https://doi.org/10.1016/j.acthis.2020.151534
17. Aslanabadi N, Entezari-Maleki T, Rezaee H, Jafarzadeh HR, Vahedpour R (2019) Curcumin for the prevention of myocardial injury following elective percutaneous coronary intervention; a pilot ran-

domized clinical trial. Eur J Pharmacol 858:172471. https://doi.org/10.1016/j.ejphar.2019.172471
18. Saberi-Karimian M, Keshvari M, Ghayour-Mobarhan M, Salehizadeh L, Rahmani S, Behnam B, Jamialahmadi T, Asgary S, Sahebkar A (2020) Effects of curcuminoids on inflammatory status in patients with non-alcoholic fatty liver disease: A randomized controlled trial. Complementary Therapies in Medicine 49, 102322. https://doi.org/10.1016/j.ctim.2020.102322
19. Alvarenga L, Salarolli R, Cardozo LFMF, Santos RS, de Brito JS, Kemp JA, Reis D, de Paiva BR, Stenvinkel P, Lindholm B, Fouque D, Mafra D (2020) Impact of curcumin supplementation on expression of inflammatory transcription factors in hemodialysis patients: a pilot randomized, double-blind, controlled study. Clin Nutr. S0261-5614(20)30110-2. https://doi.org/10.1016/j.clnu.2020.03.007
20. Soltani A, Salmaninejad A, Jalili-Nik M, Soleimani A, Javid H, Hashemy SI et al (2019) 5′-adenosine monophosphate-activated protein kinase: a potential target for disease prevention by curcumin. J Cell Physiol 234(3):2241–2251
21. Saberi-Karimian M, Katsiki N, Caraglia M, Boccellino M, Majeed M, Sahebkar A (2019) Vascular endothelial growth factor: an important molecular target of curcumin. Crit Rev Food Sci Nutr 59(2):299–312
22. Momtazi AA, Derosa G, Maffioli P, Banach M, Sahebkar A (2016) Role of microRNAs in the therapeutic effects of curcumin in non-cancer diseases. Mol Diagn Ther 20(4):335–345
23. Prasad S, Aggarwal BB (2011) Turmeric, the golden spice: from traditional medicine to modern medicine. In: Benzie IFF, Wachtel-Galor S (eds) Herbal medicine: biomolecular and herbal medicine: biomolecular and clinical aspects, 2nd edn. CRC Press/Taylor & Francis, Boca Raton, FL, pp 263–288
24. Rāzi MZ (1968) In: Khān A (ed) Al-Ḥāwi fi'l-Ṭibb (comprehensive book of medicine, in Arabic), vol 31. Osmania Oriental Publications Bureau, Osmania University, Hyderabad
25. Ibn Sinā A (1987) Al-Qānun fi'l-Ṭibb (canon of medicine, in Arabic), vol 2. I.H.M.M.R. Printing Press, New Delhi
26. Herawi M (1967) In: Bahmānyar A (ed) Al-Abniyah an Ḥaqāyeq al-Adwiyah (basics of realities on drugs, in Persian). Tehran University Publications, Tehran, Iran
27. Ibn Jazlah Y (1631) Minhāj al-Bayān fīmā Yastaʿmiluhū al-Insān (in Arabic). League of Arab States, Bulaq, Cairo, Egypt
28. Ibn Beyṭār Z (2001) Al-Jameʿe le-Mofradāt al-Adwiah wal-Aghdhiyah (comprehensive book in simple drugs and foods in Arabic), vol 2. Dār al-Kūtūb al-ʿElmiyah, Beirut, Lebanon
29. Anṭāki D (2000) In: Shams a'ddin A (ed) Tadhkereh Oulūl-albāb (memorandum book, in Arabic). Dār-al-Kūtūb al-ʿElmiyah, Beirut, Lebanon
30. Tonekāboni MM (1959) Toḥfat al-Momenin (rarity of the faithful, in Persian). Mostafavi Press, Tehran, Iran
31. Ghasāni AM (1985) Ḥadiqat al-Azhār fi Māhiyyat al-ʿUshb wa al-ʿUqqār (Flower Garden in the Nature of Herbs and Drugs,in Arabic). Dār al-Gharb al-Islāmi, Beirut, Lebanon
32. Aqili Khorāsāni MH (1992) Makhzan al-Adwiah (Drug Treasure, in Persian). Reprinted from a copy which was printed in Calcutta dated in 1844. Enqelab-e Eslami Publishing and Educational Organization, Tehran, Iran
33. Jorjāni SI (1977) In: Saeedi Sirjani AA (ed) Dhakhireh Khārazmshāhi, (treasure of Khārazmshāh, in Persian), photoprint of the manuscript dated 1206 AD. The Iranian Culture Foundation, Tehran, Iran
34. Herawi MY (2008) Baḥr al-Jawāher (Sea of Gems, in Arabic). Institute of Revival of Natural Medicine, Qom
35. Qorashi AA (607-687 AH/1211-1288 AD) al-Shāmel (The comprehensive, in Arabic). Vol. 19. Al Majmaʿ al-Thaqāfi. Abu-Dhabi, United Arab Emirates
36. Jorjāni SI (2006) Al-Aghrāḍ al-Ṭibbia wal Mabāheth al-ʿAlāiia (Medical Purposes and al-Alâee Investigations,in Persian), vol 1. Tehran University Publication, Tehran, Iran
37. Ayati Z, Ramezani M, Amiri MS, Moghadam AT, Rahimi H, Abdollahzade A et al (2019) Ethnobotany, phytochemistry and traditional uses of *Curcuma* spp. and pharmacological profile of two important species (*C. longa* and *C. zedoaria*): a review. Curr Pharm Des 25(8):871–935
38. Vasanthi HR, Parameswari RP (2010) Indian spices for healthy heart - an overview. Curr Cardiol Rev 6(4):274–279
39. Duggi S, Harish H, Handral R, Tulsianand G, Shruthi SD (2013) Turmeric: nature's precious medicine. Asian J Pharm Clin Res 6(3):10–16
40. Samala S, Veeresham C (2017) Ethnobotany of turmeric and its medicinal importance. In: Ethnobotany of India. Volume 3: North-East India and the Andaman and Nicobar Islands. CRC Press/Taylor & Francis, Boca Raton, FL, pp 293–316
41. Bone K, Mills S (2012) Principles and practice of phytotherapy: modern herbal medicine. Principles and practice of phytotherapy, 2nd edn. Elsevier, Austin, TX, pp 900–918
42. Khanna NM (1999) Turmeric – Nature's precious gift. Curr Sci 76(10):1351–1356
43. Kawasaki K, Fukuhara I, Muroyama K, Murosaki S (2017) Effect of a combination of hot water extract of *Curcuma longa* and curcumin on serum liver enzymes, inflammatory markers, and emotional states in healthy participants with moderately high body mass index: a randomized, double-blind, placebo-controlled clinical trial. Jpn Pharmacol Ther 45(2):243–252
44. Vafadar Afshar G, Rasmi Y, Yagmayee P, Khadem-Ansari MH, Makhdomii K, Rasooli J (2020) The

effects of nano-curcumin supplementation on serum level of hs-CRP, adhesion molecules, and lipid profiles in hemodialysis patients, a randomized controlled clinical trial. Iran J Kidney Dis 14(1):52–61
45. Javadi M, Khadem Haghighian H, Goodarzy S, Abbasi M, Nassiri-Asl M (2019) Effect of curcumin nanomicelle on the clinical symptoms of patients with rheumatoid arthritis: a randomized, double-blind, controlled trial. Int J Rheum Dis 22(10):1857–1862
46. Abdolahi M, Jafarieh A, Sarraf P, Sedighiyan M, Yousefi A, Tafakhori A et al (2019) The neuromodulatory effects of ω-3 fatty acids and nano-curcumin on the COX-2/iNOS network in migraines: a clinical trial study from gene expression to clinical symptoms. Endocr Metab Immune Disord Drug Targets 19(6):874–884
47. Abdolahi M, Sarraf P, Javanbakht MH, Honarvar NM, Hatami M, Soveyd N et al (2018) A novel combination of ω-3 fatty acids and nano-curcumin modulates interleukin-6 gene expression and high sensitivity C-reactive protein serum levels in patients with migraine: a randomized clinical trial study. CNS Neurol Disord Drug Targets 17(6):430–438
48. Parohan M, Sarraf P, Javanbakht MH, Foroushani AR, Ranji-Burachaloo S, Djalali M (2019) The synergistic effects of nano-curcumin and coenzyme Q10 supplementation in migraine prophylaxis: a randomized, placebo-controlled, double-blind trial. Nutr Neurosci:1–10. https://doi.org/10.1080/1028415X.2019.1627770
49. Khayat S, Fanaei H, Kheirkhah M, Moghadam ZB, Kasaeian A, Javadimehr M (2015) Curcumin attenuates severity of premenstrual syndrome symptoms: a randomized, double-blind, placebo-controlled trial. Complement Ther Med 23(3):318–324
50. Afshariani R, Farhadi P, Ghaffarpasand F, Roozbeh J (2014) Effectiveness of topical curcumin for treatment of mastitis in breastfeeding women: a randomized, double-blind, placebo-controlled clinical trial. Oman Med J 29(5):330–334
51. Ryan JL, Heckler CE, Ling M, Katz A, Williams JP, Pentland AP et al (2013) Curcumin for radiation dermatitis: a randomized, double-blind, placebo-controlled clinical trial of thirty breast cancer patients. Radiat Res 180(1):34–43
52. Ataei-Almanghadim K, Farshbaf-Khalili A, Ostadrahimi AR, Shaseb E, Mirghafourvand M (2020) The effect of oral capsule of curcumin and vitamin E on the hot flashes and anxiety in postmenopausal women: a triple blind randomised controlled trial. Complement Ther Med 48:102267. https://doi.org/10.1016/j.ctim.2019.102267
53. Masoodi M, Mahdiabadi MA, Mokhtare M, Agah S, Kashani AHF, Rezadoost AM et al (2018) The efficacy of curcuminoids in improvement of ulcerative colitis symptoms and patients' self-reported well-being: a randomized double-blind controlled trial. J Cell Biochem 119(11):9552–9559
54. Panahi Y, Ghanei M, Bashiri S, Hajihashemi A, Sahebkar A (2014) Short-term curcuminoid supplementation for chronic pulmonary complications due to sulfur mustard intoxication: positive results of a randomized double-blind placebo-controlled trial. Drug Res 65(11):567–573
55. Panahi Y, Ghanei M, Hajhashemi A, Sahebkar A (2016) Effects of curcuminoids-piperine combination on systemic oxidative stress, clinical symptoms and quality of life in subjects with chronic pulmonary complications due to sulfur mustard: a randomized controlled trial. J Diet Suppl 13(1):93–105
56. Azhdari M, Karandish M, Mansoori A (2019) Metabolic benefits of curcumin supplementation in patients with metabolic syndrome: a systematic review and meta-analysis of randomized controlled trials. Phytother Res 33(5):1289–1301
57. Panahi Y, Rahimnia AR, Sharafi M, Alishiri G, Saburi A, Sahebkar A (2014) Curcuminoid treatment for knee osteoarthritis: a randomized double-blind placebo-controlled trial. Phytother Res 28(11):1625–1631
58. Pinsornsak P, Niempoog S (2012) The efficacy of *Curcuma longa* L. extract as an adjuvant therapy in primary knee osteoarthritis: a randomized control trial. J Med Assoc Thail 95(Suppl 1):S51–S58
59. Panahi Y, Sahebkar A, Amiri M, Davoudi SM, Beiraghdar F, Hoseininejad SL et al (2012) Improvement of Sulphur mustard-induced chronic pruritus, quality of life and antioxidant status by curcumin: results of a randomised, double-blind, placebo-controlled trial. Br J Nutr 108(7):1272–1279
60. White CM, Pasupuleti V, Roman YM, Li Y, Hernandez AV (2019) Oral turmeric/curcumin effects on inflammatory markers in chronic inflammatory diseases: a systematic review and meta-analysis of randomized controlled trials. Pharmacol Res 146:104280. https://doi.org/10.1016/j.phrs.2019.104280
61. Heshmati J, Golab F, Morvaridzadeh M, Potter E, Akbari-Fakhrabadi M, Farsi F et al (2020) The effects of curcumin supplementation on oxidative stress, Sirtuin-1 and peroxisome proliferator activated receptor γ coactivator 1α gene expression in polycystic ovarian syndrome (PCOS) patients: a randomized placebo-controlled clinical trial. Diabetes Metab Syndr 14(2):77–82
62. Fleenor BS, Carlini NA, Campbell MS (2019) Curcumin and arterial function in health and disease: impact on oxidative stress and inflammation. Curr Opin Clin Nutr Metab Care 22(6):459–464
63. Jin T (2019) Mechanisms underlying the metabolic beneficial effect of curcumin intervention: beyond anti-inflammation and anti-oxidative stress. Obes Med 13:1–5
64. Tabrizi R, Vakili S, Akbari M, Mirhosseini N, Lankarani KB, Rahimi M et al (2019) The effects of curcumin-containing supplements on biomarkers of inflammation and oxidative stress: a systematic review and meta-analysis of randomized controlled trials. Phytother Res 33(2):253–262
65. Qin S, Huang L, Gong J, Shen S, Huang J, Tang Y et al (2018) Meta-analysis of randomized controlled trials of 4 weeks or longer suggest that curcumin

may afford some protection against oxidative stress. Nutr Res 60:1–12
66. Akbari M, Lankarani KB, Tabrizi R, Ghayour-Mobarhan M, Peymani P, Ferns G et al (2019) The effects of curcumin on weight loss among patients with metabolic syndrome and related disorders: a systematic review and meta-analysis of randomized controlled trials. Front Pharmacol 10:649. https://doi.org/10.3389/fphar.2019.00649
67. Di Pierro F, Bressan A, Ranaldi D, Rapacioli G, Giacomelli L, Bertuccioli A (2015) Potential role of bioavailable curcumin in weight loss and omental adipose tissue decrease: preliminary data of a randomized, controlled trial in overweight people with metabolic syndrome. Eur Rev Med Pharmacol Sci 19(21):4195–4202
68. Jalali M, Mahmoodi M, Mosallanezhad Z, Jalali R, Imanieh MH, Moosavian SP (2020) The effects of curcumin supplementation on liver function, metabolic profile and body composition in patients with non-alcoholic fatty liver disease: a systematic review and meta-analysis of randomized controlled trials. Complement Ther Med 48:102283. https://doi.org/10.1016/j.ctim.2019.102283
69. Wei Z, Liu N, Tantai X, Xing X, Xiao C, Chen L et al (2019) The effects of curcumin on the metabolic parameters of non-alcoholic fatty liver disease: a meta-analysis of randomized controlled trials. Hepatol Int 13(3):302–313
70. Jazayeri-Tehrani SA, Rezayat SM, Mansouri S, Qorbani M, Alavian SM, Daneshi-Maskooni M et al (2019) Nano-curcumin improves glucose indices, lipids, inflammation, and Nesfatin in overweight and obese patients with non-alcoholic fatty liver disease (NAFLD): a double-blind randomized placebo-controlled clinical trial. Nutr Metab (Lond) 16:8. https://doi.org/10.1186/s12986-019-0331-1
71. Melo ISVD, Santos AFD, Bueno NB (2018) Curcumin or combined curcuminoids are effective in lowering the fasting blood glucose concentrations of individuals with dysglycemia: systematic review and meta-analysis of randomized controlled trials. Pharmacol Res 128:137–144
72. Na LX, Li Y, Pan HZ, Zhou XL, Sun DJ, Meng M et al (2013) Curcuminoids exert glucose-lowering effect in type 2 diabetes by decreasing serum free fatty acids: a double-blind, placebo-controlled trial. Mol Nutr Food Res 57(9):1569–1577
73. Baziar N, Parohan M (2020) The effects of curcumin supplementation on body mass index, body weight, and waist circumference in patients with nonalcoholic fatty liver disease: a systematic review and dose-response meta-analysis of randomized controlled trials. Phytother Res 34(3):464–474
74. Campbell MS, Ouyang A, Krishnakumar IM, Charnigo RJ, Westgate PM, Fleenor BS (2019) Influence of enhanced bioavailable curcumin on obesity-associated cardiovascular disease risk factors and arterial function: a double-blinded, randomized, controlled trial. Nutrition 62:135–139
75. Saraf-Bank S, Ahmadi A, Paknahad Z, Maracy M, Nourian M (2019) Effects of curcumin on cardiovascular risk factors in obese and overweight adolescent girls: a randomized clinical trial. Sao Paulo Med J 137(5):414–422
76. Qin S, Huang L, Gong J, Shen S, Huang J, Ren H et al (2017) Efficacy and safety of turmeric and curcumin in lowering blood lipid levels in patients with cardiovascular risk factors: a meta-analysis of randomized controlled trials. Nutr J 16(1):68. https://doi.org/10.1186/s12937-017-0293-y
77. Tabrizi R, Vakili S, Lankarani KB, Akbari M, Mirhosseini N, Ghayour-Mobarhan M et al (2018) The effects of curcumin on glycemic control and lipid profiles among patients with metabolic syndrome and related disorders: a systematic review and meta-analysis of randomized controlled trials. Curr Pharm Des 24(27):3184–3199
78. Poolsup N, Suksomboon N, Kurnianta PDM, Deawjaroen K (2019) Effects of curcumin on glycemic control and lipid profile in prediabetes and type 2 diabetes mellitus: a systematic review and meta-analysis. PLoS One 14(4):e0215840. https://doi.org/10.1371/journal.pone.0215840
79. Clark CCT, Ghaedi E, Arab A, Pourmasoumi M, Hadi A (2019) The effect of curcumin supplementation on circulating adiponectin: a systematic review and meta-analysis of randomized controlled trials. Diabetes Metab Syndr 13(5):2819–2825
80. Goodarzi R, Sabzian K, Shishehbor F, Mansoori A (2019) Does turmeric/curcumin supplementation improve serum alanine aminotransferase and aspartate aminotransferase levels in patients with non-alcoholic fatty liver disease? A systematic review and meta-analysis of randomized controlled trials. Phytother Res 33(3):561–570
81. Mousavi SM, Milajerdi A, Varkaneh HK, Gorjipour MM, Esmaillzadeh A (2020) The effects of curcumin supplementation on body weight, body mass index and waist circumference: a systematic review and dose-response meta-analysis of randomized controlled trials. Crit Rev Food Sci Nutr 60(1):171–180
82. Fan C, Song Q, Wang P, Li Y, Yang M, Yu SY (2019) Neuroprotective effects of curcumin on IL-1β-induced neuronal apoptosis and depression-like behaviors caused by chronic stress in rats. Front Cell Neurosci 12:516. https://doi.org/10.3389/fncel.2018.00516
83. Zhu LN, Mei X, Zhang ZG, Xie YP, Lang F (2019) Curcumin intervention for cognitive function in different types of people: a systematic review and meta-analysis. Phytother Res 33(3):524–533
84. Ng QX, Koh SSH, Chan HW, Ho CYX (2017) Clinical use of curcumin in depression: a meta-analysis. J Am Med Dir Assoc 18(6):503–508
85. Al-Karawi D, Al Mamoori DA, Tayyar Y (2016) The role of curcumin administration in patients with major depressive disorder: mini meta-analysis of clinical trials. Phytother Res 30(2):175–183

86. Fusar-Poli L, Vozza L, Gabbiadini A, Vanella A, Concas I, Tinacci S et al (2020) Curcumin for depression: a meta-analysis. Crit Rev Food Sci Nutr 60(15):2643–2653
87. Tayarani-Najaran Z, Tayarani N, Emami SA (2014) The history of islamic medicine at a glance. In: Watson RR, Preedy VR, Zibadi S (eds) Polyphenols in human health and disease. Academic Press, San Diego, CA, pp 17–27
88. Phoolcharoen N, Oranratanaphan S, Ariyasriwatana C, Worasethsin P (2019) Efficacy of curcuminoids for reducing postoperative pain after laparoscopic gynecologic surgery: a pilot randomized trial. J Complement Integr Med 16(4):20180224. https://doi.org/10.1515/jcim-2018-0224
89. Zoodfekr L, Matinhomai H, Tarverdizadeh B (2017) The effect of aerobic exercise and curcumin consumption on clinical signs and serum levels of sexual hormones in 18-35 year-old women with premenstrual syndrome: a randomized clinical trial. Iran J Obstet Gynecol Infertil 20(8):70–79
90. Shao N, Jia H, Li Y, Li J (2017) Curcumin improves treatment outcome of Takayasu arteritis patients by reducing TNF-α: a randomized placebo-controlled double-blind clinical trial. Immunol Res 65(4):969–974
91. Ganjali S, Sahebkar A, Mahdipour E, Jamialahmadi K, Torabi S, Akhlaghi S et al (2014) Investigation of the effects of curcumin on serum cytokines in obese individuals: a randomized controlled trial. ScientificWorldJournal 2014:898361. https://doi.org/10.1155/2014/898361
92. Panahi Y, Saadat A, Beiraghdar F, Sahebkar A (2014) Adjuvant therapy with bioavailability-boosted curcuminoids suppresses systemic inflammation and improves quality of life in patients with solid tumors: a randomized double-blind placebo-controlled trial. Phytother Res 28(10):1461–1467
93. Panahi Y, Sahebkar A, Parvin S, Saadat A (2012) A randomized controlled trial on the anti-inflammatory effects of curcumin in patients with chronic sulphur mustard-induced cutaneous complications. Annal Clin Biochem 49(6):580–588
94. Saraf-Bank S, Ahmadi A, Paknahad Z, Maracy M, Nourian M (2019) Effects of curcumin supplementation on markers of inflammation and oxidative stress among healthy overweight and obese girl adolescents: a randomized placebo-controlled clinical trial. Phytother Res 33(8):2015–2022
95. Alizadeh F, Javadi M, Karami AA, Gholaminejad F, Kavianpour M, Haghighian HK (2018) Curcumin nanomicelle improves semen parameters, oxidative stress, inflammatory biomarkers, and reproductive hormones in infertile men: a randomized clinical trial. Phytother Res 32(3):514–521
96. Nasseri E, Mohammadi E, Tamaddoni A, Qujeq D, Zayeri F, Zand H (2017) Benefits of curcumin supplementation on antioxidant status in β-thalassemia major patients: a double-blind randomized controlled clinical trial. Ann Nutr Metab 71(3–4):136–144
97. Panahi Y, Khalili N, Sahebi E, Namazi S, Karimian MS, Majeed M et al (2017) Antioxidant effects of curcuminoids in patients with type 2 diabetes mellitus: a randomized controlled trial. Inflammopharmacology 25(1):25–31
98. Panahi Y, Alishiri GH, Parvin S, Sahebkar A (2016) Mitigation of systemic oxidative stress by curcuminoids in osteoarthritis: results of a randomized controlled trial. J Diet Suppl 13(2):209–220
99. Panahi Y, Saadat A, Beiraghdar F, Hosseini Nouzari SM, Jalalian HR, Sahebkar A (2014) Antioxidant effects of bioavailability-enhanced curcuminoids in patients with solid tumors: a randomized double-blind placebo-controlled trial. J Funct Foods 6(1):615–622
100. Leite DPV, Paolillo FR, Parmesano TN, Fontana CR, Bagnato VS (2014) Effects of photodynamic therapy with blue light and curcumin as mouth rinse for oral disinfection: a randomized controlled trial. Photomed Laser Surg 32(11):627–632
101. Raghava KV, Sistla KP, Narayan SJ, Yadalam U, Bose A, Mitra K (2019) Efficacy of curcumin as an adjunct to scaling and root planing in chronic periodontitis patients: a randomized controlled clinical trial. J Contemp Dent Pract 20(7):842–846
102. Jamilian M, Foroozanfard F, Kavossian E, Aghadavod E, Shafabakhsh R, Hoseini A et al (2020) Effects of curcumin on body weight, glycemic control and serum lipids in women with polycystic ovary syndrome: a randomized, double-blind, placebo-controlled trial. Clin Nutr ESPEN 36:128–133
103. Sohaei S, Amani R, Tarrahi MJ, Ghasemi-Tehrani H (2019) The effects of curcumin supplementation on glycemic status, lipid profile and hs-CRP levels in overweight/obese women with polycystic ovary syndrome: a randomized, double-blind, placebo-controlled clinical trial. Complement Ther Med 47:102201. https://doi.org/10.1016/j.ctim.2019.102201
104. Chuengsamarn S, Rattanamongkolgul S, Phonrat B, Tungtrongchitr R, Jirawatnotai S (2014) Reduction of atherogenic risk in patients with type 2 diabetes by curcuminoid extract: a randomized controlled trial. J Nutr Biochem 25(2):144–150
105. Safarian H, Parizadeh SMR, Saberi-Karimain M, Darroudi S, Javandoost A, Mohammadi F et al (2019) The effect of curcumin on serum copper and zinc and Zn/cu ratio in individuals with metabolic syndrome: a double-blind clinical trial. J Diet Suppl 16(6):625–634
106. Hodaei H, Adibian M, Nikpayam O, Hedayati M, Sohrab G (2019) The effect of curcumin supplementation on anthropometric indices, insulin resistance and oxidative stress in patients with type 2 diabetes: a randomized, double-blind clinical trial. Diabetol Metab Syndr 11:41. https://doi.org/10.1186/s13098-019-0437-7
107. Cicero AFG, Sahebkar A, Fogacci F, Bove M, Giovannini M, Borghi C (2020) Effects of phytosomal curcumin on anthropometric parameters,

insulin resistance, cortisolemia and non-alcoholic fatty liver disease indices: a double-blind, placebo-controlled clinical trial. Eur J Nutr 59(2):477–483
108. Tamaddoni A, Nasseri E, Mohammadi E, Qujeq D, Zayeri F, Zand H et al (2019) A double-blind randomized controlled trial of curcumin for improvement in glycemic status, lipid profile and systemic inflammation in β-thalassemia major. J Herb Med 21:100324. https://doi.org/10.1016/j.hermed.2019.100324
109. SilvaTAL MDC, Medeiros G, Medeiros R, de Souza AJ, Medeiros JA et al (2019) Influence of curcumin supplementation on metabolic and lipid parameters of people living with HIV/AIDS: a randomized controlled trial. BMC Complement Altern Med 19(1):202. https://doi.org/10.1186/s12906-019-2620-7
110. Adibian M, Hodaei H, Nikpayam O, Sohrab G, Hekmatdoost A, Hedayati M (2019) The effects of curcumin supplementation on high-sensitivity C-reactive protein, serum adiponectin, and lipid profile in patients with type 2 diabetes: a randomized, double-blind, placebo-controlled trial. Phytother Res 33(5):1374–1383
111. Ferguson JJA, Stojanovski E, MacDonald-Wicks L, Garg ML (2018) Curcumin potentiates cholesterol-lowering effects of phytosterols in hypercholesterolaemic individuals. A randomised controlled trial. Metabolism 82:22–35
112. Panahi Y, Khalili N, Sahebi E, Namazi S, Reiner Ž, Majeed M et al (2017) Curcuminoids modify lipid profile in type 2 diabetes mellitus: a randomized controlled trial. Complement Ther Med 33:1–5
113. Panahi Y, Hosseini MS, Khalili N, Naimi E, Soflaei SS, Majeed M et al (2016) Effects of supplementation with curcumin on serum adipokine concentrations: a randomized controlled trial. Nutrition 32(10):1116–1122
114. Panahi Y, Khalili N, Hosseini MS, Abbasinazari M, Sahebkar A (2014) Lipid-modifying effects of adjunctive therapy with curcuminoids-piperine combination in patients with metabolic syndrome: results of a randomized controlled trial. Complement Ther Med 22(5):851–857
115. Yang YS, Su YF, Yang HW, Lee YH, Chou JI, Ueng KC (2014) Lipid-lowering effects of curcumin in patients with metabolic syndrome: a randomized, double-blind, placebo-controlled trial. Phytother Res 28(12):1770–1777
116. Mohammadi A, Sahebkar A, Iranshahi M, Amini M, Khojasteh R, Ghayour-Mobarhan M et al (2013) Effects of supplementation with curcuminoids on dyslipidemia in obese patients: a randomized crossover trial. Phytother Res 27(3):374–379
117. Panahi Y, Kianpour P, Mohtashami R, Jafari R, Simental-Mendía LE, Sahebkar A (2017) Efficacy and safety of phytosomal curcumin in non-alcoholic fatty liver disease: a randomized controlled trial. Drug Res (Stuttg) 67(4):244–251
118. Rahmani S, Asgary S, Askari G, Keshvari M, Hatamipour M, Feizi A et al (2016) Treatment of non-alcoholic fatty liver disease with curcumin: a randomized placebo-controlled trial. Phytother Res 30(9):1540–1548
119. Panahi Y, Kianpour P, Mohtashami R, Jafari R, Simental-Mendía LE, Sahebkar A (2016) Curcumin lowers serum lipids and uric acid in subjects with nonalcoholic fatty liver disease: a randomized controlled trial. J Cardiovasc Pharmacol 68(3):223–229
120. Panahi Y, Khalili N, Sahebi E, Namazi S, Simental-Mendía LE, Majeed M et al (2018) Effects of curcuminoids plus piperine on glycemic, hepatic and inflammatory biomarkers in patients with type 2 diabetes mellitus: a randomized double-blind placebo-controlled trial. Drug Res (Stuttg) 68(7):403–409
121. Mirhafez SR, Farimani AR, Dehhabe M, Bidkhori M, Hariri M, Ghouchani BF et al (2019) Effect of phytosomal curcumin on circulating levels of adiponectin and leptin in patients with non-alcoholic fatty liver disease: a randomized, double-blind, placebo-controlled clinical trial. J Gastrointestin Liver Dis 28:183–189
122. Mohammadi E, Tamaddoni A, Qujeq D, Nasseri E, Zayeri F, Zand H et al (2018) An investigation of the effects of curcumin on iron overload, hepcidin level, and liver function in β-thalassemia major patients: a double-blind randomized controlled clinical trial. Phytother Res 32(9):1828–1835
123. Nouri-Vaskeh M, Afshan H, Malek Mahdavi A, Alizadeh L, Fan X, Zarei M (2020) Curcumin ameliorates health-related quality of life in patients with liver cirrhosis: a randomized, double-blind placebo-controlled trial. Complement Ther Med 49:102351. https://doi.org/10.1016/j.ctim.2020.102351
124. Sabery Karimian M, Keshvari M, Ghayour-Mobarhan M, Salehizadeh L, Rahmani S, Behnam B et al (2020) Effects of curcuminoids on inflammatory status in patients with non-alcoholic fatty liver disease: an exploratory randomized controlled trial. Complement Ther Med 49:102322. https://doi.org/10.1016/j.ctim.2020.102322
125. Nouri-Vaskeh M, Malek Mahdavi A, Afshan H, Alizadeh L, Zarei M (2020) Effect of curcumin supplementation on disease severity in patients with liver cirrhosis: a randomized controlled trial. Phytother Res 34(6):1446–1454
126. Chashmniam S, Mirhafez SR, Dehabeh M, Hariri M, Azimi Nezhad M, Nobakht MGBF (2019) A pilot study of the effect of phospholipid curcumin on serum metabolomic profile in patients with non-alcoholic fatty liver disease: a randomized, double-blind, placebo-controlled trial. Eur J Clin Nutr 73(9):1224–1235
127. Asadi S, Gholami MS, Siassi F, Qorbani M, Khamoshian K, Sotoudeh G (2019) Nano curcumin supplementation reduced the severity of diabetic sensorimotor polyneuropathy in patients with type 2 diabetes mellitus: a randomized double-blind placebo-controlled clinical trial. Complement Ther Med 43:253–260

128. Asadi S, Gholami MS, Siassi F, Qorbani M, Sotoudeh G (2019) Beneficial effects of nano-curcumin supplement on depression and anxiety in diabetic patients with peripheral neuropathy: a randomized, double-blind, placebo-controlled clinical trial. Phytother Res 34(4):896–903
129. Ahmadi M, Agah E, Nafissi S, Jaafari MR, Harirchian MH, Sarraf P et al (2018) Safety and efficacy of nanocurcumin as add-on therapy to riluzole in patients with amyotrophic lateral sclerosis: a pilot randomized clinical trial. Neurotherapeutics 15(2):430–438
130. Small GW, Siddarth P, Li Z, Miller KJ, Ercoli L, Emerson ND et al (2018) Memory and brain amyloid and tau effects of a bioavailable form of curcumin in non-demented adults: a double-blind, placebo-controlled 18-month trial. Am J Geriatr Psychiatry 26(3):266–277
131. Chico L, Ienco EC, Bisordi C, Lo Gerfo A, Petrozzi L, Petrucci A et al (2018) Amyotrophic lateral sclerosis and oxidative stress: a double-blind therapeutic trial after curcumin supplementation. CNS Neurol Disord Drug Targets 17(10):767–779
132. Fanaei H, Khayat S, Kasaeian A, Javadimehr M (2016) Effect of curcumin on serum brain-derived neurotrophic factor levels in women with premenstrual syndrome: a randomized, double-blind, placebo-controlled trial. Neuropeptides 56:25–31
133. Lopresti AL, Maes M, Meddens MJM, Maker GL, Arnoldussen E, Drummond PD (2015) Curcumin and major depression: a randomised, double-blind, placebo-controlled trial investigating the potential of peripheral biomarkers to predict treatment response and antidepressant mechanisms of change. Eur Neuropsychopharmacol 25(1):38–50
134. Esmaeili Vardanjani SA, Sehati Shafai F, Mohebi P, Deyhimi M, Delazar A, Ghojazadeh M et al (2012) Wound healing benefits of curcumin for perineal repair after episiotomy: results of an Iranian randomized controlled trial. Life Sci J 9(4):5536–5541
135. Manifar S, Obwaller A, Gharehgozloo A, Boorboor Shirazi Kordi HR, Akhondzadeh S (2012) Curcumin gel in the treatment of minor aphthous ulcer: a randomized, placebo- controlled trial. J Med Plants 11(41):40–45
136. Piyush P, Mahajan A, Singh K, Ghosh S, Gupta S (2019) Comparison of therapeutic response of lycopene and curcumin in oral submucous fibrosis: a randomized controlled trial. Oral Dis 25(1):73–79

Targeting Nuclear Factor-Kappa B Signaling Pathway by Curcumin: Implications for the Treatment of Multiple Sclerosis

3

Meysam Gachpazan, Sadra Habbibirad, Hoda Kashani, Tannaz Jamialahmadi, Hamid Reza Rahimi, and Amirhossein Sahebkar

Abstract

Multiple sclerosis (MS) is a chronic inflammatory disease of the central nervous system, which involves an auto-immune mechanism that leads to perivascular demyelination. The role of nuclear factor-kappa B (NF-κB) signaling pathway in the pathogenesis of MS has been suggested by genome-wide association studies. Therefore, strategies targeting this pathway could be potentially beneficial. Curcumin is the active component of turmeric and a phenolic phytochemical. This phytochemical has anti-inflammatory properties and has been shown by multiple studies to downregulate NF-κB and its downstream gene targets including cyclooxygenase-2, tumor necrosis factor-α, interleukin-1, and interleukin-6. This review discusses the modulatory effects of curcumin on the NF-κB signaling pathway and its downstream effectors, and the therapeutic implications of this modulation on MS.

Keywords

Multiple sclerosis · Curcumin · NF-kappa B · MS · Nuclear factor-kappa B

M. Gachpazan
Faculty of Medicine, Department of Modern Sciences and Technologies, Mashhad University of Medical Sciences, Mashhad, Iran

Student Research Committee, Department of Modern Sciences & Technologies, School of Medicine, Mashhad University of Medical Sciences, Mashhad, Iran

S. Habbibirad · H. Kashani
Faculty of Medicine, Department of Modern Sciences and Technologies, Mashhad University of Medical Sciences, Mashhad, Iran

T. Jamialahmadi
Department of Food Science and Technology, Quchan Branch, Islamic Azad University, Quchan, Iran

Department of Nutrition, Faculty of Medicine, Mashhad University of Medical Sciences, Mashhad, Iran

H. R. Rahimi (✉)
Neurogenic Inflammation Research Center, Mashhad University of Medical Sciences, Mashhad, Iran
e-mail: rahimihr@mums.ac.ir

A. Sahebkar(✉)
Neurogenic Inflammation Research Center, Mashhad University of Medical Sciences, Mashhad, Iran

Biotechnology Research Center, Pharmaceutical Technology Institute, Mashhad University of Medical Sciences, Mashhad, Iran

Halal Research Center of IRI, FDA, Tehran, Iran
e-mail: sahebkara@mums.ac.ir

P. C. Guest (ed.), *Studies on Biomarkers and New Targets in Aging Research in Iran*, Advances in Experimental Medicine and Biology 1291, https://doi.org/10.1007/978-3-030-56153-6_3

3.1 Introduction

Multiple sclerosis (MS) is a chronic inflammatory disease of the central nervous system (CNS) [1, 2]. It is an autoimmune and demyelinating disease [3] that has a multifactorial pathogenesis including lifestyle, genetic, environmental, infective, and non-infective risk factors [2]. In the early stages of MS, autoreactive CD4+ T cell activity and T-helper 1 (Th1) production have an important role in CNS damage. In addition, $CD4^+$ Th1 overactivity has also been identified in MS active lesions [3]. $CD4^+$ $CD25^+$ T-regulatory (TR) cells contribute to the suppression of autoimmune responses [4]. Also, $CD8^+$ T cells may be involved in the CNS damage. Inflammation is a major risk factor for MS, and inflammatory cytokines play an important role in the pathogenesis of MS [5].

Chemokines and receptors, as immune molecules, perform an important function in inflammation [6]. High levels of three chemokines in the cerebrospinal fluid (CSF) have been detected during acute demyelinating events in the CNS, such as acute MS attack. These chemokines include interferon-γ-inducible protein of 10 kDa (IP-10), monokine induced by interferon-γ (Mig) and regulated on activation normal T-cell expressed and secreted (RANTES) [7]. CXC chemokine receptor 3 (CXCR3) is expressed in the majority of T cells within the CSF independent of inflammatory process in the CNS. Such molecules play a major role in the accumulation of T cells in MS patients [8].

Many clinical trials have shown that Th1, Th17, interferon-γ (IFN-γ), and interleukin-17 (IL-17) play an important role in MS pathogenesis [9]. Animal model studies have shown that the distribution of IL-17 in MS lesions can be linked to the pathogenesis of the disease. This link is characterized by the increased activity of $CD4^+$ Th-17 cells. $IL\text{-}17^+$ T cells are significantly increased in the active lesions of MS compared to non-active areas. IL-17 cellular subtypes such as $CD4^+$ and $CD8^+$ T cells play an important role in the active lesions of MS [3].

Leukocytes perform diapedesis via interaction with the endothelial adhesion molecules. The diapedesis of cells occurs through the perivascular cuff into the extra-cellular matrix before entrance into the CNS. Adhesion molecules such as α4β1 integrin have been identified to cause a relapse in MS patients while αB crystalline was found to contribute to disease remission [10–12]. Osteopontin (Opn), which has both adhesive and Th-1-like cytokine activity, as well as other cytokines secreted by T cells and antigen-presenting cells, injure the myelin-generated cells (oligodendrocytes) inside the brain [10]. An animal study showed that inflammation may be increased by Opn. This molecule inhibits Foxo3a transcription via increasing nuclear factor kappa B (NF-κB) activity, as well as changes in the expression of proapoptotic proteins such as Bim, Bak, and Bax. Thus, the survival of activated T cells can increase. Opn-excess conditions increase inflammation, while brain-infiltrating lymphocytes cells will die under low Opn conditions [13]. Overexpression of Opn is also seen in MS lesions [14].

Magnetic resonance imaging (MRI) scans carried out on MS patients have demonstrated that inhibition of α4 integrin using an antibody can significantly reduce the number of new active lesions [15]. Plasma cells produce specific antibodies against myelin, which cause injuries to the myelin sheath [10, 16]. However, T-lymphocyte apoptosis and anti-inflammatory actions can be induced by ligands of peroxisome proliferator-activated receptor-gamma (PPARγ). Inflammatory and T lymphocyte activity in glial cells can be inhibited by PPARγ ligands. Pershadsingh et al. showed that pioglitazone, as a PPARγ agonist, not only reduced the clinical symptoms, but also improved external clinical efficacy without side effects [17].

Multiple sclerosis–associated retro virus (MSRV) infection plays a role in many cases of MS. Envelope proteins of this virus (Env-ms) have a pro-inflammatory potential in MS. Therefore, these can induce leukocyte diapedesis to the regions affected in MS. The production of IL-6 and IL-8 and overexpression of intracellular adhesion molecule 1 (ICAM-1) is due to Env-ms. Duperray et al. demonstrated that recognition of Env-ms by the innate immune system is done via Toll-like receptor 4 (TLR4). Therefore, the Env-ms/TLR4 pathway has an important role in the inflammation and demyelination effects affiliated with MS [18].

3.2 NF-κB

NF-κB, as a nuclear transcription factor, plays a central role during the innate and adaptive immune responses. Inflammation is the key outcome of NF-κB activation due to the overexpression of downstream pro-inflammatory genes [17, 19, 20]. In 2004, Feng et al. demonstrated that NF-κB is essential for the development of the B cell [21]. In human astrocytes cultures, Choi et al. showed that the gene expression of cytokines and chemokines is regulated by the NF-κB signaling pathway, in both resting and activating conditions [22].

The NF-κB pathway is activated through the IκB kinase complex consisting of either canonical or non-canonical routes driven by the inhibitor of NF-κB kinase subunit beta (IKKβ) and IKKα, respectively, while the IκB protein inhibits the NF-κB pathway. Thus, NF-κB is inhibited by IκB and activated by IKKβ and IKKα [23]. For activation, the NF-κB protein should be in a heterodimeric form incorporating subunits of the Rel family (e.g., p50, Rel A or p65, p52, v-Rel, c-Rel, and Rel B) [24]. Phosphorylation of the IκB protein separates this inhibitory protein from the NF-κB complex and leads to its degradation by the proteasomes. Some known external factors including mitogens, inflammatory cytokines, ultraviolet radiation, ionizing radiation, viral proteins, bacterial lipopolysaccharides (LPS), and reactive oxygen species (ROS) could facilitate this process. Phosphorylated IκB proteins will disinhibit the NF-κB pathway and induce translocation of its free dimer to the nucleus. Subsequent binding to the cis-acting κB element leads to the transcription of genes encoding different substances such as cytokines, cell adhesion molecules, and growth factors [24].

3.2.1 NF-κB in the Nervous System

NF-κB is essential for the proper functioning of cells in the nervous system including neurons, astrocytes, microglia, and oligodendrocytes [25]. In line with this, Schneider et al. demonstrated that the NF-κB pathway is a candidate for neurological diseases including stroke [24]. The NF-κB complex regulates many functional pathways in the CNS including neuronal development, cellular responses to injury, and synaptic signaling, which contributes to learning and memory [20]. In MS, inflammatory bowel disease (IBD), and many inflammatory diseases, NF-κB plays an important role, through the regulation of various inflammatory genes. To this end, the NF-κB complex is involved in transcriptional regulation of various genes such as those in the major histocompatibility complex (MHC), as well as those encoding adhesion molecules and cytokines IL-2, IL-6, interferon-β, TGF-β, and TNF-α. In addition, genetic predisposition to some inflammatory diseases could be due to polymorphisms in the TNF gene [26, 27].

3.3 Curcumin

Many plant-derived compounds are used as multi-target approaches in clinical therapeutics. Therapeutic use of phytochemicals has a long background in several systems of traditional medicine around the world [28]. Curcumin, a polyphenolic compound that imparts the yellow color and most of the pharmacological activities of turmeric, has attracted considerable attention from the biomedical community owing to its numerous pharmacological effects [29–36]. Many studies have demonstrated the safety as well as therapeutic effects of curcumin in cancer as well as cardiovascular, inflammatory, metabolic, neurological, and skin diseases [37, 38]. Based on in vitro and in vivo models, the expression of inflammatory cytokines can be controlled by this natural compound which also has anticancer effects [39]. In 2017, Xie et al. showed that inhibition of NF-κB as well as decreased serum TNF-α may occur with larger doses of curcumin [40]. An immunomodulatory effect of curcumin has been proven now for two decades. This effect is induced via modulation of the activation of T cells, B cells, macrophages, neutrophils, natural killer cells, and dendritic cells. In addition, pro-inflammatory cytokines including TNF, IL-1,

IL-2, IL-6, IL-8, IL-12, and chemokines are downregulated by curcumin, through blockade of the transcription factor NF-κB. However, it is remarkable that the low doses of curcumin can increase the antibody responses. Ultimately, in inflammatory diseases and autoimmune disorders such as Alzheimer's disease, MS, IBD, and rheumatoid arthritis, curcumin may have beneficial effects [41].

As stated above, curcumin is a natural inhibitor of NF-κB and an anti-inflammatory agent [40]. While IKκB plays a key role in the activation of NF-κB pathway, curcumin appears to circumvent this route by blocking IκB kinase and protein kinase B (AKT) activities [23].

3.3.1 Immunoregulatory Effects of Curcumin

CD80 and CD86 play an important role in inducing T cell activity by stimulation signals. Not only does curcumin inhibit the dendritic cells and inhibit these proteins, but it also suppresses MHC class II expression. Hence, curcumin impairs pro-inflammatory cytokine (IL-12) production via inhibition of mitogen-activated protein kinase (MAPK) and NF-κB p65 translocation [43]. On the other hand, Alagawany et al. demonstrated that curcumin causes a significant increase in IgG and IgM levels in rabbits [44]. NF-κB is known as a controller of inflammatory mediators such as cyclooxygenase 2 (COX-2), inducible nitric oxide synthase (iNOS), and cytokines. Furthermore, it has high potential in the regulation of more than 400 genes, which are involved in inflammation and chronic diseases like MS [45]. As mentioned above, curcumin modulates cytokine levels via inhibition of the NF-κB signaling pathway. Curcumin prevents pancreatic leucocyte penetration and leads to decreased activation of NF-κB in T cell receptor (TCR)-stimulated in lymphocytes from nonobese diabetic (NOD) mice. Also, it reduces pro-inflammatory cytokines and nitric oxide (NO) discharge and antigen-presenting cell activity due to destruction of the T cell stimulatory function of dendritic cells [46]. In microglial cells, curcumin significantly reduces the NO release of LPS and the expression of iNOS, as well as pro-inflammatory cytokines, in a dose-dependent manner. Interestingly, curcumin prevents PI3K/AKT phosphorylation in LPS-activated microglial cells, as well as NF-κB activation [47]. Myeloid differentiation factor 88 (MyD88) is a vital protein for TLR4 activity. The activation of downstream NF-κB and pro-inflammatory cytokines production appears to depend on this adaptor protein. In 2014, Zhu et al. showed that curcumin may improve patient outcomes in traumatic brain injury (TBI) by modulating the TLR4/MyD88/NF-κB signaling pathway in microglia/macrophages [48].

One of the other signaling pathways that have critical roles is the Janus kinase/signal transducers and activators of transcription (JAK/STAT) signaling cascade. JAK/STAT signaling is involved in many physiological pathways such as cellular homeostasis, modulation of immune responses, regulation of a wide array of cytokines and growth factors of cell proliferation, differentiation, migration, and apoptosis [49]. Curcumin prevents STAT3 phosphorylation in various cell types at dosages of 20–50 μM [50, 51]. Myeloperoxidase activity (MPO; the index of leukocyte infiltration) and STAT3 phosphorylation were significantly reduced when 50 mg/kg of curcumin was administered. The DNA-binding activity of STAT3 is significantly reduced, and this was found to result in IL-1 and TNF-α downregulation [52]. Suppressors of cytokine signaling proteins (SOCS), other pro-inflammatory proteins and cytokine production mediators are JAK/STAT antagonists. Curcumin via inhibition of SOCS1 and 3 plays an important role in the suppression of LPS-induced expression of IL-6, TNF-α, and prostaglandin-endoperoxide synthase 2 (PGES2) mRNA levels in RAW murine macrophages [53]. Altogether, curcumin could play a central role in the treatment of patients with inflammatory diseases like MS via inhibiting the NF-κB signaling pathway (Table 3.1).

Table 3.1 Literature review: effects of curcumin on MS-afflicted patients

Year	Compound	Duration	Model	Outcome	Ref.
24 May 2018	Nano-curcumin contains 80 mg of curcumin in the form of nano-micelle	6 months	MS patients compared with healthy control group	↓ 1. miR-145 2. miR-132 3. STAT1 4. AP-1 5. IL-1β 6. IL-6 7. CCL2 8. TNF-α 9. NF-κB	[54]
22 May 2018	Nano-curcumin contains 80 mg of curcumin in the form of nano-micelle	6 months	MS patients compared with healthy control group	↓ 1. Th17 2. IL-17A 3. IL-23 4. RORγt	[55]
17 Sep 2018	80 mg nano-curcumin	Study started May 12, 2017	MS patients		NCT03150966
	Placebo	Last update Sep 17, 2018			
23 Jan 2017	Curcumin (500 mg 2 times/day)	Study started Apr 2012	Subjects with active relapsing MS treated		NCT01514370
	IFN beta 1a (44 mg three times/week)	Study completed Mar 2016			
13 Jul 2015	PNC/day	12 till day 29	EAE Female Lewis rats	↓ 1. Scores of disease 2. Demyelination 3. Inflammation 4. BBB breaking down 5. IL-1 6. IL-17 7. TNF-αr1 8. MCP-1 9. NF-κb 10. Oxidative stress ↑ 1. Progenitor cell markers 2. Improved remyelination	[56]
	Control (12.5 mg/kg- i.p. curcumin dissolved in PBS/day)	0 till day 29			
22 Jan 2009	Curcumin 100 or 200 mg/kg + 0.5% Methylcellulose/day	14 days	EAE Female Lewis rats	↓ 1. Proliferation of the MBP-reaction lymphocyte 2. IL-17 3. TGF-β 4. IL-6 5. IL-21 6. STAT3 7. RORγt 8. STAT3-phosphorylation inhibited 9. Inhibited differentiation and development of Th17	[57]
	0.5% Methylcellulose/day				

(continued)

Table 3.1 (continued)

Year	Compound	Duration	Model	Outcome	Ref.
26 Mar 2008	15d-PGJ2 and or curcumin	14 days	EAE SJL/J and C57BL/6 mice	↓ 1. TLR4 and TLR9 expression in CD4+ and CD8+ T cells	[58]
	Control (DMSO)				
13 Apr 2005	Luteolin, apigenin, morin, fisitin, quercetin, hesperitin, curcumin	Three days before immunization with PLP139–151 until the end of study at day 21	EAE Female SJL/J mice	1. Day 1–11, did not show difference in the development of clinical signs 2. Improvement in clinical signs during recovery phase 3. Recovered more rapidly than control group 4. Reflected by reduction in mean cumulative clinical score by 24% 5. Reduced mean maximal score	[59]
15 Jun 2002	Intervention (curcumin 50 or 100 g in DMSO/day	0–25 days	EAE Female SJL/J mice & EOC-20 mouse microglial cell line	1. IL-12 production from macrophage/microglial cells 2. Differentiation of neural Ag-specific Th1 cells 3. Inhibition of JAK-STAT pathway 4. Decreased development of paralysis only for 12 days (25% reduction) with a MMCS of 1.2 (64.3% inhibition), and decreased further to 8 days (50% inhibition) with a MMCS of 0.3 (89.3% inhibition; $p = 0.001$) 5. 50 and 100 μg curcumin also resulted in decreased CNS demyelination 6. Dose-dependent decrease in IFN-γ production 7. Inhibition in IL-12-induced tyrosine phosphorylation of STAT3 in T cells 8. Inhibition of IL-12-induced STAT4 phosphorylation 9. Inhibition of IL-12-induced tyrosine phosphorylation and activation of JAK2 and TYK2 in T cells	[60]
	Control (DMSO)				

MS multiple sclerosis, *PNC* polymerized nano-curcumin, *EAE* experimental autoimmune encephalomyelitis, *BBB* blood brain barrier, *CNS* central nervous system, *DMSO* dimethyl sulfoxide, *IL* interleukin, *IFN* interferon, *TNFα* tumor necrosis factor alpha, *TGFβ* transforming growth factor beta, *NF-κB* nuclear factor-kappa B, *JAK2* Janus kinase 2, *TYK2* tyrosine kinase 2, *miR* micro-RNA, *STAT* signal transducer and activator of transcription, *AP-1* activator protein 1, *RORγt* RAR-related orphan receptor gamma, *MCP-1* monocyte chemoattractant protein-1, *MMCS* mean maximum clinical severity

3.4 TLR 4

TLRs are a family of pathogen-recognition receptors (PRRs) that play an important role in immune system responses. TLR agonists take the lead in the design and development of vaccines [61]. Also, TLR 4 is a key regulator of the NF-κB route [62] and the innate immune system [63]. Deactivation of TLR4/NF-κB pathway is a goal of anti-inflammatory therapy in ischemic stroke and in brain injury after cerebral ischemia-reperfusion (I/R) damage, due to the inhibitory effect on inflammatory responses [64]. Myeloid differentiation factor 88 (MyD88) is a vital protein for TLR4 activity as the activation of downstream NF-κB targets and pro-inflammatory cytokine production depend on this adaptor protein. In addition, IKKα and IKKβ could inhibit the IκB protein by enhancing its phosphorylation and degradation. Moreover, IKK-related kinases (TBK1 and IKKi) are involved in activation of multiple TLRs which ultimately leads to the activation of the NF-κB pathway [65]. Zhu et al. showed that by modulating the TLR4/MyD88/NF-κB signaling pathway in microglia/macrophages in traumatic brain injury (TBI), curcumin may improve patient outcomes [48]. Lubbad et al. showed that curcumin decreases inflammation by inhibiting TLR-4, MyD88, and NF-κB proteins [66] (Fig. 3.1).

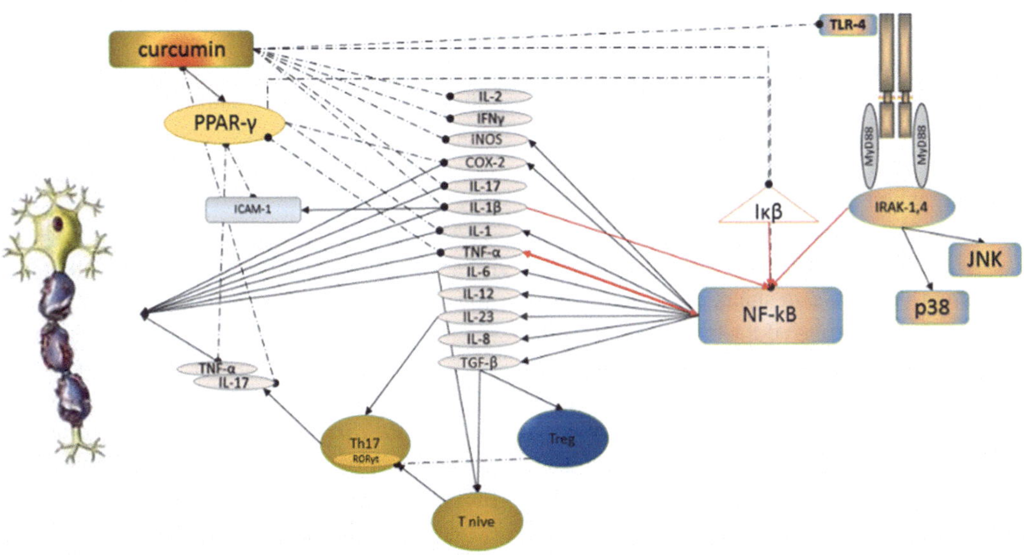

Fig. 3.1 Curcumin via activating PPARγ and inhibiting TLR4 and IKβ reduces NF-kappa B activity. PPARγ plays an important role in pro-inflammatory cytokine production and transcriptional activation of ICAM-1 and Cox-2. P38, and MMP2 plays an important role in inflammation and destruction of ECM which is suppressed by curcumin via inhibition of TLR4. IL-2, IFN-γ, IL-17, IL-1β are pro-inflammatory cytokines directly downregulated by curcumin. Activation of microglia and astrocytes in MS by infection with bacteria or viruses or other environmental factors leads to production of the pro-inflammatory cytokines via activity of NF-kappa B and AP-1. IL-6 and TGF-β and naive T cells are differentiated to T helpher 17 (Th17) cells and express retinoic acid receptor-related orphan receptor γt (RORγt), when IL-6 and TGF-β are presented to them. IL-23 and osteopontin, which stimulate Th17 to secrete IL-17 and TNF-a, are secreted with activated microglia and astrocytes. → induce, *NF-κB* NF-kappa B, *IL-2* interleukin 2, *IL-1* interleukin 1, *IL-1β* interleukin 1 beta, *IL-6* interleukin 6, *IL-8* interleukin 8, *IL-12* interleukin 12, *IL-17* interleukin 17, *IL-23* interleukin 23, *IFN-γ* interferon gamma, *iNOS* inducible nitric oxide synthase, *COX-2* cyclooxygenase-2, *TNF-α* tumor necrosis factor alpha, *TGF-β* transforming growth factor beta, *PPARγ* peroxisome proliferator-activated receptor-gamma, *TLR4* toll-like receptor 4, *IKβ* inhibitor of nuclear factor kappa-B, *ICAM-1* intracellular adhesion molecule 1, *MMP2* matrix metallopeptidase 2, *ECM* extracellular matrix, *MS* multiple sclerosis, *AP-1* activator protein 1, *Th17*, *RORγt* RAR-related orphan receptor gamma

3.5 PPARγ

Apoptosis may be activating result of PPARγ activation, which is associated with downregulation of Bcl-2 and NF-κB activity [67]. PPAR-γ nuclear receptors are increased by curcumin [68] (Fig. 3.1). In 2012, Kutsenko et al. showed that PPARγ is involved in the regulation of NF-κB activation [69]. NF-κB activity is inhibited by an interaction between PPARγ and NF-κB-p65 [70]. However, in middle cerebral artery obstruction rat models, Liu and coworkers showed that enhanced PPARγ activity by curcumin treatment reduced the neurological damage significantly, and reduced the neurological defect. Also, they showed that the mechanism of this appeared to involve the suppression of some inflammatory mediators, including TNF-α, NO, IL-1β, iNOS and COX-2, PGE2, as well as by a reduction of IκB degradation [70].

3.6 COX-2

One of the other genes that plays an important role in inflammation is COX-2, which is also known as prostaglandin-endoperoxidesynthase-2 (PTGS-2) [71, 72]. After peripheral inflammation, the expression of COX-2 in spinal cord neurons was more prominent through the activity of pro-inflammatory cytokines such as TNFα and IL1, and other factors such as TGF-β1, EGF and endotoxin [73]. In addition, the COX-2 and iNOS genes may be regulated via the NF-κB pathway [24]. Although COX-2 is expressed in healthy brain endothelial cells and neurons, increased levels of this factor can be found in inflammatory endothelial lesions [74] and this occurs via the upregulation of the NF-κB pathway [75]. In 2007, Bachmeier et al. used a reverse transcription quantitative polymerase chain reaction (qRT–PCR) technique and found that curcumin treatment of breast cancer cells can reduce COX2 and NF-κB levels. Furthermore, the promoter region of PTGS2/COX2, a classical target of NF-κB, consists of six NF-κB-responsive elements, as well as two sites involved in the 5′ promoter region of the gene encoding NF-κB [24, 76]. In patients with spinal cord MS, excess COX-2 immunoreactivity has been observed [73]. Thus, anti-inflammatory agents against COX-2 are used in the treatment of diseases such as MS [74]. Along similar lines, curcumin is able to negatively regulate COX-2 and iNOS expression, p38 MAPK activity, and colonic nitrite levels [77]. Furthermore, the expression of NF-κB and COX-2, in oral premalignant and cancer cells, has been shown to be reduced by treatment with curcumin in vitro [78].

3.7 Pro-inflammatory Cytokines

TNF-α, IL-1, IL-6, and IFN-γ are increased in MS [79, 80]. TNF-α and IL-1 are pro-inflammatory cytokines and, by downregulation of IKB inhibitory activity, promote the activation of NF-κB [81]. A study showed that the expression of IL-1β, IL-6, and TNF-α were inhibited by curcumin [82]. Since TNF-α, IL-1, and IL-6 can be produced by NF-κB [83], the mechanism of the curcumin inhibitory effect appears to occur through the NF-κB pathway through the suppression of IκB kinase and AKT [84], thus preventing the production of these pro-inflammatory chemicals.

Although TNF-α regulates PPARγ activity at pre- and post-translational levels in inflammation, PPARγ can be inhibited by TNF-α in atherosclerosis and some other pathological conditions [85]. Chiba et al. showed that in atherosclerosis, TNF-α-induced LOX-1 expression can be inhibited by PPARγ activators, and has beneficial influence on inflammatory responses [86]. Also, PPAR-γ agonists as suppressors of inflammatory cytokines could increase benefits of NSAIDs in patients with rheumatoid arthritis [87]. Finally, since PPARγ activity can be increased by curcumin, curcumin may play an important role in suppressing inflammatory cytokine production.

3.8 Conclusions

MS is a neuro-inflammatory disease [1] in which NF-κB has a critical role [20]. This appears to occur through a complex interaction of activators and inhibitors of NF-κB activity. Transcription of apoptotic proteins (e.g., Bim, Bak and Bax) is controlled by NF-κB [13]. PPARγ causes apoptosis of T-lymphocytes [88] and its ligands inhibit NF-κB activity [35]. Curcumin can inhibit NF-κB by increasing PPAR-γ [68]. T-lymphocytes' apoptosis and anti-inflammatory actions can be induced by ligands of PPARγ. PTGS2 via activating NF-κB1 and enhancing Bcl2 promotes cell survival [89]. COX-2 immunoreactivity is increased significantly in patients with spinal MS [73]. Both COX2 and NF-κB are downregulated by curcumin which can be beneficial for MS patients.

In patients with MS, TNF-α, IL-1, IL-6, and IFN-γ are increased [79, 80]. Studies have shown that NF-κB increases the production of TNF-α, IL-1 and IL-6 [42]. Curcumin inhibits IL-1β, IL-6 and TNF-α through blocking IκB kinase and AKT activity [42]. Accordingly, curcumin through inhibiting NF-kappa B signaling pathway could have an important role in the treatment of patients with inflammatory diseases like MS. In addition, a combination of natural products and pharmaceuticals offer a means of targeting multiple pathways involved in MS pathogenesis and thus may enhance the efficacy of treatment. With this in mind, curcumin represents a promising adjunct to conventional therapies for MS.

Studies on curcumin treatment in a rodent model of autoimmune encephalomyelitis (EAE) showed that demyelination, inflammation, IL-1, TNF-α, TLR4, and TLR9 expression in CD4+ and CD8+ T cells, NF-κB, and oxidative stress are decreased significantly, with a concomitant increase in remyelination [56–60]. In addition, a number of clinical studies have attempted to address the same issue. In a phase 2 study (NCT01514370), MS subjects were treated subcutaneously with IFN-β1a (44 μg three times a week) and divided into two groups receiving placebo or curcumin (500 mg 2 times a day). Another study (NCT03150966) aimed to detect the effect of oral nanocurcumin on expression levels of microRNAs and Treg Cells and Th17 cell development factors in MS patients. In other studies, 80 mg of nano-curcumin in the form of nanomicelle was tested in a comparison of MS patients with healthy controls [54, 55]. These studies showed a significant decrease in TNF-α, NF-κB, Th17, IL-17A, IL-23, and RORγt after 80 mg of nano-curcumin treatment [54, 55].

In all of these studies, the bioavailability of curcumin is a potential challenge. Curcumin has a low solubility in aqueous media, leading to low gastrointestinal absorption. Detectable plasma levels are found following the ingestion of large doses, which might not be acceptable for patients under chronic supplementation protocols [90]. With these potential issues in mind, the use of novel drug delivery systems such as nanoparticles, liposomes, and defined phospholipid complexes have been reported to increase the bioavailability of curcumin [91]. In all cases, the advantage of such fabricated delivery systems needs to be rigorously investigated in comparison with crude curcumin in direct comparative trials. Randomized controlled trials are critical to indicate if curcumin can be considered as a useful adjunct in patients suffering from MS. Future studies can focus on the combination of curcumin and other medications with immunomodulatory effects.

References

1. Lassmann H (2018) Multiple sclerosis pathology. Cold Spring Harb Perspect Med 8(3):a028936. https://doi.org/10.1101/cshperspect.a028936
2. Veroni C, Serafini B, Rosicarelli B, Fagnani C, Aloisi F (2018) Transcriptional profile and Epstein-Barr virus infection status of laser-cut immune infiltrates from the brain of patients with progressive multiple sclerosis. J Neuroinflammation 15(1):18. https://doi.org/10.1186/s12974-017-1049-5
3. Tzartos JS, Friese MA, Craner MJ, Palace J, Newcombe J, Esiri MM et al (2008) Interleukin-17 production in central nervous system-infiltrating T cells and glial cells is associated with active disease in multiple sclerosis. Am J Pathol 172(1):146–155
4. Kohm AP, Carpentier PA, Anger HA, Miller SD (2002) Cutting edge: CD4+ CD25+ regulatory T cells suppress antigen-specific autoreactive immune responses and central nervous system inflammation during active experimental autoimmune encephalomyelitis. J Immunol 169(9):4712–4716
5. Sospedra M, Martin R (2005) Immunology of multiple sclerosis. Annu Rev Immunol 23:683–747
6. Trebst C, Ransohoff RM (2001) Investigating chemokines and chemokine receptors in patients with multiple sclerosis: opportunities and challenges. Arch Neurol 58(12):1975–1980

7. Simpson JE, Newcombe J, Cuzner ML, Woodroofe MN (2000) Expression of the interferon-γ-inducible chemokines IP-10 and Mig and their receptor, CXCR3, in multiple sclerosis lesions. Neuropathol Appl Neurobiol 26(2):133–142
8. Sørensen TL, Trebst C, Kivisäkk P, Klaege KL, Majmudar A, Ravid R et al (2002) Multiple sclerosis: a study of CXCL10 and CXCR3 co-localization in the inflamed central nervous system. J Neuroimmunol 127(1–2):59–68
9. Kaskow BJ, Baecher-Allan C (2018) Effector T cells in multiple sclerosis. Cold Spring Harb Perspect Med 8(4):a029025. https://doi.org/10.1101/cshperspect.a029025
10. Steinman L (2009) A molecular trio in relapse and remission in multiple sclerosis. Nat Rev Immunol 9(6):440–447
11. Steinman L (2005) Blocking adhesion molecules as therapy for multiple sclerosis: natalizumab. Nat Rev Drug Discov 4(6):510–518
12. Sinclair C, Mirakhur M, Kirk J, Farrell M, McQuaid S (2005) Up-regulation of osteopontin and αB-crystallin in the normal-appearing white matter of multiple sclerosis: an immunohistochemical study utilizing tissue microarrays. Neuropathol Appl Neurobiol 31(3):292–303
13. Hur EM, Youssef S, Haws ME, Zhang SY, Sobel RA, Steinman L (2007) Osteopontin-induced relapse and progression of autoimmune brain disease through enhanced survival of activated T cells. Nat Immunol 8(1):74–83
14. Clemente N, Comi C, Raineri D, Cappellano G, Vecchio D, Orilieri E et al (2017) Role of anti-osteopontin antibodies in multiple sclerosis and experimental autoimmune encephalomyelitis. Front Immunol 8:321. https://doi.org/10.3389/fimmu.2017.00321
15. Tubridy N, Behan PO, Capildeo R, Chaudhuri A, Forbes R, Hawkins CP et al (1999) The effect of anti-α4 integrin antibody on brain lesion activity in MS. The UK Antegren study group. Neurology 53(3):466–466
16. Iglesias A, Bauer J, Litzenburger T, Schubart A, Linington C (2001) T-and B-cell responses to myelin oligodendrocyte glycoprotein in experimental autoimmune encephalomyelitis and multiple sclerosis. Glia 36(2):220–234
17. Hoesel B, Schmid JA (2013) The complexity of NF-κB signaling in inflammation and cancer. Mol Cancer 12(1):86. https://doi.org/10.1186/1476-4598-12-86
18. Duperray A, Barbe D, Raguenez G, Weksler BB, Romero IA, Couraud PO et al (2015) Inflammatory response of endothelial cells to a human endogenous retrovirus associated with multiple sclerosis is mediated by TLR4. Int Immunol 27(11):545–553
19. Bonizzi G, Karin M (2004) The two NF-κB activation pathways and their role in innate and adaptive immunity. Trends Immunol 25(6):280–288
20. Mattson MP, Camandola S (2001) NF-κB in neuronal plasticity and neurodegenerative disorders. J Clin Invest 107(3):247–254
21. Feng B, Cheng S, Pear WS, Liou HC (2004) NF-kB inhibitor blocks B cell development at two checkpoints. Med Immunol 3(1):1. https://doi.org/10.1186/1476-9433-3-1
22. Choi SS, Lee HJ, Lim I, Satoh J, Kim SU (2014) Human astrocytes: secretome profiles of cytokines and chemokines. PLoS One 9(4):e92325. https://doi.org/10.1371/journal.pone.0092325
23. Olivotto E, Otero M, Marcu KB, Goldring MB (2015) Pathophysiology of osteoarthritis: canonical NF-κB/IKKβ-dependent and kinase-independent effects of IKKα in cartilage degradation and chondrocyte differentiation. RMD Open 1(Suppl 1):e000061. https://doi.org/10.1136/rmdopen-2015-000061
24. Surh YJ, Chun KS, Cha HH, Han SS, Keum YS, Park KK et al (2001) Molecular mechanisms underlying chemopreventive activities of anti-inflammatory phytochemicals: down-regulation of COX-2 and iNOS through suppression of NF-κB activation. Mutat Res 480-481:243–268
25. O'Neill LA, Kaltschmidt C (1997) NF-kB: a crucial transcription factor for glial and neuronal cell function. Trends Neurosci 20(6):252–258
26. van Heel DA, Udalova IA, De Silva AP, McGovern DP, Kinouchi Y, Hull J et al (2002) Inflammatory bowel disease is associated with a TNF polymorphism that affects an interaction between the OCT1 and NF-κB transcription factors. Hum Mol Genet 11(11):1281–1289
27. Bonetti B, Stegagno C, Cannella B, Rizzuto N, Moretto G, Raine CS (1999) Activation of NF-κB and c-Jun transcription factors in multiple sclerosis lesions: implications for oligodendrocyte pathology. Am J Pathol 155(5):1433–1438
28. Nilius B, Appendino G (2013) Spices: the savory and beneficial science of pungency. Rev Physiol Biochem Pharmacol 164:1–76
29. Abdollahi E, Momtazi AA, Johnston TP, Sahebkar A (2018) Therapeutic effects of curcumin in inflammatory and immune-mediated diseases: a nature-made jack-of-all-trades? J Cell Physiol 233(2):830–848
30. Iranshahi M, Sahebkar A, Takasaki M, Konoshima T, Tokuda H (2009) Cancer chemopreventive activity of the prenylated coumarin, umbelliprenin, in vivo. Eur J Cancer Prev 18(5):412–415
31. Momtazi AA, Derosa G, Maffioli P, Banach M, Sahebkar A (2016) Role of microRNAs in the therapeutic effects of curcumin in non-cancer diseases. Mol Diagn Ther 20(4):335–345
32. Mollazadeh H, Cicero AFG, Blesso CN, Pirro M, Majeed M, Sahebkar A (2019) Immune modulation by curcumin: the role of interleukin-10. Crit Rev Food Sci Nutr 59(1):89–101
33. Hasanzadeh S, Read MI, Bland AR, Majeed M, Jamialahmadi T, Sahebkar A (2020) Curcumin: an inflammasome silencer. Pharmacol Res 159:104921.

https://doi.org/10.1016/j.phrs.2020.104921. Epub 2020 May 25. PMID: 32464325
34. Panahi Y, Kianpour P, Mohtashami R, Jafari R, Simental-Mendía LE, Sahebkar A (2017) Efficacy and safety of phytosomal curcumin in non-alcoholic fatty liver disease: a randomized controlled trial. Drug Res (Stuttg) 67(4):244–251
35. Teymouri M, Pirro M, Johnston TP, Sahebkar A (2017) Curcumin as a multifaceted compound against human papilloma virus infection and cervical cancers: A review of chemistry, cellular, molecular, and preclinical features. BioFactors 43(3):331-346
36. Bagheri H, Ghasemi F, Barreto GE, Rafiee R, Sathyapalan T, Sahebkar A (2020) Effects of curcumin on mitochondria in neurodegenerative diseases. Biofactors 46(1):5–20. https://doi.org/10.1002/biof.1566. Epub 2019 Oct 3. PMID: 31580521
37. Kunnumakkara AB, Bordoloi D, Padmavathi G, Monisha J, Roy NK, Prasad S et al (2017) Curcumin, the golden nutraceutical: multitargeting for multiple chronic diseases. Br J Pharmacol 174(11):1325–1348
38. Soleimani V, Sahebkar A, Hosseinzadeh H (2018) Turmeric (Curcuma longa) and its major constituent (curcumin) as nontoxic and safe substances: Review. Phytother Res 32(6):985–995. https://doi.org/10.1002/ptr.6054. Epub 2018 Feb 26. PMID: 29480523
39. Pan Y, Zhang X, Wang Y, Cai L, Ren L, Tang L et al (2013) Targeting JNK by a new curcumin analog to inhibit NF-kB-mediated expression of cell adhesion molecules attenuates renal macrophage infiltration and injury in diabetic mice. PLoS One 8(11):e79084. https://doi.org/10.1371/journal.pone.0079084
40. Xie YL, Chu JG, Jian XM, Dong JZ, Wang LP, Li GX et al (2017) Curcumin attenuates lipopolysaccharide/d-galactosamine-induced acute liver injury by activating Nrf2 nuclear translocation and inhibiting NF-kB activation. Biomed Pharmacother 91:70–77
41. Jagetia GC, Aggarwal BB (2007) "Spicing up" of the immune system by curcumin. J Clin Immunol 27(1):19–35
42. Sahebkar A (2014) Are curcuminoids effective C-reactive protein-lowering agents in clinical practice? Evidence from a meta-analysis. Phytother Res 28(5):633–642
43. Kim GY, Kim KH, Lee SH, Yoon MS, Lee HJ, Moon DO et al (2005) Curcumin inhibits immunostimulatory function of dendritic cells: MAPKs and translocation of NF-κB as potential targets. J Immunol 174(12):8116–8124
44. Alagawany M, Ashour EA, Reda FM (2016) 14. Effect of dietary supplementation of garlic (Allium Sativum) and turmeric (Curcuma Longa) on growth performance, carcass traits, blood profile and oxidative status in growing rabbits. Ann Anim Sci 16(2):489–505
45. Kunnumakkara AB, Sailo BL, Banik K, Harsha C, Prasad S, Gupta SC (2018) Chronic diseases, inflammation, and spices: how are they linked? J Transl Med 16(1):14. https://doi.org/10.1186/s12967-018-1381-2
46. Castro CN, Barcala Tabarrozzi AE, Winnewisser J, Gimeno ML, Antunica Noguerol M, Liberman AC et al (2014) Curcumin ameliorates autoimmune diabetes. Evidence in accelerated murine models of type 1 diabetes. Clin Exper Immunol 177(1):149–160
47. Cianciulli A, Calvello R, Porro C, Trotta T, Salvatore R, Panaro MA (2016) PI3k/Akt signalling pathway plays a crucial role in the anti-inflammatory effects of curcumin in LPS-activated microglia. Int Immunopharmacol 36:282–290
48. Zhu HT, Bian C, Yuan JC, Chu WH, Xiang X, Chen F et al (2014) Curcumin attenuates acute inflammatory injury by inhibiting the TLR4/MyD88/NF-κB signaling pathway in experimental traumatic brain injury. J Neuroinflammation 11(1):59. https://doi.org/10.1186/1742-2094-11-59
49. Rawlings JS, Rosler KM, Harrison DA (2004) The JAK/STAT signaling pathway. J Cell Sci 117(8):1281–1283
50. Bill MA, Nicholas C, Mace TA, Etter JP, Li C, Schwartz EB et al (2012) Structurally modified curcumin analogs inhibit STAT3 phosphorylation and promote apoptosis of human renal cell carcinoma and melanoma cell lines. PLoS One 7(8):e40724. https://doi.org/10.1371/journal.pone.0040724
51. Weissenberger J, Priester M, Bernreuther C, Rakel S, Glatzel M, Seifert V et al (2010) Dietary curcumin attenuates glioma growth in a syngeneic mouse model by inhibition of the JAK1, 2/STAT3 signaling pathway. Clin Cancer Res 16(23):5781–5795
52. Liu L, Liu YL, Liu GX, Chen X, Yang K, Yang YX et al (2013) Curcumin ameliorates dextran sulfate sodium-induced experimental colitis by blocking STAT3 signaling pathway. Int Immunopharmacol 17(2):314–320
53. Guimarães MR, Leite FR, Spolidorio LC, Kirkwood KL, Rossa C Jr (2013) Curcumin abrogates LPS-induced pro-inflammatory cytokines in RAW 264.7 macrophages. Evidence for novel mechanisms involving SOCS-1,-3 and p38 MAPK. Arch Oral Biol 58(10):1309–1317
54. Dolati S, Ahmadi M, Aghebti-Maleki L, Nikmaram A, Marofi F, Rikhtegar R et al (2018) Nanocurcumin is a potential novel therapy for multiple sclerosis by influencing inflammatory mediators. Pharmacol Rep 70(6):1158–1167
55. Dolati S, Ahmadi M, Rikhtegar R, Babaloo Z, Ayromlou H, Aghebati-Maleki L et al (2018) Changes in Th17 cells function after nanocurcumin use to treat multiple sclerosis. Int Immunopharmacol 61:74–81
56. Mohajeri M, Sadeghizadeh M, Najafi F, Javan M (2015) Polymerized nano-curcumin attenuates neurological symptoms in EAE model of multiple sclerosis through down regulation of inflammatory and oxidative processes and enhancing neuroprotection and myelin repair. Neuropharmacology 99:156–167
57. Xie L, Li XK, Funeshima-Fuji N, Kimura H, Matsumoto Y, Isaka Y et al (2009) Amelioration of experimental autoimmune encephalomyelitis by cur-

cumin treatment through inhibition of IL-17 production. Int Immunopharmacol 9(5):575–581
58. Chearwae W, Bright JJ (2008) 15-deoxy-Δ 12, 14-prostaglandin J 2 and curcumin modulate the expression of toll-like receptors 4 and 9 in autoimmune T lymphocyte. J Clin Immunol 28(5):558–570
59. Verbeek R, van Tol EA, van Noort JM (2005) Oral flavonoids delay recovery from experimental autoimmune encephalomyelitis in SJL mice. Biochem Pharmacol 70(2):220–228
60. Natarajan C, Bright JJ (2002) Curcumin inhibits experimental allergic encephalomyelitis by blocking IL-12 signaling through Janus kinase-STAT pathway in T lymphocytes. J Immunol 168(12):6506–6513
61. Kaur A, Kannan D, Mehta SK, Singh S, Salunke DB (2018) Synthetic toll-like receptor agonists for the development of powerful malaria vaccines: a patent review. Expert Opin Ther Pat 28(11):837–847
62. Gargiulo S, Gamba P, Testa G, Rossin D, Biasi F, Poli G et al (2015) Relation between TLR4/NF-κB signaling pathway activation by 27-hydroxycholesterol and 4-hydroxynonenal, and atherosclerotic plaque instability. Aging Cell 14(4):569–581
63. Cario E, Podolsky DK (2000) Differential alteration in intestinal epithelial cell expression of toll-like receptor 3 (TLR3) and TLR4 in inflammatory bowel disease. Infect Immun 68(12):7010–7017
64. Lan L, Tao J, Chen A, Xie G, Huang J, Lin J et al (2013) Electroacupuncture exerts anti-inflammatory effects in cerebral ischemia-reperfusion injured rats via suppression of the TLR4/NF-κB pathway. Int J Mol Med 31(1):75–80
65. Kawai T, Akira S (2007) Signaling to NF-κB by toll-like receptors. Trends Mol Med 13(11):460–469
66. Lubbad A, Oriowo M, Khan I (2009) Curcumin attenuates inflammation through inhibition of TLR-4 receptor in experimental colitis. Mol Cell Biochem 322(1–2):127–135
67. Chen GG, Lee JF, Wang SH, Chan UP, Ip PC, Lau WY (2002) Apoptosis induced by activation of peroxisome-proliferator activated receptor-gamma is associated with Bcl-2 and Nf-kB in human colon cancer. Life Sci 70(22):2631–2646
68. Siddiqui AM, Cui X, Wu R, Dong W, Zhou M, Hu M et al (2006) The anti-inflammatory effect of curcumin in an experimental model of sepsis is mediated by up-regulation of peroxisome proliferator-activated receptor-γ. Crit Care Med 34(7):1874–1882
69. Kutsenko N, Vesnina L, Kaĭdashev I (2012) Pioglitazone, an activator of PPAR-gamma, reduces the expression of kB nuclear factor and inhibits apoptosis in mononuclear cells of peripheral blood in vitro. Fiziol Zh 58(2):33–38
70. Liu ZJ, Liu W, Liu L, Xiao C, Wang Y, Jiao JS (2013) Curcumin protects neuron against cerebral ischemia-induced inflammation through improving PPAR-gamma function. Evid Based Complement Alternat Med 2013:470975. https://doi.org/10.1155/2013/470975
71. Cebola I, Custodio J, Muñoz M, Díez-Villanueva A, Paré L, Prieto P et al (2015) Epigenetics override pro-inflammatory PTGS transcriptomic signature towards selective hyperactivation of PGE 2 in colorectal cancer. Clin Epigenetics 7(1):74. https://doi.org/10.1186/s13148-015-0110-4
72. Sur D, Chakravorty R (2017) Analysis of prostaglandin-endoperoxide synthase-2 gene polymorphisms and risk of cervical cancer in an east Indian population: a case–control study. Asian J Oncol 3(1):23–27
73. Yiangou Y, Facer P, Durrenberger P, Chessell IP, Naylor A, Bountra C et al (2006) COX-2, CB2 and P2X7-immunoreactivities are increased in activated microglial cells/macrophages of multiple sclerosis and amyotrophic lateral sclerosis spinal cord. BMC Neurol 6(1):12. https://doi.org/10.1186/1471-2377-6-12
74. Miyamoto K, Miyake S, Mizuno M, Oka N, Kusunoki S, Yamamura T (2006) Selective COX-2 inhibitor celecoxib prevents experimental autoimmune encephalomyelitis through COX-2-independent pathway. Brain 129(8):1984–1992
75. Newton R, Kuitert LM, Bergmann M, Adcock IM, Barnes PJ (1997) Evidence for involvement of NF-κB in the transcriptional control of COX-2 gene expression by IL-1β. Biochem Biophys Res Commun 237(1):28–32
76. Bachmeier BE, Mohrenz IV, Mirisola V, Schleicher E, Romeo F, Höhneke C et al (2008) Curcumin downregulates the inflammatory cytokines CXCL1 and-2 in breast cancer cells via NFκB. Carcinogenesis 29(4):779–789
77. Camacho-Barquero L, Villegas I, Sánchez-Calvo JM, Talero E, Sánchez-Fidalgo S, Motilva V et al (2007) Curcumin, a Curcuma longa constituent, acts on MAPK p38 pathway modulating COX-2 and iNOS expression in chronic experimental colitis. Int Immunopharmacol 7(3):333–342
78. Sharma C, Kaur J, Shishodia S, Aggarwal BB, Ralhan R (2006) Curcumin down regulates smokeless tobacco-induced NF-κB activation and COX-2 expression in human oral premalignant and cancer cells. Toxicology 228(1):1–15
79. Lau LT, Yu ACH (2001) Astrocytes produce and release interleukin-1, interleukin-6, tumor necrosis factor alpha and interferon-gamma following traumatic and metabolic injury. J Neurotrauma 18(3):351–359
80. Merrill JE (1992) Proinflammatory and antiinflammatory cytokines in multiple sclerosis and central nervous system acquired immunodeficiency syndrome. J Immunother 12(3):167–170
81. Beg AA, Baltimore D (1996) An essential role for NF-κB in preventing TNF-α-induced cell death. Science 274(5288):782–784

82. Cho JW, Lee KS, Kim CW (2007) Curcumin attenuates the expression of IL-1β, IL-6, and TNF-α as well as cyclin E in TNF-α-treated HaCaT cells; NF-κB and MAPKs as potential upstream targets. Int J Mol Med 19(3):469–474
83. Nam NH (2006) Naturally occurring NF-kappaB inhibitors. Mini Rev Med Chem 6(8):945–951
84. Plummer SM, Holloway KA, Manson MM, Munks RJ, Kaptein A, Farrow S et al (1999) Inhibition of cyclo-oxygenase 2 expression in colon cells by the chemopreventive agent curcumin involves inhibition of NF-κB activation via the NIK/IKK signalling complex. Oncogene 18(44):6013–6020
85. Ye J (2008) Regulation of PPARγ function by TNF-α. Biochem Biophys Res Commun 374(3):405–408
86. Chiba Y, Ogita T, Ando K, Fujita T (2001) PPARγ ligands inhibit TNF-α-induced LOX-1 expression in cultured endothelial cells. Biochem Biophys Res Commun 286(3):541–546
87. Jiang C, Ting AT, Seed B (1998) PPAR-γ agonists inhibit production of monocyte inflammatory cytokines. Nature 391(6662):82–86
88. Pershadsingh HA, Heneka MT, Saini R, Amin NM, Broeske DJ, Feinstein D (2004) Effect of pioglitazone treatment in a patient with secondary multiple sclerosis. J Neuroinflammation 1(1):3. https://doi.org/10.1186/1742-2094-1-3
89. Wang D, Xia D, DuBois RN (2011) The crosstalk of PTGS2 and EGF signaling pathways in colorectal cancer. Cancers 3(4):3894–3908
90. Siviero A, Gallo E, Maggini V, Gori L, Mugelli A, Firenzuoli F, Vannacci A et al (2015) Curcumin, a golden spice with a low bioavailability. J Herbal Med 5(2):57–70
91. Anand P, Kunnumakkara AB, Newman RA, Aggarwal BB (2007) Bioavailability of curcumin: problems and promises. Mol Pharm 4(6):807–818

Carbon-based Nanomaterials and Curcumin: A Review of Biosensing Applications

4

Mohammad Mohajeri, Behzad Behnam, Aida Tasbandi, Tannaz Jamialahmadi, and Amirhossein Sahebkar

Abstract

Curcumin, the main active constituent of turmeric (*Curcuma longa* L.), is a naturally occurring phenolic compound with a wide variety of pharmacological activities. Although it has multiple pharmaceutical properties, its bioavailability and industrial usage are hindered due to rapid hydrolysis and low water solubility. Due to the growing market of curcumin, exact determination of curcumin in trade and human biological samples is important for monitoring therapeutic actions. Different nanomaterials have been suggested for sensing curcumin; and in this case, carbon-based nanomaterials (CNMs) are one of the most outstanding developments in nanomedicine, biosensing, and regenerative medicine. There are a considerable number of reports which have shown interesting potential of CNMs-based biosensors in the sensitive and selective detection of curcumin. Therefore, this review aims to increase understanding the interaction of curcumin with CNMs in the context of biosensing.

Keywords

Curcumin · Carbon nanotube · Graphene · Carbon dot · Biosensor

M. Mohajeri
Department of Medical Biotechnology and Nanotechnology, School of Medicine, Mashhad University of Medical Sciences, Mashhad, Iran

B. Behnam (✉)
Pharmaceutics Research Center, Institute of Neuropharmacology, Kerman University of Medical Sciences, Kerman, Iran

Pharmaceutical Sciences and Cosmetic Products Research Center, Kerman University of Medical Sciences, Kerman, Iran

Herbal and Traditional Medicines Research Center, Kerman University of Medical Sciences, Kerman, Iran
e-mail: behnamb@kmu.ac.ir

A. Tasbandi
Neurogenic Inflammation Research Center, Mashhad University of Medical Sciences, Mashhad, Iran

T. Jamialahmadi
Department of Food Science and Technology, Quchan Branch, Islamic Azad University, Quchan, Iran

Department of Nutrition, Mashhad University of Medical Sciences, Mashhad, Iran

A. Sahebkar (✉)
Neurogenic Inflammation Research Center, Mashhad University of Medical Sciences, Mashhad, Iran

Biotechnology Research Center, Pharmaceutical Technology Institute, Mashhad University of Medical Sciences, Mashhad, Iran

Polish Mother's Memorial Hospital Research Institute (PMMHRI), Lodz, Poland

Halal Research Center of IRI, FDA, Tehran, Iran

P. C. Guest (ed.), *Studies on Biomarkers and New Targets in Aging Research in Iran*, Advances in Experimental Medicine and Biology 1291, https://doi.org/10.1007/978-3-030-56153-6_4

4.1 Introduction

Turmeric (*Curcuma longa* L.) is a spice of the ginger family with paramount medical applications in addition to its growing popularity in the culinary world [1]. It has been widely used in Asia for medical purposes because of its antioxidant, anti-inflammatory [2], antimutagenic, antimicrobial [3, 4], and anticancer effects [5, 6]. Curcumin, also called diferuloylmethane, is the major natural polyphenol present in the turmeric rhizome [7]. It can target several signaling molecules and has activity at the cellular level, which accounts for its multiple health benefits [8–15]. Curcumin has been reported to be safe [16] in clinical studies and the numerous biological properties identified for curcumin most likely arise from its antioxidant and anti-inflammatory effects [8, 17–21].

In its natural state, curcumin has low systemic bioavailability in several species [22], which arises from low absorption in addition to metabolic conjugation and reduction [23, 24]. Furthermore, curcumin metabolites or products of the metabolic conjugation and reduction reactions appear to be much less capable of exerting the same biological actions. Despite the poor pharmacokinetic properties of curcumin, there are some reports indicating that it can exert its biological activity at certain dosages [17–19]. Therefore, detection of curcumin and its metabolites in target tissues and portal blood of patients following administration is of utmost importance. Furthermore, many evidence-based pharmacological actions of curcumin have led to the increasing global market.

Carbon-based nanomaterials (CNMs) are useful in various biomedical applications given their unique thermal and electrical conductivity, great mechanical strength, and excellent optical properties [25]. CNMs can be viewed as highly attractive biomaterials owing to their multi-functional nature. Indeed, incorporation of CNMs into biomaterials could potentially improve their functions. Accordingly, CNMs have attracted much attention in multiple areas of biomedical investigations, such as drug delivery systems [26], tissue scaffold reinforcements [27], biosensing systems [28], cardiovascular research [29], and neuroscience [30, 31]. Two of the most widely used CNMs, namely graphene and the carbon nanotube (CNT), exhibit similar electrical, optical, and thermal properties. However, their structural difference can explain their diverse characteristics. For instance, the quantum Hall effect and massless Dirac fermions result in the low-energy charge excitation at ambient temperature, as well as the optical transparency in the infrared and visible range of the spectrum for both of these CNMs [32]. Furthermore, graphene appears to be structurally robust with high flexibility, which makes it an excellent choice for engineering thin, flexible materials [33, 34]. Studies in the field of graphene in biomedical applications are emerging rapidly. Graphene oxide (GO), synthesized by the oxidation of graphite in acidic conditions, has found more common applications since it provides many advantages over pure graphene [35]. For example, it has good dispersion in aqueous media, it possesses hydrophilic functional groups necessary for chemical functionalization, and it has a wide variety of physical properties owing to its structural heterogeneity. Several biomedical applications, such as injectable cellular labeling agents, drug delivery systems, and scaffold reinforcements, have been proposed based on GO [36].

The other member of CNMs is tiny carbon dots (CDs or C-dots) which are usually less than 10 nm in size. C-dots are the youngest family of CNMs and can be regarded as very small graphenes. Therefore, there are many similarities in the physicochemical properties between them. The important key characteristic of C-dots among the other members of the CNM family is its high bioavailability. In parallel with other CNMs, there have been many applications of C-dots in biomedicine. In particular, they have been used as imaging probes in targeted drug delivery systems and in photodynamic and photothermal therapies [37, 38]. The property of CNMs in the transfer of electrons makes them useful in sensor applications, and their inclusion in biosensor platforms has led to significant progress in biosensor design. Graphenes along with CNTs have

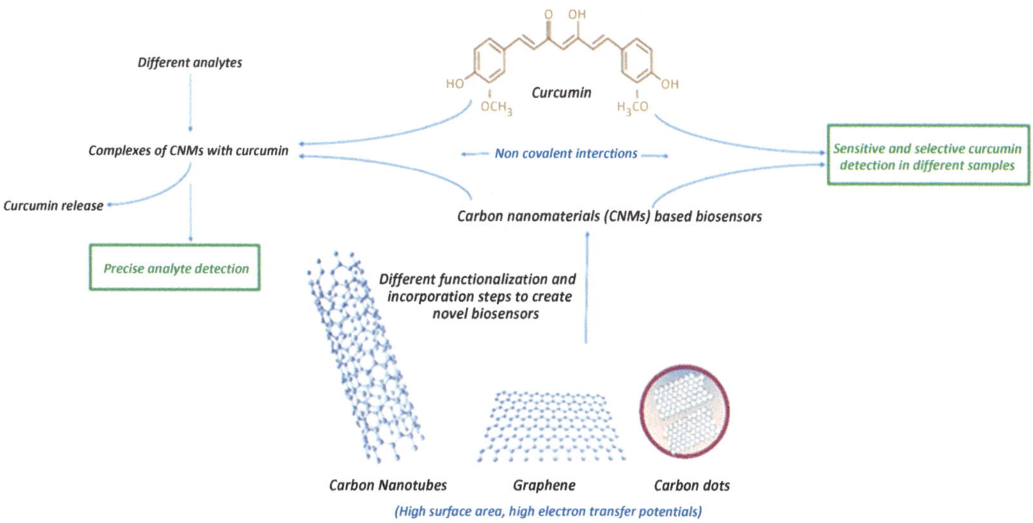

Fig. 4.1 Different structures of carbon nanomaterials and their potentials in conjugation with curcumin for bio-sensing applications

constituted indispensable elements of sensor platforms over the last decade and, presumably on account of new functionalization strategies, this will lead to new exciting developments in the field of biosensing. Additionally, C-dots hold promise as novel platforms, and development of recognition bioelements will be the future focus of research.

Accumulated literature showed that curcumin is an interesting target molecule for CNM-based biosensors. Also, curcumin-incorporated biosensors can be potentially investigated to detect other important biomolecules (Fig. 4.1). The present review provides the latest achievements in biosensor technologies based on the combination of CNMs and curcumin for medical applications.

4.2 Biosensing Applications of Carbon-based Nanomaterials

Curcumin [(1E,6E)-1,7-bis(4-hydroxy-3-methoxyphenyl)-1,6-heptadiene-3,5-dione] is an active component of the Indian spice turmeric [39, 40]. It has been traditionally used as a coloring agent, flavoring additive, and various therapeutic remedies [41]. The chemical structure of curcumin is presented in Fig. 4.1. Till now, three curcuminoids have been identified in commercial curcumin, namely curcumin (~75%), demethoxycurcumin (~20%), and bisdemethoxycurcumin (~5%) [42]. Given that curcumin, as a natural phenol, has considerable biological properties in vitro and in vivo [43–46], it is important to develop sensitive methods for its detection. More to the point, curcumin has antioxidant and pro-oxidant activities, which are dependent on its concentration [47]. The design of curcumin through addition of adjuvants, nanoparticles, and other components to increase its concentration could also have negative consequences on the body since any increase in drug concentration may cause toxicity in the system. In line with this, some well-documented harmful effects have been attributed to the use of curcumin in the body, including gastric discomfort, skin rashes, constipation, diarrhea, nausea, apoptotic alterations, necrotic cell death, and chromosomal damage [43, 44]. Hence, it is essential to measure and control the concentration of curcumin in foods and biological fluids. In addition, the sensing system should have rapid and sensitive characteristics to be able to detect this compound. The curcumin content of turmeric has also been reported to differ due to growing conditions, harvesting processes, and extracting methods [48]. It was demonstrated that the percentage of curcumin in turmeric products currently available in the market ranges from 0.5% to 5.7% [49]. Thus, the curcumin percentage is a critical component in the quality control of commercial turmeric products [48]. Additionally, turmeric powder can be, and often is, contaminated with many chemical adulterants that mimic the appearance of curcumin, due to the high cost of pure turmeric. The chief chemicals which have been used in adulteration of turmeric spices are lead chromate and the dye metanil yellow [50, 51]. Recently, the United States Food and Drug Administration (FDA) released a recall for one brand of turmeric powder imported from Bangladesh, based on the concern that the turmeric powder was contaminated with high levels of lead, which could have a negative impact on health. Besides this, children in a rural area of Bangladesh have been diagnosed with a lead concentration in the blood, which may have arisen from the contaminated turmeric [52].

Considering the dangers from lead poisoning along with other diseases that may be attributed to adulterated spices, food safety issues related to contaminated turmeric powder can have a negative impact on human health. Accordingly, the quality assurance of the commercially available turmeric products, including the measurement of the curcumin content, is critical because if the quantified amount of curcumin obtained from the products is significantly low, adulteration might be inferred. Recently, different analytical methods have been proposed to determine the curcumin concentration employing high-performance liquid chromatography (HPLC) [53], spectrophotometry [54], spectrofluorimetry [55], electrochemical techniques [56], liquid chromatography–mass spectrometry (LC-MS) [57], and square wave voltammetry [58]. At present, many important materials have been utilized in applications of nanotechnology for the development of sensor, determination, and catalytic effects [59–65]. Of note, outstanding progress has been achieved in the synthesis of carbon-based nanomaterials at non-prohibitive costs and considerable catalytic performance. Research studies addressing the catalytic activity, luminescence properties, and reductive-oxidative (redox) processes of curcumin, as well as its significant fluorescence emission in the UV–visible region, have played a central role in exploring its biological actions [66]. Lungu et al. examined the electrochemical property of curcumin and its analogue, *bis*demethoxycurcumin, by applying cyclic voltammetry (CV) and differential pulse voltammetry (DPV) techniques at an anodically activated glassy carbon electrode (GCE) in non-aqueous media. They found out that both compounds were adsorbed to the anodically activated GCE surface to varying extents [67]. A nickel-based chemically modified electrode (Ni(II)-curcumin) was employed for the electrochemical oxidation and voltammetric determination of several organic targets, such as glucose [68] and alcohol [69]. There have only been a few studies on the electrocatalytic activity of CNMs/curcumin regarding oxidation of organic compounds. The next three sections deal with the sensing systems based on CNMs for detecting curcumin or the use of CNMs/curcumin for electrochemical determination of other biomolecules. Some of these important reports are summarized in Table 4.1.

4.3 Carbon Nanotube-Based Biosensors

Nano-sized materials with distinctive properties are excellent candidates for biosensing applications. Since their discovery, CNMs have been extensively studied for their function as nano-biosensors owing to their tunable surface characteristics and good electron transfer capabilities. The structures of micro- and nanotubes are similar to those of tiny straws, outweighing other nanoparticles. CNTs exhibit many notable properties ascribed to their structure, morphology, functionality, stability, ease of modification, and suitability in hybrid materials [70], which, in turn, lead to practical applications in the electronics, photonics, renewable energy, drug delivery, and biomedical research arenas [26, 71–77].

Table 4.1 Selected reports on carbon-based nanomaterials-based biosensors for curcumin detection

Reference	Design	Structure/composition	Detection sensing method	Measurement range	Advantage
[210]	Fabrication of novel Ni-Cur/GO nano-composites for superior electrocatalytic activity in the detection of toxic p-nitrophenol	$[Ni(Cur)_2]$/graphene-oxide [GO]	Electrochemical sensor	LOD = 0.016 μM and 0.671 μA/μM/cm^2 sensitivity	Low cost, easily prepared, improved over newly reported modified electrodes by permissive practical industrial applications
[179]	Comparative evaluation of GO-based materials for electrochemical non-enzymatic sensing of Cur	Graphite ($KMnO_4$, $NaNO_3$, H2SO$_4$) Cur and H_2O_2	Voltammetric assay	Detection electrode = 0.9 pM	Successfully detected lowest concentrations of the drug approximating quantities present in the body after absorption from gut
[178]	Electrochemical behavior and voltammetric determination of Cur at electrochemically reduced GO modified glassy carbon electrode	GO/GCE and ERGO/GCE graphene-modified glassy carbon electrode	Electrochemical sensor	Linear voltammetric response from 0.2–60.0 mM Cur with LOD = 0.1 mM	GO-modified electrodes are a promising strategy for electrochemical determination of biologically important phenolic compounds
[211]	Photoluminescence and electrochemical investigation of Cur-reduced GO sheets	Cur-rGO sheets	CV and PL	Not mentioned	Charges and resonant energy from Cur molecules to GO from bonding of fluorescent Cur to Cur-rGO surfaces, through π–π stacking of aromatic rings
[56]	Electrochemical characterization of Cur and selective detection in *Curcuma longa* using a graphene-modified electrode	Graphene-modified glassy carbon electrode (GR/GCE)	CV, EIS, LSV	Cur in range of 5.0×10^{-8} to 3.010^{-6} mol/L, LOD = 3.0×10^{-8} mol/L	Good recoveries and could be applied to determine Cur in *curcuma longa* with satisfactory results
[177]	Novel determination of Cur via Ru@Au nanoparticle decorated nitrogen and sulfur-functionalized reduced graphene oxide nanomaterials	Ru@Au nanoparticles involved L-cysteine functionalized GO composite	Electrochemical determination square wave voltammetry (SWV)	Determination of Cur in plasma over linear range 0.001–0.1 nM with LOD = 2.0×10^{-13} M	Rapid, cheap, and easy to use and might be preferred to the analytical methods
[212]	Cur/MWCNT modified graphite electrode for electrochemical determination of BHA	Cur / MWCNT at graphite electrode	Electrocatalytic detection	LOD = 2.250×10^{-7} M	Modified electrode showed good catalytic response for oxidation
[213]	Carbon nanotubes hybrid hydrogels for electrically tunable release of Cur	MWCNT, gelatin, acrylamide, and PEG-DMA	Electrochemical sensor	Not mentioned	Appealing for topical treatment of skin injuries by bioactive compounds of natural origin such as Cur

(continued)

Table 4.1 (continued)

Reference	Design	Structure/composition	Detection sensing method	Measurement range	Advantage
[203]	Average 5 nm sized C-dots for sensitive detection of Cur	C-dots from DTPA	Spectrofluorimetry Cur detection through FRET	Curcumin detection in range of 0.74–5.18 μg/mL, LOD = 44.8 ng/mL	One-step synthesis of C-dots from DTPA, fast and sensitive detection of curcumin in real samples
[209]	One-step microwave-assisted C-dots from levodopa (2.3 nm)	Oxygen-rich C-dots from levodopa	Spectrofluorimetry curcumin detection based on IFE	Multifunctional biosensor for the detection of Cur with LOD = 0.4 μM, the detection of Cu^{2+}-Cur complex, the detection of bio-thiols	Rapid, eco-friendly synthesis, Cu^{2+} and Cur detection in environmental water and serum, potentials in Alzheimer disease
[204]	One-pot hydrothermal synthesis of C-dots from sodium citrate and thiourea	Prepared nitrogen and sulfur co-doped carbon dots	Spectrofluorimetry Cur detection based on IFE	Cur detection in range of 0.15–18.0 μmol/L with LOD = 0.04 μmol/ L, temperature sensitive biosensor between 5–70 °C	Simple, efficient synthesis, good photostability, high water solubility, sensitive detection of Cur in urine, reversible fluorescence quenching in the presence of increasing temperature

Cur curcumin, *GO* graphene-oxide, *rGO* reduced graphene-oxide, *CV* cyclic voltammetry, *PL* photo-luminescence, *EIS* electrochemical impedance spectroscopy, *LSV* linear sweep voltammetry, *BHA* butylated hydroxyanisole, *PEG-DMA* polyethylene glycol dimethacrylate, *DTPA* diethylenetriamine pentaacetate, *FRET* fluorescence resonance energy transfer, *MWCNT* multi-walled carbon nanotubes

They are fabricated in the form of hollow cylindrical tubes composed of carbon (graphite) with a high aspect ratio feature (~1000) that shows sp^2 hybridization. This form of hybridization is observed when one s and two p orbitals of the same atom mix to form a hybrid orbital. According to the number of graphite sheets, CNTs are categorized as single-walled nanotubes (SWNTs), double-walled nanotubes (DWNTs), and multi-walled nanotubes (MWNTs) with diameters of 1–2 nm and lengths of 50–1000 nm [78, 79]. Studies on CNTs have favorably revealed unprecedented innovations in biomedicine and nanoengineering because of their many desirable properties. The large surface area to volume ratio as well as the hollow structure pave the way toward the use of CNTs as novel carriers of drugs or bioactive compounds [80–82]. The hydrophobic character of CNTs confers them the advantages to deliver biomolecules, such as DNA [83–87], RNA [24, 88], proteins [89–91], and immune modulators [92, 93], by the capacity of entering the biological membrane with no effect on the neighboring cells.

The functionalization of CNTs enables a variety of new drug carrier systems to direct the therapeutic agents, in particular anticancer drugs, to the desired site of action [94–99]. In addition, the optical properties of CNTs hold promise as a platform for emerging technologies, including contrast agents or optical tags for photodynamic therapies, photothermal ablation, and photoacoustic imaging approaches [100–104]. Although there are CNT toxicity issues, these could be ameliorated by proper surface functionalization [105, 106]. The considerable conductive profile of CNTs allows the development of biosensors for enzymes [107, 108], genes [109, 110], and cancer [111, 112], as well as sensors for air pollutants [113, 114]. In addition, the remarkable mechanical strength of CNTs gives them the ability to modulate cellular behaviors, such as cell adhesion, proliferation, and differentiation [115–117]. Furthermore, their large surface area provides the potential for their use as a stationary phase in the separation of biomolecules (e.g. DNA and proteins) [118, 119]. However, in the light of their good antimicrobial activity, CNTs are placed in the category of antimicrobial treatments in liquid media [120–122]. In this setting, CNTs can be considered scaffolds for immobilizing biomolecules at their surface, as well as materials to transduce signals related to the identification of analytes, metabolites, or disease biomarkers. As a result of their unique structural, mechanical, electrical, and optical features, CNTs are attractive for engineering state-of-the-art probes [123–125]. Indeed, these nanomaterials can serve as scaffolds or platforms that undergo functionalization applying conjugation techniques to improve the processes of multiplex detection and signal transduction. They possess electrical conductivity almost 100 times stronger than copper wires, which enables them to transduce electric signals produced upon identification of a target. They show thermal conductivity greater than diamonds and a tensile strength around 100 times higher than steel. More importantly, their ability to cross biological membranes gives them the potential for in vivo applications with minimal invasiveness [126].

Curcumin has been integrated into multiple biosensors for the exact detection of various molecules. Zheng and Song constructed a curcumin-MWCNT-GCE modified electrode via electrodepositing curcumin at the surface of GCE modified with MWCNT. This modified electrode revealed a well-known two-electron and two-proton redox coupling with the formal potential of 0.14 V that could be attributed to the electrochemical oxidation product of curcumin. In addition, it had good electrocatalytic activity toward the oxidation of hydrazine at a reduced overpotential and enhanced peak current, compared with a curcumin-modified GCE, a MWCNT-modified GCE, or an activated GCE. The catalytic rate constant was 6.26×10^3/M/s using chronoamperometry. On the basis of amperometry, the calibration curve for the detection of hydrazine was linear in the range of 2–44 μM in phosphate buffered saline (pH 8.0). The detection limit and the sensitivity appeared to be 1.4 μM and 22.9 nA/μM, respectively [127]. More recently, Kumar et al. developed a curcumin/MWCNT electrode for the electrochemical determination of butylated hydroxyanisole (BHA), a synthetic food antioxidant. The electropolymerization of curcumin over MWCNT functionalized graphite electrode resulted in a new electrode with improved stability and reproducibility. The use of MWCNT over chemically modified electrodes enhanced the catalytic current and lowered the detection limit. More specifically, this electrode exhibited an apparent catalytic response for the oxidation of BHA with the detection limit of 2.250×10^{-7} M and excellent reproducibility with a relative standard deviation (RSD) of 2.8% [128].

To detect curcumin using CNM-based sensors, Daneshgar et al. initially fabricated two sensing systems based on MWCNT-modified electrodes (MWCNT/GCE) as well as a dysprosium nanowire carbon paste electrode (DyNW/CPE) through the technique of adsorptive stripping voltammetry in stationary solution and fast Fourier transform voltammetry at the flowing solution. Under optimized conditions, curcumin could be determined over a linear range with a detection limit of 5.0×10^{-9} mol/L and 5.0×10^{-10} mol/L for the traditional square wave and fast Fourier transform square wave voltammetry (FFTSWV) with an RSD between 0.2% and 0.5%. In this study, the anodic peak current and curcumin concentration (1.0×10^{-8} mol/L to 1.0×10^{-6} mol/L) showed a linear relationship in the ranges of 2.0×10^{-9} M to 1.0×10^{-6} M and 1.0×10^{-8} M to 1.0×10^{-6} M for the DyNW/CPE and MWCNT/GCE configurations, respectively. They concluded that although the DyNW/CPE version was slightly more sensitive than the MWCNT/GCE approach, both electrodes exerted a lower passivation effect and displayed better stability, selectivity, and reproducibility compared to unmodified electrodes. The surface of the electrodes provided a large surface area for the adsorption and oxidation of curcumin [129]. In the study by Jain et al., a simple voltammetric method was proposed to quantify curcumin in PBS at pH 2.5 in a solubilized system using a MWCNT/GCE sensor. Through testing cetyltrimethylammonium bromide (CTAB), sodium lauryl sulphate (SLS), Tween 20, and Triton X surfactants, they showed that the solubility and chemical stability of curcumin were significantly elevated in the CTAB solution. Moreover, scanning electron microscopy (SEM) and transmission electron microscopy (TEM) exhibited the formation of a nanocomposite film of MWCNT at the surface of the GCE. It was apparent that lower charge transfer ratio for MWCNT/GCE improved the electrical properties for redox processes and gave a larger surface area for electrode reactions. Under the optimized conditions, the oxidation peak current for curcumin appeared to be proportional to its concentration in the range of 4×10^{-7} g/mL to 20×10^{-7} g/mL with a detection limit of 3.28×10^{-7} g/mL and a quantification limit of 10.94×10^{-7} g/mL. Also, this sensor showed remarkable reproducibility (RSD = 4.14%), excellent precision (2.27–3.68% RSD over 4–20 μg/mL), and appropriate selectivity (the recovery of curcumin in the presence of multiple interferents was in the range of 96–104%). The results suggested that the oxidation of curcumin on the MWCNT/GCE sensor could be considered predominantly as a adsorption controlled process [130]. Additionally, Chaisiwamongkhol et al. employed adsorptive stripping voltammetry (AdsSV) for the determination and quantification of curcumin at a MWCNT-modified basal plane pyrolytic graphite (MWCNT-BPPG) electrode. In accordance with the AdsSV, this sensor was accurate and reproducible (<7% variation). The adsorption of curcumin on the electrode surface followed a Langmuir adsorption isotherm. Linear calibration for curcumin fell within the range of 2–100 μM ($R^2 = 0.999$) with a detection limit of 0.45 μM and a quantification limit of 1.49 μM. The real turmeric powder sample for the analysis was obtained following a one-step procedure. The MWCNT-BPPG electrode with AdsSV was capable of determining the curcumin equivalent in the turmeric powder sample with recoveries in the range of 92–108% [131].

4.4 Graphene-based Biosensors

Another type of CNM, graphene, is a one-atom-thick carbon sheet constructed by sp^2-bonded carbon atoms in a planar honeycomb lattice and resembles CNTs in terms of chemical composition and crystalline structure, although they are unlike CNTs in shape [132, 133]. As graphene shows unique mechanical and chemical properties besides high electrical conductivity, it offers an attractive platform for a range of applications in areas such as tissue engineering, regenerative medicine, and drug delivery [132, 134–136]. It constitutes the basic building block of all graphitic allotropes. In initial efforts, single-layer graphene was separated from graphite by micromechanical cleavage [132]. It is employed in many areas, namely, condensed-matter and high-energy physics [137–139], electronics [140–143], material science [144–147], and other new technologies [118, 148–153]. Furthermore, it has received attention from scholars for its bioelectronics, biosensing, and biological applications.

Recently, the development of sensitive biosensors has shifted toward the use of graphene-based materials due to their considerable advantages, such as increased surface loading of the target ligand by virtue of a high surface-to-volume ratio, notable conductivity, a small band gap essential for sensitive electrical and electrochemical read-outs, and tunable optical characteristics for optical read-outs (e.g., fluorescence and plasmonics). Of the many CNMs available for biosensing applications [154], graphene and its different forms, including graphene oxide, reduced graphene oxide, and graphene nanoribbons, have gained much attention in recent years. A great number of sensing mechanisms are involved in graphene-based biosensors, such as optical, electrochemical, or electrical approaches. To achieve recognition of disease biomarkers and biological analytes based on electrochemical and electrical sensing concepts, high electron transfer rates and charge-carrier mobility, and low electrical noise levels, graphene-based biosensors are helpful for highly sensitive detection [155, 156]. In addition, chemically prepared graphene derivatives show a high density of the edge-plane-like sites/defects and number of active sites for electron transfer to chemical and biological species [155]. Furthermore, the substantial optical transparency of graphene monolayers allows their use in optical-based graphene biosensors and in enhancing the sensing performance of plasmonic sensors [157]. Graphene has also been found with a strong chemical enhancement mechanism, leading to the development of surface-enhanced Raman scattering (SERS) substrates [158]. Irrespective of the detection method employed, the hydrophobic domains or π-systems on graphene-based transducers gives them perfect supporting layers for the immobilization of biomolecules [159].

The use of graphene has opened up new possibilities for the development of state-of-the-art biosensors [160–163], glucose sensors [118, 153, 160, 164, 165], single bacterium sensors and DNA transistor [166], sensitive immunosensor for cancer biomarker detection [162], and antimicrobial purposes [167–170]. In order to enhance the catalytic performance of the sensing system, graphene-based materials have become a popular approach [59, 60, 171–176]. In this regard, a graphene-modified GCE (GR/GCE) was constructed to determine the trace amounts of curcumin. Applying linear sweep voltammetry (LSV), a robust linear relationship with curcumin concentrations was achieved in the range of 5×10^{-8} to 3×10^{-6} mol/L with a detection limit of 3.0×10^{-8} mol/L. This method was further used to identify curcumin in *Curcuma longa* L. extracts. It was found that the GR/GCE responded only to curcumin in the herbal samples and there was no interference. As a voltammetric sensor, this system offered the advantages of high selectivity (<±5%), satisfactory accuracy (100–107%), excellent long-term storage stability (93.5% of the initial response), and good reproducibility (RSD = 5.2%). Kotan et al. used Ru@Au nanoparticles which involved an L-cysteine functionalized rGO composite (NSrGO/Ru@AuNPs) for the electrochemical determination of curcumin by square wave voltammetry (SWV) on GCE modified with NSrGO/Ru@AuNPs composite (NSrGO/Ru@AuNPs/GCE). The effective surface areas of rGO/GCE and NSrGO/Ru@AuNPs/GCE were 347 cm^2/mg and 1289 cm^2/mg, respectively. Furthermore, this nanocomposite can successfully determine curcumin levels in plasma. The linearity range of curcumin was 0.001–0.1 nM with a detection limit of 2.0×10^{-13} M and a quantitation limit of 1.0×10^{-12} M. Also, the prepared sensor was reusable and showed the properties of selectivity, stability, simplicity, and efficiency in target detection. A high percentage of recovery reflected the capability of the sensor to quantify curcumin without interference [177]. The study by Zhang et al. was performed on an electrochemically rGO (ERGO)-modified GCE in 100 mM KCl–10 mM PBS (pH 7.4). In order to understand the electrochemical behaviors of polyphenol curcumin better, demethoxycurcumin and bisdemethoxycurcumin were also used at bare GCE, GO/GCE, and ERGO/GCE sensors. The electrochemical behavior of curcumin at the ERGO/GCE sensor was tested using cyclic voltammetry, and this revealed a higher electrocatalytic activity toward curcumin, compared with the bare GCE and GO/GCE electrodes. Under physiological conditions, the modified electrode was found to have a linear voltammetric response from 0.2 μM to 60.0 μM for curcumin with a detection limit of 0.1 μM. Also, it was found that the electrochemical activity of curcumin at the ERGO/GCE sensor was mainly a surface-confined process. Repetitive measurements of 10.0 μM curcumin were indicative of good reproducibility with RSDs of 3.5% ($n = 11$) and 4.3% ($n = 5$). As for the practicality of the modified electrode, quantitative recovery was recorded from 97% to 100% [178].

More recently, Dey et al. compared GO and rGO electrodes modified on GCE for assay of curcumin [179]. Of these electrodes, the rGO-modified electrode had a lower detection limit of 0.9 pM with higher selectivity desirable good signal quality. This detection limit was the lowest reported thus far in the literature for curcumin for any sensing assay. However, linearity in detection was in the range from 1 nm to 100 nM for the GO-modified electrodes and 0.1–10 nM for the rGO-modified electrodes, with recoveries around 93.33% and 95.55%, respectively. Both electrodes displayed specific signals for curcumin under conditions similar to the physiology of the body, with good non-enzymatic sensing characteristics. Repeat measures 90% and 80% of the initial response were respectively obtained for rGO- and GO-modified electrodes after a 6 week interval. Having more outstanding properties over the GO-modified electrode, the rGO-modified electrode could be a favorable option for the real-time usability in sensing (RSD = 5% versus 4.5%). Curcumin detection in blood samples would be very practical as Rahimnejad et al. recently proposed a biosensor based on an rGO-modified carbon paste electrode (RGO/CPE). The obtained sensor had a limit of detection (LOD) of 3.1 μM for curcumin through the applied cyclic voltammetry (CV) and differential pulse voltammetry (DPV) techniques [180].

Curcumin has good biocompatibility and can be used for the green synthesis of graphene. For example, Hatamie et al. produced rGO from GO sheets by a novel low-temperature method [181]. GO sheets were chemically functionalized by applying polyphenol curcumin. On the basis of CV and photoluminescence spectroscopy, charges and resonant energy were transferred from curcumin molecules to the surfaces of GO sheets owing to the bonding of the fluorescence curcumin molecules to the Cur-rGO surfaces. Additionally, the curcumin-rGO appeared to be able to serve as an excellent super capacitor. The elevation of the curcumin-rGO electrical conductivity was related to the attachment of the phenolic group of curcumin on top of the graphene layers, via the π–π stacking interaction. Indeed, curcumin acts as an electron donor, repressing the fluorescence of the GO sheets and enhancing their electrochemical properties.

4.5 Carbon Dot-based Biosensors

Another nanomaterial is carbon dots (CDs), also known as fluorescent carbon dots, carbon nanoparticles, or carbon quantum dots. C-dots are composed of sp2/sp3 carbons with some heteroatoms like nitrogen and oxygen and therefore show high solubility in biological media in contrast to other CNM members [38]. Lately, their popularity is on the rise because of useful features, such as high chemical stability and inertness, outstanding fluorescence emissions, low toxicity, environmental compatibility, and low costs [182]. Therefore, a wide variety of applications have been proposed for CDs in biomedical sciences, including bioimaging [183–185], drug and gene delivery [186, 187], and chemical sensing [188]. According to the existing literature, bacteria [189], nucleic acids [190], proteins [191], small organic molecules [192, 193], pesticides [194, 195], explosives [196], drugs [197], and metal ions [194, 198] have been identified using CD-based sensors thus far. These tiny carbonaceous nanomaterials have been reported to be biocompatible with low cytotoxicity. Ostadhossein et al. showed that degradation and excretion of C-dots from the body could be mediated by the enzyme myeloperoxidase in a short time [199]. In fact, the biocompatible nature of C-dots has led to their being suggested as a potential future replacement for the currently approved dyes [200].

Another important potential of C-dots is in the biosensor field. Nano-biosensors based on C-dots could be decorated with various surface molecules and benefit from photostability, tunable fluorescence emission, high aqueous solubility, and biocompatibility. In this regard, C-dots incorporated sensors have been reported as sensors for several targets including glucose, dopamine, metal ions, pH, DNA, proteins, enzymes, and bacteria [201, 202]. Also, numerous reports have shown that C-dot based biosensors could be effectively used in the precise detection of curcumin, in which C-dot fluorescence will be quenched and compromised by the attachment of curcumin molecules mainly through the inner filter effect (IFE) or fluorescence resonance energy transfer (FRET) mechanisms. One of the first studies in this subject was the report of Shi et al. in 2015, in which they showed that 5 nm C-dots with a strong blue fluorescence were quenched by increasing concentrations of curcumin. The mechanism of quenching was said to be done by FRET. They concluded that the curcumin can be detected with a detection limit of approximately 45 ng/mL, and they concluded that their sensor could be effective for the detection of curcumin in real marketed samples. The selectivity of proposed sensor was good, as it was tested among various metals, amino, acids and sugars; and none of the interfering substances except for Fe^{+2} had considerable effects [203]. The interfering effects of iron were also stated in detail in another study in which the fluorescence of the C-dots was quenched by addition of curcumin, and this emission could be recovered upon addition of Fe^{3+} and complex formation between curcumin and Fe^{3+} [204]. Similarly, in 2015, Zhang et al. presented a green synthetic protocol for N-doped C-dots with an average size of 5.23 nm [205]. The fluorescence of the C-dots could be quenched via an inner filter effect by the presence of curcumin, and the presented system showed good selectivity among various interfering molecules, including some common metal ions, anions, some small organic molecules, and amino acids. The detection limit was reported as 8.48×10^{-8} mol/L, and the proposed system showed excellent detection of curcumin in human urine samples. Similarly, another heteroatom-doped C-dot approach was reported as boron and nitrogen co-doped carbon dots (BNCDs) by Bian et al. [206]. Their tiny BNCDs were able to detect curcumin with a detection limit of 65 nM. The reported fluorescent probe was also effective among various interfering molecules and also showed high recoveries (>96.5%) of curcumin in urine samples. The IFE phenomenon was proposed to be the mechanism by which the curcumin absorbs the excitation or emission lights of C-dots. Another interesting paper by Liu et al. reported that nitrogen and phosphorus dual-doped C-dots (NP-C-dots) is effective for curcumin detection through the IFE effect [207]. The label-free sensor from NP-C-dots showed a detection limit of 58 nM toward curcumin, and good sensitivity among multiple interfering substances including 15 amino acids and 15 drugs. It was able to recover curcumin in real food samples at high rates of >95%. Also, the sensor showed low toxicity in vitro when used effectively for cellular imaging of SiHa cells. More recently, nitrogen and sulfur co-doped fluorescence C-dots (N,S-CDs) were presented as curcumin detectors with a detection limit of 40 nM. Interestingly the C-dots were also sensitive to the temperature changes, and the authors concluded that the reported N,S-CDs could be introduced as temperature and curcumin detectors with high sensitivity [204]. New sources for C-dots production are being sought as Fahad beig et al. used chicken egg whites to obtain tiny 3.3 nm C-dots. The resultant C-dots were studied for their effective bacterial labeling (*Staphylococcus aureus* and *Escherichia coli*) and curcumin detection in complex samples [208]. A more recent study by Luo et al. dealt with simultaneous detection of curcumin, Cu^{+2} ions, and a Cu^{+2}-curcumin complex. In the presence of beta-amyloid, curcumin can form a stable complex with Cu^{+2} and ameliorate neurotoxicity [209]. Therefore, Cu^{+2}-curcumin detection may prove helpful in the treatment of Alzheimer disease. In this report, fluorescent and spherical tiny (~1.3 nm) carbon quantum dots (CQDs) were prepared from the levodopa drug as a carbon source. The levo-CQDs were capable of detecting Cu^{2+} ions with high selectivity and sensitivity in the presence of different cations with an LOD of 0.4 μM. Moreover, the absorption wavelength for curcumin had an important overlap with the excitation peak for levo-CQDs which resulted in the precise detection of curcumin through the fluorescence quenching via IFE. Moreover, the reported levo-CQDs were capable of detecting Cu^{+2}-curcumin complex formation. Also sensing of curcumin or Cu^{+2} was effectively done in real biological samples including human serum, tap water, and river water with high recovery rates.

What is evident from the above studies is that various kinds of C-dots with different structures can be synthesized, mainly with the diameters of $\approx$ 5 nm, and the emission of these tiny nanoparticles can be effectively quenched by curcumin molecules. Most of the sensors based on C-dots showed high sensitivity and selectivity among various types of elements, amino acids, foods, and drugs. Importantly some groups showed that the C-dots were useful for cellular imaging while being non-toxic to the cells.

4.6 Conclusions

The considerable growth of societal demands, especially in the arena of healthcare challenges in relation to human population aging, necessitates next-generation diagnostic devices for medical purposes. These should be reliable, rapid, simple, sensitive, selective, and cost-effective as compared to the more currently available immunological assays. Besides, these diagnostic alternatives should meet our concerns for the environment and global security. In recent years, several CNM-based biosensors have emerged with an ability to satisfy the sensitivity and selectivity aspects of biological analytes such as curcumin detection in practice. Some limited research has combined curcumin with CNMs for the recognition of micron-sized targets. Indeed, the optical, electronic, and structural properties of CNTs, GO, and rGO are well suited for curcumin detection or as combination approaches with curcumin. However, these are not enough and do not address industrial production and commercialization needs. In addition, the new designs require validation studies applying high numbers of patient specimens. Biosensors based on CNMs seem to have simple and fast production procedures, and they are less toxic while they can be highly sensitive through spectrofluorometric techniques. Although still at the laboratory stage, research on CNM-based biosensors is promising for curcumin detection or identifying the trace amounts of biomolecules as curcumin combination approaches. For future studies, more experiments along with computational simulations are required on the bio-interactions between CNMs and curcumin to further understand the mechanisms underlying their biomedical applications.

References

1. Priyadarsini KI (2014) The chemistry of curcumin: from extraction to therapeutic agent. Molecules 19(12):20091–20112
2. Lestari ML, Indrayanto G (2014) Curcumin. Profiles Drug Subst Excip Relat Methodol 39:113–204
3. Mahady GB, Pendland SL, Yun G, Lu ZZ (2002) Turmeric (Curcuma longa) and curcumin inhibit the growth of Helicobacter pylori, a group 1 carcinogen. Anticancer Res 22(6C):4179–4181
4. Reddy RC, Vatsala PG, Keshamouni VG, Padmanaban G, Rangarajan PN (2005) Curcumin

for malaria therapy. Biochem Biophys Res Commun 326(2):472–474

5. Vera-Ramirez L, Perez-Lopez P, Varela-Lopez A, Ramirez-Tortosa M, Battino M, Quiles JL (2013) Curcumin and liver disease. Biofactors 39(1):88–100
6. Wright LE, Frye JB, Gorti B, Timmermann BN, Funk JL (2013) Bioactivity of turmeric-derived curcuminoids and related metabolites in breast cancer. Curr Pharm Des 19(34):6218–6225
7. Aggarwal BB, Kumar A, Bharti AC (2003) Anticancer potential of curcumin: preclinical and clinical studies. Anticancer Res 23(1A):363–398
8. Gupta SC, Patchva S, Aggarwal BB (2013) Therapeutic roles of curcumin: lessons learned from clinical trials. AAPS J 15(1):195–218
9. Teymouri M, Pirro M, Johnston TP, Sahebkar A (2017) Curcumin as a multifaceted compound against human papilloma virus infection and cervical cancers: A review of chemistry, cellular, molecular, and preclinical features. BioFactors 43(3):331–346
10. Abdollahi E, Momtazi AA, Johnston TP, Sahebkar A (2018) Therapeutic effects of curcumin in inflammatory and immune-mediated diseases: a nature-made jack-of-all-trades? J Cell Physiol 233(2):830–848
11. Iranshahi M, Sahebkar A, Takasaki M, Konoshima T, Tokuda H (2009) Cancer chemopreventive activity of the prenylated coumarin, umbelliprenin, in vivo. Eur J Cancer Prev 18(5):412–415
12. Panahi Y, Kianpour P, Mohtashami R, Jafari R, Simental-Mendía LE, Sahebkar A (2017) Efficacy and safety of phytosomal curcumin in non-alcoholic fatty liver disease: a randomized controlled trial. Drug Res 67(4):244–251
13. Bagheri H, Ghasemi F, Barreto GE, Sathyapalan T, Jamialahmadi T, Sahebkar A (2020) The effects of statins on microglial cells to protect against neurodegenerative disorders: A mechanistic review. Biofactors. 46(3):309–325. https://doi.org/10.1002/biof.1597. Epub 2019 Dec 17. PMID: 31846136
14. Momtazi AA, Derosa G, Maffioli P, Banach M, Sahebkar A (2016) Role of microRNAs in the therapeutic effects of curcumin in non-cancer diseases. Mol Diagn Ther 20(4):335–345
15. Panahi Y, Ahmadi Y, Teymouri M, Johnston TP, Sahebkar A (2018) Curcumin as a potential candidate for treating hyperlipidemia: A review of cellular and metabolic mechanisms. J Cell Physiol 233(1):141–152. https://doi.org/10.1002/jcp.25756. Epub 2017 Jun 6. PMID: 28012169
16. Soleimani V, Sahebkar A, Hosseinzadeh H (2018) Turmeric (Curcuma longa) and its major constituent (curcumin) as nontoxic and safe substances: Review. Phytother Res 32(6):985–995. https://doi.org/10.1002/ptr.6054. Epub 2018 Feb 26. PMID: 29480523
17. Aggarwal BB, Harikumar KB (2009) Potential therapeutic effects of curcumin, the anti-inflammatory agent, against neurodegenerative, cardiovascular, pulmonary, metabolic, autoimmune and neoplastic diseases. Int J Biochem Cell Biol 41(1):40–59
18. Saberi-Karimian M, Keshvari M, Ghayour-Mobarhan M, Salehizadeh L, Rahmani S, Behnam B et al (2020) Effects of curcuminoids on inflammatory status in patients with non-alcoholic fatty liver disease: a randomized controlled trial. Complement Ther Med 49:102322. https://doi.org/10.1016/j.ctim.2020.102322
19. Karimian MS, Pirro M, Majeed M, Sahebkar A (2017) Curcumin as a natural regulator of monocyte chemoattractant protein-1. Cytokine Growth Factor Rev 33:55–63
20. Mollazadeh H, Cicero AFG, Blesso CN, Pirro M, Majeed M, Sahebkar A (2019) Immune modulation by curcumin: the role of interleukin-10. Crit Rev Food Sci Nutr 59(1):89–101
21. Panahi Y, Hosseini MS, Khalili N, Naimi E, Simental-Mendía LE, Majeed M et al (2016) Effects of curcumin on serum cytokine concentrations in subjects with metabolic syndrome: a post-hoc analysis of a randomized controlled trial. Biomed Pharmacother 82:578–582
22. Ireson C, Orr S, Jones DJ, Verschoyle R, Lim CK, Luo JL et al (2001) Characterization of metabolites of the chemopreventive agent curcumin in human and rat hepatocytes and in the rat in vivo, and evaluation of their ability to inhibit phorbol ester-induced prostaglandin E2 production. Cancer Res 61(3):1058–1064
23. Ireson CR, Jones DJ, Orr S, Coughtrie MW, Boocock DJ, Williams ML et al (2002) Metabolism of the cancer chemopreventive agent curcumin in human and rat intestine. Cancer Epidemiol Biomark Prev 11(1):105–111
24. Sharma RA, Euden SA, Platton SL, Cooke DN, Shafayat A, Hewitt HR et al (2004) Phase I clinical trial of oral curcumin: biomarkers of systemic activity and compliance. Clin Cancer Res 10(20):6847–6854
25. Teradal NL, Jelinek R (2017) Carbon nanomaterials in biological studies and biomedicine. Adv Healthc Mater 6(17). https://doi.org/10.1002/adhm.201700574
26. Mohajeri M, Behnam B, Sahebkar A (2019) Biomedical applications of carbon nanomaterials: drug and gene delivery potentials. J Cell Physiol 234(1):298–319
27. Harrison BS, Atala A (2007) Carbon nanotube applications for tissue engineering. Biomaterials 28(2):344–353
28. Wang Z, Dai Z (2015) Carbon nanomaterial-based electrochemical biosensors: an overview. Nanoscale 7(15):6420–6431
29. Rezaee M, Behnam B, Banach M, Sahebkar A (2018) The Yin and Yang of carbon nanomaterials in atherosclerosis. Biotechnol Adv 36(8):2232–2247
30. Mohajeri M, Behnam B, Barreto GE, Sahebkar A (2019) Carbon nanomaterials and amyloid-beta interactions: potentials for the detection and treatment of Alzheimer's disease? Pharmacol Res 143:186–203

31. Rauti R, Musto M, Bosi S, Prato M, Ballerini L (2018) Properties and behavior of carbon nanomaterials when interfacing neuronal cells: How far have we come? Carbon 143:430–446
32. Geim AK, Novoselov KS (2007) The rise of graphene. Nat Mater 6(3):183–191
33. Kim KS, Zhao Y, Jang H, Lee SY, Kim JM, Ahn JH et al (2009) Large-scale pattern growth of graphene films for stretchable transparent electrodes. Nature 457(7230):706–710
34. Eda G, Fanchini G, Chhowalla M (2008) Large-area ultrathin films of reduced graphene oxide as a transparent and flexible electronic material. Nat Nanotechnol 3(5):270–274
35. Dreyer DR, Park S, Bielawski CW, Ruoff RS (2010) The chemistry of graphene oxide. Chem Soc Rev 39(1):228–240
36. Wang Y, Li Z, Wang J, Li J, Lin Y (2011) Graphene and graphene oxide: biofunctionalization and applications in biotechnology. Trends Biotechnol 29(5):205–212
37. Wang J, Qiu J (2016) A review of carbon dots in biological applications. J Mater Sci 51(10): 4728–4738
38. Jaleel JA, Pramod K (2018) Artful and multifaceted applications of carbon dot in biomedicine. J Control Release 269:302–321
39. Marković Z, Kepić D, Matijašević D, Pavlović V, Jovanović S, Stanković N et al (2017) Ambient light induced antibacterial action of curcumin/graphene nanomesh hybrids. RSC Adv 7(57):36081–36092
40. Roşu MC, Páll E, Socaci C, Măgeruşan L, Pogăcean F, Coroş M et al (2017) Cytotoxicity of methylcellulose-based films containing graphenes and curcumin on human lung fibroblasts. Process Biochem 52:243–249
41. Goel A, Kunnumakkara AB, Aggarwal BB (2008) Curcumin as "Curecumin": from kitchen to clinic. Biochem Pharmacol 75(4):787–809
42. Marković ZM, Prekodravac JR, Tošić DD, Holclajtner-Antunović ID, Milosavljević MS, Dramićanin MD et al (2015) Facile synthesis of water-soluble curcumin nanocrystals. J Serb Chem Soc 80(1):63–72
43. Bengmark S (2006) Curcumin, an atoxic antioxidant and natural NFκB, cyclooxygenase-2, lipooxygenase, and inducible nitric oxide synthase inhibitor: a shield against acute and chronic diseases. J Parenter Enter Nutr 30(1):45–51
44. Hatcher H, Planalp R, Cho J, Torti F, Torti S (2008) Curcumin: from ancient medicine to current clinical trials. Cell Mol Life Sci 65(11):1631–1652
45. Huang YS, Hsieh T-J, Lu CY (2015) Simple analytical strategy for MALDI-TOF-MS and nanoU-PLC-MS/MS: quantitating curcumin in food condiments and dietary supplements and screening of acrylamide-induced ROS protein indicators reduced by curcumin. Food Chem 174:571–576
46. Mohajeri M, Behnam B, Cicero AFG, Sahebkar A (2018) Protective effects of curcumin against aflatoxicosis: A comprehensive review. J Cell Physiol 233(4):3552–3577
47. Ziyatdinova GK, Nizamova AM, Budnikov HC (2012) Voltammetric determination of curcumin in spices. J Anal Chem 67(6):591–594
48. Cheng J, Weijun K, Yun L, Jiabo W, Haitao W, Qingmiao L et al (2010) Development and validation of UPLC method for quality control of Curcuma longa Linn.: fast simultaneous quantitation of three curcuminoids. J Pharm Biomed Anal 53(1):43–49
49. Tayyem RF, Heath DD, Al-Delaimy WK, Rock CL (2006) Curcumin content of turmeric and curry powders. Nutr Cancer 55(2):126–131
50. Dhakal S, Chao K, Schmidt W, Qin J, Kim M, Chan D (2016) Evaluation of turmeric powder adulterated with Metanil yellow using FT-Raman and FT-IR spectroscopy. Foods 5(2):E36. https://doi.org/10.3390/foods5020036
51. Govindarajan VS (1980) Turmeric—chemistry, technology, and quality. Crit Rev Food Sci Nutr 12(3):199–301
52. Gleason K, Shine JP, Shobnam N, Rokoff LB, Suchanda HS, Ibne Hasan MOS et al (2014) Contaminated turmeric is a potential source of lead exposure for children in Rural Bangladesh. J Environ Public Health 2014:730636. https://doi.org/10.1155/2014/730636
53. Nasr M, Rahman A, Hamdy M (2019) Simultaneous determination of curcumin and resveratrol in lipidic nanoemulsion formulation and rat plasma using HPLC: optimization and application to real samples. J AOAC Int 102(4):1095–1101
54. Gupta NK, Nahata A, Dixit VK (2010) Development of Spectrofluorimetric Method for the determination of curcumin. Asian J Trad Med 5(1):12–18
55. Wang F, Wu X, Wang F, Liu S, Jia Z, Yang J (2006) The sensitive fluorimetric method for the determination of curcumin using the enhancement of mixed micelle. J Fluoresc 16(1):53–59
56. Li K, Li Y, Yang L, Wang L, Ye B (2014) The electrochemical characterization of curcumin and its selective detection in Curcuma using a graphene-modified electrode. Anal Methods 6(19):7801–7808
57. Ramalingam P, Ko YT (2014) A validated LC-MS/MS method for quantitative analysis of curcumin in mouse plasma and brain tissue and its application in pharmacokinetic and brain distribution studies. J Chromatogr B Analyt Technol Biomed Life Sci 969:101–108
58. Arslan E, Çakır S (2014) A novel palladium nanoparticles-polyproline-modified graphite electrode and its application for determination of curcumin. J Solid State Electrochem 18(6):1611–1620
59. Atar N, Eren T, Demirdögen B, Yola ML, Çağlayan MO (2015) Silver, gold, and silver@gold nanoparticle-anchored l-cysteine-functionalized reduced graphene oxide as electrocatalyst for methanol oxidation. Ionics 21:2285–2293
60. Atar N, Eren T, Yola ML (2015) Ultrahigh capacity anode material for lithium ion battery based on

rod gold nanoparticles decorated reduced graphene oxide. Thin Solid Films 590:156–162

61. Goyal RN, Gupta VK, Chatterjee S (2010) Voltammetric biosensors for the determination of paracetamol at carbon nanotube modified pyrolytic graphite electrode. Sensors Actuators B Chem 149(1):252–258
62. Gupta SC, Prasad S, Kim JH, Patchva S, Webb LJ, Priyadarsini IK et al (2011) Multitargeting by curcumin as revealed by molecular interaction studies. Nat Prod Rep 28(12):1937–1955
63. Gupta VK, Jain AK, Singh LP, Khurana U (1997) Porphyrins as carrier in PVC based membrane potentiometric sensors for nickel(II). Anal Chim Acta 355(1):33–41
64. Jain R, Gupta VK, Jadon N, Radhapyari K (2010) Voltammetric determination of cefixime in pharmaceuticals and biological fluids. Anal Biochem 407(1):79–88
65. Yola ML, Atar N (2014) A novel voltammetric sensor based on gold nanoparticles involved in p-aminothiophenol functionalized multi-walled carbon nanotubes: application to the simultaneous determination of quercetin and rutin. Electrochim Acta 119:24–31
66. Hatamie S, Akhavan O, Sadrnezhaad SK, Ahadian MM, Shirolkar MM, Wang HQ (2015) Curcumin-reduced graphene oxide sheets and their effects on human breast cancer cells. Mater Sci Eng C Mater Biol Appl 55:482–489
67. Lungu A, Sandu I, Boscornea C, Tomas S, Mihailciuc C (2010) Electrochemical study of curcumin and bisdemethoxycurcumin on activated glassy carbon electrode. Rev Roum Chim 55(2):109–115
68. Yousef Elahi M, Heli H, Bathaie SZ, Mousavi MF (2007) Electrocatalytic oxidation of glucose at a Ni-curcumin modified glassy carbon electrode. J Solid State Electrochem 11(2):273–282
69. Ciszewski A (1995) Catalytic oxidation of methanol on a glassy carbon electrode electrochemically modified by a conductive NiII-curcumin film. Electroanalysis 7(12):1132–1135
70. Gibson CT, Carnally S, Roberts CJ (2007) Attachment of carbon nanotubes to atomic force microscope probes. Ultramicroscopy 107(10–11):1118–1122
71. Antaris AL, Robinson JT, Yaghi OK, Hong G, Diao S, Luong R et al (2013) Ultra-low doses of chirality sorted (6, 5) carbon nanotubes for simultaneous tumor imaging and photothermal therapy. ACS Nano 7(4):3644–3652
72. Bernholc J, Brenner D, Buongiorno Nardelli M, Meunier V, Roland C (2002) Mechanical and electrical properties of nanotubes. Annu Rev Mater Res 32(1):347–375
73. Hong G, Diao S, Chang J, Antaris AL, Chen C, Zhang B et al (2014) Through-skull fluorescence imaging of the brain in a new near-infrared window. Nat Photonics 8(9):723–730
74. Liang Y, Li Y, Wang H, Dai H (2013) Strongly coupled inorganic/nanocarbon hybrid materials for advanced electrocatalysis. J Am Chem Soc 135(6):2013–2036
75. Liu Z, Robinson JT, Tabakman SM, Yang K, Dai H (2011) Carbon materials for drug delivery & cancer therapy. Mater Today 14(7–8):316–323
76. Liu Z, Tabakman S, Welsher K, Dai H (2009) Carbon nanotubes in biology and medicine: in vitro and in vivo detection, imaging and drug delivery. Nano Res 2(2):85–120
77. Wang H, Dai H (2013) Strongly coupled inorganic–nano-carbon hybrid materials for energy storage. Chem Soc Rev 42(7):3088–3113
78. Harris PJF (2009) Carbon nanotube science: synthesis, properties and applications, 1st edn. Cambridge University Press, Cambridge
79. Iijima S (1991) Helical microtubules of graphitic carbon. Nature 354(6348):56–58
80. Heister E, Neves V, Lamprecht C, Silva SRP, Coley HM, McFadden J (2012) Drug loading, dispersion stability, and therapeutic efficacy in targeted drug delivery with carbon nanotubes. Carbon 50(2):622–632
81. Klumpp C, Kostarelos K, Prato M, Bianco A (2006) Functionalized carbon nanotubes as emerging nanovectors for the delivery of therapeutics. Biochim Biophys Acta 1758(3):404–412
82. Tsai HC, Lin JY, Maryani F, Huang CC, Imae T (2013) Drug-loading capacity and nuclear targeting of multiwalled carbon nanotubes grafted with anionic amphiphilic copolymers. Int J Nanomedicine 8:4427–4440
83. Behnam B, Shier WT, Nia AH, Abnous K, Ramezani M (2013) Non-covalent functionalization of single-walled carbon nanotubes with modified polyethyleneimines for efficient gene delivery. Int J Pharm 454(1):204–215
84. Hashem Nia A, Behnam B, Taghavi S, Oroojalian F, Eshghi H, Shier WT et al (2017) Evaluation of chemical modification effects on DNA plasmid transfection efficiency of single-walled carbon nanotube–succinate– polyethylenimine conjugates as non-viral gene carriers. Med Chem Comm 8(2):364–375
85. Pantarotto D, Singh R, McCarthy D, Erhardt M, Briand JP, Prato M et al (2004) Functionalized carbon nanotubes for plasmid DNA gene delivery. Angew Chem Int Ed Engl 43(39):5242–5246
86. Shi Kam NW, Liu Z, Dai H (2006) Carbon nanotubes as intracellular transporters for proteins and DNA: an investigation of the uptake mechanism and pathway. Angew Chem Int Ed Engl 118(4):591–595
87. Wu Y, Phillips JA, Liu H, Yang R, Tan W (2008) Carbon nanotubes protect DNA strands during cellular delivery. ACS Nano 2(10):2023–2028
88. Zhang Z, Yang X, Zhang Y, Zeng B, Wang S, Zhu T et al (2006) Delivery of telomerase reverse transcriptase small interfering RNA in complex with positively charged single-walled carbon nanotubes suppresses tumor growth. Clin Cancer Res 12(16):4933–4939

89. Shi Kam NW, Dai H (2005) Carbon nanotubes as intracellular protein transporters: generality and biological functionality. J Am Chem Soc 127(16):6021–6026
90. Pantarotto D, Briand J-P, Prato M, Bianco A (2004) Translocation of bioactive peptides across cell membranes by carbon nanotubes. Chem Commun (Camb) 1:16–17
91. Pantarotto D, Partidos CD, Graff R, Hoebeke J, Briand JP, Prato M et al (2003) Synthesis, structural characterization, and immunological properties of carbon nanotubes functionalized with peptides. J Am Chem Soc 125(20):6160–6164
92. Bianco A, Hoebeke J, Godefroy S, Chaloin O, Pantarotto D, Briand JP et al (2005) Cationic carbon nanotubes bind to CpG oligodeoxynucleotides and enhance their immunostimulatory properties. J Am Chem Soc 127(1):58–59
93. Partidos CD, Hoebeke J, Wieckowski S, Chaloin O, Bianco A, Moreau E et al (2009) Immunomodulatory consequences of ODN CpG-polycation complexes. Methods 49(4):328–333
94. Chaudhuri P, Soni S, Sengupta S (2010) Single-walled carbon nanotube-conjugated chemotherapy exhibits increased therapeutic index in melanoma. Nanotechnology 21(2):025102
95. Heister E, Neves V, Tîlmaciu C, Lipert K, Beltrán VS, Coley HM et al (2009) Triple functionalisation of single-walled carbon nanotubes with doxorubicin, a monoclonal antibody, and a fluorescent marker for targeted cancer therapy. Carbon 47(9):2152–2160
96. Li H, Zhang N, Hao Y, Wang Y, Jia S, Zhang H et al (2014) Formulation of curcumin delivery with functionalized single-walled carbon nanotubes: characteristics and anticancer effects in vitro. Drug Deliv 21(5):379–387
97. Li R, Wu R, Zhao L, Wu M, Yang L, Zou H (2010) P-glycoprotein antibody functionalized carbon nanotube overcomes the multidrug resistance of human leukemia cells. ACS Nano 4(3):1399–1408
98. Samorì C, Ali-Boucetta H, Sainz R, Guo C, Toma FM, Fabbro C et al (2010) Enhanced anticancer activity of multi-walled carbon nanotube–methotrexate conjugates using cleavable linkers. Chem Commun (Camb) 46(9):1494–1496
99. Yang ST, Wang X, Jia G, Gu Y, Wang T, Nie H et al (2008) Long-term accumulation and low toxicity of single-walled carbon nanotubes in intravenously exposed mice. Toxicol Lett 181(3):182–189
100. Shiraki T, Dawn A, Le TNL, Tsuchiya Y, S-i T, Shinkai S (2011) Heat and light dual switching of a single-walled carbon nanotube/thermo-responsive helical polysaccharide complex: a new responsive system applicable to photodynamic therapy. Chem Commun (Camb) 47(25):7065–7067
101. De la Zerda A, Zavaleta C, Keren S, Vaithilingam S, Bodapati S, Liu Z et al (2008) Carbon nanotubes as photoacoustic molecular imaging agents in living mice. Nat Nanotechnol 3(9):557–562
102. Shi Kam NW, O'Connell M, Wisdom JA, Dai H (2005) Carbon nanotubes as multifunctional biological transporters and near-infrared agents for selective cancer cell destruction. Proc Natl Acad Sci U S A 102(33):11600–11605
103. Kim JW, Galanzha EI, Shashkov EV, Moon HM, Zharov VP (2009) Golden carbon nanotubes as multimodal photoacoustic and photothermal high-contrast molecular agents. Nat Nanotechnol 4(10):688–694
104. Zhou F, Xing D, Ou Z, Wu B, Resasco DE, Chen WR (2009) Cancer photothermal therapy in the near-infrared region by using single-walled carbon nanotubes. J Biomed Opt 14(2):021009. https://doi.org/10.1117/1.3078803
105. Mohammadi E, Zeinali M, Mohammadi-Sardoo M, Iranpour M, Behnam B, Mandegary A (2020) The effects of functionalization of carbon nanotubes on toxicological parameters in mice. Hum Exp Toxicol 39(9):096032711989998. https://doi.org/10.1177/0960327119899988
106. Ahmadi H, Ramezani M, Yazdian-Robati R, Behnam B, Razavi Azarkhiavi K, Hashem Nia A et al (2017) Acute toxicity of functionalized single wall carbon nanotubes: A biochemical, histopathologic and proteomics approach. Chem Biol Interact 275:196–209
107. Lin Y, Lu F, Wang J (2004) Disposable carbon nanotube modified screen-printed biosensor for amperometric detection of organophosphorus pesticides and nerve agents. Electroanalysis 16(1–2):145–149
108. Qu F, Yang M, Jiang J, Shen G, Yu R (2005) Amperometric biosensor for choline based on layer-by-layer assembled functionalized carbon nanotube and polyaniline multilayer film. Anal Biochem 344(1):108–114
109. Tang X, Bansaruntip S, Nakayama N, Yenilmez E, Chang YL, Wang Q (2006) Carbon nanotube DNA sensor and sensing mechanism. Nano Lett 6(8):1632–1636
110. Li J, Ng HT, Cassell A, Fan W, Chen H, Ye Q et al (2003) Carbon nanotube nanoelectrode array for ultrasensitive DNA detection. Nano Lett 3(5):597–602
111. Yu X, Munge B, Patel V, Jensen G, Bhirde A, Gong JD et al (2006) Carbon nanotube amplification strategies for highly sensitive immunodetection of cancer biomarkers. J Am Chem Soc 128(34):11199–11205
112. Justino CIL, Rocha-Santos TAP, Duarte AC (2013) Advances in point-of-care technologies with biosensors based on carbon nanotubes. TrAC Trends Anal Chem 45:24–36
113. Espinosa EH, Ionescu R, Chambon B, Bedis G, Sotter E, Bittencourt C et al (2007) Hybrid metal oxide and multiwall carbon nanotube films for low temperature gas sensing. Sensors Actuators B Chem 127(1):137–142
114. Van Hieu N, Thuy LTB, Chien ND (2008) Highly sensitive thin film NH3 gas sensor operating at room temperature based on SnO2/MWCNTs composite. Sensors Actuators B Chem 129(2):888–895

115. Mooney E, Dockery P, Greiser U, Murphy M, Barron V (2008) Carbon nanotubes and mesenchymal stem cells: biocompatibility, proliferation and differentiation. Nano Lett 8(8):2137–2143
116. Venkatesan J, Qian Z-J, Ryu B, Ashok Kumar N, Kim SK (2011) Preparation and characterization of carbon nanotube-grafted-chitosan – Natural hydroxyapatite composite for bone tissue engineering. Carbohydr Polym 83(2):569–577
117. Zanello LP, Zhao B, Hu H, Haddon RC (2006) Bone cell proliferation on carbon nanotubes. Nano Lett 6(3):562–567
118. Jiang F, Wang Y, Hu X, Shao N, Na N, Delanghe JR et al (2010) Carbon nanotubes-assisted polyacrylamide gel electrophoresis for enhanced separation of human serum proteins and application in liverish diagnosis. J Sep Sci 33(21):3393–3399
119. Wang Z, Luo G, Chen J, Xiao S, Wang Y (2003) Carbon nanotubes as separation carrier in capillary electrophoresis. Electrophoresis 24(24):4181–4188
120. Benincasa M, Pacor S, Wu W, Prato M, Bianco A, Gennaro R (2011) Antifungal activity of amphotericin B conjugated to carbon nanotubes. ACS Nano 5(1):199–208
121. Schiffman JD, Elimelech M (2011) Antibacterial activity of electrospun polymer mats with incorporated narrow diameter single-walled carbon nanotubes. ACS Appl Mater Interfaces 3(2):462–468
122. Vecitis CD, Schnoor MH, Rahaman MS, Schiffman JD, Elimelech M (2011) Electrochemical multiwalled carbon nanotube filter for viral and bacterial removal and inactivation. Environ Sci Technol 45(8):3672–3679
123. Holzinger M, Baur J, Haddad R, Wang X, Cosnier S (2011) Multiple functionalization of single-walled carbon nanotubes by dip coating. Chem Commun (Camb) 47(8):2450–2452
124. Mundra RV, Wu X, Sauer J, Dordick JS, Kane RS (2014) Nanotubes in biological applications. Curr Opin Biotechnol 28:25–32
125. Münzer A, Heimgreiter M, Melzer K, Weise A, Fabel B, Abdellah A et al (2013) Back-gated spray-deposited carbon nanotube thin film transistors operated in electrolytic solutions: an assessment towards future biosensing applications. J Mater Chem B 1(31):3797–3802
126. Tîlmaciu C-M, Morris MC (2015) Carbon nanotube biosensors. Front Chem 3:59. https://doi.org/10.3389/fchem.2015.00059
127. Zheng L, Song JF (2009) Curcumin multi-wall carbon nanotubes modified glassy carbon electrode and its electrocatalytic activity towards oxidation of hydrazine. Sensors Actuators B Chem 135(2):650–655
128. Krishna Kumar K, Devendiran M, Jyothithamizhanban, Narayanan SS (2014) Curcumin/MWCNT modified graphite electrode for electrochemical determination of BHA. Int J Innov Res in Sci Eng 2(special issue 1):654–659
129. Daneshgar P, Norouzi P, Moosavi-Movahedi AA, Ganjali MR, Haghshenas E, Dousty F et al (2009) Fabrication of carbon nanotube and dysprosium nanowire modified electrodes as a sensor for determination of curcumin. J Appl Electrochem 39(10):1983
130. Jain R, Haque A, Verma A (2017) Voltammetric quantification of surfactant stabilized curcumin at MWCNT/GCE sensor. J Mol Liq 230:600–607
131. Chaisiwamongkhol K, Ngamchuea K, Batchelor-McAuley C, Compton RG (2017) Multiwalled carbon nanotube modified electrodes for the adsorptive stripping voltammetric determination and quantification of curcumin in turmeric. Electroanalysis 29(4):1049–1055
132. Novoselov KS, Geim AK, Morozov SV, Jiang D, Zhang Y, Dubonos SV et al (2004) Electric field effect in atomically thin carbon films. Science 306(5696):666–669
133. Zhang Y, Petibone D, Xu Y, Mahmood M, Karmakar A, Casciano D et al (2014) Toxicity and efficacy of carbon nanotubes and graphene: the utility of carbon-based nanoparticles in nanomedicine. Drug Metab Rev 46(2):232–246
134. Geim AK (2009) Graphene: status and prospects. Science 324(5934):1530–1534
135. Kostarelos K, Novoselov KS (2014) Exploring the interface of graphene and biology. Science 344(6181):261–263
136. Novoselov KS, Fal'ko VI, Colombo L, Gellert PR, Schwab MG, Kim K (2012) A roadmap for graphene. Nature 490(7419):192–200
137. Katsnelson MI, Novoselov KS (2007) Graphene: new bridge between condensed matter physics and quantum electrodynamics. Solid State Commun 143(1):3–13
138. Novoselov KS, Geim AK, Morozov SV, Jiang D, Katsnelson MI, Grigorieva IV et al (2005) Two-dimensional gas of massless Dirac fermions in graphene. Nature 438(7065):197–200
139. Zhang Y, Tan YW, Stormer HL, Kim P (2005) Experimental observation of the quantum Hall effect and Berry's phase in graphene. Nature 438(7065):201–204
140. Gomez-Navarro C, Weitz RT, Bittner AM, Scolari M, Mews A, Burghard M et al (2007) Electronic transport properties of individual chemically reduced graphene oxide sheets. Nano Lett 7(11):3499–3503
141. Liang X, Fu Z, Chou SY (2007) Graphene transistors fabricated via transfer-printing in device active-areas on large wafer. Nano Lett 7(12):3840–3844
142. Stampfer C, Schurtenberger E, Molitor F, Güttinger J, Ihn T, Ensslin K (2008) Tunable graphene single electron transistor. Nano Lett 8(8):2378–2383
143. Standley B, Bao W, Zhang H, Bruck J, Lau CN, Bockrath M (2008) Graphene-based atomic-scale switches. Nano Lett 8(10):3345–3349
144. Bai J, Zhong X, Jiang S, Huang Y, Duan X (2010) Graphene nanomesh. Nat Nanotechnol 5(3):190–194

145. Akhavan O (2010) Graphene nanomesh by ZnO nanorod photocatalysts. ACS Nano 4(7):4174–4180
146. Wu ZS, Pei S, Ren W, Tang D, Gao L, Liu B et al (2009) Field emission of single-layer graphene films prepared by electrophoretic deposition. Adv Mater 21(17):1756–1760
147. Shang NG, Papakonstantinou P, McMullan M, Chu M, Stamboulis A, Potenza A et al (2008) Catalyst-free efficient growth, orientation and biosensing properties of multilayer graphene nanoflake films with sharp edge planes. Adv Funct Mater 18(21):3506–3514
148. Arsat R, Breedon M, Shafiei M, Spizziri PG, Gilje S, Kaner RB et al (2009) Graphene-like nano-sheets for surface acoustic wave gas sensor applications. Chem Phys Lett 467(4):344–347
149. Schedin F, Geim AK, Morozov SV, Hill EW, Blake P, Katsnelson MI et al (2007) Detection of individual gas molecules adsorbed on graphene. Nat Mater 6:652–655
150. Robinson JT, Perkins FK, Snow ES, Wei Z, Sheehan PE (2008) Reduced graphene oxide molecular sensors. Nano Lett 8(10):3137–3140
151. Wang X, Zhi L, Tsao N, Tomovic Z, Li J, Mullen K (2008) Transparent carbon films as electrodes in organic solar cells. Angew Chem Int Ed Engl 47(16):2990–2992
152. Gilje S, Han S, Wang M, Wang KL, Kaner RB (2007) A chemical route to graphene for device applications. Nano Lett 7(11):3394–3398
153. Liu Y, Yu D, Zeng C, Miao Z, Dai L (2010) Biocompatible graphene oxide-based glucose biosensors. Langmuir 26(9):6158–6160
154. Holzinger M, Le Goff A, Cosnier S (2014) Nanomaterials for biosensing applications: a review. Front Chem 2:63. https://doi.org/10.3389/fchem.2014.00063
155. Ambrosi A, Chua CK, Bonanni A, Pumera M (2014) Electrochemistry of graphene and related materials. Chem Rev 114(14):7150–7188
156. Viswanathan S, Narayanan TN, Aran K, Fink KD, Paredes J, Ajayan PM et al (2015) Graphene–protein field effect biosensors: glucose sensing. Mater Today 18(9):513–522
157. Szunerits S, Maalouli N, Wijaya E, Vilcot J-P, Boukherroub R (2013) Recent advances in the development of graphene-based surface plasmon resonance (SPR) interfaces. Anal Bioanal Chem 405(5):1435–1443
158. Sharma B, Frontiera RR, Henry A-I, Ringe E, Van Duyne RP (2012) SERS: materials, applications, and the future. Mater Today 15(1–2):16–25
159. Szunerits S, Boukherroub R (2018) Graphene-based biosensors. Interface Focus 8(3):20160132–20160132
160. Wu H, Wang J, Kang X, Wang C, Wang D, Liu J et al (2009) Glucose biosensor based on immobilization of glucose oxidase in platinum nanoparticles/graphene/chitosan nanocomposite film. Talanta 80(1):403–406
161. Zhou M, Zhai Y, Dong S (2009) Electrochemical sensing and biosensing platform based on chemically reduced graphene oxide. Anal Chem 81(14):5603–5613
162. Du D, Zou Z, Shin Y, Wang J, Wu H, Engelhard MH et al (2010) Sensitive immunosensor for cancer biomarker based on dual signal amplification strategy of graphene sheets and multi-enzyme functionalized carbon nanospheres. Anal Chem 82(7):2989–2995
163. Choi BG, Park H, Park TJ, Yang MH, Kim JS, Jang S-Y et al (2010) Solution chemistry of self-assembled graphene nanohybrids for high-performance flexible biosensors. ACS Nano 4(5):2910–2918
164. Shan C, Yang H, Song J, Han D, Ivaska A, Niu L (2009) Direct electrochemistry of glucose oxidase and biosensing for glucose based on graphene. Anal Chem 81(6):2378–2382
165. Alwarappan S, Liu C, Kumar A, Li C-Z (2010) Enzyme-doped graphene nanosheets for enhanced glucose biosensing. J Phys Chem C 114(30):12920–12924
166. Mohanty N, Berry V (2008) Graphene-based single-bacterium resolution biodevice and DNA transistor: interfacing graphene derivatives with nanoscale and microscale biocomponents. Nano Lett 8(12):4469–4476
167. Akhavan O, Ghaderi E (2010) Toxicity of graphene and graphene oxide nanowalls against bacteria. ACS Nano 4(10):5731–5736
168. Hu W, Peng C, Luo W, Lv M, Li X, Li D et al (2010) Graphene-based antibacterial paper. ACS Nano 4(7):4317–4323
169. Akhavan O, Ghaderi E (2009) Photocatalytic reduction of graphene oxide nanosheets on TiO2 thin film for photoinactivation of bacteria in solar light irradiation. J Phys Chem C 113(47):20214–20220
170. Yousefi M, Dadashpour M, Hejazi M, Hasanzadeh M, Behnam B, de la Guardia M et al (2017) Antibacterial activity of graphene oxide as a new weapon nanomaterial to combat multidrug-resistance bacteria. Mater Sci Eng C Mater Biol Appl 74:568–581
171. Gupta VK, Atar N, Yola ML, Eryılmaz M, Torul H, Tamer U et al (2013) A novel glucose biosensor platform based on Ag@AuNPs modified graphene oxide nanocomposite and SERS application. J Colloid Interface Sci 406:231–237. https://doi.org/10.1016/j.jcis.2013.06.007
172. Gupta VK, Eren T, Atar N, Yola ML, Parlak C, Karimi-Maleh H (2015) CoFe2O4@TiO2 decorated reduced graphene oxide nanocomposite for photocatalytic degradation of chlorpyrifos. J Mol Liq 208:122–129
173. Gupta VK, Jain AK, Agarwal S, Maheshwari G (2007) An iron(III) ion-selective sensor based on a μ-bis(tridentate) ligand. Talanta 71(5):1964–1968
174. Gupta VK, Yola ML, Atar N, Ustundağ Z, Solak AO (2013) A novel sensitive Cu(II) and Cd(II) nanosensor platform: Graphene oxide terminated

p-aminophenyl modified glassy carbon surface. Electrochim Acta 112:541–548

175. Gupta VK, Yola ML, Atar N, Üstündağ Z, Solak AO (2014) Electrochemical studies on graphene oxide-supported metallic and bimetallic nanoparticles for fuel cell applications. J Mol Liq 191:172–176
176. Khani H, Rofouei MK, Arab P, Gupta VK, Vafaei Z (2010) Multi-walled carbon nanotubes-ionic liquid-carbon paste electrode as a super selectivity sensor: application to potentiometric monitoring of mercury ion(II). J Hazard Mater 183(1–3):402–409
177. Kotan G, Kardaş F, Yokuş ÖA, Akyıldırım O, Saral H, Eren T et al (2016) A novel determination of curcumin via Ru@Au nanoparticle decorated nitrogen and sulfur-functionalized reduced graphene oxide nanomaterials. Anal Methods 8(2):401–408
178. Zhang D, Ouyang X, Ma J, Li L, Zhang Y (2016) Electrochemical behavior and voltammetric determination of curcumin at electrochemically reduced graphene oxide modified glassy carbon electrode. Electroanalysis 28(4):749–756
179. Dey N, Devasena T, Sivalingam T (2018) A comparative evaluation of graphene oxide based materials for electrochemical non-enzymatic sensing of curcumin. Mater Res Express 5(2). https://doi.org/10.1088/2053-1591/aaaa78
180. Rahimnejad M, Zokhtareh R, Moghadamnia AA, Asghary M (2020) An electrochemical sensor based on reduced graphene oxide modified carbon paste electrode for curcumin determination in human blood serum. Port Electrochim Acta 38(1):29–42
181. Hatamie S, Ahadian MM, Iraji zad A, Akhavan O, Jokar E (2018) Photoluminescence and electrochemical investigation of curcumin-reduced graphene oxide sheets. J Iran Chem Soc 15(2):351–357
182. Namdari P, Negahdari B, Eatemadi A (2017) Synthesis, properties and biomedical applications of carbon-based quantum dots: an updated review. Biomed Pharmacother 87:209–222
183. Desai ML, Jha S, Basu H, Singhal RK, Sharma P, Kailasa SK (2018) Chicken egg white and L-cysteine as cooperative ligands for effective encapsulation of Zn-doped silver nanoclusters for sensing and imaging applications. Colloids Surf A Physicochem Eng Asp 559:35–42
184. Li J, Zuo G, Pan X, Wei W, Qi X, Su T et al (2018) Nitrogen-doped carbon dots as a fluorescent probe for the highly sensitive detection of Ag+ and cell imaging. Luminescence 33(1):243–248
185. Zhu P, Lyu D, Shen PK, Wang X (2019) Sulfur-rich carbon dots as a novel fluorescent imaging probe for distinguishing the pathological changes of mouse-bone cells. J Lumin 207:620–625
186. Feng T, Ai X, An G, Yang P, Zhao Y (2016) Charge-convertible carbon dots for imaging-guided drug delivery with enhanced in vivo cancer therapeutic efficiency. ACS Nano 10(4):4410–4420
187. Mohammadinejad R, Dadashzadeh A, Moghassemi S, Ashrafizadeh M, Dehshahri A, Pardakhty A et al (2019) Shedding light on gene therapy: carbon dots for the minimally invasive image-guided delivery of plasmids and noncoding RNAs-A review. J Adv Res 18:81–93
188. Hormozi-Nezhad MR, Taghipour M (2016) Quick speciation of iron (II) and iron (III) in natural samples using a selective fluorescent carbon dot-based probe. Anal Methods 8(20):4064–4068
189. Yang J, Zhang X, Ma YH, Gao G, Chen X, Jia HR et al (2016) Carbon dot-based platform for simultaneous bacterial distinguishment and antibacterial applications. ACS Appl Mater Interfaces 8(47):32170–32181
190. Lim SY, Shen W, Gao Z (2015) Carbon quantum dots and their applications. Chem Soc Rev 44(1):362–381
191. Maiti S, Das K, Das PK (2013) Label-free fluorimetric detection of histone using quaternized carbon dot–DNA nanobiohybrid. Chem Commun (Camb) 49(78):8851–8853
192. Qu Y, Ren G, Yu L, Zhu B, Chai F, Chen L (2019) The carbon dots as colorimetric and fluorescent dual-readout probe for 2-nitrophenol and 4-nitrophenol detection. J Lumin 207:589–596
193. Zhao J, Pan X, Sun X, Pan W, Yu G, Wang J (2018) Detection of metronidazole in honey and metronidazole tablets using carbon dots-based sensor via the inner filter effect. Luminescence 33(4):704–712
194. Bhamore JR, Jha S, Basu H, Singhal RK, Murthy Z, Kailasa SK (2018) Tuning of gold nanoclusters sensing applications with bovine serum albumin and bromelain for detection of Hg 2+ ion and lambda-cyhalothrin via fluorescence turn-off and on mechanisms. Anal Bioanal Chem 410(11):2781–2791
195. Jiao Z, Zhang H, Jiao S, Guo Z, Zhu D, Zhao X (2019) A turn-on biosensor-based aptamer-mediated carbon quantum dots nanoaggregate for acetamiprid detection in complex samples. Food Anal Methods 12(3):668–676
196. Campos BB, Contreras-Cáceres R, Bandosz TJ, Jiménez-Jiménez J, Rodríguez-Castellón E, da Silva JCE et al (2016) Carbon dots as fluorescent sensor for detection of explosive nitrocompounds. Carbon 106:171–178
197. Yang H, Ran G, Yan J, Zhang H, Hu X (2018) A sensitive fluorescence quenching method for the detection of tartrazine with acriflavine in soft drinks. Luminescence 33(2):349–355
198. Jing N, Tian M, Wang Y, Zhang Y (2019) Nitrogen-doped carbon dots synthesized from acrylic acid and ethylenediamine for simple and selective determination of cobalt ions in aqueous media. J Lumin 206:169–175
199. Ostadhossein F, Misra SK, Mukherjee P, Ostadhossein A, Daza E, Tiwari S et al (2016) Defined host–guest chemistry on nanocarbon for sustained inhibition of cancer. Small 12(42):5845–5861
200. Yang S-T, Wang X, Wang H, Lu F, Luo PG, Cao L et al (2009) Carbon dots as nontoxic and high-

performance fluorescence imaging agents. J Phys Chem C Nanomater Interfaces 113(42):18110–18114
201. Campuzano S, Yáñez-Sedeño P, Pingarrón JM (2019) Carbon dots and graphene quantum dots in electrochemical biosensing. Nano 9(4):634
202. Wang Y, Zhu Y, Yu S, Jiang C (2017) Fluorescent carbon dots: rational synthesis, tunable optical properties and analytical applications. RSC Adv 7(65):40973–40989
203. Shi Y, Li C, Liu S, Liu Z, Zhu J, Yang J et al (2015) Facile synthesis of fluorescent carbon dots for determination of curcumin based on fluorescence resonance energy transfer. RSC Adv 5(79):64790–64796
204. Yan F, Zu F, Xu J, Zhou X, Bai Z, Ma C et al (2019) Fluorescent carbon dots for ratiometric detection of curcumin and ferric ion based on inner filter effect, cell imaging and PVDF membrane fouling research of iron flocculants in wastewater treatment. Sensors Actuators B Chem 287:231–240
205. Zhang Q, Zhang C, Li Z, Ge J, Li C, Dong C et al (2015) Nitrogen-doped carbon dots as fluorescent probe for detection of curcumin based on the inner filter effect. RSC Adv 5(115):95054–95060
206. Bian W, Wang X, Wang Y, Yang H, Huang J, Cai Z et al (2018) Boron and nitrogen co-doped carbon dots as a sensitive fluorescent probe for the detection of curcumin. Luminescence 33(1):174–180
207. Liu Y, Gong X, Dong W, Zhou R, Shuang S, Dong C (2018) Nitrogen and phosphorus dual-doped carbon dots as a label-free sensor for curcumin determination in real sample and cellular imaging. Talanta 183:61–69. https://doi.org/10.1016/j.talanta.2018.02.060
208. Baig MMF, Chen YC (2017) Bright carbon dots as fluorescence sensing agents for bacteria and curcumin. J Colloid Interface Sci 501:341–349
209. Luo T, Bu L, Peng S, Zhang Y, Zhou Z, Li G et al (2019) One-step microwave-assisted preparation of oxygen-rich multifunctional carbon quantum dots and their application for Cu2+−curcumin detection. Talanta 205:120117. https://doi.org/10.1016/j.talanta.2019.120117
210. Ragu S, Chen S-M, Ranganathan P, Rwei S-P (2016) Fabrication of a novel nickel-curcumin/graphene oxide nanocomposites for superior electrocatalytic activity toward the detection of toxic p-nitrophenol. Int J Electrochem Sci 11:9133–9144
211. Hatamie S, Ahadian MM, Akhavan O, Jokar E (2018) Photoluminescence and electrochemical investigation of curcumin-reduced graphene oxide sheets. J Iran Chem Soc 15(2):351–357
212. Kumara KK, Devendirana M, Jyothithamizhanban, Narayanan SS (2014) Curcumin/MWCNT Modified Graphite Electrode for Electrochemical Determination of BHA. International Conference on Advances in New materials (ICAN 2014). Volume 2 Special Issue 1. Organized By:- Department of Inorganic Chemistry, University of Madras
213. Cirillo G, Curcio M, Spizzirri UG, Vittorio O, Tucci P, Picci N et al (2017) Carbon nanotubes hybrid hydrogels for electrically tunable release of Curcumin. Eur Polym J 90:1–12

Effects of Curcumin and Its Analogues on Infectious Diseases

5

Reza Ranjbar, Hossein Bagheri, Faezeh Ghasemi, Paul C. Guest, and Amirhossein Sahebkar

Abstract

Infectious diseases (IDs) are life-threatening illnesses, which result from the spread of pathogenic microorganisms such as bacteria, viruses, fungi, and parasites. IDs are a major challenge for the healthcare systems around the world, leading to a wide variety of clinical manifestations and complications. Despite the capability of frontline-approved medications to partially prevent or mitigate the invasion and subsequent damage of IDs to host tissues and cells, problems such as drug resistance, insufficient efficacy, unpleasant side effects, and high expenses stand in the way of their beneficial applications. One strategy is to evaluate currently explored and available bioactive compounds as possible anti-microbial agents. The natural polyphenol curcumin has been postulated to possess various properties including anti-microbial activities. Studies have shown that it possess pleiotropic effects against bacterial- and parasitic-associating IDs including drug-resistant strains. Curcumin can also potentiate the efficacy of available anti-bacterial and anti-parasitic drugs in a synergistic fashion. In this review, we summarize the findings of these studies along with reported controversies of native curcumin and its analogues, alone and in combination, toward its application in future studies as a natural anti-bacterial and anti-parasitic agent.

Keywords

Infectious disease · Curcumin · Curcumin analogues · Antimicrobial · Anti-bacterial · Anti-parasitic · Therapy · Drug resistance

R. Ranjbar · H. Bagheri
Molecular Biology Research Center, Systems Biology and Poisonings Institute, Baqiyatallah University of Medical Sciences, Tehran, Iran

F. Ghasemi (✉)
Blood Transfusion Research Center, High Institute for Research and Education in Transfusion Medicine, Tehran, Iran

P. C. Guest
Laboratory of Neuroproteomics, Department of Biochemistry and Tissue Biology, Institute of Biology, University of Campinas (UNICAMP), Campinas, Brazil

A. Sahebkar (✉)
Neurogenic Inflammation Research Center, Mashhad University of Medical Sciences, Mashhad, Iran

Biotechnology Research Center, Pharmaceutical Technology Institute, Mashhad University of Medical Sciences, Mashhad, Iran

Polish Mother's Memorial Hospital Research Institute (PMMHRI), Lodz, Poland

Halal Research Center of IRI, FDA, Tehran, Iran
e-mail: sahebkara@mums.ac.ir

P. C. Guest (ed.), *Studies on Biomarkers and New Targets in Aging Research in Iran*, Advances in Experimental Medicine and Biology 1291, https://doi.org/10.1007/978-3-030-56153-6_5

5.1 Introduction

Plants, plant extracts, and their bioactive compounds have long been considered as potential remedial candidates to cure a wide range of conditions, disorders, and diseases. In this regard, turmeric (*Curcuma longa* L.), a bright yellow-orange spice from Zingiberaceae, has been broadly applied in traditional Indian and other medicines for its empirical-therapeutic as a remedy and for the management of various conditions [1]. The most bioactive compound of turmeric is curcumin (diferuloylmethane) which has been the subject of increasing research interest in many fields of study and medical areas. Moreover, curcumin has been authenticated by various studies as having therapeutic effects against a wide range of diseases and pathologies [2–10].

Curcumin is a naturally occurring polyphenol comprising two ferulic acid residues joined by a methylene bridge [11, 12]. Due to its hydrophobic nature, curcumin has weak pharmacokinetics and is degraded and eliminated rapidly [11, 13]. This impedes its applications using many routes of administration in vivo and in clinical studies. However, many researchers have attempted to overcome this through the design and characterization of curcumin analogues using nano-, micro-, and conjugation-based technologies, as well as other drug delivery system approaches.

Infectious diseases (IDs) are illnesses caused by the invasion of the host by pathogenic microorganisms. IDs can be spread directly/indirectly from one individual to another [14]. It should be taken into consideration that these pathogens are not restricted to specific species and can be transmitted to different species. For instance, new human pathogens often emerge from an animal reservoir [15].

Pathogenic microorganisms are classified as bacteria, viruses, parasites, or fungi. The wide variety of pathogens and IDs have constantly evolved over time and even changed the way that humans interact with their environments. Pathogens have also re-emerged possessing new adaptive features, such as multidrug resistance (e.g., drug-resistant *mycobacterium tuberculosis*), and have given rise to new epidemics and pandemics, including the 2019 novel coronavirus (COVID-19) outbreak in china [16–18] which has now spread to 210 countries and territories around the world (as of April 20, 2020). Some of the most infectious pathogens have a long history of evolution, in which they first emerge, cause epidemics/pandemics, adapt, and eventually become endemic with the potential of further outbreaks [19]. IDs reduce the life expectancy and health-related quality of life (HRQoL) in patients, and they take a heavy toll on family members and impose profound socioeconomic consequences on global healthcare systems [20–22].

For these reasons, the search for novel anti-infective agents and ways of overcoming drug resistance is now a crucial challenge facing mankind today [23, 24]. In this regard, curcumin acts as an immunomodulatory agent, regulating the host defenses in response to a disease condition and can even disrupt the drug resistance status. However, the use of curcumin in this manner has met with some controversy, despite the findings that it can mitigate and partially eradicate infections caused by some of the most damaging pathogens [25, 26].

It has been reported that native curcumin and its nanotized, micronized, micellized, hybridized forms, its derivatives, and synthetic analogues, as well as synergistic combination treatment approaches, can have protective effects against bacterial- and parasitic-associating IDs including their antibiotic and/or drug-resistant strains. The bacterial-associating IDs that have been investigated for curcumin effects include *Neisseria gonorrhoeae* (*N. gonorrhoeae*) [27], *Helicobacter pylori* (*H. pylori*) [28, 29], *Listeria monocytogenes* (*L. monocytogenes*) [30], *Staphylococcus aureus* (*S. aureus*) [31, 32], *Klebsiella pneumonia* (*K. pneumonia*) [33], *Salmonellaenterica serovar Typhimurium* (*S. typhimurium*) and *Salmonella typhi* (*S. typhi*) [34, 35], *Shigella flexneri* (*S. flexneri*), *Yersinia enterocolitica* (*Y. enterocolitica*) [25], *Mycobacterium tuberculosis* (*M. tuberculosis*) [36, 37], *Mycobacterium bovis* (*M. bovis*) [38], *Mycobacterium abscessus* (*M. abscessus*) [39], *Mycobacterium fortuitum* (*M.

fortuitum) [37], and *Mycobacterium marinum* (*M. marinum*) [36]. Furthermore, the anti-parasitic effects of curcumin that have been recently studied in parasitic-associating IDs include *Angiostrongylus cantonensis* (*A. cantonensis*) [40], *Giardia lamblia* (*G. lamblia*) [41, 42], Leishmania spp. [43–47], Plasmodium spp. [48–51], Schistosoma spp. [52–54], and *Toxoplasma gondii* (*T. gondii*) [55, 56].

In this article, we have performed an in-depth review of the anti-bacterial and anti-parasitic properties of native curcumin, its analogues, and different formulations, and combinations with other treatments, in ID-associating/generating bacteria and parasites.

5.2 Bacterial-Associating IDs

Bacteria are microscopic and unicellular prokaryote organisms, which exist in almost every environment, temperature, and climate in the world. Although most bacteria are harmless or even beneficial, many are detrimental and lead to serious health consequences and even death [57, 58]. Moreover, considering the diverse nature of bacteria, it is critical to select and develop the most appropriate therapeutic approaches to combat these effects [59, 60]. For this to be successful, it is important to understand that bacteria are under the influence of nine general factors: (1) nutrient access, (2) cell division and segregation, (3) passive dispersal, (4) surface attachment, (5) active motility, (6) polar differentiation, (7) predator escape means, (8) cellular differentiation, and (9) environmental interactions [61]. Both gram-positive and gram-negative bacteria have distinct shapes varying from rods, bacilli, spheres, and cocci to more exotic shapes which develop by the elaboration of prosthecae via polar differentiation [59, 61, 62]. Furthermore, different bacteria can produce a diversity of cellular appendages, such as pili and flagella or filaments, which also occur in various overall shapes, lengths, and thicknesses [59, 61]. After bacteria escape detection and attack from the host immune system, the next stage in the host–pathogen interaction involves its adherence via pili, flagella, or filaments to specific surfaces of host [59]. There have also been reports that antibiotic resistance can emerge from bacterial filaments through the creation of an environmental niche [63].

The cell wall structure of gram-positive and gram-negative bacteria contain toxic components which play a role in the pathogenesis of septic shock [64]. The latter is a lethal condition that can lead to collapse of the circulatory system and give rise to multiple organ system failure. Bacteria can also secrete toxins that have dire consequences on the human body, with combined impacts on the complement cascade, cytokine response, and coagulation cascade components. Some bacteria over-produce high-molecular-weight polysaccharides and exopolysaccharides as evidenced in clinical samples from patients. The extracellular sugar coating on bacteria is called a capsule which is one of the virulence factors employed by bacteria to evade and escape clearance from the infectious site by host immune system and antibiotic treatments. Moreover, it is well established that the capsule is one of the major bacterial constituents for biofilm formation, a feature which facilitates their colonization capacity [65].

Once pathogenic bacteria infect an organism, they adapt to their new environment, by subverting the host cellular ubiquitination machinery in a manner that changes the normal functions of the host [66]. In many instances, this can lead to the death of the host cells, and in other cases it can lead to maintenance of cell viability and modification of the host cells as a niche for replication. For instance, Yersinia spp., Salmonella spp., Shigella spp., *L. monocytogenes*, and *Legionella pneumophila* can cause rearrangements in the host cytoskeleton and plasma membrane that allow bacteria to penetrate the internal [67, 68]. In all cases, these bacterial manipulations of the host cellular machinery result from the secreted-bacterial proteins, which leads to modification or targeting of the specific cellular factors by either introduction into the cells or by binding to cell surface receptors and triggering the associated intracellular signaling pathways [68, 69].

One way our bodies fight bacterial infections is via macrophages engulfing bacteria into phagosomes and converting these into phagolysosomes. Furthermore, these phagolysosomes produce reactive oxygen species (ROS) and reactive nitrogen species (RNS) to kill the bacteria. Even though most bacteria are quickly eliminated and degraded in the phagolysosome, some have developed mechanisms to survive this stage, such as Salmonella spp. which has redundant mechanisms to detoxify ROS such as hydrogen peroxide [70, 71].

H. pylori is a gram-negative, invasive, flagellated, and microaerophilic bacterium, which leads to gastric infection accompanied by inflammation [72, 73]. This infection is mostly asymptomatic; but, when it affects human health, it can lead to gastritis, gastric ulcers, and duodenal ulcers. *H. pylori* infection is also a major risk factor for gastric cancer. The prevalence of *H. pylori* is high in most countries. Approximately one-third of adults in Northern European and North American populations are infected; and the prevalence in Southern and Eastesn Europe, South America, and Asia is often higher than 50% [74]. Furthermore, recurrence of *H. pylori* infection occurs in more than 4% of the cases annually [75]. Generally, *H. pylori* adhere to epithelial cells of gastric mucosa via attachment to the outer membrane and it injects cytotoxin-associated gene A (CagA) into the cell interior via a type IV secretory system, leading to alterations in cytokine signaling and cell cycle control [72, 76]. In addition, *H. pylori* produces a functional cytotoxin that can cause severe gastric injury and increase the risk for gastric cancer. Treatment or eradication of *H. pylori* not only increases the HRQoL but also decreases the risk of further gastric cancer development.

S. aureus is a gram-positive pathogen that gives rise to a variety of clinical infections, including severe pneumonia, respiratory distress, refractory hypoxemia, pulmonary edema, infective endocarditis, bacteremia, septicemia, osteoarticular, and osteomyelitis, with widespread lung inflammation and loss of epithelial and endothelial integrity [77, 78]. Pathogenic strains of *S. aureus* produce α-hemolysin, which is a pore-forming hemolytic toxin, enabling lysis of several types of cells such as erythrocytes, lymphocytes, macrophages, monocytes, and alveolar epithelial cells [79, 80]. It has been reported that *S. aureus* has colonized approximately 30% of the human population [81]. Moreover, *S. aureus* is currently treated by different antibiotics but quickly gains drug resistance features, which leads to the deficiency of effective treatments [82]. Methicillin-resistant *S. aureus* (MRSA) is resistant to most antibiotics and is responsible for high morbidity and mortality rates and increased healthcare financial burdens [83]. Therefore, there is an urgent need to develop new approaches and treatments to combat these drug-resistant strains of *S. aureus*.

Tuberculosis is an ancient disease that has affected human beings for nearly 4000 years [84]. It is responsible for more than two million deaths each year and is thought to infect one in three individuals worldwide in a latent manner. It is caused by bacillus *M. tuberculosis* and *M. bovis* [38, 85]. Furthermore, *M. abscessus*, *M. marinum, and M. fortuitum* can cause non-tuberculosis mycobacteria diseases [36, 86] and the emergence of drug-resistant strains of *Mycobacterium* species has exacerbated the death toll [85, 87]. The cell walls of *Mycobacterium* species have unique characteristics rendering them impermeable to several compounds, a feature that is responsible for the acquisition of resistance to numerous drugs. Although *M. tuberculosis* is transmitted between individuals through the air, a study of *M. abscessus* showed that transmission can also occur via cosmetic surgery, vitamin/mineral/amino acid injection mesotherapy, and intravenous infusion therapies [88]. The signs and symptoms of tuberculosis and non-tuberculosis depend on where in the body the bacteria have taken root [86, 89, 90]. For example, they can affect bone, the central nervous system (CNS), intestines, kidneys, and many other organ systems; but infection primarily leads to pulmonary disease. *M. marinum* is free-living water-borne atypical mycobacterium species that infects fish and amphibians, but can also cause a disease in humans known as "aquarium granuloma," manifested as cutaneous infections and lesions [91]. *M. fortuitum* is a rapidly growing mycobacterium that is known to form biofilms, leading to infections in multiple areas of the body including skin, bone, lymph nodes, eye after trauma and joints [92].

Salmonella is a food-borne pathogen, which gives rise to wide clinical manifestations and constitutes a significant public health problem. Salmonella spp., such as the human-specific organism *S. typhi*, leads to typhoid fever, and non-typhoidal serotypes like *S. typhimurium*, which can cause gastroenteritis and focal infections [93]. Salmonella can adapt to, reside in, and colonize gut mucosa [94]. It generates salmonellosis, one of the leading causes of morbidity and mortality around the world. Choosing the right treatment for typhoid fever patients is a challenge particularly in immunocompromised individuals [95]. In addition, diarrhea-associated salmonella infection can lead to such severe dehydrations that it requires prompt medical attention [96].

K. pneumonia is a gram-negative, rod-shaped bacillus bacterium that can be seen mostly in lung and sputum [97, 98]. High mortality and morbidity rates can occur, especially in neonates and infants. Clinical manifestations of *K. pneumonia* include high fever, pleuritic chest pain, "currant jelly" sputum, and leukopenia. In addition, drug-resistant species of *K. pneumonia* have emerged.

N. gonorrhoeae is a gram-negative bacterial pathogen, which is the causative agent of gonorrhea and different sequelae that can occur as asymptomatic infection rises within the genital tract or disseminates to other tissues. Global rates of gonorrhea are on the rise, fueled by the emergence of antibiotic resistance strains [99]. *N. gonorrhoeae* can live in the human body as they have developed novel ways to control the human-restricted mucosal niche [100–102].

S. flexneri is a gram-negative bacterium which causes bacillary dysentery in humans. It has been reported that more than one million deaths occur per year due to infections with Shigella spp., and the victims are often children. The pathogenesis of Shigella focuses on the capacity of this organism to penetrate the colonic epithelium where it can cause severe mucosal inflammation [103]. *L. monocytogenes* is a gram-negative bacterium and the causative agent of listeriosis, a rare but lethal food-borne disease. During infection, *L. monocytogenes* can cross several host barriers and invade the cytosol of a variety of cell types [104]. *Y. enterocolitica* is a gram-negative rod bacterium, which belongs to the Enterobacteriaceae family. *Y. enterocolitica* initially leads to gastrointestinal disease, although it can cause extraintestinal infections, particularly in immunocompromised individuals [105].

As has been described, bacterial-associating IDs control the host cells/tissues as a niche via production of compounds that can abuse the host cellular machinery in a manner that changes its normal functions. Moreover, treatment strategies have mainly focused on killing or eradicating the bacteria and targeting their virulence factors and toxic compounds to minimize or halt the bacterial damage, inflammation, and growth. Bacterial evolution has aligned with the acquisition of drug resistance features. These have been mostly associated with alterations in the cellular ubiquitination machinery of the hosts, making these particular strains more challenging to deal with. Therefore, the development and use of new drugs and treatment approaches that target the signaling pathway alterations and/or drug resistance features of the bacteria are urgently needed.

5.3 Mechanisms Underlying Native Curcumin in Bacterial-Associating IDs

Some plants and their bioactive compounds have been shown to have anti-bacterial properties which can attenuate or inhibit the evasion/invasion properties of bacteria and the damage caused to tissues or cells. Curcumin and curcumin combination treatments have been shown to have different bactericidal properties, combat the drug-resistant features, and eradicate some dangerous and lethal bacterial-associating IDs. The fact that the underlying pathological processes of bacterial infections are diverse may be the root cause for the controversies of curcumin effects on some bacterial-associating IDs. In this section, we discuss reports of the molecular targets of native curcumin and its combination with some common and approved treatments (Table 5.1).

Studies on *H. pylori* infection have provided novel insights into the therapeutic properties of curcumin. An *in silico* study showed that the energy binding of curcumin to *H. pylori* Cag^+ pathogenicity island is high [106]. In vitro studies reported that curcumin can inhibit the *H. pylori*-induced motogenic response along with signaling pathways such as IκBα degradation, the activity of IκB kinases α and β (IKKα and β), and NF-κB DNA-binding, which controls many genes associated with inflammatory processes [107, 108]. Moreover, it was shown that the curcumin minimum inhibitory concentration (MIC) (5–50μg/mL) was effective in inhibiting growth of *H. pylori* strains [29]. In *H. pylori*-infected C57BL/6 mice, administration of 25 mg/kg curcumin for 7 days showed efficacy in reducing gastric damage, eradicating *H. pylori* from infected mice, and healing the associated gastric damage [29]. Curcumin supplementation (200 or 600 mg/kg) in Sprague-Dawley rats for 7 days resulted in an inhibition on the NF-κB pathway [109]. Furthermore, curcumin in BALB/c mice exerted a significant increase in IL-4 serum levels, and a marked decrease in IFN-γ, lipid peroxide, and myeloperoxidase levels [110]. Curcumin administration (500 mg/kg) for 6 and 18 weeks in C57BL/6 mice inhibited the expression of inflammatory cytokines and chemokines, toll-like receptors, and the canonical adaptor for inflammatory signaling pathways, MyD88. Also, no evidence of inflammation was witnessed after curcumin treatment [111]. Curcumin treatment of *H. pylori* patients with peptic ulcer was safe and improved dyspepsia symptoms, although there was no effect on the eradication of the infection [112]. Additionally, a clinical study showed that curcumin treatment as a monotherapy cured 3 out of the 25 patients infected by *H. pylori*, with a significant decrease in overall symptom severity [28]. It was also reported that curcumin monotherapy can downregulate the production of the proinflammatory cytokine interleukin (IL)-17 via the induction of indoleamine 2,3-dioxygenase in *H. pylori*-infected human gastric mucosa [113]. Improved eradication was observed using the combination therapy of curcumin with clarithromycin, amoxicillin, and pantoprazole, as shown by a significant decrease in malondialdehyde and glutathione peroxidase markers, along with an increase in total antioxidant capacity of the gastric mucosa [114].

Owing to the global burden of *S. aureus*, several studies worked on curcumin efficacy against *S. aureus* infection and associated consequences. Wang et al. reported that curcumin direct binding to α-hemolysin blocks its hemolytic activity and reduced α-hemolysin-mediated injury of human alveolar cells infected with *S. aureus* [32]. Their results demonstrated that curcumin treatment protected C57BL/6 J mice from *S. aureus* (including MRSA strains)-induced pneumonia. Furthermore, treatment with 50 mg/kg of curcumin downregulated the keratinocyte-derived chemokine (KC), IL-1β, macrophage inflammatory protein (MIP)-2, TNF-α, IL-6 and transforming growth factor beta (TGF-β) genes in *S. aureus*-infected mice with acute lung injury. Moreover, curcumin diminished *S. aureus*-induced pneumonia, as well as other effects, and reduced activation of plasminogen activator inhibitor-1 and neutrophil infiltration [115]. The combination treatment of curcumin with erythromycin suppressed bacterial growth, mitigated bone infection, and decreased TNF-α and IL-6 expression. Compared to the respective monotherapies, the curcumin–erythromycin combination had a significantly greater impact on MRSA strain-induced osteomyelitis in rats [31].

M. tuberculosis has been known to acquire resistance to numerous drugs and antibiotics. Therefore, several studies have aimed to assess the anti-tuberculosis effects of curcumin. An in vitro study reported that curcumin can enhance *M. tuberculosis* clearance via induction of caspase-3-dependent apoptosis and autophagy in differentiated THP-1 human monocytes and primary human alveolar macrophages, and by inhibiting NF-κB activation [116]. Li et al. reported that curcumin can reduce the growth inhibition of macrophages induced by the *M. tuberculosis*

Table 5.1 Effects of curcumin on bacterial-associating IDs

Model of the study – species	Dose/route	Observations and conclusion	Reference
In vitro – *N. gonorrhoeae*	100μM	Curcumin demonstrated potential to block NF-κB-mediated cytokine release and innate immune response by inhibiting IKKα, IKKβ, and NIK. Also, it inhibited *N. gonorrhoeae*-induced signaling and completely abolished adherence of bacteria to cells in late infection	[27]
Clinical trial – *H. pylori*	30 mg b.i.d. for 7 days	Curcumin was not effective for the eradication of *H. pylori*, although 3 of 25 patients were cured of *H. pylori* infection with significant decrease in overall severity of symptoms by decreasing mucosal inflammation and diminishing the reduction of serologic signs of gastric inflammation, leading to improvement of dyspeptic symptoms	[28]
Randomized controlled trial – *H. pylori*	500 mg/day	Standard anti-*H. pylori* regimen of curcumin in patients with peptic ulcer was safe and improved dyspepsia symptoms but had no enhancing effect on the eradication of *H. pylori* infection	[112]
In vivo and in vitro – *H. pylori*	5–50μg/mL and 25 mg/kg orally fed for 7 days	Antibacterial effect of curcumin in *H. pylori*-infected C57BL/6 mice and in vitro condition showed strong therapeutic potential against *H. pylori* infection and high effectiveness in eradication of *H. pylori* from infected mice and restoration of *H. pylori*-induced gastric damage	[29]
In vitro – *H. pylori*	Up to 80μM	Curcumin inhibited IκBα degradation, activity of IKKα and β, and NF-κB DNA-binding. However, it did not affect/inhibit MAPK, ERK1/2, and p38, which are activated by *H. pylori* infection. Further, curcumin blocked *H. pylori*-induced motogenic response	[108]
Randomized controlled trial – *H. pylori*	700 mg turmeric tablet orally three times a day for 4 weeks	Triple therapy + curcumin treatment group strongly decreased malondialdehyde markers, GPx, increased total antioxidant capacity of gastric mucosa and reduced oxidative damage to DNA and significantly reduced all active, chronic, and endoscopic inflammation scores of patients at the end of the study compared to baseline and compared to triple therapy. Also the eradication rate was significantly increased compared to triple therapy alone ($P < 0.05$)	[114]
In vivo – *H. pylori*	500 mg/kg	Curcumin inhibited the expression of inflammatory cytokines and chemokines, and toll-like receptors and MyD88. Also, Immunohistochemistry showed no evidence of inflammation after curcumin treatment in C57BL/6 mice	[111]
In vivo – *H. pylori*	200 mg/kg and 600 mg/kg	Curcumin supplementation exerted its anti-inflammatory effect in male Sprague-Dawley rats against *H. pylori* infection, by reducing macromolecular leakage through the suppression of NF-κB p65 expression in gastric epithelial cells	[109]
Clinical trial – *H. pylori*	200 μM	Curcumin downregulated the IL-17 production through the induction of indoleamine 2,3-dioxygenase in *H. pylori*-infected human gastric mucosa.	[113]

(continued)

Table 5.1 (continued)

Model of the study – species	Dose/route	Observations and conclusion	Reference
In vivo – *H. pylori*	Dose-dependent	Of 5 samples of curcumin, #1 was the most active in female BALB/c mice, followed by samples #5 and #3. Overall, curcumin significantly increased IL-4 serum levels, with the highest effects with samples #1 and #3, and all samples (with the exception of #4) significantly decreased IFN-γ levels. Gastric levels of lipid peroxide were significantly decreased by samples #1, #3, and #5; and all samples (except sample #4) had decreased levels of myeloperoxidase	[110]
In vitro – *H. pylori*	Up to 80μmol/L	Curcumin downregulated *H. pylori*-induced activation-induced cytidine deaminase expression in gastric epithelial cells via inhibition of NF-κB pathway. This suggested it is a potent chemopreventive candidate against *H. pylori*-related gastric carcinogenesis	[107]
In vivo – *L.monocytogenes*	Dose- and time-dependent manner	Nano-encapsulated curcumin formulation (Eudragit L-100) showed reduction of tissue injury and compensation of alterations in the enzymes of the energetic metabolism and demonstrated a potential antioxidant effect from lipid peroxidation at a dose ten fold lower than the free form	[30]
In vivo – MRSA	50 mg/kg bw i.p. injection	Erythromycin and curcumin combination treatment significantly suppressed bacterial growth, substantially alleviated bone infection, and reduced TNF-α and IL-6 production. A combination of erythromycin and curcumin had stronger efficiency against MRSA–induced osteomyelitis in adult male Wistar rats than monotherapy	[31]
In vivo and in vitro – *S. aureus*	0–20 mg/kg and 0–25μM	Curcumin-loaded PVP nanoparticles attenuated the biofilm-induced expansion of MDSC and regulated M1- and M2-phenotypic MDSC expression. By reversing suppression of T-cell proliferation, curcumin nanoparticles downregulated cytokines and ROS. Also, the recovery of bone permeative destruction demonstrated that curcumin nanoparticles enhanced therapeutic potency of vancomycin in vivo	[120]
In vivo – *S. aureus*	50 mg/kg i.p. injection	Curcumin significantly attenuated *S. aureus*-induced pneumonia, barrier disruption, lung edema, and vascular leakage. The reduction of activation of plasminogen activator inhibitor-1 and infiltration of neutrophils by curcumin led to lower levels of proinflammatory cytokines and chemokines in female C57BL/6 mice	[115]
In vitro and in vivo – *S. aureus*	0–16μg/ml and 100 mg/kg s.c. injection	Direct binding of curcumin to α-hemolysin inhibited α-hemolysin hemolytic activity. Also, curcumin significantly attenuated α-hemolysin-mediated injury of the human alveolar cell (A549) co-cultured with *S. aureus*. The in vivo data further demonstrated that treatment with curcumin protected C57BL/6 J mice from pneumonia caused by *S. aureus*, including MRSA strains	[32]

(continued)

Table 5.1 (continued)

Model of the study – species	Dose/route	Observations and conclusion	Reference
In vivo – *S. aureus*	40 mg/kg orally gavage	CMC2.24 treatment significantly reduced mortality, total bacterial flux, lung tissue apoptosis, inflammatory mediators, NF-κB expression ($p < 0.05$), and MMPs-2, −9, −12 activities ($p < 0.05$). It was shown that FVB/N mice with surfactant protein B-C allele are more susceptible to *S. aureus* pneumonia than FVB/N mice with surfactant protein B-T allele. Also CMC2.24 attenuated lung injury and reduced mortality	[121]
In vivo – MRSA	Dose- and time-dependent IV injection	UCNPs-curcumin produced $^1O2$ after irradiation at 980 nm, and inhibited the growth of MRSA by destroying cell membranes, leading to bacterial cytoplasmic leakage. The antibacterial effect of UCNPs-curcumin also reported by control on MRSA-induced pneumonia in half–male and half-female of Kunming mice. This provided a new type of near-infrared light photosensitizer, which improved the effects of curcumin in vivo	[122]
In vivo – *K. pneumoniae*	150 mg/kg orally fed for 15 days	Curcumin alone and in combination with augmentin reduced acute lung injury associated with pneumonia observed in treated BALB/c mice. The immunomodulatory/anti-inflammatory action of curcumin resulted in decreased production of various inflammatory mediators such as TNF-α, NO, malondialdehyde, myeloperoxidase, and neutrophil recruitment	[33]
In vivo – *K. pneumoniae*	150 mg/kg orally fed for 15 days	Curcumin alone and in combination with clarithromycin showed a much more significant ($p < 0.05$) reduction in neutrophil influx along with reduced levels of various inflammatory mediators in BALB/c mice of both sexes	[119]
In vitro – *S. Typhimurium*	20μM	Curcumin increased the resistance of S. Typhimurium against antimicrobial agents like antimicrobial peptides, ROS, and nitrogen species, which might be attributed to upregulation of genes involved in resistance against antimicrobial peptides – pmrD and pmrHFIJKLM – and genes with antioxidant function – mntH, soda, and sitA	[35]
In vitro andin vivo – *S. Typhimurium* and *S. Typhi*	Up to 20 mM–80 mg/kg bw/day for 4 days	Curcumin interfered with the action of ciprofloxacin by increasing the proliferation of *S. Typhimurium* and *S. Typhi* in macrophages. In the BALB/c mice model of typhoid fever, mice fed with curcumin had increased bacterial burden in reticuloendothelial system and succumbed to death faster. The antioxidant property of curcumin was crucial in protecting salmonella against the oxidative damage induced by ciprofloxacin or interferon γ	[34]
In vitro – *S. flexneri, L. monocytogenes, S. Typhimurium, S. Typhi, S. aureus*, and *Y. enterocolitica*	10μM for 2 h	Curcumin prevented active phagosomal escape of cytosolic pathogens and enhanced active inhibition of lysosomal fusion by vacuolar pathogens. Also, curcumin retarded the escape of cytosolic pathogens, reducing their inter- and intracellular spread. The membrane-stabilizing activity of curcumin was crucial for its differential effect on the virulence of the bacteria	[25]

(continued)

Table 5.1 (continued)

Model of the study – species	Dose/route	Observations and conclusion	Reference
In vitro and in vivo – *M. tuberculosis* including drug-resistant strains	Dose- and time-dependent	CPMD-6d dihydrochloride (curcumin derivative) synergized with anti-tuberculosis drugs reduced bacterial load in lungs and spleen of female BALB/c mice at 25 mg/kg compared with ethambutol at 100 mg/kg	[37]
In vitro – *M. tuberculosis* including drug-resistant strains	10, 30, 50μM	Curcumin enhanced clearance of *M. tuberculosis* in differentiated THP-1 human monocytes and primary human alveolar macrophages. It was also an inducer of caspase-3-dependent intrinsic/extrinsic apoptosis and autophagy, and mediated anti-*M. Tuberculosis* cellular functions, partly, via inhibition of NF-κB activation	[116]
In vitro – *M. tuberculosis*	10 and 20μM	Curcumin can exert its protective effect against 19-kDa lipoprotein-inducing macrophage apoptosis secreted by the *M. tuberculosis* via regulating JNK pathway	[118]
In vitro – *M. tuberculosis*	10 or 20μM	Curcumin attenuated the inhibition of macrophage cell growth and led to increase in the expression of IL-6 and TNF-α induced by 19-kDa lipoprotein. Curcumin also reduced phosphorylation of JNK and p38. Curcumin may protect macrophages from 19-kDa lipoprotein-induced cell apoptosis via a TLR2-mediated JNK-dependent pathway	[117]
In vitro – *M. tuberculosis* including drug-resistant strains	Up to 50μM	Mono-carbonyl analogs of curcumin inhibited the growth of *M. tuberculosis* and *M. marinum*, including rifampicin-resistant strains	[36]
In vivo – *M. tuberculosis*	i.p. injection of 100 mg/kg body weight	Curcumin nanoparticles significantly reduced hepatotoxicity induced by antitubercular antibiotics during treatment in female BALB/c mice. A combination of nanoparticle-formulated curcumin plus antitubercular antibiotics reduced risk for disease reactivation and reinfection. Nanoparticle formulation of curcumin significantly ameliorated time required for antibiotic therapy, reducing the possibility of generating drug-resistant variants of the organisms	[123]
In vitro – *M. tuberculosis*	Dose-dependent	Quinolidene-based monocarbonyl curcumin analogues 3e, 3 h, 4a, and 4e exhibited very good antitubercular activity. Potency, low cytotoxicity, and selectivity of these curcumin analogs support them as valid compounds for further optimization	[38]
In vitro – antibiotic resistance of *M. abscessus*	Dose-dependent	Curcumin was a potential antibiotic resistance breaker, capable of restoring the effectiveness of failing antibiotics by reducing their minimum inhibitory concentrations. The strong effect of curcumin and amikacin on biofilm emphasized the significance of curcumin synergy association with other antibiotics	[39]

19-kDa lipoprotein, as well as IL-6 and TNF-α levels via a JNK-dependent pathway [117, 118]. Additionally, curcumin may have effects on drug-resistant strains as shown by a study on antibiotic resistance of *M. abscessus* [39]. This showed that curcumin in combination with other compounds, such as amikacin, clarithromycin, ciprofloxacin, and linezolid, markedly decreased motility at 1/8 MIC and suppressed mature biofilms at 4 × MIC. In addition, combinations of curcumin with amikacin promoted a general decrease in *M. abscessus* aggregates, suggesting that it may offer a useful approach for the treatment of antibiotic resistance strains of mycobacterium.

The contrasting effects of curcumin are more eminent in studies of salmonella infection. Marathe et al. showed that curcumin could increase pathogenicity of *S. typhimurium* by elevating the resistance against antimicrobial agents such as antimicrobial peptides, ROS and RNS [35]. In another study, they showed that curcumin interfered with the action of ciprofloxacin, leading to an increase in the proliferation of *S. Typhimurium* and *S. Typhi* in macrophages [34]. Curcumin was also found to exacerbate *S. Typhimurium, S. Typhi, S. aureus,* and *Y. enterocolitica* infection in macrophages [25].

In pneumonia, the effects of curcumin have been largely positive. A study showed that the combination of curcumin with augmentin in *K. pneumonia*-infected mice led to a significant reduction in neutrophil influx into the lungs and a large decrease in production oxidative and inflammatory biomarkers [33]. Moreover, curcumin monotherapy ameliorated lung inflammation without a significant reduction in the bacterial load. Another study found that curcumin monotherapy did not reduce the bacterial load, although its combination with clarithromycin resulted in a more significant reduction in neutrophil influx and reduced levels of various inflammatory markers and responses [119].

It has been reported that curcumin can attenuate *S. flexneri and L. monocytogenes* infections [25]. Furthermore, treatment of *N. gonorrhoeae* infection with 100μM curcumin inhibited IKKα/β and NF-κB-inducing kinase, indicating its potent activity to block NF-κB-mediated cytokine release and the innate immune response [27]. It was also able to suppress the adherence of *N. gonorrhoeae* to cells in late infection.

5.4 Mechanisms Underlying Curcumin Analogues in Bacterial-Associating IDs

Although curcumin has considerable effects on bacterial-associating IDs, its unstable status, poor intestinal absorption and low bioavailability and pharmacokinetics, hinders its in vivo and clinical trial use. Therefore, researchers have focused on designing and characterizing curcumin derivatives and analogues to compensate for these effects. In this section, we have explored previous curcumin derivatives and analogue studies on bacterial-associating IDs (Table 5.1).

Peng et al. stated that in *S. aureus* biofilm-induced periprosthetic joint infection (PJI) models of C57BL/6 mice, curcumin nanoparticles or PVP-based curcumin (curcumin-loaded PVP nanoparticles) reduced the biofilm-induced expansion of myeloid-derived suppressor cells. In addition, it downregulated cytokines and ROS and reversed the suppression of T cell proliferation. The administration of curcumin nanoparticles improved the therapeutic strength of vancomycin in C57BL/6 mice [120]. In addition, 40 mg/kg tri-ketonic chemically modified curcumin (CMC2.24) treatment of *S. aureus*-infected hTG SP-B mice led to a significant decrease in mortality, total bacterial flux, apoptosis of lung tissue cells, inflammatory mediators, NF-κB expression, and activities of matrix metalloproteinases (MMPs)-2, -9, and -12 [121]. Another study showed that up-conversion of nanoparticles conjugated with curcumin (UCNPs-curcumin) by near-infrared light led to attenuation of MRSA-induced pneumonia and decreased bacterial counts in the lungs of a mouse model [122].

Among twenty-one curcumin derivatives, CPMD-6d dihydrochloride has been found to be the most effective, with a MIC of 2μg/mL against *M. tuberculosis* and a drug-resistant strain [37]. Furthermore, CPMD-6d dihydrochloride showed a MIC of 16μg/mL against *M. fortuitum* and *M. abscessus* although this curcumin derivative was inactive at MIC concentrations >64μg/mL. Another study showed that mono-carbonyl analogues of curcumin could inhibit the in vitro growth of *M. marinum* and *M. tuberculosis,* including rifampicin-resistant strains [36]. Nevertheless, no synergistic effects were witnessed between these analogues and rifampicin on growth inhibition. Tousif et al. reported that nanoparticle-formulated curcumin led to an enhanced bioavailability in mice of approximately five-fold compared to the native molecule

[123]. Moreover, curcumin nanoparticles potently lessened the hepatotoxicity induced by anti-tuberculosis antibiotic treatment. Also, the combination treatment of nanoparticle-formulated curcumin with anti-tuberculosis antibiotics considerably decreased the risk for disease reactivation and reinfection.

Another study showed that native curcumin exhibited strong antibacterial activity, and nanocapsules containing curcumin were able to decrease the effects generated by *L. monocytogenes* regarding tissue injury and enzyme alterations related to energy metabolism, along with antioxidant effects against lipoperoxidation [30].

5.5 Parasitic-Associating IDs

There are three main classes of disease-causing parasites in humans known as protozoa, helminths, and ectoparasites. Humans serve as hosts to more than 300 species of parasitic worms and over 70 species of protozoa along with a variety of arthropods, annelida, and mollusks [124]. Some of these originated from our primate ancestors, and some were obtained from domesticated animals and form insects such as mosquitoes. Parasitic diseases are a series of IDs caused or transmitted by parasites, with symptoms ranging from itching (usually affecting the anus and vaginal area) to diarrhea, nausea, vomiting, dehydration, abdominal pain, anemia, pneumonia, and weight loss [125]. Parasitic-associating IDs predominate in warm and wet areas of tropical, subtropical, and temperate zones. The prevalence of parasitic-associating IDs, particularly in underdeveloped and developing countries, poses a threat to human life and HRQoL, and generates large global socioeconomic burdens [126]. Neglected tropical diseases (NTDs) are a group of diseases that occur in tropical and subtropical climates and are closely connected to poverty. These diseases thrive in areas with restricted access to proper sanitation, clean water, and healthcare, and which contain people in touch with different types of animals and vectors of microbial infective diseases [127]. According to the latest NTD update by the World Health Organization (WHO), 13 parasitic diseases exist in its portfolio. These include Chagas disease, dracunculiasis (Guinea-worm disease), echinococcosis, food-borne trematode infections, human African trypanosomiasis, leishmaniasis, lymphatic filariasis, onchocerciasis, scabies and other ectoparasites, schistosomiasis, intestinal worms, and taeniasis [127, 128]. This emphasizes the point that some parasitic-associating IDs have been overlooked for centuries. In addition, NTDs, such as malaria, trichomoniasis, giardiasis, cryptosporidiosis, and toxoplasmosis, are a challenge to the global healthcare system. Thus, there is an urgent need in parasitology to develop approaches to confront and eradicate these parasites including their drug resistance strains.

Considering that parasites have a close range of interactions with the host's immune system at the level of molecules involved in defense, it is not surprising that parasites have evolved ways to interact with and evade the host's immune system. This signifies that parasites have developed such strategies throughout their evolution by adapting to the host's immune system and modifying their routes of entry or transmission to improve the possibility of encounters with the next host. These strategies include passive evasion, active modulation, interference, antigenic variation, antigen masking, and molecular mimicry [129–131]. Parasites can passively or actively escape the immune system by penetrating or placing their eggs inside immune-privileged tissues like the CNS and the eye [130, 132, 133]. In antigen masking and molecular mimicry, the parasites can become invisible to the host immune system through covering their surface components [130, 134] and by changing their surface identity [130, 133, 135]. Eventually, these evasion tactics allow the parasites to disrupt or manipulate the host immune responses and/or inhibit the generation of immune memory responses [131].

Due to the vast range of parasitic-associated diseases, parasites vary in virulence factors and have distinct targets, indicating that their effects are dependent on the host species and phenotype [136]. Indeed, they are often host-specific and incapable of infecting even closely linked species [137, 138].

The pathogenesis of parasites is based on physiological, biochemical, or molecular mechanisms, which leads to harmful influences on the host, such as depletion of resources, tissue destruction, decreased fecundity, and early death [130, 139].

Malaria is a serious mosquito-borne parasitic ID, responsible for high morbidity and mortality around the world. The pathogenesis of malaria is intricate, and the malaria clinical presentation varies from severe and complicated, to moderate and uncomplicated, and there are even asymptomatic forms [140]. Individuals affected with malaria frequently experience fever, chills, and flu-like illness. Five *Plasmodium* species encompass the *Plasmodium falciparum* (*P. falciparum*), *Plasmodiumvivax*, *Plasmodiummalariae*, *Plasmodiumovale*, and *Plasmodiumknowlesi*, which are the causative agents for malaria. However, *P. falciparum* causes the most severe form of malaria and consequently higher mortality rates [141]. *P. berghei* and *P. chabaudi* infection of laboratory mouse strains is frequently utilized for in vivo studies as a model for human malaria [142, 143]. According to the WHO in 2018, there were more than 228 million cases of malaria and 405,000 malaria-related deaths worldwide. However, children under 5 years old represent the most vulnerable group affected by malaria, with more than 272,000 deaths. Furthermore, drug-resistant strains of *Plasmodium,* especially *P. falciparum*, is a recurring problem in this field [144]. Therefore, the identification and implementation of novel approaches and drugs are needed urgently.

Giardia (also known as *Giardia intestinalis*, *G. lamblia*, and *G. duodenalis*) is a microscopic parasitic protozoan that causes a diarrheal illness called giardiasis in children [145, 146] and adults [147], in urban and rural areas. *G. lamblia* is responsible for public health difficulties in most developing countries and some developed countries. It is found on surfaces or in soil, food, and water contaminated with feces from infected humans or animals. The symptoms of giardiasis are diarrhea, nausea, and malabsorption [148]. It is estimated that more than 200 million individuals are diagnosed each year with giardiasis worldwide [149]. Since 2004, giardiasis has been included in the portfolio of the WHO NTDs initiative, which has a common connection with poverty [128]. Given the current global burden, it is essential to take a comprehensive approach in the treatment of this disease. Although giardiasis treatments such as those with either metronidazole or albendazole are about 90% effective, a drug resistance feature occurs frequently, highlighting the need for new approaches [150].

Leishmaniasis is a parasitic infectious disease caused by a protozoa parasite from over 20 leishmania species, including *L. donovani, L. tropica, L. major, and L. braziliensis.* Leishmaniasis is spread by the bite of phlebotomine sandflies and found in the tropics, subtropics, and Southern Europe. There are several different types of leishmaniasis. Cutaneous leishmaniasis is the most frequent form, which causes skin sores. Visceral leishmaniasis, also known as kala-azar, is the most serious form of the disease, which affects several internal organs including spleen, liver, and bone marrow. The WHO has estimated occurrence of leishmaniasis at 700,000 to one million new cases each year, with a death toll of 26,000 to 65,000 [151]. Despite the availability of successful treatments, lesion recurrences have been reported in some cases [152]. In addition, leishmaniasis co-infection with other diseases such as HIV can occur [153].

Schistosomiasis, also known as bilharzia, is an acute and chronic parasitic-causing disease caused by blood flukes (trematode worms) of the Schistosoma genus including *S. mansoni*, *S. Japonicum*, *S. intercalatum*, and *S. mekongi* [154]. Schistosomiasis is considered by the WHO as a NTD with an estimation of at least 220 million people infected in 2017 [155]. Co-infection of schistosomiasis with salmonella, hepatitis B virus (HBV), or hepatitis C virus (HBC) can occur [154].

Toxoplasmosis is a protozoan parasitic infection caused by a single-cell microorganism called *T. gondii*. *T. gondii* is an apicomplexan, which can infect the CNS and eyes. Infection during pregnancy can lead to congenital infection with severe neurological effects. In immunocompromised people, reactivation of latent foci may cause encephalitis. *T. gondii* infection in immunocompetent individuals is generally asymptomatic and the infection may be present for the rest of their lives [156, 157]. Estimates indicate that *T. gondii* infects almost one-third of the world population, making it one of the most burgeoning parasitic organisms. Although there are some therapeutics for the treatment of *T. gondii* infections, some of these do not have good blood-brain barrier (BBB) permeability or CNS penetration [157]. In addition, many of these drugs have toxic side effects. Thus, new treatments are needed with reduced toxicity, and which can achieve better therapeutic concentrations and BBB permeability over shorter treatment durations [158, 159]. In addition, they should be safe to use in pregnancy, inexpensive, and practical. One possible candidate is a glyoxalase (Glo) inhibitor. The glyoxalase pathway consists of two enzymes, Glo1 and Glo2, which are part of a ubiquitous cellular system responsible for detoxification of methylglyoxal generated via increased glycolytic flux following infection by protozoan parasites, including *T. gondii* tachyzoites [160, 161]. A study showed that curcumin inhibited the enzymatic activity of recombinant Glo1 as well as propagation of T. gondii in vitro [55].

A. cantonensis, a parasitic nematode (roundworm), is the responsible factor involved in a human disease known as angiostrongyliasis, or rat lungworm disease, which presents as eosinophilic meningitis or meningoencephalitis [162, 163]. The symptoms vary from moderate headache to a range of neurologically induced debilitations, along with coma and, in rare incidences, death. The disease severity appears to be related to the number of worms present and their location in the body, as well as the intensity of the inflammatory responses of the host [164, 165]. Humans are accidental hosts and do not spread the infection to other humans. Although one study showed that curcumin did not reduce A. cantonensis worm number in an eosinophilic meningitis mouse model when administered alone, a combination treatment consisting of curcumin and albendazole led to improved reduction of infection-associated eosinophil counts in cerebrospinal fluid [40].

In short, parasitic-associating IDs can profoundly affect the HRQoL, with massive socioeconomic consequences on global healthcare systems. Moreover, many of the current first-line medications have significant toxicities and are expensive, and drug-resistance can lead to increased disease severity and death of the hosts. This creates a significant and urgent need for new and improved approaches for the treatment of these infections.

5.6 Mechanisms Underlying Native Curcumin in Parasitic-Associating IDs

Some plants and their bioactive compounds have been shown to possess anti-parasitic properties. In this regard, curcumin has been reported as having considerable properties on parasitic-associating IDs in both drug-sensitive and drug-resistant strains. The mechanisms underlying these effects are detailed below via a number of in vitro and in vivo studies, as well as through several clinical trials (Table 5.2).

In a study of *P. falciparum*, curcumin treatment demonstrated a degree of parasiticidal activity with an IC_{50} of 5μM [166]. These effects appeared to be due partly to a perturbing effect on the *P. falciparum* microtubules. Furthermore, Ali et al. showed that curcumin treatment of *P. berghei*-infected mice resulted in a concentration-dependent chemosuppression of parasitemia development [51]. Moreover, curcumin treatment at the highest dosage (30 mg/kg) led to suppression of more than 50% and increased median survival of the infected mice, compared to the control treatment. Curcumin treatment also led to increased phosphorylation of glycogen synthase kinase 3 beta (GSK3β), which inhibited this signaling pathway, and caused a decrease in serum levels of the proinflammatory cytokines TNF-α and IFN-γ, and an increase in the levels of the anti-inflammatory cytokines IL-10 and IL-4. The results of Martinelli and coworkers demonstrated that curcumin alone and in combination with piperine had a small antimalarial effect in *P. chabaudi*-infected Swiss albino CD1 mice [50].

Table 5.2 Effects of curcumin on parasitic-associating infectious diseases

Model of the study – species	Dose/route	Observations and conclusion	Reference
In vivo – *A. cantonensis*	20 mg/kg for 20 days	Albendazole–curcumin combination treatment alleviated symptoms in the animal model of BALB/c mice infected with A. cantonensis through reduction of worm numbers in brain and eosinophil counts in the CSF, which may reduce the risk of BBB breakdown generally found in eosinophilic meningitis	[40]
In vivo and in vitro – *G. lamblia*	Dose- and time-dependent	In vitro anti-giardia effects of 50 mg/mL curcumin extract was significant compared to controls ($P < 0.001$). In vivo effects of curcumin extract significantly reduced fecal cyst and intestinal trophozoite counts in infected Balb/c mice, and improved intestinal mucosal damage induced by Giardia infection. Curcumin extract may provide effective therapeutic alternatives with low side effects and without drug resistance in the treatment of giardiasis	[41]
In silico and in vitro – *G. lamblia*	0.3, 3, and 15μM	Curcumin inhibited proliferation and adhesion of Giardia and caused reduction of tubulin expression. By docking analysis and molecular dynamics, it was shown that curcumin had a high probability to bind at the interface of the tubulin dimer close to the vinblastine-binding site. The data indicated that curcumin might inhibit Giardia proliferation by perturbing microtubules	[42]
In vivo – *G. lamblia*	Dose-dependent	Combinations of silver, chitosan, and curcumin nanoparticles eradicated the Giardia from stool and intestine	[172]
In vitro – *L. donovani*	0, 5, 10, 15, and 20μg/mL	Curcumin analogue demonstrated robust anti-prolific activity against L. donovani. It altered general ultrastructure and mitochondrial physiology of the parasite and triggered apoptotic cell death	[43]
In vitro and in vivo – *L. donovani*	Dose-dependent	In vivo antileishmanial activity of curcumin nanoparticle in hamsters demonstrated significant suppression of parasitic replication in spleen. In vitro cytotoxicity study against the J774A.1 cell line demonstrated its comparative non-toxicity of curcumin nanoparticle toward macrophage cells. The potential of curcumin nanoparticle was confirmed by minimal observed cytotoxicity in in vivo studies	[173]
In vivo – *L. tropica*	Dose-dependent	Curcumin-loaded self-emulsifying drug delivery system in a damaged skin model suggested broad-spectrum antileishmanial and antibacterial activity and could be employed for the treatment of cutaneous leishmaniasis and associated secondary infections	[44]
In vitro – *L. major*	Dose-dependent	Indium curcumin and gallium curcumin complex demonstrated more antileishmanial activity than native curcumin and could be therapeutic candidates for further investigations	[174]
In vitro – *L. major* and *L. braziliensis*	up to 7.8μg/mL	Curcumin might be a promising alternative photosensitizer because it demonstrated no toxicity. However, further tests in co-culture with macrophages and other leishmania species should be conducted to determine better conditions before in vivo tests are performed	[46]

(continued)

Table 5.2 (continued)

Model of the study – species	Dose/route	Observations and conclusion	Reference
In vivo and in vitro – *Leishmania* spp.	Dose-dependent	Antioxidative supplements such as curcumin may compromise microbial defense	[47]
In vivo and in vitro – *L. donovani*	Dose-dependent	Nanoformulation of curcumin demonstrated significant leishmanicidal activity in both in vitro and in vivo researches	[45]
In silico and in vitro – *P. falciparum*	Dose-dependent	Parasiticidal activity was found at curcumin concentrations of 5μM. The study suggested that curcumin cellular effects are at least, in part, due to perturbing effects on microtubules of *P. falciparum*. Thus, 5-μM concentration offers opportunity to examine the effects of curcumin that may be independent of ROS generation, in particular, the effects on microtubule structure	[166]
In vitro and in vivo – *P. falciparum*	Dose-dependent	Co-encapsulated NCs showed significant reduction in P. falciparum. *C. elegans* worms exposed to NC-QC manifested higher survival and longevity and no decrease in reproductive capacity. Co-encapsulation of QN and curcumin was effective against P. falciparum, reducing toxic effects generated by chronic exposure of the free drugs in *C. elegans*	[48]
In vitro and in vivo – *P. berghei*	Dose-dependent	The percentage suppression of *P. berghei* was significantly high in CA-PLGA 5 mg/kg at day 5 (79.0%) and day 8 (72.5%). The nanotized CA-PLGA formulation was novel, simple to develop and nontoxic, and was active against malaria at low concentration. Therefore, CA-PLGA may be a novel therapeutic candidate for malaria treatment	[49]
In vitro and in vivo – *P. falciparum* and *P. berghei*	Dose-dependent	Nanotized curcumin was ten-fold more effective for growth inhibition of *P. falciparum* in vitro compared to the native counterpart. Moreover, *P. berghei*-infected C57BL/6 mice orally treated with nanotized curcumin showed increased survival by more than 2 months with complete clearance of parasites compared to untreated animals (survived for 8 days). Nanotized curcumin is a promising therapeutic candidate against malaria	[171]
In vivo – *P. berghei*	Dose-dependent	Curcumin caused decrease in serum TNF-α and IFN-γ level (by 2- and 6-fold) while IL-10 and IL-4 were elevated (by 1.4- and 1.8-fold). This showed for the first time that the antimalarial action of curcumin involved inhibition of GSK3β	[51]
In vivo – *P. chabaudi*	Dose-dependent	Curcumin alone and in combination with piperine had only a modest antimalarial effect and was not able to reverse the artemisinin-resistant phenotype, and significantly affect the growth when used in combination with artemisinin	[50]

(continued)

Table 5.2 (continued)

Model of the study – species	Dose/route	Observations and conclusion	Reference
In vivo – *P. chabaudi*	Dose-dependent	Curcumin was nontoxic in BALB/c mice. The combination of curcumin/chloroquine/piperine reduced parasitemia to 37% 7 days after treatment versus the control group's 65%, with an additive interaction. However, this combination did not show favorable drug interaction in murine model of malaria. Treatment of mice with sub-therapeutic drug doses resulted in temporary increase in genes encoding deubiquitylating enzymes indicating ubiquitin proteasome system interference. Future studies exploring suitable drug synergists for curcumin are of interest	[167]
In vivo – *S. mansoni*	1.56–100μM	Curcumin reduced viability of adult worms and killed them. The tegument of the parasite suffered morphological changes, mitochondria underwent alterations, and chromatin condensed. Curcumin-induced DNA damage, fragmentation and elevated expression of SmCASP3/7 transcripts, and activity of Caspase 3. The level of superoxide anion and activity of SOD increased, whereas that of GST, GR, and GPx decreased, which culminated in the oxidation of proteins in adult female and male worms incubated with curcumin. Altogether, curcumin generated oxidative stress followed by apoptotic-like-events in both adult female and male S. mansoni worms, eventually killing them	[52]
In vitro – *S. mansoni*	50 and 100μM	Curcumin-loaded PLGA nanoparticles caused the death of all S. mansoni and reduced motor activity and caused partial alterations in the tegument of adult S. mansoni. This study showed the first schistosomicidal activity of curcumin-loaded PLGA nanoparticles	[175]
In vivo – *S. mansoni*	Dose-dependent	Turmeric significantly decreased parasite worm burden, granuloma size, and pathology of the affected liver. However, it is still less effective than praziquantel	[168]
In vivo – *S. mansoni*	400 mg/kg body weight divided into 16 injection	Curcumin-treated infected mice had low serum levels of IL-12 and TNF-α. Curcumin augmented specific IgG and IgG1 responses against both soluble worm antigen and soluble egg antigen. In summary, curcumin modulated cellular and humoral immune responses of infected CD1 albino mice and led to significant decrease in parasite burden and liver pathology in acute murine *S. mansoni*	[169]
In vitro – *Schistosoma* spp.	Dose- and time-dependent	This showed that *S. haematobium* are more sensitive than *S. mansoni* to curcumin. Curcumin was found to affect antigenicity of surface membrane molecules of *S. haematobium*, but not *S. mansoni*. Curcumin also significantly affected *S. mansoni* eggs' hatchability and viability. Their data emphasized that curcumin is a promising schistosomicidal drug	[53]
In vitro – *S. japonicum*	Dose- and time-dependent	Curcumin demonstrated schistosomicidal effects on all life stages of *S. japonicum*. It exhibited optimal activity against adult stage and decreased motor activity of the worms without tegumental alterations. Their promising in vitro results on all stages of *S. japonicum* assures further evaluation for prophylactic and therapeutic values in the early and late schistosomiasis	[54]

(continued)

Table 5.2 (continued)

Model of the study – species	Dose/route	Observations and conclusion	Reference
In vivo – *S. mansoni*	300 mg/kg/day Oral administration	The results indicated that curcumin has potent antifibrotic activity in both suppressing and reversing *S. mansoni*-induced liver fibrosis	[170]
In vitro – *T. gondii*	Dose- and time-dependent	Curcumin inhibited enzymatic activity of the recombinant *T. gondii* Glo1 gene as well as the parasitic propagation of *T. gondii*. The inhibitory effect of curcumin on enzymatic activity of *T. gondii* Glo1 gene and parasitic propagation of *T. gondii* could be valued to explore a potential and potent drug for the treatment of toxoplasmosis	[55]
In vivo – *T. gondii*	Dose-dependent	The results showed potential of curcumin suspension and curcumin nanoemulsion in the treatment of acute and chronic toxoplasmosis in BALB/c mouse models. However, curcumin nanoemulsion was more efficient than curcumin suspension, and it seemed that curcumin nanoemulsion has the potential for the treatment of acute and chronic toxoplasmosis, especially in those with latent bradyzoites in brain	[56]

The effects of curcumin were also studied in mice infected with *P. chabaudi* clones that were resistant to the standard anti-malarial treatments, chloroquine and artemisinin [167]. This revealed that the combination of curcumin + chloroquine + piperine decreased parasitemia to 37%, compared to 65% seen for the control group. Finally, treatment of mice with therapeutic doses of the drugs resulted in a transient increase in genes encoding deubiquitylating enzymes, suggesting an inhibitory effect on the ubiquitin-proteasome system.

A study showed that administration of ginger (10 mg/kg) and curcumin (20 mg/kg) separately to giardia-infected Balb/c mice led to a significant increase in the number of dead cysts in the feces compared to giardia-infected untreated mice [41]. In addition, both treatments led to a significant improvement in the intestinal mucosal damages induced by the infection. Another study showed that high concentrations of curcumin (3 and 15μM) disrupted the *G. lamblia* cytoskelton, with a decrease in tubulin expression and changes in its distribution [42]. Curcumin also inhibited the proliferation and adhesion of *G. lamblia*.

Another study examined the photodynamic activity of curcumin on *L. major* and *L. braziliensis* promastigote, which revealed a strong effect on the viability of both parasites [46]. In addition, the morphology of promastigotes was altered. In contrast, another study found that long-term, low-dose, and oral administration of curcumin activated peroxisome proliferator-activated receptor-γ (PPARγ), deactivated the type 1 immune response, and reduced inducible nitric oxide (NO) synthase levels, thereby exacerbating pathogenesis in an *L. donovani*-infected mouse model [47]. This effect may have been due to curcumin acting as an anti-oxidant and interfering with the natural reactive radicals mounted in defense against the pathogen.

In another study, curcumin was effective in inducing apoptosis and oxidative damage in *S. mansoni* worms [52]. This occurred via increased superoxide anion levels and superoxide dismutase activity, increased caspase 3 mRNA and enzyme activity levels, and decreased glutathione-S-transferase, glutathione reductase, and glutathione reductase levels, in male and female worms. A different study showed that a combination treatment of turmeric and praziquantel reduced *S. mansoni* worm load and improved the associated pathology in infected Swiss albino mice [168]. The combination treatment also decreased granuloma size, although turmeric alone had no effect on granuloma number and was less efficient in the other effects compared to the praziquantel treatment. In a similar study, curcumin treatment of *S. mansoni* infected Swiss albino mice was effective in decreasing worm and egg loads, as well as decreasing liver granuloma volume and collagen content [169]. The curcumin treatment also restored hepatic enzymes to normal levels, improved the infection-associated hepato-splenomegaly and eosinophilia, augmented the immune response to the worms and eggs, and decreased serum levels of IL-12 and TNF-α. Another study showed that curcumin treatment killed *S. mansoni* and *S. haematobium* adult worms, with stronger effects on the latter worm type [53]. In addition, the curcumin treatment decreased hatchability and viability of *S. mansoni* eggs. Similar effects were observed in *S. japonicum* worms following curcumin treatment, with schistosomicidal effects seen at every stage of life [54]. The findings of another study suggested that curcumin may have antifibrotic activity as it was found to suppress and reverse *S. mansoni*-induced liver fibrosis in infected Swiss albino mice [170].

5.7 Mechanisms Underlying Curcumin Analogues in Parasitic-Associating IDs

Despite the therapeutic effects of curcumin, it has poor intestinal absorption and bioavailability, which hinders its in vivo and *clinical trial* applications. However, researchers have designed and tested curcumin analogues in attempts to overcome this drawback. Below, we discuss some of these studies in parasitic-associating IDs (Table 5.2).

Caenorhabditis elegans (*C. elegans*) worms exposed to *P. falciparum* and co-encapsulated quinine + curcumin-loaded polysorbate-coated polymeric nanocapsules resulted in their higher survival and longevity with no reduction in reproductive capacity compared to the corresponding monotreatments [48]. Another study showed that curcumin + artesunate-loaded poly d,l-lactic-co-glycolic acid (CA-PLGA) had improved antiplasmodial efficacy against *P. berghei* and lower toxicity in mice, compared with free curcumin and free artesunate [49]. Another study showed that nanotized curcumin was 10 times more effective than native curcumin in increasing the survival of *P. berghei*-infected mice [171].

One study showed that a combined therapy of silver, chitosan, and curcumin nanoparticles yielded better outcomes than each compound as a monotherapy on *G. lamblia*-infected rats [172]. Chauhan et al. tested the effects of the curcumin analogue trans-dibenzalacetone (DBA) on intracellular amastigotes of *L. donovani* and found a leishmanicidal effect that occurred via the arrest of cell growth at the G0/G1 phase [43]. An investigation found a cytotoxic effect of curcumin-loaded mannose-functionalized chitosan nanoparticles against the J774A.1 cell line with non-toxicity toward macrophage cells [173]. Furthermore, a study by khan et al. showed that curcumin-loaded self-emulsifying drug delivery systems had broad spectrum antileishmanial and antibacterial effects against cutaneous leishmaniasis and associated secondary infections [44]. Another study found significant growth inhibition using indium and gallium curcumin against leishmania parasites compared to curcumin and diacetylcurcumin formulations [174]. A nanoformulation of curcumin showed good leishmanicidal activity in vitro and in vivo, although there was a synergystic effect when this compound was used in combination with miltefosine [45]. This appeared to be driven by increased production of ROS and nitrogen species, elevated phagocytic activity, and increased lymphocyte proliferation. Curcumin-loaded PLGA nanoparticles also reduced motor activity, prompted partial changes in the tegument, and induced death of *S. mansoni* worms [175]. Finally, a curcumin nonoemulsion decreased the mean counts of peritoneum tachyzoites in mice inoculated with type I or type II *T. gondii* [56].

5.8 Promises and Controversies of Curcumin on Bacterial- and Parasitic-Associating IDs

As has already mentioned, native curcumin and its analogues have potential antibacterial and antiparasitic properties. Further to this, hepatoprotective effects of curcumin have been reported

[176], and it can be applied as a supplement therapy to decrease the adverse effects of some frontline licensed medications. The curcumin analogues and nano-formulations appear to show higher efficacy in purposes such as parasite suppression and reduction of hepatotoxicity compared with native curcumin [171]. This likely occurs via increased bioavailability and pharmacokinetics of the curcumin molecule. As examples, nanoparticle-formulated curcumin enhanced bioavailability in a tuberculosis-infected mouse model [123], a nano-encapsulated formulation of curcumin showed antioxidant effects at lower concentration [30], and nanotized CA-PLGA had a better efficacy on malaria [49], compared with native curcumin. Curcumin has also been shown to possess potent anti-inflammatory [1] and wound healing effects [177]. This appears to occur through reduction of inflammation and decreased damage to cells caused by microbial infections. Furthermore, curcumin can reduce the damages of oxidative stress to DNA, cells, and tissues [110, 114].

Along with these positive findings, there have also been some studies which showed mixed or negative results. For example, Although some efficacy of native curcumin was reported on eradication of *H. pylori* infection [29], a clinical trial found no effect of curcumin monotherapy [112] or its combination with lactoferrin, *N*-acetylcysteine, and pantoprazole on *H. pylori* eradication [28]. On the other hand, a clinical study reported that curcumin combination with the omeprazole, amoxicillin, and metronidazole triple therapy led to improved impact on *H. pylori* infection compared to each treatment alone [114]. Furthermore, a combination treatment consisting of silver, chitosan, and curcumin nanoparticles significantly eradicated infection from the intestines of *G. lamblia*-infected rats [172].

Failures of antimalarial approaches have been documented in some endemic countries, particularly regarding artemisinin and derivative compounds [178, 179]. However, the oxidative insult caused by artemisinin was found to be ameliorated by the administration of curcumin [180]. In contrast, the administration of a curcumin-piperine-artemisinin combination treatment did not show a desirable drug interaction a murine model of malaria, suggesting the need for further studies testing for the suitability of specific drug partners for combination therapies with curcumin [167]. Drug delivery approaches might be promising owing to the reduction of off-target activity. For example, nanotized CA-PLGA was efficacious against the malaria parasite at low dosages [49]. Although quinine has been the second-line drug of choice for the treatment of chloroquine-resistant malaria, high dosages can lead to cardiac arrhythmia, fatal hypotension, and cardiotoxicity [181]. However, the efficacy quinine on *P. falciparum* cultures was enhanced by co-encapsulation with curcumin.

In leishmania infection, macrophages generate ROS and RNS and the infected cells of the host produce NO as a defense against the parasite. However, the antioxidant activity of curcumin diminishes the activity of NO on the leishmania parasite [182]. Conversely, some curcumin analogues such as DBA have been shown to trigger apoptic death of *L. donovani* by altering the ultrastructure and physiology of mitochondria via dissipation of the proton gradient [43].

As described above, the combination therapy of silver + chitosan + curcumin nanoparticles yielded better outcomes than the corresponding monotherapies in *G. lamblia*-infected rats [172]. In addition, the combination of curcumin with the compounds *Andrographis paniculata* and *Hedyotis corymbosa*, which have been in use since ancient time as hepatoprotectants and for fever reduction, showed synergistic anti-malarial activities in vivo compared to the monotherapies [183]. Furthermore, a curcumin and erythromycin combination therapy had a greater impact on *S. aureus* infection and bacterial growth, compared to either treatment alone [31].

Some of the negative findings of curcumin treatment may be related to its antioxidant and anti-inflammatory properties interfering with the natural host response, thereby exacerbating infection [25, 34, 35, 47]. Another area which should be investigated further concern the capacity of curcumin as a microtubule destabilizer. Although this property has been found to be useful in halting *G. lamblia* [42] and *P. falciparum* [166] proliferation, this destabilizing effect may also lead to host cell damage owing to off-target effects. We suggest that further research of these phenomena should be conducted with native curcumin, analogues, and different formulations, considering the differences in efficacy.

5.9 Conclusions and Future Perspectives

Taken together, numerous studies have shown that curcumin has moderate to high anti-inflammatory anti-bacterial, and anti-parasitic activities, with efficacy in reducing the harmful effects of some bacterial- and parasitic-associating IDs. However, it is clear that further research is needed to firmly establish the benefit-risk profile of this natural compound regarding its use as an antimicrobial agent and in other therapeutic areas. Considering the potential therapeutic value and the controversies, further preclinical, safety, toxicity, and efficacy studies, as well as randomized, controlled, double-blinded clinical trials are required. These should include the testing of native curcumin, curcumin analogues, curcumin combination treatments as well as different curcumin formulations.

References

1. Ayati Z, Ramezani M, Amiri MS, Moghadam AT, Rahimi H, Abdollahzade A, Sahebkar A, Emami SA (2019) Ethnobotany, Phytochemistry and Traditional Uses of Curcuma spp. and Pharmacological Profile of Two Important Species (C. longa and C. zedoaria): A Review. Curr Pharm Des 25(8):871–935. https://doi.org/10.2174/1381612825666190402163940. PMID: 30947655
2. Khan S, Imran M, Butt TT, Ali Shah SW, Sohail M, Malik A et al (2018) Curcumin based nanomedicines as efficient nanoplatform for treatment of cancer: new developments in reversing cancer drug resistance, rapid internalization, and improved anticancer efficacy. Trends Food Sci Technol 80:8–22
3. Patel SS, Acharya A, Ray RS, Agrawal R, Raghuwanshi R, Jain P (2020) Cellular and molecular mechanisms of curcumin in prevention and treatment of disease. Crit Rev Food Sci Nutr 60(6):887–939
4. Abdollahi E, Momtazi AA, Johnston TP, Sahebkar A (2018) Therapeutic effects of curcumin in inflammatory and immune-mediated diseases: a nature-made jack-of-all-trades? J Cell Physiol 233(2):830–848
5. Iranshahi M, Sahebkar A, Takasaki M, Konoshima T, Tokuda H (2009) Cancer chemopreventive activity of the prenylated coumarin, umbelliprenin, in vivo. Eur J Cancer Prev 18(5):412–415
6. Mollazadeh H, Cicero AFG, Blesso CN, Pirro M, Majeed M, Sahebkar A (2019) Immune modulation by curcumin: the role of interleukin-10. Crit Rev Food Sci Nutr 59(1):89–101
7. Panahi Y, Ahmadi Y, Teymouri M, Johnston TP, Sahebkar A (2018) Curcumin as a potential candidate for treating hyperlipidemia: A review of cellular and metabolic mechanisms. J Cell Physiol 233(1): 141–152. https://doi.org/10.1002/jcp.25756. Epub 2017 Jun 6. PMID: 28012169.
8. Momtazi AA, Derosa G, Maffioli P, Banach M, Sahebkar A (2016) Role of microRNAs in the therapeutic effects of curcumin in non-cancer diseases. Mol Diagn Ther 20(4):335–345
9. Teymouri M, Pirro M, Johnston TP, Sahebkar A (2017) Curcumin as a multifaceted compound against human papilloma virus infection and cervical cancers: A review of chemistry, cellular, molecular, and preclinical features. BioFactors 43(3):331–346.
10. Bagheri H, Ghasemi F, Barreto GE, Rafiee R, Sathyapalan T, Sahebkar A (2019) Effects of curcumin on mitochondria in neurodegenerative diseases. Biofactors 46(1):5–20.
11. Priyadarsini KI (2014) The chemistry of curcumin: from extraction to therapeutic agent. Molecules 19(12):20091–20112
12. Priyadarsini KI (2013) Chemical and structural features influencing the biological activity of curcumin. Curr Pharm Des 19(11):2093–2100
13. Ghasemi F, Bagheri H, Barreto GE, Read MI, Sahebkar A (2019) Effects of curcumin on microglial cells. Neurotox Res 36(1):12–26
14. Understanding Emerging and Re-emerging Infectious Diseases. NIH Curriculum Supplement Series [Internet]. https://www.ncbi.nlm.nih.gov/books/NBK20370/
15. van Doorn HR (2014) Emerging infectious diseases. Medicine (Abingdon) 42(1):60–63
16. Borgdorff MW, van Soolingen D (2013) The re-emergence of tuberculosis: what have we learnt from molecular epidemiology? Clin Microbiol Infect 19(10):889–901
17. Sejvar JJ (2003) West nile virus: an historical overview. Ochsner J 5(3):6–10
18. Huang C, Wang Y, Li X, Ren L, Zhao J, Hu Y et al (2019) Clinical features of patients infected with 2019 novel coronavirus in Wuhan, China. Lancet 395(10223):497–506
19. Leventhal GE, Hill AL, Nowak MA, Bonhoeffer S (2015) Evolution and emergence of infectious

diseases in theoretical and real-world networks. Nat Commun 6:6101. https://doi.org/10.1038/ncomms7101
20. Smith KM, Machalaba CC, Seifman R, Feferholtz Y, Karesh WB (2019) Infectious disease and economics: the case for considering multi-sectoral impacts. One Health 7:100080. https://doi.org/10.1016/j.onehlt.2018.100080
21. Brown J, Capocci S, Smith C, Morris S, Abubakar I, Lipman M (2015) Health status and quality of life in tuberculosis. Int J Infect Dis 32:68–75
22. Basavaraj KH, Navya MA, Rashmi R (2010) Quality of life in HIV/AIDS. Indian J Sex Transm Dis AIDS 31(2):75–80
23. Praditya D, Kirchhoff L, Brüning J, Rachmawati H, Steinmann J, Steinmann E (2019) Anti-infective properties of the golden spice curcumin. Front Microbiol 10:912
24. Günther G (2014) Multidrug-resistant and extensively drug-resistant tuberculosis: a review of current concepts and future challenges. Clin Med (Lond) 14(3):279–285
25. Marathe SA, Sen M, Dasgupta I, Chakravortty D (2012) Differential modulation of intracellular survival of cytosolic and vacuolar pathogens by curcumin. Antimicrob Agents Chemother 56(11):5555–5567
26. Srivastava RM, Singh S, Dubey SK, Misra K, Khar A (2011) Immunomodulatory and therapeutic activity of curcumin. Int Immunopharmacol 11(3):331–341
27. Wessler S, Muenzner P, Meyer TF, Naumann M (2005) The anti-inflammatory compound curcumin inhibits Neisseria gonorrhoeae-induced NF-kappaB signaling, release of pro-inflammatory cytokines/chemokines and attenuates adhesion in late infection. Biol Chem 386(5):481–490
28. Di Mario F, Cavallaro LG, Nouvenne A, Stefani N, Cavestro GM, Iori V et al (2007) A curcumin-based 1-week triple therapy for eradication of Helicobacter pylori infection: something to learn from failure? Helicobacter 12(3):238–243
29. De R, Kundu P, Swarnakar S, Ramamurthy T, Chowdhury A, Nair GB et al (2009) Antimicrobial activity of curcumin against helicobacter pylori isolates from India and during infections in mice. Antimicrob Agents Chemother 53(4):1592–1597
30. Jaguezeski AM, Souza CF, Perin G, Reis JH, Gomes TMA, Baldissera MD et al (2019) Effect of free and nano-encapsulated curcumin on treatment and energetic metabolism of gerbils infected by Listeria monocytogenes. Microb Pathog 134:103564. https://doi.org/10.1016/j.micpath.2019.103564
31. Zhou Z, Pan C, Lu Y, Gao Y, Liu W, Yin P et al (2017) Combination of erythromycin and curcumin alleviates Staphylococcus aureus induced osteomyelitis in rats. Front Cell Infect Microbiol 7:379. https://doi.org/10.3389/fcimb.2017.00379
32. Wang J, Zhou X, Li W, Deng X, Deng Y, Niu X (2016) Curcumin protects mice from Staphylococcus aureus pneumonia by interfering with the self-assembly process of α-hemolysin. Sci Rep 6:28254. https://doi.org/10.1038/srep28254
33. Bansal S, Chhibber S (2010) Curcumin alone and in combination with augmentin protects against pulmonary inflammation and acute lung injury generated during Klebsiella pneumoniae B5055-induced lung infection in BALB/c mice. J Med Microbiol 59(Pt 4):429–437
34. Marathe SA, Kumar R, Ajitkumar P, Nagaraja V, Chakravortty D (2013) Curcumin reduces the antimicrobial activity of ciprofloxacin against Salmonella typhimurium and Salmonella typhi. J Antimicrob Chemother 68(1):139–152
35. Marathe SA, Ray S, Chakravortty D (2010) Curcumin increases the pathogenicity of Salmonella enterica serovar Typhimurium in murine model. PLoS One 5(7):e11511. https://doi.org/10.1371/journal.pone.0011511
36. Baldwin PR, Reeves AZ, Powell KR, Napier RJ, Swimm AI, Sun A et al (2015) Monocarbonyl analogs of curcumin inhibit growth of antibiotic sensitive and resistant strains of Mycobacterium tuberculosis. Eur J Med Chem 92:693–699
37. Singh AK, Yadav P, Karaulia P, Singh VK, Gupta P, Puttrevu SK et al (2017) Biological evaluation of novel curcumin-pyrazole-mannich derivative active against drug-resistant Mycobacterium tuberculosis. Future Microbiol 12:1349–1362
38. Subhedar DD, Shaikh MH, Nawale L, Sarkar D, Khedkar VM, Shingate BB (2017) Quinolidene based monocarbonyl curcumin analogues as promising antimycobacterial agents: synthesis and molecular docking study. Bioorg Med Chem Lett 27(4):922–928
39. Marini E, Di Giulio M, Magi G, Di Lodovico S, Cimarelli ME, Brenciani A et al (2018) Curcumin, an antibiotic resistance breaker against a multiresistant clinical isolate of Mycobacterium abscessus. Phytother Res 32(3):488–495
40. Shyu LY, Chang HH, Hsu JD, Lin DP, Teng YH, Lee HH (2012) Curcumin alleviates eosinophilic meningitis through reduction of eosinophil count following albendazole treatment against Angiostrongylus cantonensis in mice. Parasitology 139(3):358–365
41. Dyab AK, Yones DA, Ibraheim ZZ, Hassan TM (2016) Anti-giardial therapeutic potential of dichloromethane extracts of Zingiber officinale and Curcuma longa in vitro and in vivo. Parasitol Res 115(7):2637–2645
42. Gutiérrez-Gutiérrez F, Palomo-Ligas L, Hernández-Hernández JM, Pérez-Rangel A, Aguayo-Ortiz R, Hernández-Campos A et al (2017) Curcumin alters the cytoskeleton and microtubule organization on trophozoites of Giardia lamblia. Acta Trop 172:113–121
43. Chauhan IS, Rao GS, Shankar J, Chauhan LKS, Kapadia GJ, Singh N (2018) Chemoprevention of Leishmaniasis: in-vitro antiparasitic activity of dibenzalacetone, a synthetic curcumin analog leads to apoptotic cell death in Leishmania donovani. Parasitol Int 67(5):627–636

44. Khan M, Ali M, Shah W, Shah A, Yasinzai MM (2019) Curcumin-loaded self-emulsifying drug delivery system (cu-SEDDS): a promising approach for the control of primary pathogen and secondary bacterial infections in cutaneous leishmaniasis. Appl Microbiol Biotechnol 103(18):7481–7490
45. Tiwari B, Pahuja R, Kumar P, Rath SK, Gupta KC, Goyal N (2017) Nanotized curcumin and miltefosine, a potential combination for treatment of experimental visceral Leishmaniasis. Antimicrob Agents Chemother 61(3):e01169-16. https://doi.org/10.1128/AAC.01169-16
46. Pinto JG, Fontana LC, de Oliveira MA, Kurachi C, Raniero LJ, Ferreira-Strixino J (2016) In vitro evaluation of photodynamic therapy using curcumin on Leishmania major and Leishmania braziliensis. Lasers Med Sci 31(5):883–890
47. Adapala N, Chan MM (2008) Long-term use of an anti-inflammatory, curcumin, suppressed type 1 immunity and exacerbated visceral leishmaniasis in a chronic experimental model. Lab Investig 88(12):1329–1339
48. Velasques K, Maciel TR, de Castro Dal Forno AH, Teixeira FEG, da Fonseca AL, Varotti FP et al (2018) Co-nanoencapsulation of antimalarial drugs increases their in vitro efficacy against Plasmodium falciparum and decreases their toxicity to Caenorhabditis elegans. Eur J Pharm Sci 118:1–12
49. Oyeyemi O, Morenkeji O, Afolayan F, Dauda K, Busari Z, Meena J et al (2018) Curcumin-artesunate based polymeric nanoparticle; antiplasmodial and toxicological evaluation in murine model. Front Pharmacol 9:562. https://doi.org/10.3389/fphar.2018.00562
50. Martinelli A, Rodrigues LA, Cravo P (2008) Plasmodium chabaudi: efficacy of artemisinin + curcumin combination treatment on a clone selected for artemisinin resistance in mice. Exp Parasitol 119(2):304–307
51. Ali AH, Sudi S, Basir R, Embi N, Sidek HM (2017) The antimalarial effect of curcumin is mediated by the inhibition of glycogen synthase kinase-3beta. J Med Food 20(2):152–161
52. de Paula AD, Brunetto Moreira Moscardini M, Rezende Morais E, Graciano de Paula R, Ferreira PM, Afonso A et al (2016) Curcumin generates oxidative stress and induces apoptosis in adult Schistosoma mansoni worms. PLoS One 11(11):e0167135. https://doi.org/10.1371/journal.pone.0167135
53. Abou El Dahab MM, Shahat SM, Mahmoud SSM, Mahana NA (2019) In vitro effect of curcumin on Schistosoma species viability, tegument ultrastructure and egg hatchability. Exp Parasitol 199:1–8
54. Chen YQ, Xu QM, Li XR, Yang SL, Zhu-Ge HX (2012) In vitro evaluation of Schistosomicidal potential of curcumin against Schistosoma japonicum. J Asian Nat Prod Res 14(11):1064–1072
55. Goo YK, Yamagishi J, Ueno A, Terkawi MA, Aboge GO, Kwak D et al (2015) Characterization of toxoplasma gondii glyoxalase 1 and evaluation of inhibitory effects of curcumin on the enzyme and parasite cultures. Parasit Vectors 8:654. https://doi.org/10.1186/s13071-015-1268-5
56. Azami SJ, Teimouri A, Keshavarz H, Amani A, Esmaeili F, Hasanpour H et al (2018) Curcumin nanoemulsion as a novel chemical for the treatment of acute and chronic toxoplasmosis in mice. Int J Nanomedicine 13:7363–7374
57. Haque M, Sartelli M, McKimm J, Abu Bakar M (2018) Health care-associated infections – an overview. Infect Drug Resist 11:2321–2333
58. Lyons NA, Kolter R (2015) On the evolution of bacterial multicellularity. Curr Opin Microbiol 24:21–28
59. Yang DC, Blair KM, Salama NR (2016) Staying in shape: the impact of cell shape on bacterial survival in diverse environments. Microbiol Mol Biol Rev 80(1):187–203
60. Kamaruzzaman NF, Kendall S, Good L (2017) Targeting the hard to reach: challenges and novel strategies in the treatment of intracellular bacterial infections. Br J Pharmacol 174(14):2225–2236
61. Young KD (2006) The selective value of bacterial shape. Microbiol Mol Biol Rev 70(3):660–703
62. Silhavy TJ, Kahne D, Walker S (2010) The bacterial cell envelope. Cold Spring Harb Perspect Biol 2(5):a000414. https://doi.org/10.1101/cshperspect.a000414
63. Bos J, Zhang Q, Vyawahare S, Rogers E, Rosenberg SM, Austin RH (2015) Emergence of antibiotic resistance from multinucleated bacterial filaments. Proc Natl Acad Sci U S A 112(1):178–183
64. Wilson JW, Schurr MJ, LeBlanc CL, Ramamurthy R, Buchanan KL, Nickerson CA (2002) Mechanisms of bacterial pathogenicity. Postgrad Med J 78(918):216
65. Wang H, Wilksch JJ, Strugnell RA, Gee ML (2015) Role of capsular polysaccharides in biofilm formation: an AFM nanomechanics study. ACS Appl Mater Interfaces 7(23):13007–13013
66. Alomairi J, Bonacci T, Ghigo E, Soubeyran P (2015) Alterations of host cell ubiquitination machinery by pathogenic bacteria. Front Cell Infect Microbiol 5:17. https://doi.org/10.3389/fcimb.2015.00017
67. Ireton K, Payrastre B, Cossart P (1999) The Listeria monocytogenes protein InlB is an agonist of mammalian phosphoinositide 3-kinase. J Biol Chem 274(24):17025–17032
68. Di Paolo G, De Camilli P (2006) Phosphoinositides in cell regulation and membrane dynamics. Nature 443(7112):651
69. Lu R, Goldberg MB (2010) Bacterial exploitation of host cell signaling. Sci Transl Med 2(51):51ps48
70. Slauch JM (2011) How does the oxidative burst of macrophages kill bacteria? Still an open question. Mol Microbiol 80(3):580–583
71. Garin J, Diez R, Kieffer S, Dermine JF, Duclos S, Gagnon E et al (2001) The phagosome proteome:

insight into phagosome functions. J Cell Biol 152(1):165–180
72. Peek RM Jr, Crabtree JE (2006) Helicobacter infection and gastric neoplasia. J Pathol 208(2):233–248
73. Kotilea K, Bontems P, Touati E (2019) Epidemiology, diagnosis and risk factors of helicobacter pylori infection. Adv Exp Med Biol 1149:17–33
74. Eusebi LH, Zagari RM, Bazzoli F (2014) Epidemiology of helicobacter pylori infection. Helicobacter 19(Suppl 1):1–5. https://doi.org/10.1111/hel.12165
75. Sjomina O, Pavlova J, Niv Y, Leja M (2018) Epidemiology of helicobacter pylori infection. Helicobacter 23(Suppl 1):e12514. https://doi.org/10.1111/hel.12514
76. Huang Y, Wang QL, Cheng DD, Xu WT, Lu NH (2016) Adhesion and invasion of gastric mucosa epithelial cells by helicobacter pylori. Front Cell Infect Microbiol 6:159. https://doi.org/10.3389/fcimb.2016.00159
77. Tong SYC, Davis JS, Eichenberger E, Holland TL, Fowler VG Jr (2015) Staphylococcus aureus infections: epidemiology, pathophysiology, clinical manifestations, and management. Clin Microbiol Rev 28(3):603–661
78. Johnson ER, Matthay MA (2010) Acute lung injury: epidemiology, pathogenesis, and treatment. J Aerosol Med Pulm Drug Deliv 23(4):243–252
79. Nguyen VT, Kamio Y (2004) Cooperative assembly of beta-barrel pore-forming toxins. J Biochem 136(5):563–567
80. Song L, Hobaugh MR, Shustak C, Cheley S, Bayley H, Gouaux JE (1996) Structure of staphylococcal alpha-hemolysin, a heptameric transmembrane pore. Science 274(5294):1859–1866
81. Wertheim HF, Melles DC, Vos MC, van Leeuwen W, van Belkum A, Verbrugh HA et al (2005) The role of nasal carriage in Staphylococcus aureus infections. Lancet Infect Dis 5(12):751–762
82. Haussler S, Parsek MR (2010) Biofilms 2009: new perspectives at the heart of surface-associated microbial communities. J Bacteriol 192(12):2941–2949
83. Maeda Y, Kenny F, Coulter WA, Loughrey A, Nagano Y, Goldsmith CE et al (2007) Bactericidal activity of denture-cleaning formulations against planktonic health care-associated and community-associated methicillin-resistant Staphylococcus aureus. Am J Infect Control 35(9):619–622
84. Jordao L, Vieira OV (2011) Tuberculosis: new aspects of an old disease. Int J Cell Biol 2011:403623. https://doi.org/10.1155/2011/403623
85. Khan MK, Islam MN, Ferdous J, Alam MM (2019) An overview on epidemiology of tuberculosis. Mymensingh Med J 28(1):259–266
86. Lee MR, Sheng WH, Hung CC, Yu CJ, Lee LN, Hsueh PR (2015) Mycobacterium abscessus complex infections in humans. Emerg Infect Dis 21(9):1638–1646
87. Hett EC, Rubin EJ (2008) Bacterial growth and cell division: a mycobacterial perspective. Microbiol Mol Biol Rev 72(1):126
88. Liu R, To KK, Teng JL, Choi GK, Mok KY, Law KI et al (2013) Mycobacterium abscessus bacteremia after receipt of intravenous infusate of cytokine-induced killer cell therapy for body beautification and health boosting. Clin Infect Dis 57(7):981–991
89. Zaman K (2010) Tuberculosis: a global health problem. J Health Popul Nutr 28(2):111–113
90. Smith I (2003) Mycobacterium tuberculosis pathogenesis and molecular determinants of virulence. Clin Microbiol Rev 16(3):463–496
91. Rallis E, Koumantaki-Mathioudaki E (2007) Treatment of Mycobacterium marinum cutaneous infections. Expert Opin Pharmacother 8(17):2965–2978
92. Tulloch LG (2012) 24 – Nontuberculous mycobacterial skin infections. In: Jong EC, Stevens DL (eds) Netter's infectious diseases. Saunders, Philadelphia, PA, pp 117–123
93. Sanderson KE, Liu SL, Tang L, Johnston RN (2015) Chapter 71 – Salmonella Typhi and Salmonella Paratyphi A. In: Tang Y-W et al (eds) Molecular medical microbiology, 2nd edn. Academic Press, Cambridge, MA, pp 1275–1306
94. Graziani C, Losasso C, Luzzi I, Ricci A, Scavia G, Pasquali P (2017) Chapter 5 – Salmonella. In: Dodd C, Aldsworth T, Stein R (eds) Foodborne diseases, 3rd edn. Academic Press, Cambridge, MA, pp 133–169
95. Parry CM, Hien TT, Dougan G, White NJ, Farrar JJ (2002) Typhoid fever. N Engl J Med 347(22):1770–1782
96. Crum-Cianflone NF (2008) Salmonellosis and the gastrointestinal tract: more than just peanut butter. Curr Gastroenterol Rep 10(4):424–431
97. Farver CF (2018) 10 – Bacterial diseases. In: Zander DS, Farver CF (eds) Pulmonary pathology, 2nd edn. Elsevier, Amsterdam, Netherlands, pp 163–200
98. Brabb T, Newsome D, Burich A, Hanes M (2012) Chapter 23 – Infectious Diseases. In: Suckow MA, Suckow M, Stevens K, Wilson R (eds) The laboratory rabbit, guinea pig, hamster, and other rodents. Academic Press, Cambridge, MA, pp 637–683
99. McSheffrey GG, Gray-Owen SD (2015) Chapter 82 – Neisseria gonorrhoeae. In: Tang YW, Saila A (eds) Molecular medical microbiology, 2nd edn. Academic Press, Cambridge, MA, pp 1471–1485
100. Kaushic C, Jerse AE, Beagley KW (2015) Chapter 107 – Animal models of immunity to female genital tract infections and vaccine development. In: Mestecky J, Strober W, Russell M, Cheroutre H, Lambrecht BN, Kelsall B (eds) Mucosal immunology, 4th edn. Academic Press, Cambridge, MA, pp 2059–2096
101. Quillin SJ, Seifert HS (2018) Neisseria gonorrhoeae host adaptation and pathogenesis. Nat Rev Microbiol 16(4):226–240
102. Goldstein E, Kirkcaldy RD, Reshef D, Berman S, Weinstock H, Sabeti P et al (2012) Factors related to increasing prevalence of resistance to ciprofloxacin and other antimicrobial drugs in Neisseria

gonorrhoeae, United States. Emerg Infect Dis 18(8):1290–1297
103. Philpott DJ, Edgeworth JD, Sansonetti PJ (2000) The pathogenesis of Shigella flexneri infection: lessons from in vitro and in vivo studies. Philos Trans R Soc Lond Ser B Biol Sci 355(1397):575–586
104. David DJV, Cossart P (2017) Recent advances in understanding Listeria monocytogenes infection: the importance of subcellular and physiological context. F1000Res 6:F1000 Faculty Rev-1126. https://doi.org/10.12688/f1000research
105. Magill AJ, Strickland GT, Maguire JH, Hill DR, Ryan ET, Solomon T (2013) 47 – Miscellaneous bacterial enteritides. In: Hunter's tropical medicine and emerging infectious disease, 9th edn. Saunders, Philadelphia, PA, pp 471–475; ASIN: B00QCLR9N2
106. Srivastava AK, Tewari M, Shukla HS, Roy BK (2015) In silico profiling of the potentiality of curcumin and conventional drugs for CagA oncoprotein inactivation. Arch Pharm 348(8):548–555
107. Zaidi SF, Yamamoto T, Refaat A, Ahmed K, Sakurai H, Saiki I et al (2009) Modulation of activation-induced cytidine deaminase by curcumin in Helicobacter pylori-infected gastric epithelial cells. Helicobacter 14(6):588–595
108. Foryst-Ludwig A, Neumann M, Schneider-Brachert W, Naumann M (2004) Curcumin blocks NF-kappaB and the motogenic response in Helicobacter pylori-infected epithelial cells. Biochem Biophys Res Commun 316(4):1065–1072
109. Sintara K, Thong-Ngam D, Patumraj S, Klaikeaw N, Chatsuwan T (2010) Curcumin suppresses gastric NF-kappaB activation and macromolecular leakage in Helicobacter pylori-infected rats. World J Gastroenterol 16(32):4039–4046
110. Team E (2017) Erratum to effects of curcumin on Helicobacter pylori infection. Ann Transl Med 5(6):153. https://doi.org/10.21037/atm.2017.03.13
111. Santos AM, Lopes T, Oleastro M, Gato IV, Floch P, Benejat L et al (2015) Curcumin inhibits gastric inflammation induced by Helicobacter pylori infection in a mouse model. Nutrients 7(1):306–320
112. Khonche A, Biglarian O, Panahi Y, Valizadegan G, Soflaei SS, Ghamarchehreh ME et al (2016) Adjunctive therapy with curcumin for peptic ulcer: a randomized controlled trial. Drug Res (Stuttg) 66(8):444–448
113. Larussa T, Gervasi S, Liparoti R, Suraci E, Marasco R, Imeneo M et al (2018) Downregulation of interleukin- (IL-) 17 through enhanced indoleamine 2,3-dioxygenase (IDO) induction by curcumin: a potential mechanism of tolerance towards helicobacter pylori. J Immunol Res 2018:3739593. https://doi.org/10.1155/2018/3739593
114. Judaki A, Rahmani A, Feizi J, Asadollahi K, Hafezi Ahmadi MR (2017) Curcumin in combination with triple therapy regimes ameliorates oxidative stress and histopathologic changes in chronic gastritis-associated helicobacter pylori infection. Arq Gastroenterol 54(3):177–182
115. Xu F, Diao R, Liu J, Kang Y, Wang X, Shi L (2015) Curcumin attenuates staphylococcus aureus-induced acute lung injury. Clin Respir J 9(1):87–97
116. Bai X, Oberley-Deegan RE, Bai A, Ovrutsky AR, Kinney WH, Weaver M et al (2016) Curcumin enhances human macrophage control of Mycobacterium tuberculosis infection. Respirology 21(5):951–957
117. Li M, Wu Z, Niu W, Wan Y, Zhang L, Shi G et al (2014) The protective effect of curcumin against the 19kDa Mycobacterium tuberculosis protein-induced inflammation and apoptosis in human macrophages. Mol Med Rep 10(6):3261–3267
118. Li MY, Wang HL, Huang J, Shi GC, Wan YG, Wang JX et al (2014) Curcumin inhibits 19-kDa lipoprotein of Mycobacterium tuberculosis induced macrophage apoptosis via regulation of the JNK pathway. Biochem Biophys Res Commun 446(2):626–632
119. Bansal S, Chhibber S (2014) Phytochemical-induced reduction of pulmonary inflammation during Klebsiella pneumoniae lung infection in mice. J Infect Dev Ctries 8(7):838–844
120. Peng KT, Chiang YC, Huang TY, Chen PC, Chang PJ, Lee CW (2019) Curcumin nanoparticles are a promising anti-bacterial and anti-inflammatory agent for treating periprosthetic joint infections. Int J Nanomedicine 14:469–481
121. Xu Y, Ge L, Abdel-Razek O, Jain S, Liu Z, Hong Y et al (2016) Differential susceptibility of human SP-B genetic variants on lung injury caused by bacterial pneumonia and the effect of a chemically modified curcumin. Shock 45(4):375–384
122. Ye Y, Li Y, Fang F (2014) Upconversion nanoparticles conjugated with curcumin as a photosensitizer to inhibit methicillin-resistant Staphylococcus aureus in lung under near infrared light. Int J Nanomedicine 9:5157–5165
123. Tousif S, Singh DK, Mukherjee S, Ahmad S, Arya R, Nanda R et al (2017) Nanoparticle-formulated curcumin prevents posttherapeutic disease reactivation and reinfection with Mycobacterium tuberculosis following isoniazid therapy. Front Immunol 8:739. https://doi.org/10.3389/fimmu.2017.00739
124. Cox FE (2002) History of human parasitology. Clin Microbiol Rev 15(4):595–612
125. Kiani H, Haghighi A, Rostami A, Azargashb E, Tabaei SJS, Solgi A et al (2016) Prevalence, risk factors and symptoms associated to intestinal parasite infections among patients with gastrointestinal disorders in Nahavand, Western Iran. Rev Inst Med Trop Sao Paulo 58:42. https://doi.org/10.1590/S1678-9946201658042
126. Wang Y (2017) Introduction to parasitic disease. In: Li H (ed) Radiology of parasitic diseases: a practical approach. Springer, Amsterdam, Netherlands, pp 3–3
127. Engels D, Zhou X-N (2020) Neglected tropical diseases: an effective global response to local poverty-

related disease priorities. Infect Dis Poverty 9(1):10. https://doi.org/10.1186/s40249-020-0630-9
128. World Health Organization. Neglected tropical diseases. https://www.who.int/neglected_diseases/diseases/en/
129. Moore J (2002) Parasites and the behavior of animals, Oxford series in Ecology and Evolution. Oxford University Press, Oxford
130. Schmid-Hempel P (2009) Immune defence, parasite evasion strategies and their relevance for 'macroscopic phenomena' such as virulence. Philos Trans R Soc Lond Ser B Biol Sci 364(1513):85–98
131. Schmid-Hempel P (2008) Parasite immune evasion: a momentous molecular war. Trends Ecol Evol 23(6):318–326
132. Bhopale GM (2003) Pathogenesis of toxoplasmosis. Comp Immunol Microbiol Infect Dis 26(4):213–222
133. Cortés A, Sotillo J, Muñoz-Antolí C, Molina-Durán J, Esteban JG, Toledo R (2017) Antibody trapping: a novel mechanism of parasite immune evasion by the trematode Echinostoma caproni. PLoS Negl Trop Dis 11(7):e0005773. https://doi.org/10.1371/journal.pntd.0005773
134. Bloom BR (1979) Games parasites play: how parasites evade immune surveillance. Nature 279(5708):21–26
135. Frank SA (2002) Immunology and evolution of infectious disease. Princeton University Press, Princeton, NJ
136. Poulin R (2011) Evolutionary ecology of parasites. Princeton University Press, Princeton, NJ
137. Roberts L, Janovy J Jr (1996) Foundations of parasitology, 5th edn. Wm. C. Brown, Co., Dubuque, IA, p 659. ASIN: B000UUO8CQ
138. Nelson FBL, Brown GP, Shilton C, Shine R (2015) Host–parasite interactions during a biological invasion: the fate of lungworms (Rhabdias spp.) inside native and novel anuran hosts. Int J Parasitol Parasites Wildl 4(2):206–215
139. Graham AL, Allen JE, Read AF (2005) Evolutionary causes and consequences of immunopathology. Annu Rev Ecol Evol Syst 36(1):373–397
140. Laishram DD, Sutton PL, Nanda N, Sharma VL, Sobti RC, Carlton JM et al (2012) The complexities of malaria disease manifestations with a focus on asymptomatic malaria. Malar J 11(1):29
141. de Koning-Ward TF, Dixon MW, Tilley L, Gilson PR (2016) Plasmodium species: master renovators of their host cells. Nat Rev Microbiol 14(8):494–507
142. Craig AG, Grau GE, Janse C, Kazura JW, Milner D, Barnwell JW et al (2012) The role of animal models for research on severe malaria. PLoS Pathog 8(2):e1002401. https://doi.org/10.1371/journal.ppat.1002401
143. Stephens R, Culleton RL, Lamb TJ (2012) The contribution of Plasmodium chabaudi to our understanding of malaria. Trends Parasitol 28(2):73–82
144. World Health Organization (2020) Malaria. https://www.who.int/malaria/en/
145. Noor Azian MY, San YM, Gan CC, Yusri MY, Nurulsyamzawaty Y, Zuhaizam AH et al (2007) Prevalence of intestinal protozoa in an aborigine community in Pahang, Malaysia. Trop Biomed 24(1):55–62
146. Dib HH, Lu SQ, Wen SF (2008) Prevalence of Giardia lamblia with or without diarrhea in South East, South East Asia and the Far East. Parasitol Res 103(2):239–251
147. Nyarango RM, Aloo PA, Kabiru EW, Nyanchongi BO (2008) The risk of pathogenic intestinal parasite infections in Kisii Municipality, Kenya. BMC Public Health 8:237. https://doi.org/10.1186/1471-2458-8-237
148. Farthing MJ (1997) The molecular pathogenesis of giardiasis. J Pediatr Gastroenterol Nutr 24(1):79–88
149. Hooshyar H, Rostamkhani P, Arbabi M, Delavari M (2019) Giardia lamblia infection: review of current diagnostic strategies. Gastroenterol Hepatol Bed Bench 12(1):3–12
150. Leitsch D (2015) Drug resistance in the microaerophilic parasite Giardia lamblia. Curr Trop Med Rep 2(3):128–135
151. Organization WH (2019) Leishmaniasis. World Health Organization
152. Weigel MM, Armijos RX (2001) The traditional and conventional medical treatment of cutaneous leishmaniasis in rural Ecuador. Rev Panam Salud Publica 10(6):395–404
153. Kumar Bhat N, Ahuja V, Dhar M, Ahmad S, Pandita N, Gupta V et al (2017) Changing epidemiology: a new focus of Kala-azar at high-altitude Garhwal region of north India. J Trop Pediatr 63(2):104–108
154. Barsoum RS, Esmat G, El-Baz T (2013) Human schistosomiasis: clinical perspective: review. J Adv Res 4(5):433–444
155. World Health Organization (2019) Schistosomiasis. https://www.who.int/schistosomiasis/en/
156. Dubey JP, Lindsay DS, Speer CA (1998) Structures of toxoplasma gondii tachyzoites, bradyzoites, and sporozoites and biology and development of tissue cysts. Clin Microbiol Rev 11(2):267–299
157. Halonen SK, Weiss LM (2013) Toxoplasmosis. Handb Clin Neurol 114:125–145
158. Katlama C, De Wit S, O'Doherty E, Van Glabeke M, Clumeck N (1996) Pyrimethamine-clindamycin vs. pyrimethamine-sulfadiazine as acute and long-term therapy for toxoplasmic encephalitis in patients with AIDS. Clin Infect Dis 22(2):268–275
159. Alday PH, Doggett JS (2017) Drugs in development for toxoplasmosis: advances, challenges, and current status. Drug Des Devel Ther 11:273–293
160. Hasim S, Hussin NA, Alomar F, Bidasee KR, Nickerson KW, Wilson MA (2014) A glutathione-independent glyoxalase of the DJ-1 superfamily plays an important role in managing metabolically generated methylglyoxal in Candida albicans. J Biol Chem 289(3):1662–1674

161. Urscher M, Alisch R, Deponte M (2011) The glyoxalase system of malaria parasites–implications for cell biology and general glyoxalase research. Semin Cell Dev Biol 22(3):262–270
162. Alicata JE (1965) Biology and distribution of the rat lungworm, angiostrongylus cantonensis, and its relationship to eosinophilic meningoencephalitis and other neurological disorders of man and animals. In: Advances in parasitology, vol 3. Academic Press, Cambridge, MA, pp 223–248
163. Rosen L, Chappell R, Laqueur GL, Wallace GD, Weinstein PP (1962) Eosinophilic meningoencephalitis caused by a metastrongylid lung-worm of rats. JAMA 179(8):620–624
164. Kliks MM, Kroenke K, Hardman JM (1982) Eosinophilic radiculomyeloencephalitis: an angiostrongyliasis outbreak in American Samoa related to ingestion of Achatina fulica snails. Am J Trop Med Hyg 31(6):1114–1122
165. Hollingsworth R, Cowie R, Joshi R, Sebastian L (2006) Global advances in ecology and management of golden apple snails. Philippine Rice Research Institute, Maligaya, Muñoz, Nueva Ecija; Philippines
166. Chakrabarti R, Rawat PS, Cooke BM, Coppel RL, Patankar S (2013) Cellular effects of curcumin on Plasmodium falciparum include disruption of microtubules. PLoS One 8(3):e57302. https://doi.org/10.1371/journal.pone.0057302
167. Neto Z, Machado M, Lindeza A, do Rosario V, Gazarini ML, Lopes D (2013) Treatment of plasmodium chabaudi parasites with curcumin in combination with antimalarial drugs: drug interactions and implications on the ubiquitin/proteasome system. J Parasitol Res 2013:429736. https://doi.org/10.1155/2013/429736
168. Hussein A, Rashed S, El Hayawan I, El-Sayed R, Ali H (2017) Evaluation of the anti-schistosomal effects of turmeric (Curcuma longa) versus praziquantel in Schistosoma mansoni infected mice. Iran J Parasitol 12(4):587–596
169. Allam G (2009) Immunomodulatory effects of curcumin treatment on murine schistosomiasis mansoni. Immunobiology 214(8):712–727
170. El-Agamy DS, Shebl AM, Said SA (2011) Prevention and treatment of Schistosoma mansoni-induced liver fibrosis in mice. Inflammopharmacology 19(6):307–316
171. Ghosh A, Banerjee T, Bhandary S, Surolia A (2014) Formulation of nanotized curcumin and demonstration of its antimalarial efficacy. Int J Nanomedicine 9:5373–5387
172. Said DE, Elsamad LM, Gohar YM (2012) Validity of silver, chitosan, and curcumin nanoparticles as anti-Giardia agents. Parasitol Res 111(2):545–554
173. Chaubey P, Mishra B, Mudavath SL, Patel RR, Chaurasia S, Sundar S et al (2018) Mannose-conjugated curcumin-chitosan nanoparticles: efficacy and toxicity assessments against Leishmania donovani. Int J Biol Macromol 111:109–120
174. Fouladvand M, Barazesh A, Tahmasebi R (2013) Evaluation of in vitro antileishmanial activity of curcumin and its derivatives "gallium curcumin, indium curcumin and diacethyle curcumin". Eur Rev Med Pharmacol Sci 17(24):3306–3308
175. Luz PP, Magalhaes LG, Pereira AC, Cunha WR, Rodrigues V, Andrade ESML (2012) Curcumin-loaded into PLGA nanoparticles: preparation and in vitro schistosomicidal activity. Parasitol Res 110(2):593–598
176. Farzaei MH, Zobeiri M, Parvizi F, El-Senduny FF, Marmouzi I, Coy-Barrera E et al (2018) Curcumin in liver diseases: a systematic review of the cellular mechanisms of oxidative stress and clinical perspective. Nutrients 10(7):E855. https://doi.org/10.3390/nu10070855
177. Tejada S, Manayi A, Daglia M, Nabavi SF, Sureda A, Hajheydari Z et al (2016) Wound healing effects of curcumin: a short review. Curr Pharm Biotechnol 17(11):1002–1007
178. Lim P, Alker AP, Khim N, Shah NK, Incardona S, Doung S et al (2009) Pfmdr1 copy number and arteminisin derivatives combination therapy failure in falciparum malaria in Cambodia. Malar J 8:11. https://doi.org/10.1186/1475-2875-8-11
179. Dondorp AM, Nosten F, Yi P, Das D, Phyo AP, Tarning J et al (2009) Artemisinin resistance in Plasmodium falciparum malaria. New Engl J Med 361(5):455–467
180. Desai KR, Rajput DK, Patel PB, Highland HN (2015) Ameliorative effects of curcumin on artesunate-induced subchronic toxicity in testis of swiss albino male mice. Dose Response 13(2):1559325815592393. https://doi.org/10.1177/1559325815592393
181. Achan J, Talisuna AO, Erhart A, Yeka A, Tibenderana JK, Baliraine FN et al (2011) Quinine, an old anti-malarial drug in a modern world: role in the treatment of malaria. Malar J 10:144. https://doi.org/10.1186/1475-2875-10-144
182. Chan MM, Adapala NS, Fong D (2005) Curcumin overcomes the inhibitory effect of nitric oxide on Leishmania. Parasitol Res 96(1):49–56
183. Mishra K, Dash AP, Swain BK, Dey N (2009) Anti-malarial activities of andrographis paniculata and Hedyotis corymbosa extracts and their combination with curcumin. Malar J 8:26. https://doi.org/10.1186/1475-2875-8-26

Protective Effects of Curcumin on Endothelium: An Updated Review

6

Mona Alidadi, Luca Liberale, Fabrizio Montecucco, Muhammed Majeed, Khalid Al-Rasadi, Maciej Banach, Tannaz Jamialahmadi, and Amirhossein Sahebkar

Abstract

Endothelial dysfunction is the common early stage of most cardiovascular afflictions. The endothelium is considered the main mediator of vascular homeostasis via its vasodilator, anti-inflammatory and anticoagulant properties. Among the different endothelial-derived mediators, nitric oxide is produced by nitric oxide synthase and has a critical role in regulating endothelial function. Physiological and pathological processes such as aging and diabetes mellitus are associated with disturbances of endothelial function which, at least at the earliest stage, can be reversed by lifestyle and pharmacological intervention to reduce the risk of incident cardiovascular diseases. Among dietary strategies, curcumin is a cheap

M. Alidadi
School of Medicine, Mashhad University of Medical Sciences, Mashhad, Iran

L. Liberale
Center for Molecular Cardiology, University of Zürich, Schlieren, Switzerland

First Clinic of Internal Medicine, Department of Internal Medicine, University of Genoa, Genoa, Italy

F. Montecucco
IRCCS Ospedale Policlinico San Martino Genoa – Italian Cardiovascular Network, Genoa, Italy

First Clinic of Internal Medicine, Department of Internal Medicine and Centre of Excellence for Biomedical Research (CEBR), University of Genoa, Genoa, Italy

M. Majeed
Sabinsa Corporation, East Windsor, NJ, USA

K. Al-Rasadi
Medical Research Centre, Sultan Qaboos University, Muscat, Oman

M. Banach
Department of Hypertension, WAM University Hospital in Lodz, Medical University of Lodz, Lodz, Poland

Polish Mother's Memorial Hospital Research Institute (PMMHRI), Lodz, Poland

T. Jamialahmadi
Biotechnology Research Center, Pharmaceutical Technology Institute, Mashhad University of Medical Sciences, Mashhad, Iran

Department of Nutrition, Faculty of Medicine, Mashhad University of Medical Sciences, Mashhad, Iran

A. Sahebkar (✉)
Biotechnology Research Center, Pharmaceutical Technology Institute, Mashhad University of Medical Sciences, Mashhad, Iran

Neurogenic Inflammation Research Center, Mashhad University of Medical Sciences, Mashhad, Iran

Halal Research Center of IRI, FDA, Tehran, Iran

Polish Mother's Memorial Hospital Research Institute (PMMHRI), Lodz, Poland
e-mail: sahebkara@mums.ac.ir

P. C. Guest (ed.), *Studies on Biomarkers and New Targets in Aging Research in Iran*, Advances in Experimental Medicine and Biology 1291, https://doi.org/10.1007/978-3-030-56153-6_6

and safe nutraceutical polyphenol with proven antioxidant and anti-inflammatory properties. Given the important role of such processes in the development of endothelium dysfunction, a role for curcumin in the prevention or treatment of this condition has been hypothesized. This review summarizes the available literature on the beneficial role of curcumin on vascular endothelial function.

Keywords

Endothelium · Endothelial dysfunction · Arterial dysfunction · Nitric oxide bioavailability · Curcumin · Curcuminoid · Atherosclerosis · FMD · Adhesion molecule

6.1 Introduction

The endothelium is the inner single layer of the vessel wall, which is directly connected with blood. It is a paracrine organ which controls the vascular tone by releasing several mediators in response to the hemodynamic forces and blood-borne signals. Furthermore it also interacts with platelets and leukocytes to modulate thrombogenicity and inflammation [1]. Also, the endothelium acts as a semi-selective barrier between all organs and blood flow to facilitate the exchange of macromolecules and blood gases [2]. Any alteration of its homeostatic regulatory properties is described as endothelial dysfunction, which can be caused by several conditions and lead to leukocyte adherence, platelet activation and activation of the coagulation cascade, release of pro-oxidant and pro-inflammatory mediators, increase of vascular permeability, vasoconstriction, and vascular smooth muscle cell proliferation [3]. Endothelial dysfunction represents the early phase and is, therefore, a sensible predictor of most types of cardiovascular disease (CVD), the leading cause of mortality in the world [4, 5]. Most of the traditional and nontraditional risk factors of CVD onset including aging, hyperlipidemia, hypertension, diabetes mellitus, metabolic syndrome, obesity, and smoking associate with endothelial dysfunction by reducing vascular nitric oxide (NO) bioavailability and ultimately lead to deterioration of vascular function [6–8]. Being the most important endothelium-derived vaso-relaxant, anti-oxidant, and anti-inflammatory factor, NO is the key mediator of endothelial homeostatic function and plays a crucial role in the modulation of vascular tone and reactivity [9]. Vascular aging is characterized by reduced expression and activity of endothelial nitric oxide synthase (eNOS), leading to decreased NO bioavailability, overproduction of reactive oxygen species (ROS), and enhanced synthesis of endothelium-derived vasoconstrictor factors such as endothelin-1 and cyclooxygenase-derived prostanoids [10–12].

In recent decades different invasive and non-invasive techniques have been described in order to assess endothelial function. Among these, the measurement of the arterial dilation in response to blood flow increase (i.e., flow-mediated dilationr, FMD) is a non-invasive ultrasound-based technique method that has shown high correlation with more invasive measurements together with good repeatability and reproducibility when assessed in a standardized manner. Accordingly, FMD is widely used as a clinical marker of vascular endothelial function with high predictive ability on CVD onset [13]. Non-functional assessment of endothelial health includes the measurement of different circulating biomarkers of endothelial activation, inflammation, and oxidative stress, including soluble intercellular adhesion molecule 1 (sICAM-1), soluble vascular cell adhesion molecule 1 (sVCAM-1), E-selectin, asymmetric dimethylarginine (ADMA), thrombomodulin (TM), tumor necrosis factor-α (TNFα), monocyte chemoattractant protein-1 (MCP-1), and C-reactive protein (CRP) [7, 14–23]. However, due to their low specificity, only some of these biomarkers have potential clinical use and none are currently included in any guidelines [7].

According to recent investigations, the use of cardiovascular protective drugs (such as the statins) and lifestyle interventions including smoking cessation, weight management, change in dietary habits, and increase in physical activity can be effective strategies to improve endothelial function and even revert the early stages of vascular dysfunction [13, 24–29]. Emerging data also suggest that dietary phytochemicals can contribute to the cardiovascular health, especially in patients not responding optimally to conventional therapies [30, 31]. Curcumin, or diferuloylmethane, is a polyphenolic compound derived from the rhizomes of *Curcuma longa*, a native plant of South and Southeast Asia, and is broadly used as dietary spice, food-coloring agent, and herbal remedy in these regions [32, 33]. Different studies suggest that curcumin holds beneficial protective effects on different human organs and systems including hepatoprotective, analgesic, lipid-lowering, reno-, cardio-, and pulmuno-protective effects. Mechanisms underlying such properties have been only partially investigated and may include antioxidant and anti-inflammatory functions as well as immunomodulatory, anti-tumor, neuroprotective and hepatoprotective effects [33–43]. Given the role of oxidative stress and inflammation in the development of endothelial dysfunction, it has been suggested that curcumin may suppress or reverse this deleterious process, thereby reducing CVD risk. The aim of the present review article is to give an overview of the available experimental and clinical evidence of the effects of curcumin supplementation on endothelial function.

6.2 In Vitro Studies and Ex Vivo Experiments with Curcumin Addiction

Several in vitro studies support a role for curcumin on the preservation of endothelial function (Table 6.1). Tham et al. investigated the effect of a synthetic curcuminoid analogue [2,6-bis-(4-hydroxyl-3-methoxybenzylidine) cyclohexanone or BHMC] on endothelial function in human umbilical vein endothelial cells (HUVECs). In this study endothelial dysfunction was induced by lipopolysaccharide (LPS) treatment and BHMC was able to blunt the expression of ICAM-1 and VCAM-1, as well as the production of MCP-1 at the transcriptional level in a concentration-dependent way [44]. Consistent with these results, many studies demonstrated that pre-treatment with curcumin significantly reduces TNFα-mediated upregulation of ICAM-1, VCAM-1, and MCP-1 as well as E selectin at both transcriptional and translational levels in HUVECs [45–47]. Also, curcumin significantly blunted the reduction in the TM protective levels due to TNFα treatment by acting at the transcriptional level [48]. Again, similar dose-dependent effects of curcumin supplementation on the expression of adhesion molecules have been obtained in another study employing a different in vitro model based on methylglyoxal (MGO) treatment of endothelial cells [49]. Furthermore, the elevation of VCAM-1 expression induced in endothelial cells by poly-L-lactic acid (PLLA) was similarly reduced by curcumin treatment. Of interest in this study, curcumin was also able to blunt PLLA-induced eNOS downregulation although levels of NO have not been directly investigated [50]. Senescent endothelial cells are characterized by a certain degree of dysfunctionality. In an in vitro model of stress-induced endothelial senescence by hydrogen peroxide treatment, pre-treatment with different concentrations of curcumin was found to partially reverse the reduction of eNOS phosphorylation and levels of NO found in the culture medium of senescent cells [51].

However, the effect of curcumin treatment on endothelial disturbances induced by different stressors has been reported to vary based on the concentration used as well as on its application pre- or post-induction of the damage. Such differences may partially explain some apparently discordant reports. In HUVECs activated by TNFα, curcumin treatment (5 μM for 20 h) reduced the MCP-1 expression significantly, although no effect was observed in the expression of VCAM-1 [52], while another study indicated that 1 h of curcumin pretreatment at different concentrations (1, 5, 10, and 25 μM) substantially diminished the production of MCP-1, ICAM-1, and VCAM-1 in TNFα-activated HUVECs [53]. Of note, most of the previously mentioned studies point at nuclear factor kappa B (NF-κB) as the intracellular mediator of cur-

Table 6.1 List of in vitro and ex vivo studies reporting the effects of curcumin supplementation on indices of endothelial function

Reference	Dose	Treatment duration	Model	Method	Main findings
[50]	5 μM	48 h	Human aortic endothelial cells	VCAM-1 eNOS	Curcumin significantly reduced poly-L-lactic acid-dependent eNOS downregulation and VCAM-1 expression
[52]	1 μM	20 h	Human cell line (EA.hy926)	VCAM-1 MCP-1	Curcumin blunted TNFα-induced expressions of MCP-1. No effect on VCAM-1 expression
[56]	0.5 μM 1 μM	3 h	HUVECs	VCAM-1 ICAM-1 MCP-1 Permeability	Curcumin prevented TNFα-dependent increase in endothelial permeability without modulating VCAM-1, ICAM-1, MCP-1expressions
[54]	1 μM 10 μM 100 μM	1 h	HUVECs	VCAM-1 ICAM-1 E-selectin	Curcumin reduced the expression of adhesion molecules at lower concentrations
[49]	10^{-7} M 10^{-6} M 10^{-5} M	48 h	HUVECs	ICAM-1	Curcumin dose-dependently attenuated methylglyoxal-induced ICAM-1 expression
[111]	10 μM	24 h	HUVECs	No, eNOS	Pretreatment with curcumin prevented endothelial senescence-associated downregulation of eNOS and NO levels in response to H_2O_2
[45]	2.5 μM 5 μM	1 h	Human cardiovascular endothelial cell line (EAhy926)	ICAM-1	Pretreatment by curcumin diminished surface expression of ICAM-1 in endothelial cells undergoing TNF-induced activation
[44]	0.8 μM 3.1 μM 12.6 μM (synthetic curcuminoid BHMC)	24 h	HUVECs	VCAM-1 ICAM-1 MCP-1	BHMC dose-dependently reduced the expression of adhesion molecules and MCP-1 stimulated by LPS
[55]	0.25 μM 0.5 μM 1 μM 2.5 μM	24 h	HUVECs	Morphology Permeability	Curcumin prevented MGO-induced endothelial hyperpermeability and ameliorated cytoskeleton rearrangement
[58]	10 μM 30 μM (curcumin and Tetrahydro-curcumin)	Acute	Aorta from male Wistar rats	Carbachol-induced EDD	Curcumin and tetrahydrocurcumin effectively ameliorated endothelial dysfunction
[110]	0.1 μM 0.5 μM 1 μM	3 h	Aorta from $ApoE^{-}/^{-}$ mice	VCAM-1, ICAM-1	Higher concentrations of curcumin downregulated the expression of VCAM-1
[112]	1 μM 5 μM 10 μM 25 μM	1 h 12 h	HUVECs	VCAM-1 ICAM-1 MCP-1 NO	Treatment with curcumin reduced the expression of MCP-1 and adhesion molecules while increasing NO production in TNFα-stimulated endothelial cells

(continued)

Table 6.1 (continued)

Reference	Dose	Treatment duration	Model	Method	Main findings
[57]	10^{-13} M 10^{12} M 10^{-11} M	2 h	Aorta from male Sprague-Dawley rats	ACh-induced EDD	Curcumin reversed high-glucose-induced vasodilation impairment
[46]	3 μM 10 μM 30 μM	1 h	HUVECs	ICAM-1, MCP-1	Curcumin attenuated the expression of ICAM-1 and MCP-1 in TNFα-stimulated HUVECs
[60]	5 μM 10 μM	24 h	Porcine coronary arteries	eNOS, Bradykinin-induced EDD	Curcumin at 10 μM notably improved eNOS expression and ameliorated vasodilation
[59]	5 μM	24 h	Porcine coronary arteries	Bradykinin-induced EDD	Curcumin significantly reversed homocysteine-induced the impairment of endothelium-dependent dilation
[47]	40 μM	1 h	HUVECs	VCAM-1 ICAM-1 E-selectin	In TNF-treated HUVECs, expression of adhesion molecules decreased with curcumin pretreatment

cumin effect in endothelial cells. An interesting recent comprehensive study evaluated the potential protective role of curcumin against endothelial dysfunction induced by polluting agents such as particulate material with an aerodynamic diameter of ≤10 μm (PM10) and titanium dioxide nanoparticles (TiO_2-NPs) [54]. In this study, pretreatment with curcumin could blunt PM10- and TiO_2-NPs-induced endothelial expression of adhesion molecules and monocyte adhesion to the endothelial layer. Also curcumin was able to reduce the oxidative stress due to the polluting agents as assed by 2,7-dichlorodihydrofluorescein (H2DCF) oxidation [54].

Functionally, pretreatment of HUVECs with curcumin was reported to prevent MGO-induced endothelial hyperpermeability and improve cytoskeleton rearrangement through reduced actin cytoskeletal rearrangement and increased gap junctional intercellular communication via connexin [45] and zonulin-1 expression [55]. Consistent with this study, TNFα-induced hyperpermeability was found to be decreased after pretreatment of HUVECs with curcumin [56]. Also, the protective effects of curcumin on NO production under stress conditions have been confirmed to have functional relevance in different studies which added curcumin in the media of myograph experiments on aortic rings [57, 58]. Ex vivo incubation of aortic rings in the presence of high glucose was shown to impair acetylcholine (Ach)-induced endothelial-dependent dilation (EDD), thus mirroring endothelial dysfunction induced by diabetes mellitus. Here, the addition of curcumin at different concentration in the high glucose media was able to improve ACh-induced EDD in a dose-dependent manner. Sodium nitroprusside (SNP)–induced dilation did not change, thereby proving the endothelial-specific effects [57]. These results suggest a possible role for curcumin supplementation in preventing or ameliorating diabetes-induced endothelial dysfunction and deserve further investigation in clinical trials. In humans, hyper-homocysteinemia is associated with impaired endothelial function and is thought to increase the risk of CVD. A study investigated the putative role of curcumin in blunting such a deleterious effect by assessing bradykinin-dependent EDD in porcine coronary arteries incubated with control media, homocysteine, curcumin, and homocysteine. As expected, arteries treated with homocysteine showed a significant reduced vasodilatory ability while curcumin treatment could blunt homocysteine-induced impairment of EDD. Again, SNP experiments were used to show the endothelial-dependent effect [59]. Of interest, functional studies also demonstrated a role for curcumin in reducing impaired EDD due to exogenous chemicals such as the human immunodeficiency virus (HIV) protease inhibitor ritonavir [60].

6.3 Animal Studies

Curcumin supplementation was shown to have beneficial effects on many animal models of disease including diabetes mellitus, metabolic syndrome, hypertension, and myocardial infarction. Some of those effects were mediated by its protective role on endothelial function as detailed in this chapter (Table 6.2).

Age is among the most important determinant of endothelial dysfunction, eNOS activation and NO levels are importantly reduced in aged arteries, and this is often accompanied with increased circulating levels of vascular activation biomarkers. Fleenor et al. investigated whether curcumin diet supplementation (4 weeks) could prevent age-dependent endothelial dysfunction [61]. In their study, curcumin supplementation restored carotid EDD of old animals to levels of young animals by acting on vascular redox status as assessed by arterial nitro-tyrosine levels, superoxide production, NADPH oxidase p67 subunit, and manganese superoxide dismutase expression [61]. Endothelial dysfunction is generally considered the earliest step of arterial stiffening, a recognized CV risk factor. In this study, curcumin supplementation also ameliorated age-associated large elastic artery stiffening as assessed by aortic pulse wave velocity (aPWV) by reducing advanced glycation end-products (AGEs) and collagen aortic contents and preserving vascular elastin levels [61].

High glucose and carbohydrate, as found in the Western diet, are known to induce endothelial dysfunction in both animals and humans. A recent study showed that curcumin has the potential to revert vascular activation and inflammation induced by such deleterious dietary regimens. Here, Sprague-Dawley male rats were fed on high sucrose, high-fat diet (HSF diet), for 8 weeks to induce obesity and vascular dysfunction. In a second phase, rats were fed with control or curcumin-supplemented diet for 4 weeks. At the end of the study, levels of CRP, ICAM-1, and VCAM-1 in the curcumin-treated group were significantly decreased by 14.3%, 11.6%, and 21.8%, respectively, compared to control group. Also, NO levels significantly improved by 118.7% in the curcumin group compared to HSF diet group [62]. Similarly, in another endothelial dysfunction model induced by diabetes, curcumin administration at the dose of 300 mg/kg for 12 weeks improved the expression of ICAM-1 compared to the untreated group [63].

Systemic inflammation has been recently identified as an important risk factor for the development of CVD. Circulating cytokines and immune cells can activate the endothelium increasing its expression of ROS, procoagulants, and inflammatory mediators, thus further fueling the process and causing endothelial dysfunction. In a study conducted by Zhang et al., male C57BL/6 mice were treated with a single dose of TNFα to induce vascular inflammation after 2-week feeding with standard diet (SD), SD + curcumin, SD + lutein (a carotenoid with proven anti-oxidative effects), or SD + curcumin + lutein [52]. At the end of the experiment, curcumin and lutein alone did not significantly abolish TNFα-induced expression of VCAM-1 and MCP-1, but the combination of curcumin and lutein synergistically diminished the expression of VCAM-1 and MCP-1 as compared to the control groups [52]. Consistent with this study, Rana et al. showed that pretreatment with curcumin at the dosage of 100 mg/kg and 300 mg/kg ameliorated lipopolysaccharide (LPS)-induced vascular endothelial dysfunction compared to LPS alone, with a dose-dependent effect [64]. Different CV risk factors are known to induce a chronic low-grade systemic inflammation at the onset of endothelial dysfunction and vascular disease. In a rat model of metabolic syndrome (MetS), curcumin administration in low, middle, and high doses led to a dose-dependent reduction in the circulating levels of TNFα and associated with better lipid and glucose profiles as well as reduced intima/media thickness [65]. Mechanistically, inducible NOS (iNOS) was found to be responsible for the protective effect [65]. Recently, another study reinforced these

Table 6.2 List of animal models reporting the effects of curcumin supplementation on indices of endothelial function

Reference	Dose	Duration	Population	Method	Main finding
[52]	500 mg/kg	2 weeks	Male C57BL/6 mice	VCAM-1 MCP-1	Curcumin did not significantly inhibit TNFα-induced expression of VCAM-1 and MCP-1
[65]	Low dose Middle dose High dose	4 weeks	Sprague-Dawley rats	TNFα eNOS	Curcumin decreased the level of TNFα and increased eNOS expression
[62]	300 mg/kg	4 weeks	Male Sprague-Dawley rats	ICAM-1, VCAM-1 CRP NO	Treatment with curcumin blunted Western diet-induced effects on ICAM-1, VCAM-1, CRP, and NO
[70]	100 mg/kg	4 weeks	Stroke-prone spontaneously hypertensive rats	ACh-induced EDD, NO	Curcumin reduced stroke incidence and increased rat survival by improving NO level and endothelial-dependent relaxation of carotid artery
[71]	50 mg/kg	10 weeks	Male Wistar rats	ACh-induced EDD eNOS	Curcumin increased eNOS expression and ameliorated endothelial dysfunction
[67]	200 mg/kg	3 days	Male wild-type C57BL/6 mice	IL-6 TNFα	TNF- α and IL-6 exhibited significantly lower circulating levels in curcumin-treated mice after zymosan injection
[77]	120 mg/kg	4 weeks	Male Wistar rats	eNOS MCP-1	Curcumin suppressed the overexpression of MCP-1 and increased the expression of eNOS in a rat model of chronic kidney disease
[74]	100 mg 200 mg 400 mg	4 weeks	Male Wistar rats	ACh-induced EDD	Curcumin supplementation at two higher dosages blunted methotrexate-induced endothelial dysfunction
[75]	300 mg/kg (curcuma oil)	4 weeks (hamsters) 2 weeks (rats)	Male Golden Syrian hamsters and Wistar rats	ACh-induced EDD IL-6 CRP	Curcumin blunted endothelial dysfunction induced by high fructose diet in hamsters and rats. CRP and IL-6 levels decreased significantly in rats supplemented with curcumin
[64]	100 mg/kg 300 mg/kg	10 days	Male Swiss albino mice	ACh-induced EDD, IL-6, TNFα	Curcumin dose-dependently prevented LPS-induced endothelial dysfunction and reduced IL-6 and TNFα circulating levels

(continued)

Table 6.2 (continued)

Reference	Dose	Duration	Population	Method	Main finding
[63]	300 mg/kg	12 weeks	Male Wistar rats	ICAM-1	Curcumin reduced ICAM-1 expression in diabetes-related endothelial dysfunction
[69]	50 mg/kg 100 mg/kg	6 weeks	Male Sprague-Dawley rats	ACh-induced EDD, eNOS	Curcumin effectively increased endothelial-dependent vasorelaxation in a dose-dependent manner by upregulating eNOS expression
[78]	200 mg/kg	15 days	Male Sprague-Dawley rats	Morphological and ultrastructural alterations	Curcumin prevented cyclosporin A-induced endothelial dysfunction
[68]	50 mg/kg 100 mg/kg (tetrahydrocurcumin)	8 weeks	Male ICR mice	eNOS	Curcumin supplementation increased eNOS expression in cadmium-induced hypertension model
[61]	Curcumin-supplemented food at a dose (0.2%)	4 weeks	Young and old male C57BL6/N mice	ACh-induced EDD eNOS	Curcumin supplementation improved age-associated arterial stiffness, endothelial dysfunction, oxidative stress, and increases in collagen and AGEs
[76]	15 mg/kg 30 mg/kg 60 mg/kg	4 weeks	Male Sprague-Dawley rats	NO	Renal NO levels were increased by curcumin supplementation in fructose-fed rats
[73]	30 mg/kg 300 mg/kg	12 weeks	Male Wistar rats	ACh-induced EDD	Curcumin dose-dependently improved endothelial dysfunction in STZ-induced diabetic rats

findings by showing that curcumin could blunt the exaggerated vasoconstriction in a rat model of MetS through reduced AGEs formation and endothelial dysfunction [66]. Again, ROS scavenging together with direct vasodilatory effects were reported as responsible for curcumin's beneficial effects [66]. Systemic inflammation reaches its maximal expression during multiple organ dysfunction syndrome (MODS). In a mouse model of zymosan-induced MODS, 3 days of curcumin treatment (200 mg/kg) was able to reduce the levels of systemic inflammation (circulating TNFα and IL-6) as compared to the control group [67]. Whether such an effect was at least partially mediated by the effect of curcumin on vascular endothelium was not specifically investigated.

Endothelial dysfunction was reported to precede and accompany arterial hypertension development. In a mouse model of hypertension induced by cadmium administration, curcumin at the dose of 50 or 100 mg/kg significantly ameliorated hypertension, endothelium dysfunction, and arterial stiffness through increased eNOS and NO bioavailability [68]. In addition, 6 weeks of curcumin treatment improved the hemodynamic performance in a rat model of hypertension by reducing arterial pressure and facilitating blood flow through reduced vascular resistance [69]. Also, EED in response to acetylcholine was significantly improved in aortic rings isolated from the hypertensive rats treated with curcumin. Again, the curcumin effect was endothelial-specific as it did not affect the endothelial-independent response induced by SNP. In another study, daily curcumin administration in stroke-prone spontaneously hypertensive rats was able to significantly delay the onset of stroke and increase the animal survival rate by blunting oxidative stress and improving endothelial dependent relaxation of carotid arteries [70]. Also, the curcumin-mediated ROS reduction and NO increase were blocked by the inhibition of a mitochondrial uncoupling protein, thus underlining the importance of such a process in mediating the vascular effects of curcumin [70].

Several other studies have also shown that treatment with curcumin increased ACh-induced endothelium-dependent vasorelaxation in vascular damage animal models, including rats, mice, and hamsters [61, 71–75] (Table 6.2). Curcumin supplementation at the dose of 15, 30, or 60 mg/kg for 4 weeks dose-dependently improved NO production and endothelial dysfunction in a rat model of hyperuricemia and renal endothelial dysfunction [76]. Consistently, Tapia et al. reported that curcumin treatment at the dose of 120 mg/kg for 4 weeks suppressed the overexpression of MCP-1 and increased the expression of eNOS in renal cortex of chronic kidney disease rat model [77]. Finally, a protective effect of curcumin supplementation was also observed for cyclosporin A-induced endothelial dysfunction in rats [78].

6.4 Clinical Trials

The above-mentioned promising results obtained in the experimental setting prompted the design of different randomized clinical trials (RCTs) aimed at assessing the effects of curcumin intake on vascular function. Curcumin is safe and has a good tolerability and toxicity profile in humans. Indeed even high doses (up to 12 g/day) are well tolerated and only minor side effects have been reported, namely, diarrhea. To date, seven RCTs (Table 6.3) have assessed endothelial function in patients after curcumin-based dietary intervention [79–85]. In one double-blind, randomized, controlled trial, 59 apparently healthy young individuals were randomized to receive capsules containing placebo or curcumin at 250 or 1000 mg/day. After 8 weeks, curcumin supplementation resulted in a dose-dependent improvement of endothelial function as assessed by FMD. Although a trend was reported for the lower dosage, only the higher supplementation strategy was able to significantly increase FMD although only by 3.0% with respect to placebo [83]. In another randomized, double-blind, controlled trial, 39 healthy middle-aged men and postmenopausal women were randomized to receive curcumin at the high dose of 2 g/day or placebo. In this study, endothelial function was assessed by both FMD and forearm blood flow response to acetylcholine infusions (FBFAch) before and at the end of the 12 weeks of intervention. Curcumin treatment enhanced brachial artery FMD and FBF Ach (36% and 37%, respectively), whereas FBF response to SNP did not change, confirming the endothelial-specific effects of curcumin [84]. Of interest, in this study curcumin treatment partially reverted the acute reduction in FBFACh induced by treatment with a NO synthase inhibitor, indicating that improvements in artery endothelial function were mediated by an increase in NO bioavailability [84]. Yet, curcumin supplementation did not modulate circulating biomarkers of oxidative stress and inflammation such as oxidized lipoprotein, CRP, and TNFα, supporting the case that these may not reflect the local vascular endothelial state [84]. Another RCT enrolled postmenopausal women and randomized them to 3 groups: control, moderate aerobic exercise, and curcumin supplementation (150 mg/day). After 8 weeks, FMD increased significantly and equally in both the exercise and curcumin groups as compared to placebo [79].

In contrast to the studies mentioned above, other trials failed to show a protective effect of curcumin on endothelial dysfunction. In young obese men, a 12-week intervention with an enhanced bioavailable curcumin formulation at the dose of 500 mg/day showed an improvement

Table 6.3 List of clinical trials reporting the effects of curcumin supplementation on indices of endothelial function

Reference	Study design	Control group	Dose	Duration	Population	Number	Method	Main finding
[82]	Randomized, controlled, double-blind	Mix of corn, olive oil, and 20 mg fish oil	160 mg/day	16 weeks	Older sedentary overweight/obese adults	74	CRP, transcranial Doppler ultrasound	No changes observed between groups
[85]	Randomized, controlled, double-blind	Placebo capsules	1200 mg/day	3 months	Type 2 diabetic patients	88	ICAM-1, VCAM	There were no statistical demonstrable differences between the groups
[80]	Randomized, crossover, double-blind	Microcrystalline cellulose	5 g (single dose)	2 h	Healthy smokers	18	FMD, RHI	RHI and FMD did not significantly change
[84]	Randomized, controlled, double-blind	Placebo capsules	2 g/day	12 weeks	Healthy middle-aged men and postmenopausal women	39	FMD, FBF	FMD and FBF_{Ach} improved in the curcumin group (36% and 37%, respectively)
[83]	Randomized, controlled, double-blind	Corn starch	250 mg/day 1000 mg/day	8 weeks	Apparently healthy adults	59	FMD	Curcumin dose-dependently improved FMD
[81]	Randomized, controlled, double-blind	Fenugreek fiber	500 mg/day	12 weeks	Young, obese men	22	RHI	No significant improvement was observed
[79]	Randomized, controlled, double-blind	Placebo	150 mg/day	8 weeks	Postmenopausal women	21	FMD	FMD improved in the curcumin group

in homocysteine and high-density lipoprotein (HDL) levels but did not affect blood pressure and the reactive hyperemia index (RHI), a measure of endothelial function [81]. Recently, another RCT investigated curcumin supplementation (160 mg/day) on cerebrovascular function of overweight/obese adults and failed to show any effect of such intervention on cerebrovascular responsiveness to hypercapnia, carotid arterial stiffness blood pressure, or markers of inflammation [82]. Similarly, in an Indian cohort of diabetes mellitus patients, 1200 mg/day curcumin supplementation for 3 months was able to reduce different measurements of arterial stiffness despite non-significant changes in the levels of ICAM-1 and VCAM [85]. Unfortunately, no functional technique to assess endothelial function has been employed in this RCT. Therefore, the authors could not draw clear conclusion on this aspect [85]. Lastly, in healthy smokers, the acute administration of curcumin (5 g) 2 h before RHI and FMD measurements did not lead to any improvement in endothelial function [80]. Yet, a subgroup analysis based on gender and cardiovascular risk score showed a significant effect of curcumin on FMD in women and individuals with lower CVD risk [80].

The apparently contrasting results may result from different explanations, including the use of different cohorts and different techniques employed to estimate endothelial function. Also, the dosage and therapeutic scheme of curcumin supplementation differed among the seven RCTs. In addition, All RCTs enrolled a low number of patients. Therefore larger studies employing state-of-art techniques for endothelial function assessment and involving heterogeneous populations are needed to allow for solid subgroup analysis and to draw solid conclusions.

6.5 Mechanisms of Action

Although the exact mechanism by which curcumin improves vascular endothelial function still remains unclear, all data point toward the antioxidant and anti-inflammatory properties of this polyphenol as mediators of its effect [86]. Of interest, such effects maybe mediated by different ROS-generating and scavenging enzymes. As mentioned previously, eNOS is one of the vital molecules involved in the regulation of endothelium-dependent dilation. Several studies have indicated that curcumin improved vascular function by upregulating eNOS expression or functionality in endothelium cells [60, 61, 71–75]. Other scavenging enzymes such as manganese superoxide dismutase (MnSOD) may partially account for curcumin effects on endothelial function [61]. Also, the available data have led to the proposal that curcumin may directly scavenge free radicals [59–61]. As is the case for other polyphenols [87], curcumin may act through the increased synthesis and function of the deacetylase sirtuin 1 (SIRT1). Indeed, the antioxidant effects of SIRT1 are known and well described [12, 88–92]. In an endothelial dysfunction model induced by H_2O_2, curcumin repressed the negative effect of H_2O_2 on eNOS phosphorylation by improving protein expression and enzymatic activity of SIRT1 [51]. Lastly, the heme oxygenase-1 (HO-1) pathway is probably one of the mechanisms through which curcumin acts on endothelial dysfunction. HO-1, as an inducible isoform of HO, is a powerful antioxidant enzyme that catalyzes the degradation of heme [93, 94]. HO-1 plays a critical role in preventing the expression of adhesion molecules, and subsequently vascular inflammation [93, 95, 96]. Curcumin increases HO-1 activity dose-dependently [57, 97, 98], acting as a potent stimulator [97].

NADPH oxidase (NOX) is among the most important enzymes involved in ROS generation. During endothelial dysfunction, NOX is often upregulated in the vascular tissue, leading to increased oxidative stress [99–101]. Different studies suggest that curcumin attenuates the level of oxidative stress by reducing the expression of NOX subunits (p67, gp91phox, and gp47phox) or by directly affecting their activity [61, 69, 99, 102, 103].

Oxidative stress interacts with inflammation to initiate a vascular vicious circle that drives the onset endothelial dysfunction [104–106]. As shown in Fig. 6.1, the transcription factor NF-κB is a main hub of this interaction being directly stimulated by

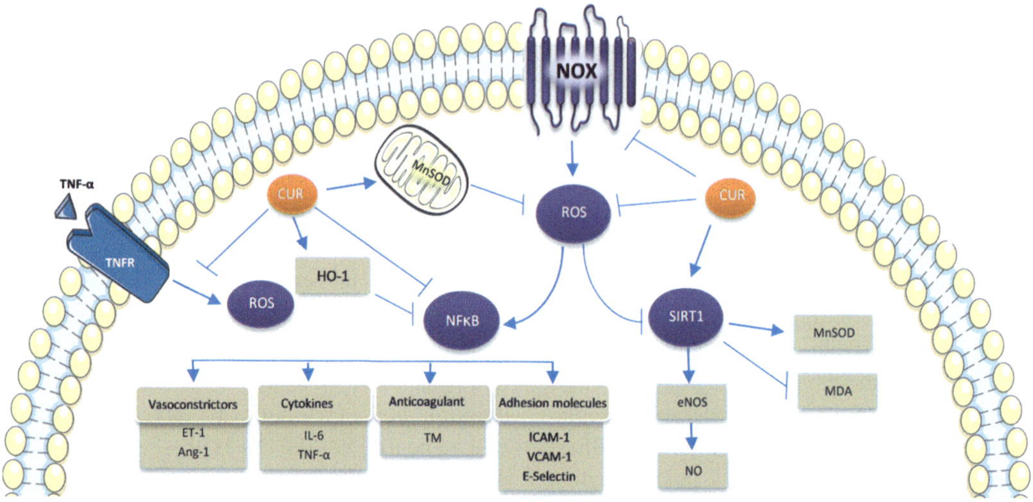

Fig. 6.1 Mechanisms of action of curcumin to improve endothelial dysfunction. *Ang-1* angiotensin 1; *CUR* curcumin, *eNOS* endothelial nitric oxide synthase, *ET-1* endothelin 1, *HO-1* heme oxygenase 1, *ICAM-1* intercellular adhesion molecule 1, *IL* interleukin, *MDA* malondialdehyde, *MnSOD* manganese superoxide dismutase, *NO* nitric oxide, *NOX* NADPH oxidase, *ROS* reactive oxygen species, *SIRT* sirtuin 1, *TM* thrombomodulin, *TNFα* tumor necrosis factor α, *TNFR* TNFα receptor, *VCAM-1* vascular cell adhesion molecule 1

ROS and regulating the expression of pro-oxidants, pro-coagulants, and pro-inflammatory cytokines such as TNFα [107–109]. Different studies have shown that curcumin improves TNFα-induced vascular damage, by suppressing the expression of NF-κB [47, 52, 67, 110].

6.6 Conclusions

Given that endothelium dysfunction is an initial sign of CVD, interventions improving endothelial function may help to prevent this condition. Among the dietary interventions which have shown promising results in improving the functionality of the inner layer of blood vessels, the accessible and safe nutraceutical curcumin has gained much attention in the last decades. Preclinical studies have demonstrated the efficacy of curcumin on vascular endothelial function preservation through its antioxidant and anti-inflammatory effects. Yet, the clinical evidence from RCTs remains weak and this is mirrored by the absence of indication for curcumin supplementation in CVD prevention guidelines. Although the scientific evidence suggests a potential role for curcumin in preventing endothelial dysfunction, adequate clinical trials are required to confirm this hypothesis and pave the way for this inexpensive and safe compound in the field of CVD prevention.

References

1. Aversa A, Bruzziches R, Francomano D, Greco EA, Migliaccio S, Lenzi A (2010) The role of steroids in endothelial function in the ageing male. Eur Endocrinol 7(2):115–118
2. Verma S, Buchanan MR, Anderson TJ (2003) Endothelial function testing as a biomarker of vascular disease. Circulation 108(17):2054–2059
3. Verma S, Anderson TJ (2002) Fundamentals of endothelial function for the clinical cardiologist. Circulation 105(5):546–549
4. World Health Organization. Cardiovascular diseases (CVDs). https://www.who.int/news-room/fact-sheets/detail/cardiovascular-diseases-(cvds)#

5. Barthelmes J, Nägele MP, Ludovici V, Ruschitzka F, Sudano I, Flammer AJ (2017) Endothelial dysfunction in cardiovascular disease and Flammer syndrome-similarities and differences. EPMA J 8(2):99–109
6. Bonetti PO, Lerman LO, Lerman A (2003) Endothelial dysfunction: a marker of atherosclerotic risk. Arterioscler Thromb Vasc Biol 23(2):168–175
7. Park KH, Park WJ (2015) Endothelial dysfunction: clinical implications in cardiovascular disease and therapeutic approaches. J Korean Med Sci 30(9):1213–1225
8. Gimbrone MA Jr, García-Cardeña G (2016) Endothelial cell dysfunction and the pathobiology of atherosclerosis. Circ Res 118(4):620–636
9. Levine AB, Punihaole D, Levine TB (2012) Characterization of the role of nitric oxide and its clinical applications. Cardiology 122(1):55–68
10. Herrera MD, Mingorance C, Rodríguez-Rodríguez R, de Sotomayor MA (2010) Endothelial dysfunction and aging: an update. Ageing Res Rev 9(2):142–152
11. Smith AR, Visioli F, Hagen TM (2006) Plasma membrane-associated endothelial nitric oxide synthase and activity in aging rat aortic vascular endothelia markedly decline with age. Arch Biochem Biophys 454(1):100–105
12. Liberale L, Camici GG (2012) The role of vascular aging in atherosclerotic plaque development and vulnerability. Curr Pharm Des 25(29):3098–3111
13. Zhang L, Gong D, Li S, Zhou X (2012) Meta-analysis of the effects of statin therapy on endothelial function in patients with diabetes mellitus. Atherosclerosis 223(1):78–85
14. Dzikowska-Diduch O, Domienik-Karłowicz J, Górska E, Demkow U, Pruszczyk P, Kostrubiec M (2017) E-selectin and sICAM-1, biomarkers of endothelial function, predict recurrence of venous thromboembolism. Thromb Res 157:173–180
15. Fang Y, Li C, Shao R, Yu H, Zhang Q (2018) The role of biomarkers of endothelial activation in predicting morbidity and mortality in patients with severe sepsis and septic shock in intensive care: a prospective observational study. Thromb Res 171:149–154
16. Martínez-Urbistondo D, Beltrán A, Beloqui O, Huerta A (2016) The neutrophil-to-lymphocyte ratio as a marker of systemic endothelial dysfunction in asymptomatic subjects. Nefrología 36(4):397–403
17. Oyama J, Nagatomo D, Yoshioka G, Yamasaki A, Kodama K, Sato M et al (2016) The relationship between neutrophil to lymphocyte ratio, endothelial function, and severity in patients with obstructive sleep apnea. J Cardiol 67(3):295–302
18. Wang H, Luo W, Wang J, Guo C, Wang X, Wolffe SL et al (2012) Obesity-induced endothelial dysfunction is prevented by deficiency of P-selectin glycoprotein ligand-1. Diabetes 61(12):3219–3227
19. Xue J, Huang W, Chen X, Li Q, Cai Z, Yu T et al (2017) Neutrophil-to-lymphocyte ratio is a prognostic marker in acute ischemic stroke. J Stroke Cerebrovasc Dis 26(3):650–657
20. Santos HO, Izidoro LFM (2018) Neutrophil-lymphocyte ratio in cardiovascular disease risk assessment. Int J Cardiovasc Sci 31(5):532–537
21. Sen N, Afsar B, Ozcan F, Buyukkaya E, Isleyen A, Akcay AB et al (2013) The neutrophil to lymphocyte ratio was associated with impaired myocardial perfusion and long term adverse outcome in patients with ST-elevated myocardial infarction undergoing primary coronary intervention. Atherosclerosis 228(1):203–210
22. Balta S, Celik T, Mikhailidis DP, Ozturk C, Demirkol S, Aparci M et al (2016) The relation between atherosclerosis and the neutrophil–lymphocyte ratio. Clin Appl Thromb Hemost 22(5):405–411
23. Devaraj S, Kumaresan PR, Jialal I (2011) C-reactive protein induces release of both endothelial microparticles and circulating endothelial cells in vitro and in vivo: further evidence of endothelial dysfunction. Clin Chem 57(12):1757–1761
24. Vafeiadou K, Weech M, Sharma V, Yaqoob P, Todd S, Williams CM et al (2012) A review of the evidence for the effects of total dietary fat, saturated, monounsaturated and n-6 polyunsaturated fatty acids on vascular function, endothelial progenitor cells and microparticles. Br J Nutr 107(3):303–324
25. Desch S, Sonnabend M, Niebauer J, Sixt S, Sareban M, Eitel I et al (2010) Effects of physical exercise versus rosiglitazone on endothelial function in coronary artery disease patients with prediabetes. Diabetes Obes Metab 12(9):825–828
26. Gokce N, Vita JA, McDonnell M, Forse AR, Istfan N, Stoeckl M et al (2005) Effect of medical and surgical weight loss on endothelial vasomotor function in obese patients. Am J Cardiol 95(2):266–268
27. Johnson HM, Gossett LK, Piper ME, Aeschlimann SE, Korcarz CE, Baker TB et al (2010) Effects of smoking and smoking cessation on endothelial function: 1-year outcomes from a randomized clinical trial. J Am Coll Cardiol 55(18):1988–1995
28. Konduracka E, Galicka-Latala D, Cieslik G, Rostoff P, Fedak D, Sieradzki J et al (2008) Effect of atorvastatin on endothelial function and inflammation in long-duration type 1 diabetic patients without coronary heart disease and arterial hypertension. Diabetes Obes Metab 10(9):719–725
29. Economides PA, Caselli A, Tiani E, Khaodhiar L, Horton ES, Veves A (2004) The effects of atorvastatin on endothelial function in diabetic patients and subjects at risk for type 2 diabetes. J Clin Endocrinol Metab 89(2):740–747
30. Alidadi M, Jamialahmadi T, Cicero AFG, Bianconi V, Pirro M, Banach M, Sahebkar A (2020) The potential role of plantderived natural products in improving arterial stiffness: A review of dietary intervention studies. Trends in Food Science and Technology 99:426–440
31. Banach M, Patti AM, Giglio RV, Cicero AFG, Atanasov AG, Bajraktari G, Bruckert E, Descamps O, Djuric DM, Ezhov M, Fras Z, von Haehling S, Katsiki N, Langlois M, Latkovskis G, Mancini

GBJ, Mikhailidis DP, Mitchenko O, Moriarty PM, Muntner P, Nikolic D, Panagiotakos DB, Paragh G, Paulweber B, Pella D, Pitsavos C, Reiner Ž, Rosano GMC, Rosenson RS, Rysz J, Sahebkar A, Serban MC, Vinereanu D, Vrablík M, Watts GF, Wong ND, Rizzo M (2018) International Lipid Expert Panel (ILEP). The Role of Nutraceuticals in Statin Intolerant Patients. J Am Coll Cardiol 72(1):96–118. https://doi.org/10.1016/j.jacc.2018.04.040. PMID: 29957236
32. Chuengsamarn S, Rattanamongkolgul S, Luechapudiporn R, Phisalaphong C, Jirawatnotai S (2012) Curcumin extract for prevention of type 2 diabetes. Diabetes Care 35(11):2121–2127
33. Aggarwal BB, Kumar A, Aggarwal MS, Shishodia S (2005) Curcumin derived from turmeric (Curcuma longa): a spice for all seasons. Phytopharm Cancer Chemoprevent 23:351–387
34. Miriyala S, Panchatcharam M, Rengarajulu P (2007) Cardioprotective effects of curcumin. Adv Exp Med Biol 595:359–377
35. Forouzanfar F, Read MI, Barreto GE, Sahebkar A (2020) Neuroprotective effects of curcumin through autophagy modulation. IUBMB Life 72(4):652–664. https://doi.org/10.1002/iub.2209. Epub 2019 Dec 5. PMID: 31804772
36. Abdollahi E, Momtazi AA, Johnston TP, Sahebkar A (2018) Therapeutic effects of curcumin in inflammatory and immune-mediated diseases: a nature-made jack-of-all-trades? J Cell Physiol 233(2):830–848
37. Iranshahi M, Sahebkar A, Takasaki M, Konoshima T, Tokuda H (2009) Cancer chemopreventive activity of the prenylated coumarin, umbelliprenin, in vivo. Eur J Cancer Prev 18(5):412–415
38. Hasanzadeh S, Read MI, Bland AR, Majeed M, Jamialahmadi T, Sahebkar A (2020) Curcumin: an inflammasome silencer. Pharmacol Res 159:104921. https://doi.org/10.1016/j.phrs.2020.104921. Epub 2020 May 25. PMID: 32464325
39. Mollazadeh H, Cicero AFG, Blesso CN, Pirro M, Majeed M, Sahebkar A (2019) Immune modulation by curcumin: the role of interleukin-10. Crit Rev Food Sci Nutr 59(1):89–101
40. Momtazi AA, Sahebkar A (2016) Difluorinated curcumin: a promising curcumin analogue with improved anti-tumor activity and pharmacokinetic profile. Curr Pharm Des 22(28):4386–4397
41. Panahi Y, Kianpour P, Mohtashami R, Jafari R, Simental-Mendía LE, Sahebkar A (2017) Efficacy and safety of phytosomal curcumin in non-alcoholic fatty liver disease: a randomized controlled Trial. Drug Res 67(4):244–251
42. Teymouri M, Pirro M, Johnston TP, Sahebkar A (2017) Curcumin as a multifaceted compound against human papilloma virus infection and cervical cancers: a review of chemistry, cellular, molecular, and preclinical features. BioFactors 43(3):331–346
43. Momtazi AA, Derosa G, Maffioli P, Banach M, Sahebkar A (2016) Role of microRNAs in the Therapeutic Effects of Curcumin in Non-Cancer Diseases. Molecular Diagnosis and Therapy 20(4): 335–345.
44. Tham CL, Hazeera Harith H, Wai Lam K, Joong Chong Y, Singh Cheema M, Roslan Sulaiman M et al (2015) The synthetic curcuminoid BHMC restores endotoxin-stimulated HUVEC dysfunction: Specific disruption on enzymatic activity of p38 MAPK. Eur J Pharmacol 749:1–11
45. Kam A, Li KM, Razmovski-Naumovski V, Nammi S, Chan K, Grau GE et al (2015) Curcumin reduces tumour necrosis factor-enhanced annexin v-positive microparticle release in human vascular endothelial cells. J Pharm Pharm Sci 18(4):424–433
46. Kim YS, Ahn Y, Hong MH, Joo SY, Kim KH, Sohn IS et al (2006) Curcumin attenuates nuclear factor-κB, c-Jun N-terminal kinase and p38 in tumor necrosis factor-α-stimulated endothelial cells. Korean Circulation Journal 36(7):482–489
47. Kumar A, Dhawan S, Hardegen NJ, Aggarwal BB (1998) Curcumin (diferuloylmethane) inhibition of tumor necrosis factor (TNF)-mediated adhesion of monocytes to endothelial cells by suppression of cell surface expression of adhesion molecules and of nuclear factor-κB activation. Biochem Pharmacol 55(6):775–783
48. Nan B, Lin P, Lumsden AB, Yao Q, Chen C (2005) Effects of TNF-α and curcumin on the expression of thrombomodulin and endothelial protein C receptor in human endothelial cells. Thromb Res 115(5):417–426
49. Sun YP, Gu JF, Tan XB, Wang CF, Jia XB, Feng L et al (2016) Curcumin inhibits advanced glycation end product-induced oxidative stress and inflammatory responses in endothelial cell damage via trapping methylglyoxal. Mol Med Rep 13(2):1475–1486
50. Chen D, Weng L, Chen C, Zheng J, Wu T, Zeng S et al (2019) Inflammation and dysfunction in human aortic endothelial cells associated with poly-l-lactic acid degradation in vitro are alleviated by curcumin. J Biomed Mater Res A 107(12):2756–2763
51. Usui F, Shirasuna K, Kimura H, Tatsumi K, Kawashima A, Karasawa T et al (2015) Inflammasome activation by mitochondrial oxidative stress in macrophages leads to the development of angiotensin II-induced aortic aneurysm. Arterioscler Thromb Vasc Biol 35(1):127–136
52. Zhang L, Wang X, Zhang L, Virgous C, Si H (2019) Combination of curcumin and luteolin synergistically inhibits TNF-α-induced vascular inflammation in human vascular cells and mice. J Nutr Biochem 73:108222. https://doi.org/10.1016/j.jnutbio.2019.108222
53. Kim YS, Ahn Y, Hong MH, Joo SY, Kim KH, Sohn IS et al (2007) Curcumin attenuates inflammatory responses of TNF-alpha-stimulated human endothelial cells. J Cardiovasc Pharmacol 50(1):41–49
54. Montiel-Dávalos A, Silva Sánchez GJ, Huerta-García E, Rueda-Romero C, Chafre GS, Mitre-Aguilar IB, Alfaro-Moreno E, Pedraza-Chaverri

J, López-Marure R (2017) Curcumin inhibits activation induced by urban particulate material or titanium dioxide nanoparticles in primary human endothelial cells. PLoS One 12(12)
55. Hu TY, Liu CL, Chen JY, Hu ML (2013) Curcumin ameliorates methylglyoxal-induced alterations of cellular morphology and hyperpermeability in human umbilical vein endothelial cells. J Funct Foods 5(2):745–754
56. Monfoulet LE, Mercier S, Bayle D, Tamaian R, Barber-Chamoux N, Morand C et al (2017) Curcumin modulates endothelial permeability and monocyte transendothelial migration by affecting endothelial cell dynamics. Free Radic Biol Med 112:109–120
57. Fang XD, Yang F, Zhu L, Shen YL, Wang LL, Chen YY (2009) Curcumin ameliorates high glucose-induced acute vascular endothelial dysfunction in rat thoracic aorta. Clin Exp Pharmacol Physiol 36(12):1177–1182
58. Tep-Areenan P, Suksamrarn A (2012) Curcumin and tetrahydrocurcumin restore the impairment of endothelium-dependent vasorelaxation induced by homocysteine thiolactone in rat aortic rings. Int J Pharmacol 8(2):128–133
59. Ramaswami G, Chai H, Yao Q, Lin PH, Lumsden AB, Chen C (2004) Curcumin blocks homocysteine-induced endothelial dysfunction in porcine coronary arteries. J Vasc Surg 40(6):1216–1222
60. Chai H, Yan S, Lin P, Lumsden AB, Yao Q, Chen C (2005) Curcumin blocks HIV protease inhibitor ritonavir-induced vascular dysfunction in porcine coronary arteries. J Am Coll Surg 200(6):820–830
61. Fleenor BS, Sindler AL, Marvi NK, Howell KL, Zigler ML, Yoshizawa M et al (2013) Curcumin ameliorates arterial dysfunction and oxidative stress with aging. Exp Gerontol 48(2):269–276
62. Tsai IJ, Chen CW, Tsai SY, Wang PY, Owaga E, Hsieh RH (2018) Curcumin supplementation ameliorated vascular dysfunction and improved antioxidant status in rats fed a high-sucrose, high-fat diet. Appl Physiol Nutr Metab 43(7):669–676
63. Wongeakin N, Bhattarakosol P, Patumraj S (2014) Molecular mechanisms of curcumin on diabetes-induced endothelial dysfunctions: Txnip, ICAM-1, and NOX2 expressions. Biomed Res Int 2014:161346. https://doi.org/10.1155/2014/161346
64. Rana M, Reddy SS, Maurya P, Singh V, Chaturvedi S, Kaur K et al (2014) Turmerone enriched standardized Curcuma longa extract alleviates LPS induced inflammation and cytokine production by regulating TLR4-IRAK1-ROS-MAPK-NFκB axis. J Funct Foods 16:152–163
65. Yan LJ, Wang SB, Wang XQ, Cao XM (2019) Protective effect and mechanism of curcumin on aorta in rats with metabolic syndrome. Zhongguo Zhong Yao Za Zhi 44(21):4685–4690
66. Ahmed OAA, El-Bassossy HM, Azhar AS, Tarkhan MM, El-Mas MM (2020) Interference with AGEs formation and AGEs-induced vascular injury mediates curcumin vascular protection in metabolic syndrome. Sci Rep 10(1):315. https://doi.org/10.1038/s41598-019-57268-z
67. Liu S, Zhang J, Pang Q, Song S, Miao R, Chen W et al (2016) The protective role of curcumin in zymosan-induced multiple organ dysfunction syndrome in mice. Shock 45(2):209–219
68. Sangartit W, Kukongviriyapan U, Donpunha W, Pakdeechote P, Kukongviriyapan V, Surawattanawan P et al (2014) Tetrahydrocurcumin protects against cadmium-induced hypertension, raised arterial stiffness and vascular remodeling in mice. PloS One 9(12):e114908. https://doi.org/10.1371/journal.pone.0114908
69. Boonla O, Kukongviriyapan U, Pakdeechote P, Kukongviriyapan V, Pannangpetch P, Prachaney P et al (2014) Curcumin improves endothelial dysfunction and vascular remodeling in 2K-1C hypertensive rats by raising nitric oxide availability and reducing oxidative stress. Nitric Oxide 42:44–53
70. Lan C, Chen X, Zhang Y, Wang W, Wang WE, Liu Y et al (2018) Curcumin prevents strokes in stroke-prone spontaneously hypertensive rats by improving vascular endothelial function. BMC Cardiovasc Disord 18(1):43. https://doi.org/10.1186/s12872-018-0768-6
71. Bai XJ, Hao JT, Wang J, Zhang WF, Yan CP, Zhao JH et al (2018) Curcumin inhibits cardiac hypertrophy and improves cardiovascular function via enhanced Na+/Ca2+ exchanger expression after transverse abdominal aortic constriction in rats. Pharmacol Rep 70(1):60–68
72. Manhas A, Khanna V, Prakash P, Goyal D, Malasoni R, Naqvi A et al (2014) Curcuma oil reduces endothelial cell-mediated inflammation in postmyocardial ischemia/reperfusion in rats. J Cardiovasc Pharmacol 64(3):228–236
73. Rungseesantivanon S, Thenchaisri N, Ruangvejvorachai P, Patumraj S (2010) Curcumin supplementation could improve diabetes-induced endothelial dysfunction associated with decreased vascular superoxide production and PKC inhibition. BMC Complement Altern Med 10:57. https://doi.org/10.1186/1472-6882-10-57
74. Sankrityayan H, Majumdar AS (2015) Curcumin and folic acid abrogated methotrexate induced vascular endothelial dysfunction. Can J Physiol Pharmacol 94(1):89–96
75. Singh V, Jain M, Misra A, Khanna V, Prakash P, Malasoni R et al (2015) Curcuma oil ameliorates insulin resistance & associated thrombotic complications in hamster & rat. Indian J Med Res 141(June):823–832
76. Zhang DM, Li YC, Xu D, Ding XQ, Kong LD (2012) Protection of curcumin against fructose-induced hyperuricaemia and renal endothelial dysfunction involves NO-mediated JAK-STAT signalling in rats. Food Chem 134(4):2184–2193
77. Tapia E, García-Arroyo F, Silverio O, Rodríguez-Alcocer AN, Jiménez-Flores AB, Cristobal M et al

(2016) Mycophenolate mofetil and curcumin provide comparable therapeutic benefit in experimental chronic kidney disease: role of Nrf2-Keap1 and renal dopamine pathways. Free Radic Res 50(7):781–792
78. Sagiroglu T, Kanter M, Yagci MA, Sezer A, Erboga M (2014) Protective effect of curcumin on cyclosporin A-induced endothelial dysfunction, antioxidant capacity, and oxidative damage. Toxicol Ind Health 30(4):316–327
79. Akazawa N, Choi Y, Miyaki A, Tanabe Y, Sugawara J, Ajisaka R et al (2012) Curcumin ingestion and exercise training improve vascular endothelial function in postmenopausal women. Nutr Res 32(10):795–799
80. Barber-Chamoux N, Milenkovic D, Verny MA, Habauzit V, Pereira B et al (2018) Substantial variability across individuals in the vascular and nutrigenomic response to an acute intake of curcumin: a randomized controlled trial. Mol Nutr Food Res 62(5). https://doi.org/10.1002/mnfr.201700418
81. Campbell MS, Ouyang A, Krishnakumar IM, Charnigo RJ, Westgate PM, Fleenor BS (2019) Influence of enhanced bioavailable curcumin on obesity-associated cardiovascular disease risk factors and arterial function: A double-blinded, randomized, controlled trial. Nutrition 62:135–139
82. Kuszewski JC, Wong RHX, Wood LG, Howe PRC (2020) Effects of fish oil and curcumin supplementation on cerebrovascular function in older adults: a randomized controlled trial. Nutr Metab Cardiovasc Dis 30(4):625–633
83. Oliver JM, Stoner L, Rowlands DS, Caldwell AR, Sanders E, Kreutzer A et al (2017) Novel form of curcumin improves endothelial function in young, healthy individuals: a double-blind placebo controlled Study. J Nutr Metab 2016:1089653. https://doi.org/10.1155/2016/1089653
84. Santos-Parker JR, Strahler TR, Bassett CJ, Bispham NZ, Chonchol MB, Seals DR (2017) Curcumin supplementation improves vascular endothelial function in healthy middle-aged and older adults by increasing nitric oxide bioavailability and reducing oxidative stress. Aging (Albany NY) 9(1):187–208
85. Srinivasan A, Selvarajan S, Kamalanathan S, Kadhiravan T, Prasanna Lakshmi NC, Adithan S (2019) Effect of Curcuma longa on vascular function in native Tamilians with type 2 diabetes mellitus: a randomized, double-blind, parallel arm, placebo-controlled trial. Phytother Res 33(7):1898–1911
86. Hallajzadeh J, Milajerdi A, Kolahdooz F, Amirani E, Mirzaei H, Asemi Z (2019) The effects of curcumin supplementation on endothelial function: a systematic review and meta-analysis of randomized controlled trials. Phytother Res 33(11):2989–2995
87. Liberale L, Bonaventura A, Montecucco F, Dallegri F, Carbone F (2019) Impact of red wine consumption on cardiovascular health. Curr Med Chem 26(19):3542–3566
88. Da Cunha MDSB, Arruda SF (2017) Tucum-do-Cerrado (Bactris setosa Mart.) may promote anti-aging effect by upregulating SIRT1-Nrf2 pathway and attenuating oxidative stress and inflammation. Nutrients 9(11):E1243. https://doi.org/10.3390/nu9111243
89. Ren Z, He H, Zuo Z, Xu Z, Wei Z, Deng J (2019) The role of different SIRT1-mediated signaling pathways in toxic injury. Cell Mol Biol Lett 24:36. https://doi.org/10.1186/s11658-019-0158-9
90. Singh CK, Chhabra G, Ndiaye MA, Garcia-Peterson LM, Mack NJ, Ahmad N (2018) The role of sirtuins in antioxidant and redox signaling. Antioxid Redox Signal 28(8):643–661
91. Zhang X, Wu Q, Lu Y, Wan J, Dai H, Zhou X et al (2018) Cerebroprotection by salvianolic acid B after experimental subarachnoid hemorrhage occurs via Nrf2-and SIRT1-dependent pathways. Free Radic Biol Med 124:504–516
92. Corbi G, Conti V, Russomanno G, Rengo G, Vitulli P, Ciccarelli AL et al (2012) Is physical activity able to modify oxidative damage in cardiovascular aging? Oxid Med Cell Longev 2012:728547. https://doi.org/10.1155/2012/728547
93. Araujo JA, Zhang M, Yin F (2012) Heme oxygenase-1, oxidation, inflammation, and atherosclerosis. Front Pharmacol 3:119. https://doi.org/10.3389/fphar.2012.00119
94. Otterbein LE, Soares MP, Yamashita K, Bach FH (2003) Heme oxygenase-1: unleashing the protective properties of heme. Trends Immunol 24(8):449–455
95. Suganya N, Bhakkiyalakshmi E, Sarada D, Ramkumar K (2016) Reversibility of endothelial dysfunction in diabetes: role of polyphenols. Br J Nutr 116(2):223–246
96. Soares MP, Seldon MP, Gregoire IP, Vassilevskaia T, Berberat PO, Yu J et al (2004) Heme oxygenase-1 modulates the expression of adhesion molecules associated with endothelial cell activation. J Immunol 172(6):3553–3563
97. Motterlini R, Foresti R, Bassi R, Green CJ (2000) Curcumin, an antioxidant and anti-inflammatory agent, induces heme oxygenase-1 and protects endothelial cells against oxidative stress. Free Radic Biol Med 28(8):1303–1312
98. Olszanecki R, Gebska A, Korbut R (2007) The role of haem oxygenase-1 in the decrease of endothelial intercellular adhesion molecule-1 expression by curcumin. Basic Clin Pharmacol Toxicol 101(6):411–415
99. Derochette S, Franck T, Mouithys-Mickalad A, Ceusters J, Deby-Dupont G, Lejeune JP et al (2013) Curcumin and resveratrol act by different ways on NADPH oxidase activity and reactive oxygen species produced by equine neutrophils. Chem Biol Interact 206(2):186–193
100. Paravicini TM, Touyz RM (2008) NADPH oxidases, reactive oxygen species, and hypertension: clinical implications and therapeutic possibilities. Diabetes Care 31(Suppl 2):S170–S180

101. Fleenor BS, Seals DR, Zigler ML, Sindler AL (2012) Superoxide-lowering therapy with TEMPOL reverses arterial dysfunction with aging in mice. Aging Cell 11(2):269–276
102. Durrant JR, Seals DR, Connell ML, Russell MJ, Lawson BR, Folian BJ et al (2009) Voluntary wheel running restores endothelial function in conduit arteries of old mice: direct evidence for reduced oxidative stress, increased superoxide dismutase activity and down-regulation of NADPH oxidase. J Physiol 587(13):3271–3285
103. Yu W, Wu J, Cai F, Xiang J, Zha W, Fan D et al (2012) Curcumin alleviates diabetic cardiomyopathy in experimental diabetic rats. PLoS One 7(12):e52013. https://doi.org/10.1371/journal.pone.0052013
104. Liberale L, Montecucco F, Tardif JC, Libby P, Camici GG (2020) Inflamm-ageing: the role of inflammation in age-dependent cardiovascular disease. Eur Heart J 41(31):2974–2982. https://doi.org/10.1093/eurheartj/ehz961
105. Carbone F, Liberale L, Bonaventura A, Cea M, Montecucco F (2016) Targeting inflammation in primary cardiovascular prevention. Curr Pharm Des 22(37):5662–5675
106. Montecucco F, Liberale L, Bonaventura A, Vecchie A, Dallegri F, Carbone F (2017) The role of inflammation in cardiovascular outcome. Curr Atheroscler Rep 19(3):11. https://doi.org/10.1007/s11883-017-0646-1
107. Bonaventura A, Montecucco F, Dallegri F, Carbone F, Luscher TF, Camici GG et al (2019) Novel findings in neutrophil biology and their impact on cardiovascular disease. Cardiovasc Res 115(8):1266–1285
108. Bonaventura A, Liberale L, Carbone F, Vecchie A, Diaz-Canestro C, Camici GG et al (2018) The pathophysiological role of neutrophil extracellular traps in inflammatory diseases. Thromb Haemost 118(1):6–27
109. Liberale L, Dallegri F, Montecucco F, Carbone F (2017) Pathophysiological relevance of macrophage subsets in atherogenesis. Thromb Haesmost 117(1):7–18
110. Coban D, Milenkovic D, Chanet A, Khallou-Laschet J, Sabbe L, Palagani A et al (2012) Dietary curcumin inhibits atherosclerosis by affecting the expression of genes involved in leukocyte adhesion and transendothelial migration. Mol Nutr Food Res 56(8):1270–1281
111. Sun Y, Hu X, Hu G, Xu C, Jiang H (2015) Curcumin attenuates hydrogen peroxide-induced premature senescence via the activation of SIRT1 in human umbilical vein endothelial cells. Biol Pharm Bull 38(8):1134–1141
112. Lee HS, Lee MJ, Kim H, Choi SK, Kim JE, Moon HI et al (2010) Curcumin inhibits TNFalpha-induced lectin-like oxidised LDL receptor-1 (LOX-1) expression and suppresses the inflammatory response in human umbilical vein endothelial cells (HUVECs) by an antioxidant mechanism. J Enzyme Inhib Med Chem 25(5):720–729

7 The Effect of Curcumin Supplemsentation on Anthropometric Indices in Overweight and Obese Individuals: A Systematic Review of Randomized Controlled Trials

Zahra Safari, Mohammad Bagherniya, Gholamreza Askari, Thozhukat Sathyapalan, and Amirhossein Sahebkar

Abstract

Background: Curcumin is an active molecule present in turmeric and is the main therapeutic compound. There is growing evidence that curcumin could affect various anthropometric indices. We performed a systematic review to evaluate the efficacy of curcumin supplementation on anthropometric indices in obese and overweight individuals.

Methods: A comprehensive search was conducted in PubMed, Scopus, Web of Science, and Google Scholar from inception up to February 2020 to identify randomized controlled trials investigating the effect of curcumin supplementation on anthropometric indices including body weight, body mass index (BMI), waist circumference (WC), hip circumference (HC), arm circumference (AC), waist-hip ratio (WHR), total body fat (TBF), and visceral fat (VF) in obese and overweight individuals. The Jadad scale was used to assess the quality of the included studies.

Result: Twenty-eight randomized controlled trials, comprising 2168 participants, were included in the systematic review. The results of 16 papers indicated that curcumin reduced at least one of the anthropometric indices among individuals with a BMI $\geq$ 25 kg/m^2. Nevertheless, 12 articles showed that curcumin supplementation was not effective in

Z. Safari
Students' Research Committee, Department of Community Nutrition, School of Nutrition and Food Science, Isfahan University of Medical Sciences, Isfahan, Iran

M. Bagherniya · G. Askari
Food Security Research Center, Department of Community Nutrition, School of Nutrition and Food Science, Isfahan University of Medical Sciences, Isfahan, Iran

T. Sathyapalan
Department of Academic Diabetes, Endocrinology and Metabolism, Hull York Medical School, University of Hull, Hull, UK

A. Sahebkar (✉)
Biotechnology Research Center, Pharmaceutical Technology Institute, Mashhad University of Medical Sciences, Mashhad, Iran

Neurogenic Inflammation Research Center, Mashhad University of Medical Sciences, Mashhad, Iran

Halal Research Center of IRI, FDA, Tehran, Iran

Polish Mother's Memorial Hospital Research Institute (PMMHRI), Lodz, Poland
e-mail: sahebkara@mums.ac.ir

P. C. Guest (ed.), *Studies on Biomarkers and New Targets in Aging Research in Iran*, Advances in Experimental Medicine and Biology 1291, https://doi.org/10.1007/978-3-030-56153-6_7

any of the measured anthropometric factors. The included trials exhibited substantial heterogeneity in terms of the treatment protocol, follow-up duration, curcumin dosage, and background diseases of the participants.

Conclusion: Clinical trials that have independently examined the effects of curcumin in obese or overweight individuals are limited. However, available studies indicate that curcumin has beneficial impacts on various anthropometric indices. Further trials with longer duration of interventions are needed to confirm these findings.

Keywords

Curcumin · Anthropometric indices · Obesity · Overweight · Systematic review

7.1 Introduction

Obesity has currently become one of the most important public health threats in both industrial nations and developing countries [1, 2]. Obesity is referred to as the presence of excess fat in the body, which is mainly the result of an imbalance between calories consumed and calories expended [3]. This imbalance results in the increase of adipose tissue hypertrophy and adipose tissue dysfunction [4]. Generally, people with a body mass index (BMI) of 25 kg/m^2 or more are considered overweight and those with a BMI of 30 or more are classified as obese [5]. The etiology of obesity is complex. In addition to energy imbalances, genetic, physiological, environmental, psychological, social, economic, and even political factors are involved in varying degrees of obesity [6]. According to the World Health Organization (WHO), overweight and obesity directly lead to the death of at least 2.8 million adults each year worldwide [7]. Obesity is a risk factor for chronic diseases such as cardiovascular disease, stroke, diabetes mellitus, osteoarthritis, and hypertension [8]. In addition, it has been associated with a higher rate of morbidity and mortality in the elderly [9].

The initial option to prevent and treat obesity is a comprehensive lifestyle modification, which includes diet, exercise, and behavioral therapy [10]. However, there are several challenges to follow long-term weight loss diets and vigorous physical activity programs resulting in poor adherence to these types of treatment [11]. Existing therapeutic strategies for the treatment of obesity and overweight include various pharmacological agents and bariatric surgery which are more expensive for the patients and may have serious adverse effects. More recently, medicinal plants and bioactive herbal compounds without major side effects have been introduced as potential new candidates which might be helpful to manage obesity [12, 13].

Recent evidence has suggested that dietary compounds such as spices may play a key role in prevention and treatment of obesity. One of these spices that has recently attracted significant attention as a natural product with anti-obesity properties is turmeric [14]. Curcumin is a safe and bioactive molecule present in the turmeric and is the main compound responsible for its therapeutic activities [15–20]. Studies have shown that curcumin has antioxidant, anti-inflammatory, antimicrobial, cardioprotective, nephroprotective, antineoplastic, hepatoprotective, immunomodulatory, hypoglycaemic, and anti-rheumatic properties [21–27]. Some studies have shown that curcumin has an important role in reducing obesity and its risk factors, mainly due to its anti-inflammatory and antioxidant properties [28, 29].

The results of studies on the effect of curcumin on anthropometric indices have been inconsistent. Some studies have shown a beneficial effect of curcumin on anthropometric indices [30, 31], while others found no significant effect [32, 33]. However, to the best of our knowledge, no systematic reviews have explored the impact of curcumin consumption on anthropometric indices such as body weight, BMI, waist circumference, hip circumference, arm circumference, waist-hip ratio, total body fat, and visceral fat in obese or overweight individuals. Therefore, we conducted a systematic review of randomized controlled trials on curcumin treatment of obese or overweight people to assess the efficacy of curcumin in modifying anthropometric indices.

7.2 Methods

Our systematic review was designed and reported based on the guidelines of the preferred reporting items for systematic reviews and meta-analyses (PRISMA) [34].

7.2.1 Search Strategy

An online search was carried out using the following databases: PubMed (http://www.ncbi.nlm.nih.gov/pubmed), Scopus (http://www.scopus.com), ISI Web of Science (http://www.webofscience.com), and Google Scholar up to 28 February 2020. We used text words and medical subject headings (MeSH) to identify qualified articles. Search words included: ("curcumin" OR "curcuma" OR "curcuminoid" OR "curcuma longa" OR "tumeric" OR "turmeric" OR "diferuloyme") AND ("waist circumference" OR "WC" OR "Waist-Hip Ratio" OR "Waist to hip ratio" OR "WHR" OR "WHtR" OR "Body Mass Index" OR "Anthropometry" OR "Body Constitution" OR "Body Weight Changes" OR "Obesity" OR "weight loss" OR "overweight" OR "metabolic syndrome" OR "BMI" OR "Intra-Abdominal Fat" OR "Visceral fat" OR "Subcutaneous fat" OR "Subcutaneous Fat, Abdominal" OR "total fat"). The language and time constraints were not considered in search of these databases.

7.2.2 Study Selection

Title and abstract of all papers found in the early search were reviewed independently by two authors (ZS and MB). Articles that did not meet the inclusion criteria were excluded using a screen form with a hierarchical approach based on study design, population, exposure, and outcome. Reference lists of relevant review articles were reviewed to explore additional studies. If the citation was considered eligible, the full text of the article was reviewed. Any disagreements were discussed and agreed.

7.2.3 Inclusion and Exclusion Criteria

Relevant articles were included if (a) they examined the effects of curcumin on anthropometric indices, (b) their target population is obese and overweight (BMI $\geq$ 25 kg/m^2), (c) they applied a clinical trial design, and (d) they provided sufficient information on anthropometric indices in both curcumin and control groups. Studies were excluded if they used the mixture of curcumin with other herbs, and if they were experimental studies or duplicate studies, or reviews, letters, editorial articles, or case reports. When multiple publications reported the same or overlapping data, we used the most recent or the publication with the largest population.

7.2.4 Data Extraction

ZS and MB extracted the following required data: first author's name, publication year, the country of investigation, sample size, target population, gender, mean age of subjects, mean BMI of subjects, study design, dose and type of curcumin, type of placebo, and duration of intervention. Also, the final effect of curcumin in the intervention group compared with the placebo group was extracted.

7.2.5 Quality Assessment

A systematic assessment of bias in the included studies was done in accordance with the Jadad criteria. This scale consists of the following items: randomization (0–2 points), blinding (0–2 points), and dropouts and withdrawals (0–1 point). According to this scale, the overall score of a study ranges between 0 and 5, with higher scores indicative of better quality [35]. Studies with Jadad scores of $\leq$2 and $\geq$ 3 were considered as low and high quality, respectively [36]. Two authors (ZS and MB) independently assessed the quality of the studies using the following criteria: randomization, blinding, and an account of all participants.

7.3 Results

7.3.1 Search Results and Study Selection

In the initial search of the database, 1288 articles were identified, of which 441 were duplicates. After reading the titles and abstracts, 769 papers were deleted and finally, at this stage, 78 articles remained. The full texts of these 78 articles were reviewed, and according to our inclusion and exclusion criteria, 50 articles were omitted due to the following reasons: 12 included participants with a BMI < 25 kg/m^2, 14 articles only mentioning baseline data, 6 articles lacked sufficient information about anthropometric indices, 3 articles lacked a control or placebo group, 8 articles included other compounds in addition to curcumin, the full text of 4 articles were not available, 2 articles were duplicates, and one article was not a clinical trial (Fig. 7.1). A final total of 28 trials were included in the systematic review.

7.3.2 Characteristics of the Included Studies

Included studies were published between 2012 and 2020. Most of the studies were conducted in Iran (21 articles) [30, 32, 33, 37–54], and the rest in Thailand (2 articles) [31, 55], Pakistan (1 article) [30], Italy (1 article) [56], Germany (1 article) [57], China (1 article) [58], and Taiwan (1 article) [59]. The total number of participants was 2168, ranging from 30 to 235 per study. The target population in most studies was type 2 diabetes (9 studies) [31, 37, 38, 40, 45, 46, 48, 55, 58] and non-alcoholic fatty liver disease (NAFLD) (8 studies) [39, 42, 43, 47, 49, 50, 56, 60], and the rest of the studies were done on metabolic syndrome (4 studies) [30, 32, 53, 59], obesity (3 studies) [33, 51, 52], polycystic ovary syndrome (2 studies) [41, 54], dyslipidemia (1 study) [44], and moderately hyperlipidemic individuals (1 study) [57]. With the exception of 3 studies targeting women only [41, 52, 54], the rest of the studies included both genders. The average age of participants ranged from 16 ± 1.63 to 59.36 ± 1.07 years. All of the studies included adult participants except for one study of adolescents [52]. All studies were conducted on people with a BMI ≥ 25 kg/m^2. The design of the most articles was parallel, and only three papers had crossover designs [33, 44, 57]. Four studies used turmeric powder (at doses of 2100, 2400, 3000 and 3500 mg/day) for intervention [30, 37, 39, 60], 8 studies applied curcumin (500 to 1500 mg/day) [33, 38, 40, 41, 52, 54, 55, 59], 5 studies used curcuminoid (250 to 1000 mg/day) [31, 44, 51, 57, 58], 2 studies used phytosomal curcumin (800 and 1000 mg/day) [47, 56], 2 studies applied nano-curcumin (80 mg/day) [42, 48], 1 study used phospholipid curcumin (50–250 mg/day curcumin) [43], 1 study used absorption-enhanced curcumin (1000 mg/day) [32], 2 studies used curcumin with piperine (1000 mg/day) [45, 46], 1 study used curcumin formulation (500 mg/day) [49], 1 study used BIO-curcumin® (1500 mg) [50], and one study used a phosphatidylcholine complex of curcumin (1000 mg/day) [53]. The duration of interventions varied from 4 weeks to 36 weeks (Table 7.1).

7.3.3 Quality Assessment

All trials used randomization, and 21 articles (75%) described their randomization method. Blindness was used in all studies, and 25 articles (89.2%) described how their study was blinded. Twenty-six articles (92.8%) reported the number and cause of falling participants, and only 2 studies did not address these (Table 7.2). All of the included studies were categorized as high-quality studies.

7.3.4 Effect of Curcumin on Body Weight

Out of the 28 eligible studies, 23 articles examined the effect of curcumin on weight loss. The total number of participants in these 23 articles

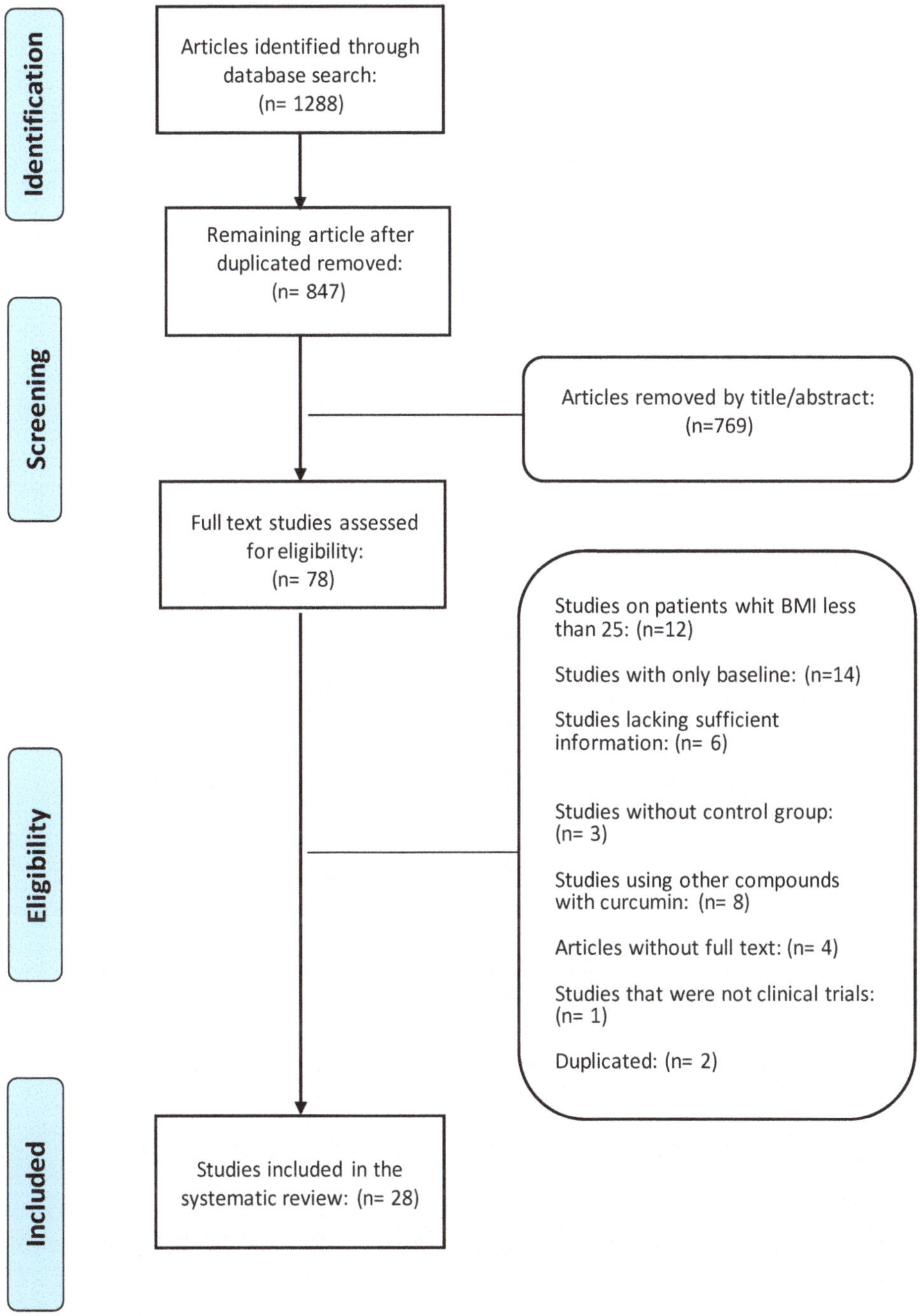

Fig. 7.1 Flow chart showing the process of the study selection

was 1661 (from 30 to 235 per study). Of these trials, 10 articles showed the effectiveness of curcumin on weight loss, and 13 showed that curcumin had no significant effect on weight.

Six parallel studies [37, 38, 40, 45, 46, 55] of participants with type 2 diabetes evaluated the effects of curcumin supplementation on body weight. The duration of intervention in 2 studies

Table 7.1 Characteristics of included randomized controlled trials (RCTs) (arranged alphabetically by first author's last name)

First author, year, country	Sample size (intervention/control)	Target population	Gender (male (M)/female(F))	Age range (years) or mean ± SD	BMI at baseline (Kg/m^2) range or mean ± SD	Study design	Intervention	Control	Duration (weeks)	Results
Adab (2019) Iran [37]	75 (36/39)	Type II diabetes	M/F	55.19 ± 7.3	28.9 ± 4.3	Parallel	Powdered rhizome of turmeric 2100 mg/day	Corn starch flour	8	Weight ↔ BMI↓
Adibian (2019), Iran [38]	44 (21/23)	Type II diabetes	M/F	59.04 ± 5.10	Intervention group ≈ 29.2/ control group ≈ 28.2[a]	Parallel	Curcumin 1500 mg/day	Rice flour	10	Weight ↓ BMI↓
Amin (2015), Pakistan [30]	126 (63/63)	Metabolic syndrome	M/F	44 ± 13.3	27.4 ± 3.9	Parallel	Turmeric 2400 mg/day	Ispaghula	8	Weight ↓ BMI↓ WC↓ HC↓ TBF↓
Chuengsamarn (2012), Thailand [55]	235 (119/116)	Type II diabetes	M/F	57.43 ± 1.2	26.64 ± 0.4	Parallel	Curcumin 1500 mg/day	Placebo	36	Weight ↓ WC↓
Chuengsamarn (2014), Thailand [31]	213 (107/106)	Type II diabetes	M/F	59.36 ± 1.07	26.96 ± 0.4	Parallel	Curcuminoid 250 mg/day	Starch	24	BMI↔ VF↓ TBF↓ WC↔
Cicero (2019), Italy [56]	80 (40/40)	NAFLD	M/F	53.5 ± 4.12	27 ± 1.84	Parallel	Phytosomal curcumin 800 mg/day	Placebo	8	BMI↓ WC↓
Esmaily (2015), Iran [33]	30 (15/15)	Obese individuals	M/F	38.37 ± 11.5	33.30 ± 4.2	Crossover	Curcumin 1000 mg/day	Bioperine	4	Weight↔ BMI↔
Ghaffari (2018), Iran [39]	46 (23/23)	NAFLD	M/F	20–60	24.9–40	Parallel	Turmeric 3000 mg/day	Corn starch	12	BMI↓
Hodaei (2019), Iran [40]	44 (21/23)	Type II diabetes	M/F	59.04 ± 7.47	28.67 ± 3.1	Parallel	Curcumin 1500 mg/day	Cooked rice Flour	10	Weight↓ BMI↔ WC↔ HC↓

First author, year, country	Sample size (intervention/control)	Target population	Gender (male (M)/female(F))	Age range (years) or mean ± SD	BMI at baseline (Kg/m^2) range or mean ± SD	Study design	Intervention	Control	Duration (weeks)	Results
Jamilian (2020), Iran [41]	50 (24/26)	Polycystic ovary syndrome	F	27.87 ± 4.09	26.88 ± 3.8	Parallel	Curcumin 500 mg/day	Starch	12	Weight ↓ BMI↓
Jazayeri-tehrani (2019), Iran [42]	84 (42/42)	NAFLD	M/F	42.15 ± 5.8	30.65 ± 2.2	Parallel	Nano-curcumin 80 mg/day	Placebo	12	Weight↔ BMI↔ WC↓
Kocher (2016), Germany [57]	42	Moderately hyperlipidemic	M/F	51.19 ± 17.51	26.59 ± 4.5	Crossover	Curcuminoids 294 mg/day	Placebo	6	Weight↔
Mirhafez (2019), Iran [43]	61 (32/29)	NAFLD	M/F	43.04 ± 11.51	29.05 ± 5.91	Parallel	Phospholipidated curcumin (capsule composed of 250 mg/day equivalent to 50 mg/day pure curcumin)	Placebo	8	Weight↔ BMI↔
Mohammadi (2017), Iran [32]	109 (37, 36/36)	Metabolic syndrome	M/F	38.73 ± 10.05	30.84 ± 4.4	Parallel	A: Absorption-enhanced curcumin (1000 mg/day equivalent to 200-mg pure curcumin/day) B: Unformulated curcumin (1000 mg/day)	Lactose and starch	6	Weight↔ BMI↔ WC↔

(continued)

Table 7.1 (continued)

First author, year, country	Sample size (intervention/control)	Target population	Gender (male (M)/female(F))	Age range (years) or mean ± SD	BMI at baseline (Kg/m²) range or mean ± SD	Study design	Intervention	Control	Duration (weeks)	Results
Mohammadi (2012), Iran [44]	30 (15/15)	Dyslipidemia	M/F	38.43 ± 10.84	32.60 ± 3.58	Crossover	Curcuminoids 1000 mg/day	Piperine	4	Weight↔ BMI↔ WC↔ HC↔ Ac↔ Body fat↔
Na (2012), China [58]	100 (50/50)	Type II diabetes	M/F	55.07 ± 7.4	27.27 ± 2.6	Parallel	Curcuminoids 300 mg/day	Starch	12	BMI↔
Navekar (2017), Iran [60]	42 (21/21)	NAFLD	M/F	41.23 ± 8.25	32.36 ± 4.6	Parallel	3500 mg/day turmeric	Starch	12	Weight ↓ BMI↓
Panahi (2016), Iran [45]	100 (50/50)	Type II diabetes	M/F	42 ± 7.54	26.93 ± 2.01	Parallel	Curcuminoids (1000 mg/day co-administered with piperine 10 mg/day)	Microcrystalline cellulose	8	Weight ↓ BMI↓
Panahi (2017), Iran [46]	100 (50/50)	Type II diabetes	M/F	42 ± 7.54	26.93 ± 2.01	Parallel	Curcuminoids (1000 mg/day co-administered with piperine 10 mg/day)	Microcrystalline cellulose	12	Weight ↓ BMI↓
Panahi (2017), Iran [47]	87 (44/43)	NAFLD	M/F	46.08 ± 11.5	29.01 ± 3.43	Parallel	Curcumin (phytosomal form; 1000 mg/day in 2 divided doses)	Placebo	8	Weight ↓ BMI↓
Rahimi (2016), Iran [48]	70 (35/35)	Type II diabetes	M/F	58.64 ± 11.13	27.09 ± 3.16	Parallel	Curcumin (as nano-micelle 80 mg/day)	Placebo	12	BMI↓
Rahmani (2016), Iran [49]	77 (37/40)	NAFLD	M/F	47.71 ± 10.68	31.10 ± 5.09	Parallel	Curcumin formulation (500 mg/day equivalent to 70-mg curcumin)	Placebo	8	Weight ↓ BMI↓

First author, year, country	Sample size (intervention/control)	Target population	Gender (male (M)/female(F))	Age range (years) or mean ± SD	BMI at baseline (Kg/m²) range or mean ± SD	Study design	Intervention	Control	Duration (weeks)	Results
Saadati (2019), Iran [50]	50 (27/23)	NAFLD	M/f	45.70 ± 11.13	32.33 ± 4.72	Parallel	BIO-curcumin® 1500 mg (3*500 mg)	Maltodexterin	12	Weight↔ BMI↔ WC↔ HC↔ WC/HC ↔
Sahebkar (2013), Iran [51]	30 (15/15)	Obese individuals	M/F	38.43 ± 10.84	32.6 ± 3.58	Crossover	Curcuminoids (1000 mg/day)	Piperine	4	Weight↔ BMI↔
Saraf (2019), Iran [52]	60 (30/30)	Overweight and obese adolescents	F	16.00 ± 1.63	30.85 ± 2.87	Parallel	Standardized 95% turmeric extract (500 mg curcumin)	Placebo	10	Weight↔ BMI↔ WC↔ HC↔ WHR ↔
Shirmohammadi (2019), Iran [53]	73 (37/36)	Metabolic syndrome	M/F	39.33 ± 10.33	30.93 ± 4.84	Parallel	Phosphatidylcholine complex of curcumin (2 capsules contained 500 mg)	Placebo	6	Weight↔ BMI↔ WC↔
Sohaei (2019), Iran [54]	51 (27/24)	Polycystic ovary syndrome	F	29.48 ± 5.12	30.45 ± 4.20	Parallel	Curcumin (500 mg/day) twice daily	Placebo	6	Weight↔ BMI↔ Body fat↔ Visceral fat↔
Yang (2014), Taiwan [59]	59 (30/29)	Metabolic syndrome	M/F	59.3 ± 12.0	30.06 ± 4.12	Parallel	Curcumin extract capsule (630 mg thrice daily)	Placebo	12	Weight↔ BMI↔

[a]Data were shown by Figures in the original article; thus, it is not the exact values
The reported results showed a significant difference between the intervention and placebo groups after the end of the intervention

Table 7.2 Quality assessment of the included studies using Jadad Scale

First author (publication year)	Randomization	Blinding	Account of all patients	Total points
Adab (2019) [37]	2	2	1	5
Adibian (2019) [38]	2	2	1	5
Amin (2015) [30]	2	2	1	5
Chuengsamarn (2012) [55]	2	2	0	4
Chuengsamarn (2014) [31]	2	2	0	4
Cicero (2019) [56]	2	2	1	5
Esmaily (2015) [33]	2	2	1	5
Ghaffari (2018) [39]	2	2	1	5
Hodaei (2019) [40]	2	2	1	5
Jamilian (2020) [41]	2	2	1	5
Jazayeri-Tehrani (2019) [42]	2	2	1	5
Kocher (2016) [57]	1	1	1	3
Mirhafez (2019) [43]	2	2	1	5
Mohammadi (2017) [32]	1	2	1	4
Mohammadi (2012) [44]	1	2	1	4
Na (2012) [58]	2	1	1	4
Navekar (2017) [60]	2	2	1	5
Panahi (2016) [45]	1	2	1	4
Panahi (2017) [46]	1	2	1	4
Panahi (2017) [47]	2	2	1	5
Rahimi (2016) [48]	2	1	1	4
Rahmani (2016) [49]	1	2	1	4
Saadati (2019) [50]	2	2	1	5
Sahebkar (2013) [51]	1	2	1	4
Saraf (2019) [52]	2	2	1	5
Shirmohammadi (2019) [53]	2	2	1	5
Sohaei (2019) [54]	2	2	1	5
Yang (2014) [59]	2	2	1	5

was 10 weeks, in 2 studies 8 weeks, in 1 study 12 weeks, and in 1 study 36 weeks. One study [37] used 2100 mg/day turmeric powder, 3 studies [38, 40, 55] used 1500 mg/day curcumin, and 2 studies [45, 46] used 1000 mg/day curcuminoids with 10 mg piperine to increase curcumin bioavailability. Among these clinical trials, 5 studies [38, 40, 45, 46, 55] showed the efficacy of curcumin in weight loss in comparison to placebo and 1 study [37] of turmeric consumption did not lead to weight loss in patients with type 2 diabetes.

On the other hand, 6 parallel studies [42, 43, 47, 49, 50, 60] were conducted among people with NAFLD. The intervention period was 8 weeks in 3 studies and 12 weeks in the other 3 studies. In terms of the type of supplements, curcumin (1500 mg/day) [50], nano-curcumin (80 mg/day) [42], curcumin formulation (500 mg/day) [49], curcumin phytosomal (1000 mg/day) [47], phospholipidated curcumin (50–250 mg/day) pure curcumin [43], and turmeric 3500 mg/day [60] were used. Three studies [47, 49, 60] showed that curcumin supplementation led to weight loss in NAFLD patients whereas 3 studies [42, 43, 50] showed that curcumin consumption was not effective in weight loss in people with NAFLD.

Four parallel trials [30, 32, 53, 59] examined the efficacy of curcumin in weight loss in individuals with metabolic syndrome. Interventions lasted 6, 8, and 12 weeks (two studies) and turmeric powder (2400 mg/day) [30], phosphatidylcholine complex of curcumin (1000 mg/day) [53], absorption-enhanced curcumin (1 g/day equivalent to 200 mg pure curcumin/day) [32], and curcumin (630 mg/day) [59] were used in these studies. Three studies [32, 53, 59] showed that curcumin was ineffective, and 1 study [30] showed the effectiveness of curcumin on weight loss compared with placebo in people with metabolic syndrome.

Three studies [33, 51, 52] were conducted on obese and overweight individuals, in which the target population of 1 study [52] was adolescent girls. There were 2 cross-sectional design studies and 1 parallel design study. The duration of interventions was 4 (2 studies) and 10 (1 study) weeks. In these studies, curcumin (500 and 1000 mg/day) [33, 52] and curcuminoid (1000 mg/day) [51] were used in the intervention group. The results of all 3 studies showed that curcumin had no effect on weight loss.

Two trials [41, 54] were conducted on people with polycystic ovary syndrome, which had parallel and cross-sectional designs. Interventions lasted 6 and 12 weeks and, in both studies, 500 mg/day curcumin was used. The longer intervention study [41] showed that curcumin supplementation led to weight loss, while the other study [54] showed that curcumin supplementation had no effect on weight loss in women with polycystic ovary syndrome.

Two crossover [44, 57] studies were performed on patients with hyperlipidemia and dyslipidemia. The duration of interventions in these studies were 4 and 6 weeks and curcuminoids were used at doses of 1000 mg/day [44] and 249 mg/day [57], respectively. The results of these two studies did not confirm the efficacy of curcumin in weight loss.

7.3.5 Effect of Curcumin on BMI

Out of the 28 selected trials, 25 studies assessed the efficacy of curcumin on BMI. The total number of samples in these 25 articles was 1861 (from 30 to 213 per study). Of these trials, 12 studies showed the effectiveness of curcumin on reducing BMI, and 13 showed that curcumin had no effect.

Of these 25 trials, eight parallel studies [31, 37, 38, 40, 45, 46, 48, 58] were conducted on people with type 2 diabetes. Interventions lasted from 8 weeks to 24 weeks in various studies. The supplements used in these studies were curcumin (1500 mg/day) [38, 40], turmeric (2100 mg/day) [37], curcuminoids at doses of 250 and 300 mg/day [31, 58], nano-curcumin (80 mg/day) [48], and curcumin-piperine (1000 mg/day) [45, 46]. The results of these eight trials showed that 5 studies [37, 38, 45, 46, 48] confirmed the efficacy of curcumin in reducing BMI, while the other 3 studies [31, 40, 58] showed that curcumin was ineffective.

Eight parallel articles [39, 42, 43, 47, 49, 50, 56, 60] were published among people with NAFLD. The duration of intervention in these 8 trials was 8 weeks (4 studies) and 12 weeks (4 studies). The type of supplement used in these studies was phytosomal curcumin at doses of 800 and 1000 mg/day [47, 56], turmeric powder at doses of 3000 and 3500 mg/day [39, 60], nano-curcumin (80 mg/day) [42], phospholipid curcumin (250 mg/day) [43], curcumin formulation (500 mg/day) [49], and 1500 mg/day curcumin [50]. The results of 5 studies [39, 47, 49, 56, 60] showed that curcumin consumption significantly reduced BMI, whereas three studies [42, 43, 50] showed no effect.

The target population for 4 parallel trials [30, 32, 53, 59] was individuals with metabolic syndrome. The interventions in these studies had 6- (two studies), 8-, and 12-week durations. The 4 trials used turmeric (2400 mg/day) [30], curcumin (630 mg/day) [59], phosphatidylcholine complex of curcumin 500 mg/day [53], and absorption-enhanced curcumin (200-mg pure curcumin/day) [32]. Apart from one study [30], the other trials [32, 53, 59] showed that curcumin had no effect on BMI.

Two parallel studies [41, 54] were performed on women with polycystic ovary syndrome. The duration of the intervention was 6 and 12 weeks, and the supplement was 500 mg curcumin. The longer intervention study [41] confirmed the efficacy of curcumin, while the other study [54] indicated that it was ineffective.

Two crossover and parallel studies [51, 52] were performed on obese individuals. The intervention period was 4 and 10 weeks, and the test groups received curcumin (500 mg/day) [52] and curcuminoid (1000 mg/day) [51]. These studies did not confirm the efficacy of curcumin in reducing BMI.

A crossover trial [44] was performed on people with dyslipidemia. The duration of the intervention was 4 weeks, and the curcuminoid supplement was 1000 mg/day. The results of this study showed that curcumin was ineffective in reducing BMI.

7.3.6 Effect of Curcumin on Waist Circumference

Eleven articles assessed the effect of curcumin on waist circumference. The total number of participants was 1104 (from 30 to 235 per study). Of these trials, 4 showed that curcumin was effective on waist circumference loss, and 7 studies showed that curcumin had no effect.

Of these 11 trials, three parallel papers [31, 40, 55] studies individuals with type 2 diabetes. The duration of intervention was 10, 24, and 36 weeks and 3 trials used curcumin (1500 mg/day; 2 studies) [40, 55] and curcuminoid (250 mg/day) [31]. The longer intervention study of 36 weeks [55] confirmed the efficacy of curcumin, while the other 2 studies [31, 40] showed curcumin to be ineffective in reducing waist circumference in people with type 2 diabetes.

Three parallel studies [30, 32, 53] were conducted on individuals with metabolic syndrome. The intervention duration was 6 (two studies) and 12 weeks. The type of supplement used was 500 mg/day phosphatidylcholine curcumin complex [53], 2400 mg/day turmeric powder [30], and 1000 mg/day absorption-enhanced curcumin [32]. The results of 2 studies [32, 53] showed that curcumin was ineffective, and 1 study [30] showed that it had an effect in reducing waist circumference.

In addition, three parallel studies [42, 50, 56] were performed on people with NAFLD. The intervention period was 8 and 12 weeks (2 studies). The type of curcumin used was nano-curcumin (80 mg/day) [42], phytosomal curcumin (800 mg/day) [56], and curcumin (1500 mg/day) [50]. Two studies [42, 56] showed that curcumin consumption led to a decrease in waist circumference, and 1 study [50] showed that curcumin consumption had no effect on waist circumference in people with NAFLD.

A parallel study [52] was conducted on obese and overweight adolescent girls. The duration of intervention was 10 weeks, and the intervention group consumed 1500 mg/day curcumin. The results of this study did not confirm any efficacy of curcumin on reducing waist circumference.

A crossover study [44] was conducted on patients with dyslipidemia with a 4-week intervention period. The intervention group consumed 1000 mg/day of curcuminoid. The results of this study did not indicate any beneficial effects of curcumin on reducing waist circumference.

7.3.7 Effect of Curcumin on Hip Circumference

There were only five studies evaluating the efficacy of curcumin on hip circumference. The total number of participants in these five articles was 310 people. Results of the two studies showed a decrease in hip circumference, whereas three studies showed no effect of curcumin on hip circumference.

The five trials were conducted on people with type 2 diabetes [40], metabolic syndrome [30], dyslipidemia [44], NAFLD [50], and overweight or obese adolescents [52]. The design of most studies was parallel, except for one that had a crossover design. Interventions lasted from 4 weeks to 12 weeks. The supplement used in the intervention group was curcumin at doses of 500 mg/day [52] and 1500 mg/day [40, 50], turmeric powder at 2400 mg/day [30], and curcuminoid at 1000 mg/day [44]. Finally, the results of 2 studies [9, 30] confirmed the efficacy of curcumin and 3 studies [44, 50, 52] showed that curcumin was ineffective in hip circumference reduction.

7.3.8 Effect of Curcumin on Total Body Fat

Four of the 28 studies indicated the efficacy of curcumin on total body fat. The total number of participants in these 4 studies was 420 people. Of these 4 trials, 3 had a parallel design and 1 study was of crossover design.

These trials were performed on people with metabolic syndrome [30], type 2 diabetes [31], dyslipidemia [55], and polycystic ovary syndrome [54]. The duration of interventions in these trials were 4, 6, 8, and 24 weeks. The supplements used in these 4 studies were curcuminoids at doses of 1000 mg/day [44] and 250 mg/day [31], curcumin at 500 mg/day [54], and turmeric powder at 2400 mg/day [30]. Two studies [30, 31] confirmed the efficacy of curcumin, and 2 [44, 54] showed that curcumin was not effective in reducing total body fat.

7.3.9 Effect of Curcumin on Visceral Fat

Of the 28 articles, 2 studies measured the amount of visceral fat. The total number of participants in these 2 studies was 264 people. These studies were performed on patients with polycystic ovary syndrome [54] and type 2 diabetes [31], and the intervention duration was 6 and 24 weeks. The supplements used in these studies were curcumin (500 mg/day) [54] and curcuminoid (250 mg/day) [31]. The longer intervention study [31] confirmed the efficacy of curcumin, while the other study [54] showed that curcumin was not effective in reducing visceral fat.

7.3.10 Effect of Curcumin on Waist-Hip Ratio

Two parallel trials measured waist-hip ratio in people with NAFLD [50] and in overweight or obese adolescents [52]. The number of participants in these 2 studies was 110 people. The duration of intervention was 10 and 12 weeks. The supplement used in the intervention group was curcumin at doses of 500 mg/day [52] and 1500 mg/day [50]. The results of these studies showed that curcumin consumption did not have any significant effect on waist-hip ratio.

7.3.11 Effect of Curcumin on Arm Circumference

Only one crossover study [44] measured arm circumference. This study was performed on 30 participants with dyslipidemia. The duration of the intervention was 4 weeks and the supplement used in the intervention group was 1000 mg/day curcuminoid. The results of this study showed that curcumin did not significantly reduce AC.

7.4 Discussion

The main findings of the current systematic reviews were that, among a total of 28 included studies, 16 studies indicated that curcumin had a significant effect on reducing at least one of the anthropometric indices among individuals with a BMI $\geq$ 25 kg/m^2. Nevertheless, 12 articles showed that curcumin supplementation was not effective in any of the measured anthropometric parameters. To the best of our knowledge, this is the first systematic review study investigating the effects of curcumin supplementation on anthropometric indices in overweight and obese individuals.

Due to the limitations of synthetic pharmacological agents and bariatric surgical procedures to manage obesity today, herbal medicines and natural herbal bio-active compounds such as curcumin, which are reliable, safe, and inexpensive, have attracted significant attention to combat obesity and its comorbidities [12]. In obesity, like many other chronic diseases, inflammation plays a major role [61]. Studies have shown that curcumin can inhibit a number of signaling pathways and molecular targets associated with inflammation [62]. Possible mechanisms of curcumin to reduce inflammation include reduction of tumor necrosis factor expression in different tissues, suppression of nuclear factor kappa-light-chain-enhancer of activated B cells (NF-κB) activity through inhibition of IκBα degradation, and inhibition of IκB kinase activation. These processes will eventually lead to the suppression of the levels of inflammatory markers such as cyclooxygenase-2 (COX-2) and vascular endothelial growth factor, along with decreased expression of proinflammatory adipocytokines including chemokines [such as monocyte chemoattractant protein-1 (MCP)-1, MCP-4, and eotaxin] and interleukins (IL-1, IL-6, and IL-8), inhibition of the Wnt/β-catenin pathway that is closely linked to obesity, and increased expression of adiponectin which negatively controls obesity [62].

Another contributing factor to obesity is oxidative stress. Studies have shown that obese people are more susceptible to oxidative stress [63]. Oxidative stress has also been associated with an increase in BMI. Many studies have reported the antioxidant properties of curcumin [28, 64]. Animal studies have shown that curcumin increased antioxidant activity of glutathione peroxidase enzyme and reduced oxidative stress. In addition, curcumin has been found to increase the levels of antioxidant vitamins that are destroyed due to the oxidative stress caused by a high-fat diet [28].

A meta-analysis of Akbari et al., which measured the supplementary effect of curcumin on patients with metabolic syndrome and related diseases, showed that curcumin significantly reduced BMI, body weight, and waist circumference [65]. In this study, it is proposed that curcumin might affect obesity through mechanisms including suppression of angiogenesis in adipose tissue and reduction in body fat, and regulation of phosphorylated AMP-activated protein kinase (AMPK) gene expression that may subsequently act as a sirtuin 1 activator. In adipose tissue, sirtuin 1 reduces the secretion of inflammatory cytokines and macrophage infiltration and improves insulin sensitivity [65].

In a meta-analysis, Baziar et al. examined the effect of curcumin consumption on patients with NAFLD. The results of this meta-analysis showed that curcumin consumption reduced BMI and waist circumference but had no significant effect on body weight. One of the reasons why curcumin might not have been effective in this meta-analysis was that fewer studies have measured weight [66]. The meta-analysis found that trials using curcumin with increased bioavailability had better effects on waist circumference and obesity. It is suggested that these effects of curcumin might be due to its role in increasing lipolysis, brown fat tissue, adipocyte energy metabolism and apoptosis, serum levels of nesfatin, and via probiotic-like effects [66].

The present study is the first systematic review of the effect of curcumin on anthropometric parameters in overweight and obese individuals. However, several limitations of this study should be acknowledged. First, most of the trials included in this systematic review were conducted among Iranian participants, so the results may not be generalizable to individuals of other ethnic origins. Second, in most of the included studies, curcumin efficacy on anthropometric indices was evaluated only as a secondary outcome, as most investigated its effects on a particular disease. Third, in each study, only some of the anthropometric indices were assessed, and studies did not focus on the same parameters. Some studies evaluated only BMI while others assessed other indices. This made it difficult to draw firm conclusions in terms of efficacy of curcumin supplementation on all anthropometry indices, using data of all the included studies. Finally, heterogeneity of the included studies in terms of different doses of curcumin, different types of curcumin, varied populations, and different treatment durations made it difficult to judge the definitive effect of curcumin on anthropometric indices in obese or overweight people, thus requiring further trials in this areas.

7.5 Conclusion

Clinical trials that have independently examined the supplementary effect of curcumin in obese or overweight individuals have been limited. However, a number of existing studies have shown some efficacy of curcumin on anthropometric indices. From a total of 28 studies, 16 papers indicated that curcumin had a significant effect on reducing at least one anthropometric indices among individuals with a BMI ≥ 25 kg/ m^2. Regardless of the type of curcumin, the longer duration of intervention studies had better results on anthropometric indices. More clinical studies with longer interventions and different forms of curcumin are needed to demonstrate the effect of this natural compound on anthropometric parameters.

References

1. Swinburn BA, Sacks G, Hall KD, McPherson K, Finegood DT, Moodie ML, Gortmaker SL (2011) The global obesity pandemic: shaped by global drivers and local environments. Lancet 378(9793):804–814
2. Bagherniya M, Sharma M, Mostafavi F, Keshavarz SA (2015) Application of social cognitive theory in predicting childhood obesity prevention behaviors in overweight and obese Iranian adolescents. Int Q Community Health Educ 35(2):133–147
3. Sarokhani D, Sarokhani M, Hasanpour Dehkordi A, Ghanei Gheshlagh R, Fakhri M (2020) Prevalence of obesity and overweight in Iranian students: a systematic review and meta-analysis. J Pediatr Endocrinol Metab 33(4):453–468
4. Sun K, Tordjman J, Clement K, Scherer PE (2013) Fibrosis and adipose tissue dysfunction. Cell Metab 18(4):470–477
5. Flegal KM, Carroll MD, Kit BK, Ogden CL (2012) Prevalence of obesity and trends in the distribution of body mass index among US adults, 1999-2010. JAMA 307(5):491–497
6. Aronne LJ, Nelinson DS, Lillo JL (2009) Obesity as a disease state: a new paradigm for diagnosis and treatment. Clin Cornerstone 9(4):9–25; discussion 26-29
7. WHO (2012) Obesity and overweight
8. Bradford PG (2013) Curcumin and obesity. Biofactors 39(1):78–87
9. Kalish VB (2016) Obesity in older adults. Prim Care 43(1):137–144. ix
10. Wadden TA, Webb VL, Moran CH, Bailer BA (2012) Lifestyle modification for obesity: new developments in diet, physical activity, and behavior therapy. Circulation 125(9):1157–1170
11. Bagherniya M, Nobili V, Blesso CN, Sahebkar A (2018) Medicinal plants and bioactive natural compounds in the treatment of non-alcoholic fatty liver disease: a clinical review. Pharmacol Res 130:213–240
12. Kazemipoor M, Radzi CWJWM, Cordell GA, Yaze I (2012) Potential of traditional medicinal plants for treating obesity: a review. 2012 International Conference on Nutrition and Food Sciences. IPCBEE 39; IACSIT Press, Singapore
13. Hasani-Ranjbar S, Jouyandeh Z, Abdollahi M (2013) A systematic review of anti-obesity medicinal plants - an update. J Diabetes Metab Disord 12(1):28. https://doi.org/10.1186/2251-6581-12-28
14. Kochhar KP (2008) Dietary spices in health and diseases (II). Indian J Physiol Pharmacol 52(4): 327–354
15. Lee WH, Loo CY, Bebawy M, Luk F, Mason RS, Rohanizadeh R (2013) Curcumin and its derivatives: their application in neuropharmacology and neuroscience in the 21st century. Curr Neuropharmacol 11(4):338–378
16. Pivari F, Mingione A, Brasacchio C, Soldati L (2019) Curcumin and type 2 diabetes mellitus: prevention and treatment. Nutrients 11(8):E1837. https://doi.org/10.3390/nu11081837
17. Mollazadeh H, Cicero AFG, Blesso CN, Pirro M, Majeed M, Sahebkar A (2019) Immune modulation by curcumin: the role of interleukin-10. Crit Rev Food Sci Nutr 59(1):89–101
18. Momtazi AA, Derosa G, Maffioli P, Banach M, Sahebkar A (2016) Role of microRNAs in the therapeutic effects of curcumin in non-cancer diseases. Mol Diagn Ther 20(4):335–345
19. Bagheri H, Ghasemi F, Barreto GE, Rafiee R, Sathyapalan T, Sahebkar A (2020) Effects of curcumin on mitochondria in neurodegenerative diseases. Biofactors 46(1):5–20. https://doi.org/10.1002/biof.1566. Epub 2019 Oct 3. PMID: 31580521.
20. Soleimani V, Sahebkar A, Hosseinzadeh H (2018) Turmeric (Curcuma longa) and its major constituent (curcumin) as nontoxic and safe substances: Review. Phytother Res 32(6):985–995. https://doi.org/10.1002/ptr.6054. Epub 2018 Feb 26. PMID: 29480523
21. Hasanzadeh S, Read MI, Bland AR, Majeed M, Jamialahmadi T, Sahebkar A (2020) Curcumin: an inflammasome silencer. Pharmacol Res 159:104921. https://doi.org/10.1016/j.phrs.2020.104921. Epub 2020 May 25. PMID: 32464325.
22. Derosa G, Maffioli P, Simental-Mendia LE, Bo S, Sahebkar A (2016) Effect of curcumin on circulating interleukin-6 concentrations: a systematic review and meta-analysis of randomized controlled trials. Pharmacol Res 111:394–404
23. Strimpakos AS, Sharma RA (2008) Curcumin: preventive and therapeutic properties in laboratory studies and clinical trials. Antioxid Redox Signal 10(3):511–546
24. Aggarwal BB, Harikumar KB (2009) Potential therapeutic effects of curcumin, the anti-inflammatory agent, against neurodegenerative, cardiovascular, pulmonary, metabolic, autoimmune and neoplastic diseases. Int J Biochem Cell Biol 41(1):40–59
25. Abdollahi E, Momtazi AA, Johnston TP, Sahebkar A (2018) Therapeutic effects of curcumin in inflammatory and immune-mediated diseases: a nature-made jack-of-all-trades? J Cell Physiol 233(2):830–848

26. Iranshahi M, Sahebkar A, Takasaki M, Konoshima T, Tokuda H (2009) Cancer chemopreventive activity of the prenylated coumarin, umbelliprenin, in vivo. Eur J Cancer Prev 18(5):412–415
27. Teymouri M, Pirro M, Johnston TP, Sahebkar A (2017) Curcumin as a multifaceted compound against human papilloma virus infection and cervical cancers: A review of chemistry, cellular, molecular, and preclinical features. BioFactors 43(3):331-346.
28. Alappat L, Awad AB (2010) Curcumin and obesity: evidence and mechanisms. Nutr Rev 68(12):729–738
29. Sadeghian M, Rahmani S, Jamialahmadi T, Johnston TP, Sahebkar A (2020) The effect of oral curcumin supplementation on health-related quality of life: A systematic review and meta-analysis of randomized controlled trials. J Affect Disord 29;278:627–636. https://doi.org/10.1016/j.jad.2020.09.091. Epub ahead of print. PMID: 33038707.
30. Amin F, Islam N, Anila N, Gilani AH (2015) Clinical efficacy of the co-administration of turmeric and black seeds (Kalongi) in metabolic syndrome - a double blind randomized controlled trial - TAK-MetS trial. Complement Ther Med 23(2):165–174
31. Chuengsamarn S, Rattanamongkolgul S, Phonrat B, Tungtrongchitr R, Jirawatnotai S (2014) Reduction of atherogenic risk in patients with type 2 diabetes by curcuminoid extract: a randomized controlled trial. J Nutr Biochem 25(2):144–150
32. Mohammadi A, Sadeghnia HR, Saberi-Karimian M, Safarian H, Ferns GA, Ghayour-Mobarhan M et al (2017) Effects of curcumin on serum vitamin E concentrations in individuals with metabolic syndrome. Phytother Res 31(4):657–662
33. Esmaily H, Sahebkar A, Iranshahi M, Ganjali S, Mohammadi A, Ferns G, Ghayour-Mobarhan M (2015) An investigation of the effects of curcumin on anxiety and depression in obese individuals: a randomized controlled trial. Chin J Integr Med 21(5):332–338
34. Moher D, Liberati A, Tetzlaff J, Altman DG (2009) Preferred reporting items for systematic reviews and meta-analyses: the PRISMA statement. Ann Intern Med 151(4):264–269
35. Jadad AR, Moore RA, Carroll D, Jenkinson C, Reynolds DJM, Gavaghan DJ et al (1996) Assessing the quality of reports ofrandomized clinical trials: is blinding necessary? Control Clin Trials 17(1):1–12
36. Moher D, Cook D, Jadad A, Tugwell P, Moher M, Jones A et al (1999) Assessing the quality of reports of randomised trials: implications for the conduct of meta-analyses. Health Technol Assess 3(12):i–iv, 1–98
37. Adab Z, Eghtesadi S, Vafa MR, Heydari I, Shojaii A, Haqqani H et al (2019) Effect of turmeric on glycemic status, lipid profile, hs-CRP, and total antioxidant capacity in hyperlipidemic type 2 diabetes mellitus patients. Phytother Res 33(4):1173–1181
38. Adibian M, Hodaei H, Nikpayam O, Sohrab G, Hekmatdoost A, Hedayati M (2019) The effects of curcumin supplementation on high-sensitivity C-reactive protein, serum adiponectin, and lipid profile in patients with type 2 diabetes: a randomized, double-blind, placebo-controlled trial. Phytother Res 33(5):1374–1383
39. Ghaffari A, Rafraf M, Navekar R, Asghari-Jafarabadi M (2018) Effects of turmeric and chicory seed supplementation on antioxidant and inflammatory biomarkers in patients with non-alcoholic fatty liver disease (NAFLD). Adv Integr Med 5(3):89–95
40. Hodaei H, Adibian M, Nikpayam O, Hedayati M, Sohrab G (2019) The effect of curcumin supplementation on anthropometric indices, insulin resistance and oxidative stress in patients with type 2 diabetes: a randomized, double-blind clinical trial. Diabetol Metab Syndr 11:41. https://doi.org/10.1186/s13098-019-0437-7
41. Jamilian M, Foroozanfard F, Kavossian E, Aghadavod E, Shafabakhsh R, Hoseini A, Asemi Z (2020) Effects of curcumin on body weight, glycemic control and serum lipids in women with polycystic ovary syndrome: a randomized, double-blind, placebo-controlled trial. Clin Nutr ESPEN 36:128–133
42. Jazayeri-Tehrani SA, Rezayat SM, Mansouri S, Qorbani M, Alavian SM, Daneshi-Maskooni M et al (2019) Nano-curcumin improves glucose indices, lipids, inflammation, and Nesfatin in overweight and obese patients with non-alcoholic fatty liver disease (NAFLD): a double-blind randomized placebo-controlled clinical trial. Nutr Metab (Lond) 16:8. https://doi.org/10.1186/s12986-019-0331-1
43. Mirhafez SR, Farimani AR, Dehhabe M, Bidkhori M, Hariri M, Ghouchani BFNM et al (2019) Effect of phytosomal curcumin on circulating levels of adiponectin and leptin in patients with non-alcoholic fatty liver disease: a randomized, double-blind, placebo-controlled clinical trial. J Gastrointestin Liver Dis 28:183–189
44. Mohammadi A, Sahebkar A, Iranshahi M, Amini M, Khojasteh R, Ghayour-Mobarhan M et al (2013) Effects of supplementation with curcuminoids on dyslipidemia in obese patients: a randomized crossover trial. Phytother Res 27(3):374–379
45. Panahi Y, Khalili N, Sahebi E, Namazi S, Karimian MS, Majeed M, Sahebkar A (2017) Antioxidant effects of curcuminoids in patients with type 2 diabetes mellitus: a randomized controlled trial. Inflammopharmacology 25(1):25–31
46. Panahi Y, Khalili N, Sahebi E, Namazi S, Reiner Z, Majeed M et al (2017) Curcuminoids modify lipid profile in type 2 diabetes mellitus: a randomized controlled trial. Complement Ther Med 33:1–5
47. Panahi Y, Kianpour P, Mohtashami R, Jafari R, Simental-Mendía LE, Sahebkar A (2017) Efficacy and safety of phytosomal curcumin in non-alcoholic fatty liver disease: a randomized controlled trial. Drug Res (Stuttg) 67(4):244–251
48. Rahimi HR, Mohammadpour AH, Dastani M, Jaafari MR, Abnous K, Mobarhan MG et al (2016) The effect of nano-curcumin on HbA1c, fasting blood glucose, and lipid profile in diabetic subjects: a randomized clinical trial. Avicenna J Phytomed 6(5):567–577

49. Rahmani S, Asgary S, Askari G, Keshvari M, Hatamipour M, Feizi A, Sahebkar A (2016) Treatment of non-alcoholic fatty liver disease with curcumin: a randomized placebo-controlled trial. Phytother Res 30(9):1540–1548
50. Saadati S, Hekmatdoost A, Hatami B, Mansour A, Zahra Z, Hedayati M et al (2018) Comparing different non-invasive methods in assessment of the effects of curcumin on hepatic fibrosis in patients with non-alcoholic fatty liver disease. Gastroenterol Hepatol Bed Bench 11(Suppl 1):S8–S13
51. Sahebkar A, Mohammadi A, Atabati A, Rahiman S, Tavallaie S, Iranshahi M et al (2013) Curcuminoids modulate pro-oxidant–antioxidant balance butnot the immune response to heat shock protein 27 and oxidized LDL in obese individuals. Phytother Res 27(12):1883–1888
52. Saraf-Bank S, Ahmadi A, Paknahad Z, Maracy M, Nourian M (2019) Effects of curcumin on cardiovascular risk factors in obese and overweight adolescent girls: a randomized clinical trial. Sao Paulo Med J 137(5):414–422
53. Shirmohammadi L, Ghayour-Mobarhan M, Saberi-Karimian M, Iranshahi M, Tavallaie S, Emamian M et al (2019) Effect of curcumin on serum cathepsin D in patients with metabolic syndrome. Cardiovasc Hematol Disord Drug Targets. https://doi.org/10.2174/1871529X19666190919110652. [Epub ahead of print]
54. Sohaei S, Amani R, Tarrahi MJ, Ghasemi-Tehrani H (2019) The effects of curcumin supplementation on glycemic status, lipid profile and hs-CRP levels in overweight/obese women with polycystic ovary syndrome: a randomized, double-blind, placebo-controlled clinical trial. Complement Ther Med 47:102201. https://doi.org/10.1016/j.ctim.2019.102201
55. Chuengsamarn S, Rattanamongkolgul S, Luechapudiporn R, Phisalaphong C, Jirawatnotai S (2012) Curcumin extract for prevention of type 2 diabetes. Diabetes Care 35(11):2121–2127
56. Cicero AF, Sahebkar A, Fogacci F, Bove M, Giovannini M, Borghi C (2020) Effects of phytosomal curcumin on anthropometric parameters, insulin resistance, cortisolemia and non-alcoholic fatty liver disease indices: a double-blind, placebo-controlled clinical trial. Eur J Nutr 59(2):477–483
57. Kocher A, Bohnert L, Schiborr C, Frank J (2016) Highly bioavailable micellar curcuminoids accumulate in blood, are safe and do not reduce blood lipids and inflammation markers in moderately hyperlipidemic individuals. Mol Nutr Food Res 60(7):1555–1563
58. Na LX, Li Y, Pan HZ, Zhou XL, Sun DJ, Meng M et al (2013) Curcuminoids exert glucose-lowering effect in type 2 diabetes by decreasing serum free fatty acids: a double-blind, placebo-controlled trial. Mol Nutr Food Res 57(9):1569–1577
59. Yang YS, Su YF, Yang HW, Lee YH, Chou JI, Ueng KC (2014) Lipid-lowering effects of curcumin in patients with metabolic syndrome: a randomized, double-blind, placebo-controlled trial. Phytother Res 28(12):1770–1777
60. Navekar R, Rafraf M, Ghaffari A, Asghari-Jafarabadi M, Khoshbaten M (2017) Turmeric supplementation improves serum glucose indices and leptin levels in patients with nonalcoholic fatty liver diseases. J Am Coll Nutr 36(4):261–267
61. Aggarwal BB (2010) Targeting inflammation-induced obesity and metabolic diseases by curcumin and other nutraceuticals. Annu Rev Nutr 30:173–199
62. Shehzad A, Ha T, Subhan F, Lee YS (2011) New mechanisms and the anti-inflammatory role of curcumin in obesity and obesity-related metabolic diseases. Eur J Nutr 50(3):151–161
63. Marseglia L, Manti S, D'Angelo G, Nicotera A, Parisi E, Di Rosa G et al (2015) Oxidative stress in obesity: a critical component in human diseases. Int J Mol Sci 16(1):378–400
64. Fu Y, Zheng S, Lin J, Ryerse J, Chen A (2008) Curcumin protects the rat liver from CCl4-caused injury and fibrogenesis by attenuating oxidative stress and suppressing inflammation. Mol Pharmacol 73(2):399–409
65. Akbari M, Lankarani KB, Tabrizi R, Ghayour-Mobarhan M, Peymani P, Ferns G et al (2019) The effects of curcumin on weight loss among patients with metabolic syndrome and related disorders: a systematic review and meta-analysis of randomized controlled trials. Front Pharmacol 10:649. https://doi.org/10.3389/fphar.2019.00649
66. Baziar N, Parohan M (2020) The effects of curcumin supplementation on body mass index, body weight, and waist circumference in patientswith nonalcoholic fatty liver disease: a systematic review and dose–response meta-analysis of randomized controlled trials. Phytother Res 34(3):464–474

Effect of Curcumin on Glycemic Control in Patients with Type 2 Diabetes: A Systematic Review of Randomized Clinical Trials

8

Atena Mahdavi, Sajjad Moradi, Gholamreza Askari, Bijan Iraj, Thozhukat Sathyapalan, Paul C. Guest, Mohammad Bagherniya, and Amirhossein Sahebkar

Abstract

Type 2 diabetes mellitus (T2DM) is a chronic metabolic condition, which carries considerable morbidity and mortality. There is growing evidence that curcumin could modulate glucose homeostasis and improve vascular risk in patients with T2DM. The aim of this systematic review was to study the effect of curcumin on glycemic indices in patients with diabetes. A comprehensive search was conducted in PubMed, Scopus, Embase, and Google Scholar up to March 5, 2020, to identify randomized control trials investigating the effect of curcumin supplementation on glycemic indices including fasting blood glucose (FBS), hemoglobin A1c (HbA1C), and the homeostatic model assessment for insulin

A. Mahdavi · G. Askari · M. Bagherniya (✉)
Food Security Research Center, Department of Community Nutrition, School of Nutrition and Food Sciences, Isfahan University of Medical Sciences, Isfahan, Iran

S. Moradi
Halal Research Center of IRI, FDA, Tehran, Iran

Nutritional Sciences Department, School of Nutritional Sciences and Food Technology, Kermanshah University of Medical Sciences, Kermanshah, Iran

B. Iraj
Isfahan Endocrine and Metabolism Research Center, Isfahan University of Medical Sciences, Isfahan, Iran

T. Sathyapalan
Department of Academic Diabetes, Endocrinology and Metabolism, Hull York Medical School, University of Hull, Hull, UK

P. C. Guest
Laboratory of Neuroproteomics, Department of Biochemistry and Tissue Biology, Institute of Biology, University of Campinas (UNICAMP), Campinas, Brazil

A. Sahebkar (✉)
Biotechnology Research Center, Pharmaceutical Technology Institute, Mashhad University of Medical Sciences, Mashhad, Iran

Neurogenic Inflammation Research Center, Mashhad University of Medical Sciences, Mashhad, Iran

Polish Mother's Memorial Hospital Research Institute (PMMHRI), Lodz, Poland

Halal Research Center of IRI, FDA, Tehran, Iran
e-mail: sahebkara@mums.ac.ir

P. C. Guest (ed.), *Studies on Biomarkers and New Targets in Aging Research in Iran*, Advances in Experimental Medicine and Biology 1291, https://doi.org/10.1007/978-3-030-56153-6_8

resistance (HOMA-IR). Eleven articles comprising 1131 individuals with T2DM were included in the study. Treatment with curcumin significantly reduced the level of FBS and HbA1c in 8 and 7 studies, respectively. HOMA-IR was evaluated in five studies, and this was reduced significantly by curcumin supplementation in three of those studies. Patients who took curcumin supplementation over longer periods (≥12 weeks) showed a significant reduction in glycemic indices. The current systematic review showed that curcumin can improve glycemic control in patients with T2DM. However, further studies are required to determine the optimum conditions for these effects of curcumin, particularly regarding readouts of insulin resistance.

Keywords

Curcumin · Turmeric · Type 2 diabetes mellitus · Glucose · Insulin resistance

8.1 Introduction

Type 2 diabetes mellitus (T2DM) is a chronic metabolic condition associated mainly with an increase in insulin resistance (IR) [1, 2]. T2DM is a major risk factor for the development of cardiovascular disorders and carries significant morbidity and mortality [1, 2]. It has been estimated that 463 million adults are living with diabetes worldwide, and this has been projected to increase to 578 million adults by 2030 [3], with Asia as a major epicenter [4].

It has been demonstrated that long-term hyperglycemia is a major risk factor for cardiovascular disease, blindness, and renal failure [5, 6]. Thus, appropriate management of T2DM is important for managing cardiovascular disorders [7]. Currently, cardiovascular disorders are a leading cause of morbidity and mortality in patients with T2DM [4]. Recent developments in the pharmaceutical have focused on the discovery of new drugs for T2DM and also on the development of newer agents with fewer side effects. There is a trend toward the use of natural therapeutic agents to improve metabolic health [8]. Synthetic medications and surgery are therapeutic strategies for the treatment of patients with T2DM, but these can have side effects and constitute expensive alternatives for the patients and healthcare systems. Therefore, efficacious medicinal plants and polyphenol compounds with fewer side effects could potentially be beneficial for patients [9–16].

Curcumin is a natural polyphenol responsible for the yellow color and several pharmacological effects of turmeric (*Curcuma longa* L.), a popular spice in Asian cuisine [17–23]. Several studies have shown that the levels of fasting blood glucose (FBG), glycated hemoglobin (HbA1c), and IR are reduced by curcumin, thus enhancing glucose metabolism [24–26]. These reports recommend that curcuminoids may influence not only glucose homeostasis but also the vascular complications in patients with T2DM [27]. However, the results of studies in terms of efficacy of curcumin supplementation on glycemic control in patients with T2DM have not been consistent. One study demonstrated that curcumin had a significant reduction in FBS and HbA1c [28], but another found no effect on these parameters [29]. Therefore, this systematic review was undertaken to investigate the effect of curcumin or combined curcuminoids on glucose metabolism in patients with T2DM based on randomized controlled trials (RCTs).

8.2 Methods

We conducted this systematic review according to the guidelines of the Preferred Reporting Items for Systematic Reviews and Meta-analyses (PRISMA) [30].

8.2.1 Search Strategy

A systematic and comprehensive search was performed on the online databases PubMed (http://www.ncbi.nlm.nih.gov/pubmed), Scopus (http://www.scopus.com), Embase (https://www.

embase.com/) and Google Scholar and was carried out up to March 5, 2020, without any limitations on date of publication. The following search terms were applied in titles and abstracts to identify the potential articles of interest: "Diabetes Mellitus" AND "Curcumin" OR "Turmeric" OR "Turmeric Yellow" OR "Diferuloylmethane." We used Medical Subject Headings (MESH) terms whenever possible. Finally, a manual search was performed using the reference lists of the included original papers and review articles to identify other relevant studies.

8.2.2 Study Selection

Using End-Note software (version X8; Thomson Reuters; https://endnote.com/), duplicate publications were removed from the initial electronic search results. Next, the titles and abstracts of the remaining records were evaluated by two independent researchers (MB and SM). As required, the full-text versions of remaining publications were assessed. Any disagreements were resolved by discussion and a final agreement made. Finally, all randomized controlled trials in humans that examined the effects of curcumin supplementation on diabetes mellitus were included.

8.2.3 Inclusion and Exclusion Criteria

Relevant articles were included if they.

(a) Included a target population of patients with T2DM.
(b) Applied a clinical trial design.
(c) Used curcumin or turmeric supplementation.
(d) Compared the effects of curcumin/turmeric supplementation with placebo.

Studies were excluded if they:

(a) Used a mixture of curcumin with other natural compounds.
(b) Were non-English language records.
(c) Had no control or placebo group were experimental studies.
(d) Were duplicate studies.
(e) Were reviews, letters, editorial articles, or case reports.

8.2.4 Data Extraction

The following data were extracted from the full text of eligible studies:

(a) First author's last name.
(b) Publication year.
(c) Location of the study.
(d) Subjects
(e) Study design.
(f) Number of participants.
(g) Gender.
(h) Age.
(i) Baseline body mass index (BMI).
(j) Intervention.
(k) Control.
(l) Treatment duration.
(m) Main outcomes.
(n) Jadad score.

Data extraction was conducted by AM and MB.

8.2.5 Quality Assessment of Studies

The Jadad scale was used to assess the quality of the included studies. This scale is constructed of the following items:

(a) Randomization (0–2 points).
(b) Blinding (0–2 points).
(c) Dropouts and withdrawals (0–1 point).

Therefore, the overall score of a study can range between 0 and 5, with higher scores indicting higher quality [31, 32]. Studies achieving Jadad scores of less than or equal to 2 and greater than or equal to 3 were considered low quality and high quality, respectively.

8.3 Results

8.3.1 Study Characteristics

The search led to identification of a total of 2886 articles, of which 1055 were duplicates and these were removed, leaving 1831 records. After reading the title and abstract, 1793 irrelevant records were omitted, leaving 38 articles which were analyzed in full. Twenty-seven articles were excluded, which either included duplicated data ($n = 4$), studied other conditions such as metabolic syndrome, non-alcoholic fatty liver disease, hyperlipidemia, or cardiovascular risk factors ($n = 9$), were surveys assessing the effect of curcumin on antioxidant or redox status ($n = 4$), investigated glyburide therapy or vascular function ($n = 2$), were experimental studies ($n = 4$), constituted a meta-analysis ($n = 1$), were not in English ($n = 2$), or did not include a placebo ($n = 1$). This resulted in a total of 11 articles which were included in the final analysis (Fig. 8.1).

The principal features of the selected studies are reviewed in Table 8.1. All clinical trials were published between 2008 and 2019. The sample size differed from 32 to 240 and consisted of a total of 1131 individuals with T2DM. The studies had been performed in various countries, including Iraq [33], Iran [34–38], Thailand [4, 29], India [39], China [28], and Australia [40], and approximately 91% of the data were obtained from other Asian countries ($n = 10$). The participant ages ranged from 27 to 70 years old, and the study durations ranged between 1 month and 9 months. Two studies used curcumin (1000, 15,000 mg/day) [33, 38], one used curcumin nanomicelle (80 mg/day) [36], five utilized curcuminoids (250 and 1000 mg/day) [4, 28, 29, 34, 39], one used curcumin (Meriva®) (2 × 500 mg tablets) [40], another study used 2.1 g powdered turmeric rhizomes per day [35], and another studied nano-curcumin (80 mg/day) [37]. All of the studies used the American Diabetes Association (ADA) practice guidelines to diagnose patients with T2DM.

8.3.2 Quality Assessment

All 11 trials applied randomization, but 3 did not report a randomization method [33, 34, 39] (Table 8.2). A total of nine studies used a blinded approach, two studies did not blind [33, 39], and eight studies reported how the study was blinded [4, 28, 29, 34, 35, 37, 38, 40]. Eight articles described the number and cause of loss to follow-up, but three articles had no description regarding this [4, 33, 39]. In total, nine articles were categorized as high quality [4, 28, 29, 34–38, 40], and two studies were classified as being of low quality using the Jadad scale [33, 39].

8.3.3 Effect of Curcumin on FBS, HbA1c, and HOMA-IR

All included studies assessed FBS and HbA1c as their main outcomes, and five studies reported HOMA-IR as their outcome [4, 28, 29, 35, 38]. In one study, the turmeric rhizome (2100 mg/day) was used although none of the glycemic indices (FBS, HbA1c, and HOMA-IR) were significantly altered in comparison to the control group [35]. Curcumin was used in the other ten studies and in eight of these, supplementation significantly reduced FBS levels [4, 28, 29, 33, 34, 36–38]. However, two studies which used curcumin described no effect [39, 40]. The results of seven studies showed that curcumin reduced HbA1c significantly [4, 28, 29, 33, 34, 36, 37], but there was no effect on HbA1c levels in three studies [38–40]. In three studies, curcumin supplementation resulted in significantly decreased HOMA-IR [4, 28, 29], while in one study, HOMA-IR did not significantly change in response to curcumin consumption [38]. In three studies, the use of curcuminoids (250–300 mg/day) led to a significant reduction in all three glycemic indices in comparison to the control group [4, 28, 29]. In four studies that used either curcumin (1000 mg/day) [33], curcuminoids (1000 mg/day plus piperine 10 mg/day) [34], curcumin (nano-micelle 80 mg/day) [36], and nano-curcumin (80 mg/day) [37],

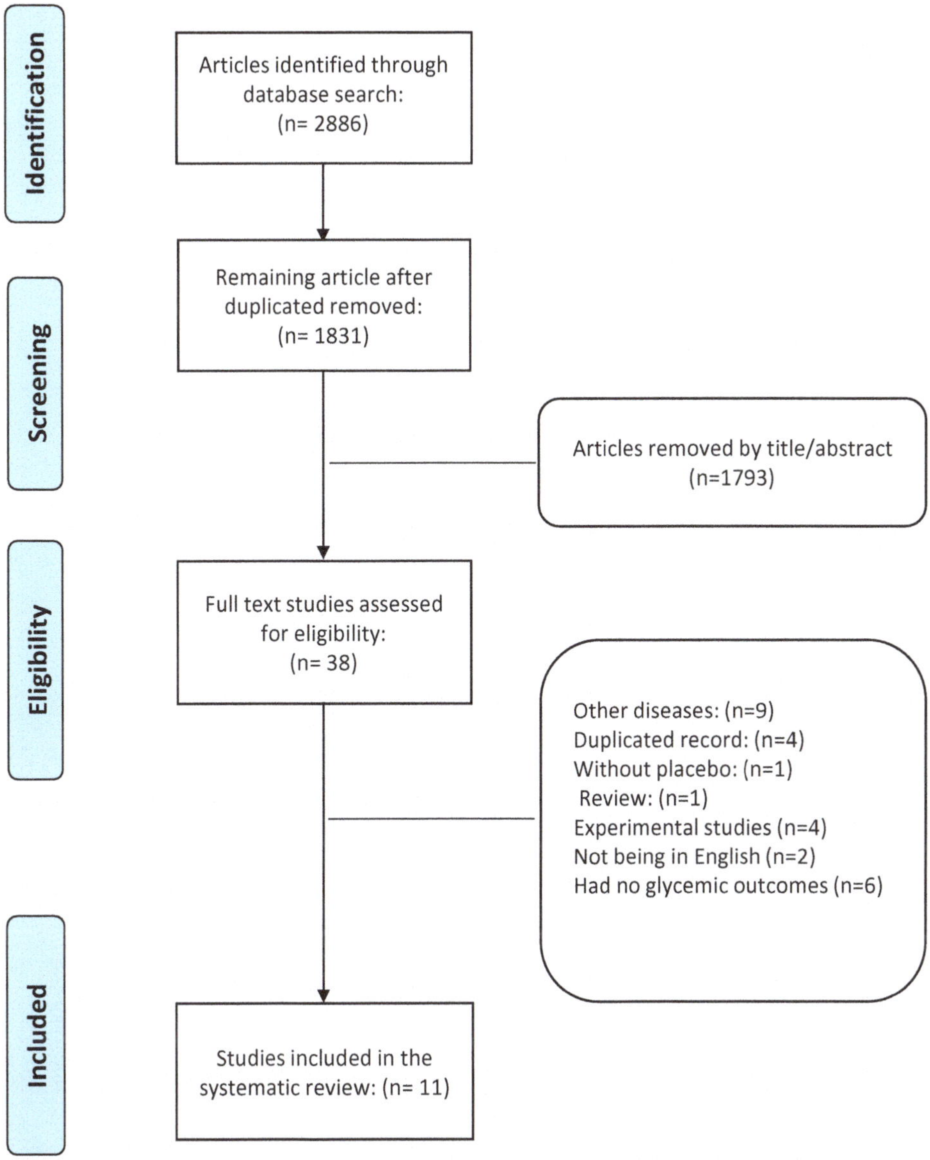

Fig. 8.1 Flow chart showing the study selection procedure

both FBS and HbA1c were significantly reduced. In eight studies out of ten, which used curcumin to study glycemic indices, at least one of these indices was markedly reduced compared with placebo [4, 28, 29, 33, 34, 36–38]. In terms of treatment duration, patients who took supplementation for longer periods ($\geq$ 3 months) generally showed better results compared to studies of shorter duration.

8.4 Discussion

The main finding of the current systematic review is that curcumin-related supplementation significantly reduced at least one of the glycemic indices in 8 out of the 11 studies, which passed the selection criteria. Considering the fact that T2DM is an increasing worldwide public health problem

Table 8.1 Summary of the studies included in the systematic review. All studies include males and females

First author, year, country	Patient group	Study design	Participant number (treatment/control)	Age range (years) or mean ± SD	Baseline BMI (kg/m^2) Range or mean ± SD	Intervention	Control	Treatment duration	Main outcomes (compared to placebo group)			Jadad score
									FBS	HbA1c	HOMA-IR	
Abbood (2017), Iraq [33]	Diabetes with dys-lipidemia	RCT	32 (16/16)	46.76 ± 7.89	NR	Curcumin (1000 mg/day)	Starch	12 weeks	↓	↓	NR	1
Panahi (2017), Iran [34]	Diabetes	RCT double-blind placebo	100 (50/50)	42 ± 7.54	26.93 ± 2.01	Curcuminoids (1000 mg/day plus piperine 10 mg/day)	Cellulose	12 weeks	↓	↓	NR	4
Chuengsamarn (2014), Thailand [29]	Diabetes	RCT double-blind placebo	240	59.16 ± 11.03	27.09 ± 5.37	Curcuminoids (250 mg/day)	Placebo	6 months	↓	↓	↓	5
Chuengsamarn (2012), Thailand [4]	Prediabetes	RCT double-blind placebo	240	56.95 ± 12.04	26.66 ± 5.25	Curcuminoids (250 mg/day)	Placebo	9 months	↓	↓	↓	4
Adab (2019), Iran [35]	Diabetes with dyslipidemia	RCT double-blind	80 (40/40)	54.76 ± 6	28.98 ± 3.68	Rhizome turmeric (2100 mg/day)	Placebo	8 weeks	No effect	No effect	No effect	5
Rahimi (2016), Iran [36]	Diabetes	RCT double-blind	70 (35/35)	27.09 ± 3.16	58.64 ± 11.13	Curcumin (nanomicelle 80 mg/day)	Placebo	3 months	↓	↓	NR	4
Asadi (2019), Iran [37]	Diabetes	Parallel RCT double-blind placebo	80 (40/40)	53.3 ± 6.5	31.1 ± 4.2	Nano-curcumin (80 mg/day)	Placebo	8 weeks	↓	↓	NR	5
Na (2013), China [28]	Diabetes	RCT double-blind placebo	100 (49/51)	55.07 ± 7.4	27.27 ± 2.6	Curcuminoids 300 mg/day	Starch	3 month	↓	↓	↓	5
Usharanirt (2008), India [39]	Diabetes	RCT, parallel group, placebo digital volume plethysmography	72 (21 placebo/23 atorvastatin/23 curcumin)	55.52 ± 10.76	24.66 ± 2.42	Curcuminoids 600 mg/day	Placebo	8 weeks	No effect	No effect	NR	1

First author, year, country	Patient group	Study design	Participant number (treatment/control)	Age range (years) or mean ± SD	Baseline BMI (kg/m²) Range or mean ± SD	Intervention	Control	Treatment duration	Main outcomes (compared to placebo group)			Jadad score
									FBS	HbA1c	HOMA-IR	
Hodaei (2019), Iran [38]	Diabetes	RCT double-blind placebo	53 (25/28)	58 ± 8	29.2 ± 3.76	Curcumin (1500 mg/day)	Placebo	10 weeks	↓	No effect	No effect	5
Thota (2019), Australia [40]	Diabetes	2 × 2 factorial RCT double-blind placebo	64 (26/38)	55 ± 1.3	31.1 ± 0.7	Curcumin (Meriva®) (2 × 500mg tablets)	Placebo	12 weeks	No effect	No effect	NR	5

NR not reported, *RCT*: randomized controlled trial, *FBS* fasting blood glucose, *HbA1C* hemoglobin A1c, *HOMA-IR* homeostatic model assessment for insulin resistance

Table 8.2 Quality of studies assessed by the JADAD scale

	Randomization	Blinding	Account of patients	Total point
Abbood (2017)	1	0	0	1
Panahi (2017)	1	2	1	4
Chuengsamarn (2014)	2	2	1	5
Chuengsamarn (2012)	2	2	0	4
Adab (2019)	2	2	1	5
Rahimi (2016)	2	1	1	4
Asadi (2019)	2	2	1	5
Na (2013)	2	2	1	5
Usharanirt (2008)	1	0	0	1
Hodaei (2019)	2	2	1	5
Thota (2019)	2	2	1	5

associated with considerable morbidity, including loss of kidney function, CVD, and early death [41], the findings of this study show some promise. We observed significant reductions in FBG, HbA1c, and HOMA-IR levels following supplementation with curcumin or curcumin-related compounds in most of the studies in which these indices were measured as outcomes. All three of these indices have been shown significant associations in T2DM patients with atherosclerosis, which is a risk factor for life-threatening conditions such as heart attack and stroke [29].

Most of the involved studies were of high quality, and the majority measured FBG and HbA1c. Our analysis showed that curcumin-related treatments decreased FBS significantly in 8 of the 11 studies in patients with T2DM [4, 28, 29, 33, 34, 36–38], while three of the studies showed no effect [35, 39, 40]. Seven studies confirmed that supplementation with curcumin-related compounds reduced HbA1c significantly [4, 28, 29, 33, 34, 36, 37], but there was no effect in four studies [35, 38–40]. In addition, the curcumin-related treatments led to a decrease in HOMA-IR in three of the five studies in which it was assessed [4, 28, 29], with no effect in two of the studies [35, 38]. Therefore, the question as to whether or not curcumin has an effect on insulin resistance requires further investigation.

We found that longer treatment duration with curcumin-related compounds was associated with a greater reduction in the levels of FBG, HbA1c, and HOMA-IR. This suggests that longer treatment durations may be required for curcumin to induce significant beneficial changes in glucose metabolism. This is consistent with the findings of a recent meta-analysis by Huang et al., which found that longer treatment durations with curcumin or combined curcuminoids were associated with more significant reductions in FBG, HbA1c, and HOMA-IR in T2DM patients [42].

Several strategies such as having a healthy diet, use of hypoglycemic agents, and insulin administration have been adopted to manage diabetes [43], although most of the pharmacological agents in use have some side effects [44]. Thus, natural agents such as medicinal plants and herbal remedies may have utility in the management of metabolic conditions such as T2DM without significant adverse effects [4]. Curcumin has been shown to have a beneficial effect on the management of metabolic conditions due to its anti-inflammatory and antidiabetic properties [26, 45, 46]. It has also been shown to have some beneficial effects in conditions associated with heightened inflammation, such as metabolic syndrome [47], hyperlipidemia [48], hypertension [49], and nonalcoholic fatty liver diseases [50]. Curcumin supplementation was found to inhibit the production of pro-inflammatory cytokines in obesity and metabolic syndrome [51], and it has been shown to prevent the development of T2DM and decrease insulin resistance in animal models [52, 53].

The pathophysiology of T2DM is due to a combination of insulin resistance in peripheral tissues and/or insulin deficiency at the level of pancreatic β cells [54, 55]. Research findings have suggested that curcumin intervention in a prediabetic population can prevent the development of T2DM by enhancing insulin sensitivity at target tissues and by improving β-cell function [4, 29, 37]. In addition, another study found that curcumin supplementation led to a significant increase in the levels of the adipose hormone adiponectin [4], and high levels of this hormone have been associated with a reduced risk of developing T2DM [56].

The mechanism of action of curcumin in improving biomarkers in T2DM may also be due to its antioxidant effects, which includes inhibiting the formation of free radicals and stimulating antioxidant enzymes such as glutathione peroxidase, superoxide dismutase, and catalase [57]. Increasing evidence has suggested that oxidative stress plays a major role in diabetes pathogenesis [58], and the antioxidant effects of curcumin may therefore be beneficial in the management of diabetes by improving insulin resistance and pancreatic β-cell function [59]. In addition, curcumin supplementation has been shown to lower circulating levels of free fatty acids [60, 61], which are known risk factors of metabolic disorders such as obesity and T2DM [62].

The limitations of this systematic review include the points that the number of studies available on this topic was comparatively low and sample size was small in most of these. Also, some studies were eliminated due to being written in a language other than English. In addition, the available articles showed heterogeneity with respect to the dosage and duration of intervention. Finally, most of the studies were conducted in Asian countries, so generalization of these results to other population is not certain. More randomized controlled trials should be conducted with larger sample sizes and longer durations to clarify the effects of curcumin and curcumin-related compounds on T2DM.

8.5 Conclusions

The findings of our systematic review showed a beneficial effect of curcumin and curcumin-related compounds in control of hyperglycemia in T2DM patients. This study showed that consumption of curcumin supplements could lead to a reduction in fasting blood glucose and glycosylated hemoglobin levels in patients with T2DM, compared to placebo controls. The findings with regard to effects on insulin resistance were less certain as reductions in this index were found in only three out of the five studies that measured HOMA-IR. In terms of treatment duration, participants who took supplementation for longer periods generally showed better results compared to studies that used shorter treatment durations. It can be concluded that more clinical trial studies are needed to verify the effects of curcumin using standardized protocols to allow more accurate comparisons across studies. Finally, although curcumin is generally safe and well-tolerated according ot the currently avilable clinical studies [63], evidence from long-term trials is still lacking and mertits further investigations.

References

1. Donnelly R, Davis KR (2000) Type 2 diabetes and atherosclerosis. Diabetes Obes Metab 2:S21–S30
2. Marfella R, Di Filippo C, D'Amico M, Paolisso G (2007) Diabetes, ubiquitin proteasome system and atherosclerotic plaque rupture. Circ Res 100(10):e84–e85
3. IDF Diabetes Atlas (2019) 9th edition. https://diabetesatlas.org/en/sections/worldwide-toll-of-diabetes.html
4. Zheng Y, Ley SH, Hu FB (2018) Global aetiology and epidemiology of type 2 diabetes mellitus and its complications. Nat Rev Endocrinol 14(2):88–98
5. Mealey BL, Ocampo GL (2007) Diabetes mellitus and periodontal disease. Periodontology 44(1):127–153
6. Verma H, Garg R (2017) Effect of magnesium supplementation on type 2 diabetes associated cardiovascular risk factors: a systematic review and meta-analysis. J Hum Nutr Diet 30(5):621–633
7. Knowler WC, Barrett-Connor E, Fowler SE, Hamman RF, Lachin JM, Walker EA et al (2002) Reduction in the incidence of type 2 diabetes with lifestyle intervention or metformin. N Engl J Med 346(6):393–403
8. Apostolidis E, Kwon YI, Shetty K (2006) Potential of select yogurts for diabetes and hypertension management. J Food Biochem 30(6):699–717
9. Bagherniya M, Nobili V, Blesso CN, Sahebkar A (2018) Medicinal plants and bioactive natural compounds in the treatment of non-alcoholic fatty liver disease: a clinical review. Pharmacol Res 130:213–240
10. Kooti W, Farokhipour M, Asadzadeh Z, Ashtary-Larky D, Asadi-Samani M (2016) The role of medicinal plants in the treatment of diabetes: a systematic review. Electron Physician 8(1):1832–1842

11. Talebi S, Bagherniya M, Atkin SL, Askari G, Orafai HM, Sahebkar A (2020) The beneficial effects of nutraceuticals and natural products on small dense LDL levels, LDL particle number and LDL particle size: a clinical review. Lipids Health Dis 19:1–21
12. Bagherniya M, Johnston TP, Sahebkar A (2020) Regulation of apolipoprotein B by natural products and nutraceuticals: a comprehensive review. Curr Med Chem. https://doi.org/10.2174/0929867327666200427092114. [Epub ahead of print]
13. Mollazadeh H, Cicero AFG, Blesso CN, Pirro M, Majeed M, Sahebkar A (2019) Immune modulation by curcumin: the role of interleukin-10. Crit Rev Food Sci Nutr 59(1):89–101
14. Momtazi AA, Derosa G, Maffioli P, Banach M, Sahebkar A (2016) Role of microRNAs in the therapeutic effects of curcumin in non-cancer diseases. Mol Diagn Ther 20(4):335–345
15. Banach M, Patti AM, Giglio RV, Cicero AFG, Atanasov AG, Bajraktari G, Bruckert E, Descamps O, Djuric DM, Ezhov M, Fras Z, von Haehling S, Katsiki N, Langlois M, Latkovskis G, Mancini GBJ, Mikhailidis DP, Mitchenko O, Moriarty PM, Muntner P, Nikolic D, Panagiotakos DB, Paragh G, Paulweber B, Pella D, Pitsavos C, Reiner Ž, Rosano GMC, Rosenson RS, Rysz J, Sahebkar A, Serban MC, Vinereanu D, Vrablík M, Watts GF, Wong ND, Rizzo M (2018) International Lipid Expert Panel (ILEP). The Role of Nutraceuticals in Statin Intolerant Patients. J Am Coll Cardiol 72(1):96–118. https://doi.org/10.1016/j.jacc.2018.04.040. PMID: 29957236
16. Iranshahi M, Sahebkar A, Takasaki M, Konoshima T, Tokuda H (2009) Cancer chemopreventive activity of the prenylated coumarin, umbelliprenin, in vivo. Eur J Cancer Prev 18(5):412–415
17. Anand P, Thomas SG, Kunnumakkara AB, Sundaram C, Harikumar KB, Sung B et al (2008) Biological activities of curcumin and its analogues (congeners) made by man and mother nature. Biochem Pharmacol 76(11):1590–1611
18. Abdollahi E, Momtazi AA, Johnston TP, Sahebkar A (2018) Therapeutic effects of curcumin in inflammatory and immune-mediated diseases: a nature-made jack-of-all-trades? J Cell Physiol 233(2):830–848
19. Bagheri H, Ghasemi F, Barreto GE, Rafiee R, Sathyapalan T, Sahebkar A (2020) Effects of curcumin on mitochondria in neurodegenerative diseases. Biofactors. Jan;46(1):5–20. https://doi.org/10.1002/biof.1566. Epub 2019 Oct 3. PMID: 31580521.
20. Momtazi AA, Sahebkar A (2016) Difluorinated curcumin: a promising curcumin analogue with improved anti-tumor activity and pharmacokinetic profile. Curr Pharm Des 22(28):4386–4397
21. Hasanzadeh S, Read MI, Bland AR, Majeed M, Jamialahmadi T, Sahebkar A (2020) Curcumin: an inflammasome silencer. Pharmacol Res 159:104921. https://doi.org/10.1016/j.phrs.2020.104921. Epub 2020 May 25. PMID: 32464325.
22. Teymouri M, Pirro M, Johnston TP, Sahebkar A (2017) Curcumin as a multifaceted compound against human papilloma virus infection and cervical cancers: A review of chemistry, cellular, molecular, and preclinical features. BioFactors 43(3):331–346.
23. Mohajeri M, Bianconi V, Ávila-Rodriguez MF, Barreto GE, Jamialahmadi T, Pirro M, Sahebkar A. (2020) Curcumin: a phytochemical modulator of estrogens and androgens in tumors of the reproductive system. Pharmacol Res Jun;156:104765. https://doi.org/10.1016/j.phrs.104765. Epub 2020 Mar 23. PMID: 32217147.
24. Kuroda M, Mimaki Y, Nishiyama T, Mae T, Kishida H, Tsukagawa M et al (2005) Hypoglycemic effects of turmeric (Curcuma longa L. rhizomes) on genetically diabetic KK-Ay mice. Biol Pharm Bull 28(5):937–939
25. Shao W, Yu Z, Chiang Y, Yang Y, Chai T, Foltz W et al (2012) Curcumin prevents high fat diet induced insulin resistance and obesity via attenuating lipogenesis in liver and inflammatory pathway in adipocytes. PLoS One 7(1):e28784. https://doi.org/10.1371/journal.pone.0028784
26. Weisberg SP, Leibel R, Tortoriello DV (2008) Dietary curcumin significantly improves obesity-associated inflammation and diabetes in mouse models of diabesity. Endocrinology 149(7):3549–3558
27. Katsiki N, Yovos JG, Gotzamani-Psarrakou A, Karamitsos DT (2011) Adipokines and vascular risk in type 2 diabetes mellitus. Angiology 62(8):601–604
28. Na LX, Li Y, Pan HZ, Zhou XL, Sun DJ, Meng M et al (2013) Curcuminoids exert glucose-lowering effect in type 2 diabetes by decreasing serum free fatty acids: a double-blind, placebo-controlled trial. Mol Nutr Food Res 57(9):1569–1577
29. Chuengsamarn S, Rattanamongkolgul S, Phonrat B, Tungtrongchitr R, Jirawatnotai S (2014) Reduction of atherogenic risk in patients with type 2 diabetes by curcuminoid extract: a randomized controlled trial. J Nutr Biochem 25(2):144–150
30. Moher D, Liberati A, Tetzlaff J, Altman DG (2009) Preferred reporting items for systematic reviews and meta-analyses: the PRISMA statement. Ann Intern Med 151(4):264–269
31. Jadad AR, Moore RA, Carroll D, Jenkinson C, Reynolds DJM, Gavaghan DJ et al (1996) Assessing the quality of reports of randomized clinical trials: is blinding necessary? Control Clin Trials 17(1):1–12
32. Moher D, Cook D, Jadad A, Tugwell P, Moher M, Jones A et al (1999) Assessing the quality of reports of randomised trials: implications for the conduct of meta-analyses. Health Technol Assess 3(12):i–iv, 1–98
33. Abbood MS (2017) Hypolipidaemic and anti-inflammatory effects of curcumin versus atorvastatin in type 2 diabetic patients. Int J Pharm Sci Rev Res 49(1):1–7
34. Panahi Y, Khalili N, Sahebi E, Namazi S, Reiner Ž, Majeed M et al (2017) Curcuminoids modify lipid profile in type 2 diabetes mellitus: a randomized controlled trial. Complement Ther Med 33:1–5
35. Adab Z, Eghtesadi S, Vafa MR, Heydari I, Shojaii A, Haqqani H et al (2019) Effect of turmeric on glycemic status, lipid profile, hs-CRP, and total antioxidant capacity in hyperlipidemic type 2 diabetes mellitus patients. Phytother Res 33(4):1173–1181

36. Rahimi HR, Mohammadpour AH, Dastani M, Jaafari MR, Abnous K, Mobarhan MG et al (2016) The effect of nano-curcumin on HbA1c, fasting blood glucose, and lipid profile in diabetic subjects: a randomized clinical trial. Avicenna J Phytomed 6(5):567–577
37. Asadi S, Gholami MS, Siassi F, Qorbani M, Khamoshian K, Sotoudeh G (2019) Nano curcumin supplementation reduced the severity of diabetic sensorimotor polyneuropathy in patients with type 2 diabetes mellitus: a randomized double-blind placebo-controlled clinical trial. Complement Ther Med 43:253–260
38. Hodaei H, Adibian M, Nikpayam O, Hedayati M, Sohrab G (2019) The effect of curcumin supplementation on anthropometric indices, insulin resistance and oxidative stress in patients with type 2 diabetes: a randomized, double-blind clinical trial. Diabetol Metab Syndr 11:41. https://doi.org/10.1186/s13098-019-0437-7
39. Usharani P, Mateen A, Naidu M, Raju Y, Chandra N (2008) Effect of NCB-02, atorvastatin and placebo on endothelial function, oxidative stress and inflammatory markers in patients with type 2 diabetes mellitus. Drugs R D 9(4):243–250
40. Thota RN, Acharya SH, Garg ML (2019) Curcumin and/or omega-3 polyunsaturated fatty acids supplementation reduces insulin resistance and blood lipids in individuals with high risk of type 2 diabetes: a randomised controlled trial. Lipids Health Dis 18(1):31. https://doi.org/10.1186/s12944-019-0967-x
41. Levey AS, Eckardt KU, Tsukamoto Y, Levin A, Coresh J, Rossert J et al (2005) Definition and classification of chronic kidney disease: a position statement from kidney disease: improving global outcomes (KDIGO). Kidney Int 67(6):2089–2100
42. Huang J, Qin S, Huang L, Tang Y, Ren H, Hu H (2019) Efficacy and safety of Rhizoma curcumea longae with respect to improving the glucose metabolism of patients at risk for cardiovascular disease: a meta-analysis of randomised controlled trials. J Hum Nutr Diet 32(5):591–606
43. Tan K (2004) Dyslipidaemia, inflammation and endothelial dysfunction in diabetes mellitus. Int Congr Ser 1262:511–514
44. Arun N, Nalini N (2002) Efficacy of turmeric on blood sugar and polyol pathway in diabetic albino rats. Plant Food Hum Nutr 57(1):41–52
45. Kochhar K (2008) Dietary spices in health and diseases: I. Indian J Physiol Pharmacol 52(2):106–122
46. Jacob A, Wu R, Zhou M, Wang P (2007) Mechanism of the anti-inflammatory effect of curcumin: PPAR-γ activation. PPAR Res 2007:1
47. Sahebkar A (2013) Why it is necessary to translate curcumin into clinical practice for the prevention and treatment of metabolic syndrome? Biofactors 39(2):197–208
48. Qin S, Huang L, Gong J, Shen S, Huang J, Ren H et al (2017) Efficacy and safety of turmeric and curcumin in lowering blood lipid levels in patients with cardiovascular risk factors: a meta-analysis of randomized controlled trials. Nutr J 16(1):68. https://doi.org/10.1186/s12937-017-0293-y
49. Hadi A, Pourmasoumi M, Ghaedi E, Sahebkar A (2019) The effect of curcumin/turmeric on blood pressure modulation: a systematic review and meta-analysis. Pharmacol Res 150:104505. https://doi.org/10.1016/j.phrs.2019.104505
50. Wei Z, Liu N, Tantai X, Xing X, Xiao C, Chen L et al (2019) The effects of curcumin on the metabolic parameters of non-alcoholic fatty liver disease: a meta-analysis of randomized controlled trials. Hepatol Int 13(3):302–313
51. Aggarwal BB (2010) Targeting inflammation-induced obesity and metabolic diseases by curcumin and other nutraceuticals. Annu Rev Nutr 30:173–199
52. Seo KI, Choi MS, Jung UJ, Kim HJ, Yeo J, Jeon SM et al (2008) Effect of curcumin supplementation on blood glucose, plasma insulin, and glucose homeostasis related enzyme activities in diabetic db/db mice. Mol Nutr Food Res 52(9):995–1004
53. Jang EM, Choi MS, Jung UJ, Kim MJ, Kim HJ, Jeon SM et al (2008) Beneficial effects of curcumin on hyperlipidemia and insulin resistance in high-fat–fed hamsters. Metabolism 57(11):1576–1583
54. Scheen A (2003) Pathophysiology of type 2 diabetes. Acta Clin Belg 58(6):335–341
55. Guest PC (2019) Biogenesis of the insulin secretory granule in health and disease. Adv Exp Med Biol 1134:17–32
56. Yokoyama H, Emoto M, Mori K, Araki T, Teramura M, Koyama H et al (2006) Plasma adiponectin level is associated with insulin-stimulated nonoxidative glucose disposal. J Clin Endocrinol Metab 91(1):290–294
57. Oyenihi AB, Ayeleso AO, Mukwevho E, Masola B (2015) Antioxidant strategies in the management of diabetic neuropathy. Biomed Res Int 2015:515042. https://doi.org/10.1155/2015/515042
58. Maritim AC, Sanders RA, Watkins JB 3rd (2003) Diabetes, oxidative stress, and antioxidants: a review. J Biochem Mol Toxicol 17(1):24–38
59. Jeenger MK, Shrivastava S, Yerra VG, Naidu V, Ramakrishna S, Kumar A (2015) Curcumin: a pleiotropic phytonutrient in diabetic complications. Nutrition 31(2):276–282
60. Na LX, Zhang YL, Li Y, Liu LY, Li R, Kong T et al (2011) Curcumin improves insulin resistance in skeletal muscle of rats. Nutr Metab Cardiovasc Dis 21(7):526–533
61. Boden G (2003) Effects of free fatty acids (FFA) on glucose metabolism: significance for insulin resistance and type 2 diabetes. Exp Clin Endocrinol Diabetes 111(3):121–124
62. Saponaro C, Gaggini M, Carli F, Gastaldelli A (2015) The subtle balance between lipolysis and lipogenesis: a critical point in metabolic homeostasis. Nutrients 7(11):9453–9474
63. Soleimani V, Sahebkar A, Hosseinzadeh H (2018) Turmeric (Curcuma longa) and its major constituent (curcumin) as nontoxic and safe substances: Review. Phytother Res 32(6):985–995. https://doi.org/10.1002/ptr.6054. Epub 2018 Feb 26. PMID: 29480523

9 Antidiabetic Properties of Curcumin: Insights on New Mechanisms

Elahe Mohammadi, Behzad Behnam, Reza Mohammadinejad, Paul C. Guest, Luis E. Simental-Mendía, and Amirhossein Sahebkar

Abstract

Plant extracts have been used to treat a wide range of human diseases. Curcumin, a bioactive polyphenol derived from *Curcuma longa* L., exhibits therapeutic effects against diabetes while only negligible adverse effects have been observed. Antioxidant and anti-inflammatory properties of curcumin are the main and well-recognized pharmacological effects that might explain its antidiabetic effects. Additionally, curcumin may regulate novel signaling molecules and enzymes involved in the pathophysiology of diabetes, including glucagon-like peptide-1, dipeptidyl peptidase-4, glucose transporters, alpha-glycosidase, alpha-amylase, and peroxisome proliferator-activated receptor gamma (PPARγ). Recent findings from in vitro and in vivo studies on novel signaling pathways involved in the potential beneficial effects of curcumin for the treatment of diabetes are discussed in this review.

E. Mohammadi
Student Research Committee, School of Pharmacy, Kerman University of Medical Sciences, Kerman, Iran

B. Behnam (✉)
Herbal and Traditional Medicines Research Center, Kerman University of Medical Sciences, Kerman, Iran

Pharmaceutics Research Center, Institute of Neuropharmacology, Kerman University of Medical Sciences, Kerman, Iran

Pharmaceutical Sciences and Cosmetic Products Research Center, Kerman University of Medical Sciences, Kerman, Iran
e-mail: behnamb@kmu.ac

R. Mohammadinejad
Pharmaceutics Research Center, Institute of Neuropharmacology, Kerman University of Medical Sciences, Kerman, Iran

P. C. Guest
Laboratory of Neuroproteomics, Department of Biochemistry and Tissue Biology, Institute of Biology, University of Campinas (UNICAMP), Campinas, Brazil

L. E. Simental-Mendía
Biomedical Research Unit, Mexican Social Security Institute, Durango, Mexico

A. Sahebkar (✉)
Biotechnology Research Center, Pharmaceutical Technology Institute, Mashhad University of Medical Sciences, Mashhad, Iran

Neurogenic Inflammation Research Center, Mashhad University of Medical Sciences, Mashhad, Iran

Polish Mother's Memorial Hospital Research Institute (PMMHRI), Lodz, Poland

Halal Research Center of IRI, FDA, Tehran, Iran
e-mail: sahebkara@mums.ac.ir

P. C. Guest (ed.), *Studies on Biomarkers and New Targets in Aging Research in Iran*, Advances in Experimental Medicine and Biology 1291, https://doi.org/10.1007/978-3-030-56153-6_9

Keywords

Curcumin · Diabetes · Dipeptidyl peptidase-4 · Glucagon-like peptide-1 · Glucose transporter · Alpha-glycosidase · Alpha-amylase · PPARγ

9.1 Introduction

Diabetes is a growing public health problem, which exerts a heavy economic cost for many countries and world healthcare systems [1]. Diabetes is a constellation of metabolic disturbances characterized by chronic high glucose and poor glucose regulation. This hyperglycemia may induce long-term microvascular (retinopathy, neuropathy, and nephropathy) and macrovascular (cardiovascular, cerebrovascular, and peripheral artery diseases) complications through numerous mechanisms such as production of advanced glycation products, oxidative stress, inflammation, and endoplasmic reticulum stress [2]. In addition to lifestyle changes and dietary modifications, pharmacological treatment is indicated for most of diabetes patients. Antidiabetic therapies include the following classes of oral and injectable drugs: biguanides, sulfonylureas, meglitinides, thiazolidinediones, α-glucosidase inhibitors, dipeptidyl peptidase-4 (DPP4) inhibitors, sodium-glucose co-transporter-2 (SGLT2) inhibitors, glucagon-like peptide-1 (GLP-1) agonists, and numerous insulin formulations [3]. Although new drugs showing positive outcomes for the treatment of diabetes have been approved in recent years, there is still a need for alternative or adjunct antidiabetic drugs with greater efficacy and minimal adverse effects [4]. Therefore, natural and herbal products with glucoregulatory properties have emerged as potential therapeutic options, with better safety profiles [5, 6].

Curcumin is one of the predominant natural products, which has a beneficial effect on glucose metabolism. It is the major component of *Curcuma longa* L., which has shown therapeutic potential for a variety of diseases including diabetes [7]. Previously, several biological effects of curcumin have been reported, which could be salutary against a wide range of pathological conditions [8–17]. Additionally, curcumin has been widely examined for its glucose-lowering properties [7]. Evidence from two meta-analyses supported the antidiabetic actions of curcumin and showed that the compound had only mild side effects [18, 19]. Some studies found that curcumin has a wide range of molecular targets including numerous intracellular elements, pathways, and key enzymes involved in glucose metabolism [20, 21]. Several studies have shown that curcumin reduces glucose concentrations by activating AMP-activated protein kinase and inhibiting phosphoenolpyruvate carboxykinase (PEPCK) and glucose 6-phosphatase (G6Pase) activities, as well as increasing insulin levels [22–24]. In addition, curcumin has been demonstrated to have antioxidant and anti-inflammatory activities which appear to cause decreased metabolic (triglycerides, total cholesterol, uric acid, and total body fat), inflammation [TNF-α, IL6, intracellular adhesion molecule 1 (ICAM-1), vascular cell adhesion protein 1 (VCAM-1), and transforming growth factor beta 1 (TGF-β1)], and apoptosis (Bax/Bcl2 ratio, JNK phosphorylation, and caspase 3 cleavage) biomarkers, along with an increase in antioxidant biomarkers [superoxide dismutase (SOD) and glutathione (GSH)] [25]. Moreover, beneficial effects of curcumin on glucose metabolism such as increased glycolysis and glycogen synthesis, as well as decreased gluconeogenesis, have been observed [26]. On the other hand, recent studies have revealed that curcumin has hypoglycemic activity through the regulation of incretins, α-glucosidase, amylase, glucose transporters, and peroxisome proliferator-activated receptor gamma (PPARγ) (Fig. 9.1).

This review has focused on identification of the molecular targets and biomarkers involved in the antidiabetic effects of curcumin.

9.2 Diabetes

Diabetes is a group of metabolic disorders characterized by persistent hyperglycemia as a result of impaired insulin secretion, insulin resistance, or both. This insufficiency is caused by β-cell dysfunction and/or insulin resistance in peripheral tissues [27, 28]. Diabetes is currently classified into the following categories: type 1 diabetes (T1D), type 2 diabetes (T2D), gestational diabetes, and diabetes due to other causes. T1D com-

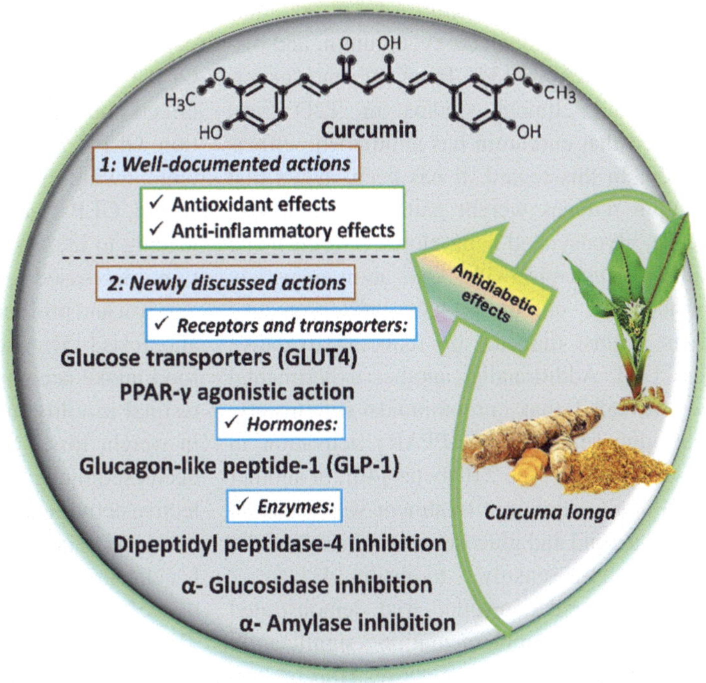

Fig. 9.1 Major pathways regarding the antidiabetic and glucose-lowering actions of curcumin

prises 5–10% of the diabetes cases and is caused by autoimmune destruction of the pancreatic β cells. This type of diabetes is characterized by the presence of autoimmune markers such as autoantibodies that target islet cell proteins, such as glutamic acid decarboxylase-65, insulin, tyrosine phosphatases IA-2 and IA-2β, and zinc transporter 8. Furthermore, T1D has a strong genetic predisposition mainly with human leukocyte antigens, particularly linked to the major histocompatibility complex class II DQ alpha 1 (DQA) and DQ beta 1 (DQB) genes. The rate of beta-cell destruction is generally rapid in children and slow in adults. Patients with T1D often require insulin therapy early in the illness [4, 29]. T2D accounts 90–95% of all types of diabetes and is distinguished by β-cell dysfunction and peripheral insulin resistance [4, 30, 31]. Most of these patients exhibit overweight or obesity, and at least initially, they do not need insulin treatment. Excess weight induces deficient insulin action leading to an increase in insulin secretion as a compensatory response. T2D is more frequent in adults older than 45 years. However, due to the increasing prevalence of obesity, this disease is becoming increasingly observed in young adults, even children and adolescents.

Gestational diabetes is defined as hyperglycemia developed during pregnancy [4, 30, 31]. Usually, it occurs in 7% of all pregnancies and is detected between 24 and 28 weeks of gestation. Gestational diabetes is accompanied by β-cell dysfunction resulting in an increased risk of developing type 2 diabetes later. Additionally, the offspring of patients with gestational diabetes have an elevated risk of congenital malformations and macrosomia. There are other types of diabetes due to different causes such as monogenic diabetes, diseases of the exocrine pancreas, posttransplantation diabetes, and drug-induced diabetes [4, 21, 30]. These forms are rare.

9.3 Curcumin and Type 2 Diabetes

Curcuma longa (Turmeric; Zingiberaceae) is a humid climate plant and native to Asia. This plant is used on food, perfumes, and a substance of herbal medicine [31]. C. longa has a wide range of therapeutic effects including hypoglycemic, hypolipidemic, anti-inflammatory, antioxidant, antihypertensive, and anticancer activities [32, 33]. Curcumin is an extracted polyphenol from

C. longa, which is a mixture of curcuminoids such as curcumin, demethoxycurcumin, and bisdemethoxycurcumin [34]. In recent years, experimental and clinical studies in T2D have suggested that curcumin has antidiabetic effects [26, 35]. In this regard, it has been shown that curcumin reduces weight gain, fasting plasma glucose, glycosylated hemoglobin (HbA1c), and insulin resistance [35–38]. Also, an in vivo study reported that curcumin has cardioprotective effects against diabetes by reducing oxidative stress [39]. Additionally, another experimental study revealed that curcuminoids significantly lower glucose levels via PPARγ activation in mouse models of type 2 diabetes [40]. In another animal study, curcumin treatment was found to ameliorate lipid and glucose metabolism disorder and improve the sensitivity to insulin [41].

A randomized controlled trial demonstrated that curcumin supplementation significantly decreased fasting plasma glucose, HbA1c, and insulin resistance by reducing serum free fatty acids in T2D patients [42]. Another clinical trial found that curcumin treatment decreased total body fat, visceral fat, insulin resistance, leptin levels, and increased serum adiponectin concentrations in T2D patients [43]. Previous meta-analyses observed that curcumin therapy significantly reduced fasting plasma glucose, HbA1c, and insulin resistance in T2D patients [18, 19]. It follows from these preclinical and clinical studies that curcumin might be a natural therapeutic option for the treatment of T2D. Different antidiabetic mechanisms of curcumin have been described including increased insulin release, improved insulin sensitivity in target tissues, decreased gluconeogenesis, and increased glycolysis and glycogen synthesis in the liver, increased glucose uptake in skeletal muscles, induced regeneration of β cells, mitigation of oxidative stress, increased glucagon-like peptide-1 (GLP-1) secretion, inhibition of dipeptidyl peptidase-4 (DPP-4), and upregulation of glucose transporters (GLUTs) and PPAR-γ [44].

9.3.1 The Incretin Pathway

Incretins are a group of gastrointestinal hormones released from the gut, which play a pivotal role in glucose metabolism. Among them, glucose-dependent insulinotropic polypeptide (GIP) and GLP-1 exhibit glucose-dependent insulinotropic effects and regulate insulin and glucagon production. GLP-1 is currently an attractive therapeutic target owing to its antidiabetic properties. In this regard, GLP-1 acts directly on pancreatic β cells leading to increased insulin synthesis and secretion, decreased glucagon secretion, increased survival and proliferation of β cells, and decreased apoptosis [45]. Also, GLP-1 reduces appetite and food intake accompanied by decreased gastrointestinal motility and gastric emptying, resulting in weight loss. Other beneficial effects such as decreased hepatic gluconeogenesis, cardioprotective activities, and reduced inflammation have been attributed to GLP-1. Enteroendocrine L-cells are responsible for basal and meal secretion of GLP-1 to maintain glucose homeostasis. The main obstacle to developing GLP-1-based drugs is that this hormone is rapidly inactivated after release by enzymatic cleavage by the DPP-4 enzyme [46]. Therefore, two routes have been proposed for the use of this pharmacological category: long-acting GLP-1 analogues or inhibition of the DPP-4 enzyme. For the first group, several injectable drugs have been approved by the FDA including liraglutide, semaglutide, lixisenatide, albiglutide, and dulaglutide as either daily or long-acting weekly injections. The second approach based on DPP-4 inhibition includes several marketed oral drugs such as sitagliptin, linagliptin, alogliptin, and saxagliptin [46–48]. Moreover, it has been reported that several phytochemicals may induce GLP-1 secretion including berberine, resveratrol, green tea extract plus epigallocatechin gallate, coffee, sweet potato, and curcumin [49].

In vitro and in vivo studies have demonstrated that the curcumin enhances GLP-1 secretion, which could be a component of its antidiabetic actions [50–54]. The effects of curcumin on GLP-1 and the DPP-4 enzyme are shown in Table 9.1. Kato et al. reported that the curcumin formulation, theracurmin, improves glucose tolerance via stimulation of GLP-1 secretion associated with induction of insulin secretion in rats after intraperitoneal glucose injection [53]. Also the binding efficiency of various phytochemicals to the GLP-1 receptor has been investigated in

Table 9.1 Summary of the curcumin on glucagon-like peptide-1 and dipeptidyl peptidase-4 (DPP-4) with basic findings

Extract/active constituent	Dosage and regimes	Major findings	Reference
Curcuma longa root extract	STC-1 cells (gut-derived cell line)	The nondigested turmeric extracts increased GLP-1 concentrations and digested extracts decreased GLP-1 secretion	[51]
Curcumin, demethoxycurcumin (DMC), bisdemethoxycurcumin (BMC), and tetrahydrocurcumin (THC)	Murine GLUTag L cell line (25 μM)	All curcumin derivatives increased GLP-1 secretion by the Ca2+–Ca2+/calmodulin-dependent kinase II pathway	[52]
Novel curcumin derivative (NCD)	Streptozotocin-treated rat islets in vitro (10 μM)	NCD significantly elevated gene GLP-1 expression	[50]
Highly absorptive curcumin dispersed with colloidal nanoparticles (Theracurmin)	Rat & Enteroendocrine L cell line (GLUTag cells) (5 mg/kg)	Curcumin improved glucose tolerance and induced GLP-1 secretion	[53]
Turmeric-based beverage	5 men, 7 women Drink contained 185 mg GAE (gallic acid equivalents)	Turmeric-based beverage consumption did not change the perceived appetite sensations after either medium- or high-fat meals	[59]
Curcumin, synthetic analogues, curcumin-glucuronide	Mouse STC-1 cells (10 mM)	Curcumin induced GLP-1 secretion and its oxidative degradation products were associated with activation of PKC, ERK, and CaM kinase II	[54]
Curcumin	Caco-2, C2C12, and AR42J cells, Con: 10, 20, and 30 μM ICR mice Oral gavage Con: 50 mg/kg 5 weeks	Curcumin inhibited DPP IV activity and ERK phosphorylation, and reduced glucose levels and HbA1c concentration in diabetic mice	[56]
Different (arene) Ru(II) curcumin complexes	–	Curcumin complexes suppressed DPP 4 activities	[57]

rats [55]. This study tested ten plant-based anti-hyperglycemic compounds and found that myricetin had the best interaction with GLP-1 receptor and curcumin had comparable results.

A number of reports have reported inhibitory effects of curcumin against the DPP-4 enzyme. Recently a comprehensive study included molecular docking and in vitro and in vivo analyses to evaluate the DPP-4 inhibition activity of some natural phenolic compounds [56]. They reported that curcumin, syringic acid, and resveratrol had affinity toward DPP-4, and the in vitro and enzymatic assays showed an inhibitory effect of curcumin. Also the in vivo data showed that curcumin supplementation was effective in controlling blood glucose levels in mice, as well as hemoglobin A1c (Hb1Ac) levels, in long-term studies [56]. There are also other molecular docking studies alone or in combination with in vitro tests which showed that curcumin and curcuma-organometallic complexes can be regarded as the lead compounds for DPP-4 inhibition [57, 58].

9.3.2 Alpha-Glucosidase and Alpha-Amylase Inhibition

Alpha-glucosidase, an enzyme found in the epithelium membrane of small intestine, plays a key role in carbohydrate digestion. Alpha-glucosidase inhibitors are a class of oral antidiabetic agents (e.g., voglibose, miglitol, and acarbose) that directly reduce the intestinal absorption of carbohydrates by inhibiting their conversion into monosaccharides, resulting in decreased postprandial glycemia [60]. Alpha-amylase catalyzes hydrolysis of the α-(1,4)-D-glycosidic linkages of starch and other glucose polymers. Thus, inhibition of alpha-amylase with antidiabetic drugs, such as acarbose, leads to suppression of starch digestion, which lowers glucose levels [61]. The inhibitory effects of curcumin against alpha-glycosidase and alpha-amylase are summarized in Table 9.2. Molecular docking studies revealed that curcumin exhibited a high binding affinity for both enzymes [62, 63]. It has also been demonstrated that curcumin exerts an effective inhibition of both enzymes in a manner similar to acarbose [63–67]. Additionally, curcumin decreases amylase activity in rat pancreatic cancer cells [68]. Gülçubuk et al. showed that serum amylase levels were significantly reduced after treatment with curcumin in rats with acute pancreatitis [69]. Another experimental study observed that curcumin administration significantly lowered amylase levels in the pancreas of rats [70]. In contrast, in vitro and in vivo studies showed that curcumin increased amylase activity [71, 72]. Du et al. reported that natural curcuminoids (curcumin, demethoxycurcumin, and bisdemethoxycurcumin) and synthetic curcumin analogs exhibit a significant α-glucosidase inhibitory effect in vitro [73]. Additionally, it was observed that turmeric ethyl acetate extract had a higher α-glucosidase and α-amylase inhibitory activity than acarbose in vitro [74]. Consistently, in vitro studies showed that a novel curcumin-loaded mixed polymeric micelle exerted a significant α-amylase inhibitory effect by increasing the stability and solubility of curcumin [75].

9.3.3 Regulation of Glucose Transporters

Glucose transport across the plasma membranes of all mammalian cells is carried out by facilitated diffusion mechanism through the GLUT family of proteins. GLUTs comprise several isoforms such as GLUT1, GLUT2, GLUT3, GLUT4, and GLUT5, which maintain glucose homeostasis and are also molecular targets of antidiabetic drugs [84]. GLUT8, a recently identified member of the GLUT family, is a high-affinity transporter of glucose and has been identified in several tissues including kidney, brain, testis, adipocytes, and muscle cells [85]. In the liver, GLUT2 and GLUT4 are the major GLUT forms which modulate hepatic glucose metabolism through glucose transportation into hepatocytes [86]. Due to its low affinity and high capacity, GLUT2 transports glucose over a large range of physiological glucose concentrations, whereas GLUT4 function is extensively regulated by insulin-activated phosphoinositide 3-kinases (PI3K) [87, 88]. GLUT4 is one of the most important insulin-regulated glucose transporters located on the surfaces of adipocytes, cardiac and skeletal muscle cells, and brain neuronal cells [89]. Besides, GLUT1 acts as a glucose sensor in β cells and is associated with the bidirectional transport of glucose regulated by hormones in hepatocytes [90]. GLUT3 is mainly found in the brain [90]. GLUT5 is present on cells of small intestine, kidneys, and testes and acts primarily as a fructose transporter. The effects of curcumin on the GLUTs are shown in Table 9.3.

Extracellular glucose is rapidly transported into liver cells by GLUT2. Lin et al. demonstrated that curcumin reduces hyperglycemia-induced hepatic stellate cell activation by inhibition of GLUT2 membrane translocation through suppression of p38 mitogen-activated protein kinase (MAPK) signaling [91]. Also, curcumin suppresses GLUT2 expression by PPAR activation and de novo glutathione synthesis [91]. An in vitro study reported that nanocurcumin, an improved curcumin formulation, increased GLUT-4 translocation to the surface of skeletal muscle cells through increased AKT phosphorylation and glycogen synthase kinase (GSK)-3β downregulation [92]. Furthermore, in vivo and in vitro experimental results found increased levels of GLUT2 after curcumin administration in diabetic rats and pancreatic islets [93]. However, because evidence is controversial and still scarce,

Table 9.2 Summary of the curcumin on alpha-amylase and alpha-glucosidase with basic findings

α- amylase and α-glucosidase			
Extract/active constituent	Dosage and regimes	Major findings	Reference
Nisha-Amalaki, an ayurvedic anti-diabetic formulation prepared from Curcuma longa and Emblicaofficinalis	–	Nisha-Amalaki inhibited α-amylase (65%), α-glucosidase (93%) and reduced starch digestion (93%)	[67]
Curcumin	50–200 μg/mL	Curcumin analogs showed a potent α-amylase and α-glucosidase inhibitory activities	[76]
Curcumin	Caco-2 cells 10 or 40 μM Mice 0.3 mg/kg	Curcumin exhibited a potential inhibition of α-glucosidase and α-amylase via molecular docking	[77]
Curcumin, demethoxycurcumin and bisdemethoxycurcumin synthetic curcumin analogs	–	Natural curcuminoids and curcumin analogs had a significant α-glucosidase inhibitory activity in vitro	[73]
Curcumin	Male Wistar albino rats Con: 100 mg/kg since day 20 prior to the initiation of acute pancreatitis	Curcumin treatment decreased serum amylase levels in induced acute pancreatitis	[69]
Curcumin	Pancreas and intestinal mucosal of adult female Wistar rats Con: 5.0, 50 (mg/mL)	Curcumin increased the activity of pancreatic amylase	[71]
Curcumin	Male Sprague-Dawley rats Intravenous, 35 mg/kg.h during the 6-h (total of 200 mg/kg)	Curcumin decreased serum amylase levels	[78]
Curcumin	Female Wistar rats Dietary study (Ds): 0.5 g Single dose study (SDs): 100 mg	Curcumin increased pancreatic amylase levels	[72]
Curcumin	Male Sprague-Dawley rats Intraperitoneally Con: 20 mg/kg (pretreatment)	Curcumin decreased plasma levels of amylase in ischemia-reperfusion in the pancreas of rats	[70]
Curcumin	Male Wistar rats Con: 0.5 g%	Curcumin enhanced the activity of pancreatic amylase	[79]
Curcumin	Male albino rats Pre and posttreated group Intraperitoneally (daily) 50 mg/kg 6 day	Curcumin significantly reduced serum amylase levels	[80]
Turmeric (curcuma longaL) rhizome	C2C12 cells 10–100 μg	Turmeric ethyl acetate extract significantly inhibits α-glucosidase and α-amylase compared with acarbose	[74]
Curcumin	Male Sprague-Dawley rats Intraperitoneally Con: 0, 50, 100, and 150 mg/kg	Curcumin reduced amylase activity	[81]

(continued)

Table 9.2 (continued)

α- amylase and α-glucosidase			
Extract/active constituent	Dosage and regimes	Major findings	Reference
Curcumin encapsulated with pluronic-based mixed polymeric micelles	In vitro Agar diffusion test	Curcumin-loaded mixed polymeric micelles inhibited α-amylase activity	[75]
Curcumin	Female Sprague Dawley rats (acute pancreatitis) Intraperitoneally 200 mg/kg of body weight AR42J rat acinar cell line	Curcumin decreased amylase activity in rat pancreatic cancer	[68]
Curcumin	Male ICR mice Intraperitoneally Con: 50, 200 mg/kg Daily for 3 days	Curcumin significantly decreased serum amylase levels	[82]
Curcumin	Male rats Intraperitoneally 150 mg/kg	Curcumin decreased amylase activity	[83]

Table 9.3 Summary of the curcumin on glucose transporters with basic findings

Extract/active constituent	Dosage and regimes	Major findings	Reference
Curcumin	Hepatic stellate cells (HSCs) 0–30 μM	Curcumin decreased hyperglycemia by inhibiting membrane translocation and gene expression of GLUT2	[91]
Curcumin nanoparticle system with chitosan (CS-NC)	Rat L6 skeletal muscle cells Con: 25 μM	CS-NC increased the GLUT4 translocation to the cell surface, increased phosphorylation of AKT (Ser-473) and its downstream target GSK-3β (Ser-9)	[92]
Curcumin	3 T3-L1 adipocytes Con: 5, 10, & 20 μM	Curcumin reduced GLUT-1 expression during hypoxia in adipocytes	[94]
Curcumin	Human corneal–limbal epithelial (HCLE) cell line, L929 mouse fibroblast cells and HK2 (human kidney) cells Con: 0–200 μM	Curcumin inhibited GLUT1 activity The results propose a plausible pathway in which intestinal epithelial GLUTs are inhibited during curcumin administration and led to glucose reduction	[95]
Curcumin (was provided with olive oil)	Male Wistar rat Dose: 100 mg/kg 8 weeks Oral gavages Islet 352cell 20 μM	Curcumin administration increased GLUT2 levels	[93]
Novel curcumin derivative (NCD)	Male Wistar rats Dose: 60 mg/kg	NCD significantly increased GLUT2 expression in pancreatic islets	[50]

future studies are mandatory in order to elucidate the potential beneficial effects of curcumin on glucose transporters.

9.3.4 Increased Expression of Peroxisome Proliferator-Activated Receptor Gamma (PPAR-γ)

PPARs are a group of ligand-activated transcription factors of nuclear hormone receptors which regulate metabolism and energy homeostasis via signaling to the cell nucleus. In mammals, there are three isoforms of PPARs including PPAR-α, PPAR-β/δ, and PPAR-γ. PPAR-γ is often found in adipose tissue and is involved in adipocyte differentiation and modulates energy stores [96]. Furthermore, this subclass of PPARs regulates metabolism and inflammation in various types of tissues and cells. PPAR-γ plays a pivotal role in lipid and glucose homeostasis through fatty acid oxidation and adipogenesis [97, 98]. Also, PPAR-γ activation improves glucose metabolism by acting as an insulin sensitizer and mitigation of inflammatory responses [99]. The thiazolidinediones are a class of antidiabetic drugs with high affinity for PPAR-γ, which improve systemic insulin sensitivity and glucose control [100]. A summary of the curcumin effects on PPAR-γ is presented in Table 9.4. It has been reported that curcumin improves insulin secretion, lipid metabolism, and free fatty acid receptor expression by upregulation of PPAR-γ. Blanquicett et al. demonstrated that curcumin treatment inhibits H_2O_2-induced reductions in PPARγ expression [101]. An experimental study showed that curcumin exerts beneficial effects on streptozotocin-induced dementia through the activation of PPAR-γ [102]. Chen et al. reported that curcuminoids and curcuminoid derivatives induce adipocyte differentiation and decrease glucose levels by PPAR-γ ligand-binding activity [40]. In contrast, Narala et al. indicated that curcumin is not a PPAR-γ ligand and that this polyphenol does not induce PPAR-γ-dependent differentiation of preadipocytes. They suggested that curcumin must have indirect effects through endogenous ligands or receptor expressions that are regulated by other pathways [103]. This appears to be in conflict with other molecular docking studies which revealed that curcumin has PPAR-γ protein binding sites [104], and another study showed that curcumin exhibits molecular interactions with PPAR-γ as an ago-

Table 9.4 Effects of curcumin on peroxisome proliferator-activated receptor gamma

Extract/active constituent	Dosage and regimes	Major findings	Reference
Curcumin -encapsulated nanoparticles	Male white albino diabetic rats orally Dose: 20 mg/kg.b.wt/day	Curcumin-encapsulated nanoparticles significantly increased PPARγ mRNA levels	[106]
Curcumin	Monolayers of human umbilical vein or bovine aortic endothelial cells (HUVECs or BAECs) (10 μM)	Curcumin treatment inhibited H_2O_2-induced reductions in PPARγ expression	[101]
Curcumin, demethoxycurcumin, and bisdemethoxycurcumin	Female genetically T2DM KK-Ay/Ta mice Dose: 0.1, 0.5 g of hexane extraction (H-ext), and 0.5 g of hexane-extraction (HE-ext)/100 g of diet, or 0.2, 1.0 g of ethanol extraction (E-ext)/100 g of diet 4 weeks	Turmeric extracts induce adipocyte differentiation and decrease glucose levels by PPAR-γ ligand-binding activity	[40]
Curcumin	Male and female swiss albino mice Dose: 20 mg/kg Orally daily for 14 days	Curcumin improves streptozotocin-induced dementia through the activation of PPAR-γ	[102]

nist in the treatment of insulin resistance [105]. Given that these results are inconclusive, further research is need to clarify these points.

9.4 Conclusions and Future Perspectives

T2D has become one of the main health problems with severe negative impacts on quality of life worldwide. Therefore, management of hyperglycemia in diabetes is one of the most important challenges in overall public health, and more effective therapies for glucose control are a current need in this field. Although there are several groups of drugs for the treatment of diabetes, all have shown adverse effects and imperfect efficacy in achieving long-term glycemic control [107]. This has created considerable space for new efficacious compounds with lower adverse effect profiles. With this in mind, herbal products with minimal side effects and potential benefits on glycemic parameters have emerged as therapeutic alternatives for the treatment of diabetes [108]. Curcumin is one of the frontrunners in this objective as a number of studies have now demonstrated that it has positive effects in treatment of diabetes and its complications by reducing the effects of hyperglycemia, insulin resistance, and islet apoptosis and necrosis [109]. In fact, curcumin poses its antidiabetic actions through a wide range of molecular targets and has an improved safety profile [110]. In terms of mechanism of action, this review has highlighted the capacity of curcumin as an inhibitor of enzymes such as alpha-glucosidase, alpha-amylase, and DPP-4. The stimulatory effects on the insulin secretagogue GLP-1 have also been discussed. Also some studies have clearly shown that curcumin can regulate the glucose transporters (GLUTs). There are reports that showed GLUT2 can be upregulated in response to curcumin administration, while in contrast some reports showed that it can be reduced. Also it might be plausible that GLUT1 inhibition in the intestinal cells by curcumin can lead to the reduced glucose levels. Also upregulation of GLUT4 in muscle cells can reduce the glucose concentration. Moreover, the other important pathway in the antidiabetic actions of curcumin might be through tPPAR-γ as some studies have indicated that curcumin has PPAR-γ agonist activity.

In conclusion, curcumin administration has positive effects in several aspects involved in the pathogenesis of diabetes. In addition to the well-recognized antioxidant and anti-inflammatory activities of curcumin, the new mechanisms discussed in this review provide promising evidence as antidiabetic therapy. However, some of the studies have shown conflicting results. Thus, further in vitro and in vivo studies are required to resolve these issues prior to testing in clinical trials.

References

1. Dall TM, Yang W, Gillespie K, Mocarski M, Byrne E, Cintina I et al (2019) The economic burden of elevated blood glucose levels in 2017: diagnosed and undiagnosed diabetes, gestational diabetes mellitus, and prediabetes. Diabetes Care 42(9):1661–1668
2. Rahman F, McEvoy JW, Ohkuma T, Marre M, Hamet P, Harrap S et al (2019) Effects of blood pressure lowering on clinical outcomes according to baseline blood pressure and cardiovascular risk in patients with type 2 diabetes mellitus: the ADVANCE trial. Hypertension 73(6):1291–1299
3. Hu W, Jiang C, Guan D, Dierickx P, Zhang R, Moscati A et al (2019) Patient adipose stem cell-derived adipocytes reveal genetic variation that predicts antidiabetic drug response. Cell Stem Cell 24(2):299–308. e296
4. Association AD (2020) 2. Classification and diagnosis of diabetes: standards of medical care in diabetes—2020. Diabetes Care 43(Suppl 1):S14–S31
5. Xu YXZ, Xi S, Qian X (2019) Evaluating traditional Chinese medicine and herbal products for the treatment of gestational diabetes mellitus. J Diabetes Res 2019:9182595. https://doi.org/10.1155/2019/9182595
6. Panahi Y, Khalili N, Sahebi E, Namazi S, Karimian MS, Majeed M et al (2017) Antioxidant effects of curcuminoids in patients with type 2 diabetes mellitus: a randomized controlled trial. Inflammopharmacology 25(1):25–31
7. Parsamanesh N, Moossavi M, Bahrami A, Butler AE, Sahebkar A (2018) Therapeutic potential of curcumin in diabetic complications. Pharmacol Res 136:181–193
8. Abdollahi E, Momtazi AA, Johnston TP, Sahebkar A (2018) Therapeutic effects of curcumin in inflammatory and immune-mediated diseases: a nature-made jack-of-all-trades? J Cell Physiol 233(2):830–848

9. Saberi-Karimian M, Keshvari M, Ghayour-Mobarhan M, Salehizadeh L, Rahmani S, Behnam B et al (2020) Effects of curcuminoids on inflammatory status in patients with non-alcoholic fatty liver disease: a randomized controlled trial. Complement Ther Med 49:102322. https://doi.org/10.1016/j.ctim.2020.102322
10. Abbas Momtazi A, Sahebkar A (2016) Difluorinated curcumin: a promising curcumin analogue with improved anti-tumor activity and pharmacokinetic profile. Curr Pharm Des 22(28):4386–4397
11. Amel Zabihi N, Pirro MP, Johnston T, Sahebkar A (2017) Is there a role for curcumin supplementation in the treatment of non-alcoholic fatty liver disease? The data suggest yes. Curr Pharm Des 23(7):969–982
12. Hasanzadeh S, Read MI, Bland AR, Majeed M, Jamialahmadi T, Sahebkar A (2020) Curcumin: an inflammasome silencer. Pharmacol Res 159:104921. https://doi.org/10.1016/j.phrs.2020.104921. Epub 2020 May 25. PMID: 32464325.
13. Mohajeri M, Behnam B, Cicero AFG, Sahebkar A (2018) Protective effects of curcumin against aflatoxicosis: a comprehensive review. J Cell Physiol 233(4):3552–3577
14. Iranshahi M, Sahebkar A, Hosseini ST, Takasaki M, Konoshima T, Tokuda H (2010) Cancer chemopreventive activity of diversin from ferula diversivittata in vitro and in vivo. Phytomedicine 17(3–4):269–273
15. Mollazadeh H, Cicero AFG, Blesso CN, Pirro M, Majeed M, Sahebkar A (2019) Immune modulation by curcumin: the role of interleukin-10. Crit Rev Food Sci Nutr 59(1):89–101
16. Momtazi AA, Derosa G, Maffioli P, Banach M, Sahebkar A (2016) Role of microRNAs in the therapeutic effects of curcumin in non-cancer diseases. Mol Diagn Ther 20(4):335–345
17. Bagheri H, Ghasemi F, Barreto GE, Sathyapalan T, Jamialahmadi T, Sahebkar A (2020) The effects of statins on microglial cells to protect against neurodegenerative disorders: A mechanistic review. Biofactors 46(3):309–325. https://doi.org/10.1002/biof.1597. Epub 2019 Dec 17. PMID: 31846136.
18. Poolsup N, Suksomboon N, Kurnianta PDM, Deawjaroen K (2019) Effects of curcumin on glycemic control and lipid profile in prediabetes and type 2 diabetes mellitus: a systematic review and meta-analysis. PLoS One 14(4):e0215840. https://doi.org/10.1371/journal.pone.0215840
19. Huang J, Qin S, Huang L, Tang Y, Ren H, Hu H (2019) Efficacy and safety of Rhizoma curcumea longae with respect to improving the glucose metabolism of patients at risk for cardiovascular disease: a meta-analysis of randomised controlled trials. J Hum Nutr Diet 32(5):591–606
20. Chapman K, Scorgie FE, Ariyarajah A, Stephens E, Enjeti AK, Lincz LF (2019) The effects of tetrahydrocurcumin compared to curcuminoids on human platelet aggregation and blood coagulation in vitro. Thromb Res 179:28–30
21. Liu X, Zhang W, Wang L, Wang S, Yu Y, Chen S et al (2019) Male patients with diabetes undergoing coronary artery bypass grafting have increased major adverse cerebral and cardiovascular events. Interact Cardiovasc Thorac Surg 28(4):607–612
22. Fujiwara H, Hosokawa M, Zhou X, Fujimoto S, Fukuda K, Toyoda K et al (2008) Curcumin inhibits glucose production in isolated mice hepatocytes. Diabetes Res Clin Pract 80(2):185–191
23. Aziz MTA, El-Asmar MF, El-Ibrashy IN, Rezq AM, Al-Malki AL, Wassef MA et al (2012) Effect of novel water soluble curcumin derivative on experimental type-1 diabetes mellitus (short term study). Diabetol Metab Syndr 4(1):30. https://doi.org/10.1186/1758-5996-4-30
24. Kim T, Davis J, Zhang AJ, He X, Mathews ST (2009) Curcumin activates AMPK and suppresses gluconeogenic gene expression in hepatoma cells. Biochem Biophys Res Commun 388(2):377–382
25. Zheng J, Cheng J, Zheng S, Feng Q, Xiao X (2018) Curcumin, a polyphenolic curcuminoid with its protective effects and molecular mechanisms in diabetes and diabetic cardiomyopathy. Front Pharmacol 9:472. https://doi.org/10.3389/fphar.2018.00472
26. Wojcik M, Krawczyk M, Wojcik P, Cypryk K, Wozniak LA (2018) Molecular mechanisms underlying curcumin-mediated therapeutic effects in type 2 diabetes and cancer. Oxidative Med Cell Longev 2018:9698258. https://doi.org/10.1155/2018/9698258
27. Yaribeygi H, Katsiki N, Behnam B, Iranpanah H, Sahebkar A (2018) MicroRNAs and type 2 diabetes mellitus: molecular mechanisms and the effect of antidiabetic drug treatment. Metabolism 87:48–55
28. Ashrafizadeh M, Yaribeygi H, Atkin SL, Sahebkar A (2019) Effects of newly introduced antidiabetic drugs on autophagy. Diabetes Metab Syndr 3(4):2445–2449
29. Katsarou A, Gudbjörnsdottir S, Rawshani A, Dabelea D, Bonifacio E, Anderson BJ et al (2017) Type 1 diabetes mellitus. Nat Rev Dis Primers 3(1):1–17
30. Chatterjee S, Khunti K, Davies MJ (2017) Type 2 diabetes. Lancet 389(10085):2239–2251
31. Ramesh M, Muthuraman A (2018) Flavoring and coloring agents: health risks and potential problems. In: Natural and artificial flavoring agents and food dyes. Elsevier, Amsterdam, Netherlands, pp 1–28
32. Raj R, Sahay S, Tripathi J (2016) Medications of diabetes mellitus and antidiabetic medicinal plants: a review. Drugs 1(1):19–28
33. Vamanu E, Gatea F, Sârbu I, Pelinescu D (2019) An in vitro study of the influence of Curcuma longa extracts on the microbiota modulation process, in patients with hypertension. Pharmaceutics 11(4):191. https://doi.org/10.3390/pharmaceutics11040191
34. Osorio-Tobón JF, Carvalho PI, Barbero GF, Nogueira GC, Rostagno MA, de Almeida Meireles MA (2016) Fast analysis of curcuminoids from turmeric (Curcuma longa L.) by high-performance liquid chromatography using a fused-core column. Food Chem 200:167–174

35. Hodaei H, Adibian M, Nikpayam O, Hedayati M, Sohrab G (2019) The effect of curcumin supplementation on anthropometric indices, insulin resistance and oxidative stress in patients with type 2 diabetes: a randomized, double-blind clinical trial. Diabetol Metab Syndr 11(1):41. https://doi.org/10.1186/s13098-019-0437-7
36. Rungseesantivanon S, Thenchaisri N, Ruangvejvorachai P, Patumraj S (2010) Curcumin supplementation could improve diabetes-induced endothelial dysfunction associated with decreased vascular superoxide production and PKC inhibition. BMC Complement Altern Med 10(1):57. https://doi.org/10.1186/1472-6882-10-57
37. Asadi S, Gholami MS, Siassi F, Qorbani M, Khamoshian K, Sotoudeh G (2019) Nano curcumin supplementation reduced the severity of diabetic sensorimotor polyneuropathy in patients with type 2 diabetes mellitus: a randomized double-blind placebo-controlled clinical trial. Complement Ther Med 43:253–260
38. Onyenibe NS, Nathaniel NA, Udogadi NS, Iyanu OO (2019) Hypoglycemic and antioxidant capacity of curcuma longa and Viscum album in alloxan induced diabetic male Wistar rats. Int J Diabetes Endocrinol 4(1):26. https://doi.org/10.11648/j.ijde.20190401.15
39. Ibrahim ZS, Soliman MM, Mahmoud S, Shukry M (2017) Cardioprotective effects of curcumin against diabetes and nicotine combined oxidative stress. Afr J Tradit Complement Altern Med 14(6):64–71
40. Nishiyama T, Mae T, Kishida H, Tsukagawa M, Mimaki Y, Kuroda M et al (2005) Curcuminoids and sesquiterpenoids in turmeric (Curcuma longa L.) suppress an increase in blood glucose level in type 2 diabetic KK-ay mice. J Agric Food Chem 53(4):959–963
41. Su LQ, Chi HY (2017) Effect of curcumin on glucose and lipid metabolism, FFAs and TNF-α in serum of type 2 diabetes mellitus rat models. Saudi J Biol Sci 24(8):1776–1780
42. Na LXLY, Pan HZ, Zhou XL, Sun DJ, Meng M, Li XX, Sun CH (2013) Curcuminoids exert glucose-lowering effect in type 2 diabetes by decreasing serum free fatty acids: a double-blind, placebo-controlled trial. Mol Nutr Food Res 57(9):1569–1577
43. Chuengsamarn SRS, Phonrat B, Tungtrongchitr R, Jirawatnotai S (2014) Reduction of atherogenic risk in patients with type 2 diabetes by curcuminoid extract: a randomized controlled trial. J Nutr Biochem 25(2):144–150
44. Wojcik M, Krawczyk M, Wozniak LA (2018) Antidiabetic activity of curcumin: insight into its mechanisms of action. In: Bagchi D, Nair S (eds) Nutritional and therapeutic interventions for diabetes and metabolic syndrome. Elsevier, Amsterdam, Netherlands, pp 385–401
45. Campbell JE, Drucker DJ (2013) Pharmacology, physiology, and mechanisms of incretin hormone action. Cell Metab 17(6):819–837
46. Müller TD, Finan B, Bloom S, D'Alessio D, Drucker DJ, Flatt P et al (2019) Glucagon-like peptide 1 (GLP-1). Mol Metab 30:72–130
47. Drucker DJ (2018) Mechanisms of action and therapeutic application of glucagon-like peptide-1. Cell Metab 27(4):740–756
48. Yaribeygi H, Maleki M, Sathyapalan T, Jamialahmadi T, Sahebkar A (2020) Anti-inflammatory potentials of incretin-based therapies used in the management of diabetes. Life Sci 241:117152. https://doi.org/10.1016/j.lfs.2019.117152
49. Tsuda T (2015) Possible abilities of dietary factors to prevent and treat diabetes via the stimulation of glucagon-like peptide-1 secretion. Mol Nutr Food Res 59(7):1264–1273
50. Aziz MA, El-Asmar MF, Rezq AM, Wassef MA, Fouad H, Roshdy NK et al (2014) Effects of a novel curcumin derivative on insulin synthesis and secretion in streptozotocin-treated rat pancreatic islets in vitro. Chin Med 9(1):3. https://doi.org/10.1186/1749-8546-9-3
51. Planes-Muñoz D, López-Nicolás R, González-Bermúdez CA, Ros-Berruezo G, Frontela-Saseta C (2018) In vitro effect of green tea and turmeric extracts on GLP-1 and CCK secretion: the effect of gastrointestinal digestion. Food Funct 9(10):5245–5250
52. Takikawa M, Kurimoto Y, Tsuda T (2013) Curcumin stimulates glucagon-like peptide-1 secretion in GLUTag cells via Ca2+/calmodulin-dependent kinase II activation. Biochem Biophys Res Commun 435(2):165–170
53. Kato M, Nishikawa S, Ikehata A, Dochi K, Tani T, Takahashi T et al (2017) Curcumin improves glucose tolerance via stimulation of glucagon-like peptide-1 secretion. Mol Nutr Food Res 61(3):1600471. https://doi.org/10.1002/mnfr.201600471
54. Alli-Oluwafuyi AM, Luis PB, Nakashima F, Giménez-Bastida JA, Presley SH, Duvernay MT et al (2019) Curcumin induces secretion of glucagon-like peptide-1 through an oxidation-dependent mechanism. Biochimie 165:250–257
55. Sharma G, Hora S, Aeri V, Katare DP (2018) Glucagon like peptide (Glp–1) receptor: a promising therapeutic target for screening of herbal antidiabetic compounds. Asian J Pharm 12(2):S512–S518
56. Huang PK, Lin SR, Chang CH, Tsai MJ, Lee DN, Weng CF (2019) Natural phenolic compounds potentiate hypoglycemia via inhibition of Dipeptidyl peptidase IV. Sci Rep 9(1):1–11
57. Antonyan A, De A, Vitali L, Pettinari R, Marchetti F, Gigliobianco M et al (2014) Evaluation of (arene) Ru (II) complexes of curcumin as inhibitors of dipeptidyl peptidase IV. Biochimie 99:146–152
58. Istyastono EP (2009) Docking studies of curcumin as a potential lead compound to develop novel dipeptydyl peptidase-4 inhibitors. Indones J Chem 9(1):132–136
59. Zanzer YC, Batista ÂG, Dougkas A, Tovar J, Granfeldt Y, Östman E (2019) Difficulties in translating appetite sensations effect of turmeric-based

beverage when given prior to isoenergetic medium- or high-fat meals in healthy subjects. Nutrients 11(4):736. https://doi.org/10.3390/nu11040736
60. Dash RP, Babu RJ, Srinivas NR (2018) Reappraisal and perspectives of clinical drug–drug interaction potential of α-glucosidase inhibitors such as acarbose, voglibose and miglitol in the treatment of type 2 diabetes mellitus. Xenobiotica 48(1):89–108
61. Taslimi P, Aslan HE, Demir Y, Oztaskin N, Maraş A, Gulçin İ et al (2018) Diarylmethanon, bromophenol and diarylmethane compounds: discovery of potent aldose reductase, α-amylase and α-glycosidase inhibitors as new therapeutic approach in diabetes and functional hyperglycemia. Int J Biol Macromol 119:857–863
62. Jhong CH, Riyaphan J, Lin SH, Chia YC, Weng CF (2015) Screening alpha-glucosidase and alpha-amylase inhibitors from natural compounds by molecular docking in silico. Biofactors 41(4):242–251
63. Rasouli H, Hosseini-Ghazvini SM-B, Adibi H, Khodarahmi R (2017) Differential α-amylase/α--glucosidase inhibitory activities of plant-derived phenolic compounds: a virtual screening perspective for the treatment of obesity and diabetes. Food Funct 8(5):1942–1954
64. Nampoothiri SV, Lekshmi P, Venugopalan V, Menon AN (2012) Antidiabetic and antioxidant potentials of spent turmeric oleoresin, a by-product from curcumin production industry. Asian Pac J Trop Dis 2:S169–S172
65. Hasimun P, Adnyana I, Valentina R, Lisnasari E (2016) Potential alpha-glucosidase inhibitor from selected Zingiberaceae family. Asian J Pharm Clin Res 9(1):164–167
66. Yousefi A, Yousefi R, Panahi F, Sarikhani S, Zolghadr AR, Bahaoddini A et al (2015) Novel curcumin-based pyrano [2, 3-d] pyrimidine anti-oxidant inhibitors for α-amylase and α-glucosidase: implications for their pleiotropic effects against diabetes complications. Int J Biol Macromol 78:46–55
67. Butala MA, Kukkupuni SK, Venkatasubramanian P, Vishnuprasad CN (2018) An ayurvedic anti-diabetic formulation made from Curcuma longa L. and Emblica officinalis L. inhibits α-amylase, α-glucosidase, and starch digestion, in vitro. Starch-Stärke 70(5–6):1700182. https://doi.org/10.1002/star.201700182
68. Wang Y, Bu C, Wu K, Wang R, Wang J (2019) Curcumin protects the pancreas from acute pancreatitis via the mitogenactivated protein kinase signaling pathway. Mol Med Rep 20(4):3027–3034
69. Gülçubuk A, Sönmez K, Gürel A, Altunatmaz K, Gürler N, Aydın S et al (2005) Pathologic alterations detected in acute pancreatitis induced by sodium taurocholate in rats and therapeutic effects of curcumin, ciprofloxacin and metronidazole combination. Pancreatology 5(4–5):345–353
70. Chen K, Chao D, Liu C, Chen C, Wang D (2010) Curcumin attenuates airway hyperreactivity induced by ischemia-reperfusion of the pancreas in rats. Transplant Proc 42(3):744–747
71. Ramakrishna Rao R, Platel K, Srinivasan K (2003) In vitro influence of spices and spice-active principles on digestive enzymes of rat pancreas and small intestine. Food Nahrung 47(6):408–412
72. Platel K, Srinivasan K (2000) Influence of dietary spices and their active principles on pancreatic digestive enzymes in albino rats. Food Nahrung 44(1):42–46
73. Du ZY, Liu RR, Shao WY, Mao XP, Ma L, Gu LQ et al (2006) α-Glucosidase inhibition of natural curcuminoids and curcumin analogs. Eur J Med Chem 41(2):213–218
74. Lekshmi P, Arimboor R, Nisha V, Menon AN, Raghu K (2014) In vitro antidiabetic and inhibitory potential of turmeric (Curcuma longa L) rhizome against cellular and LDL oxidation and angiotensin converting enzyme. Int J Food Sci 51(12):3910–3917
75. Akbar MU, Zia KM, Nazir A, Iqbal J, Ejaz SA, Akash MSH (2018) Pluronic-based mixed polymeric micelles enhance the therapeutic potential of curcumin. AAPS PharmSciTech 19(6):2719–2739
76. Banuppriya G, Sribalan R, Fathima SAR, Padmini V (2018) Synthesis of β-ketoamide curcumin analogs for anti-diabetic and AGE s inhibitory activities. Chem Biodivers 15(8):e1800105. https://doi.org/10.1002/cbdv.201800105
77. Riyaphan J, Jhong CH, Lin SR, Chang CH, Tsai MJ, Lee DN et al (2018) Hypoglycemic efficacy of docking selected natural compounds against α-glucosidase and α-amylase. Molecules 23(9):2260. https://doi.org/10.3390/molecules23092260
78. Gukovsky I, Reyes CN, Vaquero EC, Gukovskaya AS, Pandol SJ (2003) Curcumin ameliorates ethanol and nonethanol experimental pancreatitis. Am J Physiol Gastrointest Liver Physiol 284(1):G85–G95
79. Prakash UN, Srinivasan K (2012) Fat digestion and absorption in spice-pretreated rats. J Sci Food Agric 92(3):503–510
80. Shafik NM, Abou-Fard GM (2016) Ameliorative effects of curcumin on fibrinogen-like protein-2 gene expression, some oxido-inflammatory and apoptotic markers in a rat model of l-arginine-induced acute pancreatitis. J Biochem Mol Toxicol 30(6):302–308
81. Yu S, Wang M, Guo X, Qin R (2018) Curcumin attenuates inflammation in a severe acute pancreatitis animal model by regulating TRAF1/ASK1 signaling. Med Sci Monit 24:2280–2286
82. Siriviriyakul P, Chingchit T, Klaikeaw N, Chayanupatkul M, Werawatganon D (2019) Effects of curcumin on oxidative stress, inflammation and apoptosis in L-arginine induced acute pancreatitis in mice. Heliyon 5(8):e02222. https://doi.org/10.1016/j.heliyon.2019.e02222
83. Ragy MM, Ali FF, Toni ND (2019) Comparing the preventive effect of sodium hydrosulfide, leptin, and curcumin against L-arginine induced acute pancreatitis in rats: role of corticosterone and inducible nitric oxide synthase. Endocr Regul 53(4):221–230

84. Stanirowski PJ, Szukiewicz D, Pyzlak M, Abdalla N, Sawicki W, Cendrowski K (2017) Impact of pre-gestational and gestational diabetes mellitus on the expression of glucose transporters GLUT-1, GLUT-4 and GLUT-9 in human term placenta. Endocrine 55(3):799–808
85. Schiffer M, Susztak K, Ranalletta M, Raff AC, Bottinger EP, Charron MJ (2005) Localization of the GLUT8 glucose transporter in murine kidney and regulation in vivo in nondiabetic and diabetic conditions. Am J Physiol Gastrointest Liver Physiol 289(1):F186–F193
86. Rathinam A, Pari L (2016) Myrtenal ameliorates hyperglycemia by enhancing GLUT2 through Akt in the skeletal muscle and liver of diabetic rats. Chem Biol Interact 256:161–166
87. Leturque A, Brot-Laroche E, Le Gall M, Stolarczyk E, Tobin V (2005) The role of GLUT2 in dietary sugar handling. Cell Physiol Biochem 61(4):529–537
88. Zhao FQ, Keating AF (2007) Functional properties and genomics of glucose transporters. Curr Genomics 8(2):113–128
89. Rudich A, Konrad D, Török D, Ben-Romano R, Huang C, Niu W et al (2003) Indinavir uncovers different contributions of GLUT4 and GLUT1 towards glucose uptake in muscle and fat cells and tissues. Diabetologia 46(5):649–658
90. Navale AMPA (2016) Glucose transporters: physiological and pathological roles. Biophys Rev 8(1):5–9
91. Lin J, Chen A (2011) Curcumin diminishes the impacts of hyperglycemia on the activation of hepatic stellate cells by suppressing membrane translocation and gene expression of glucose transporter-2. Mol Cell Endocrinol 333(2):160–171
92. Chauhan P, Tamrakar AK, Mahajan S, Prasad G (2018) Chitosan encapsulated nanocurcumin induces GLUT-4 translocation and exhibits enhanced anti-hyperglycemic function. Life Sci 213:226–235
93. Rashid K, Sil PC (2015) Curcumin enhances recovery of pancreatic islets from cellular stress induced inflammation and apoptosis in diabetic rats. Toxicol Appl Pharmacol 282(3):297–310
94. Priyanka A, Shyni G, Anupama N, Raj PS, Anusree S, Raghu K (2017) Development of insulin resistance through sprouting of inflammatory markers during hypoxia in 3T3-L1 adipocytes and amelioration with curcumin. Eur J Pharmacol 812:73–81
95. Gunnink LK, Alabi OD, Kuiper BD, Gunnink SM, Schuiteman SJ, Strohbehn LE et al (2016) Curcumin directly inhibits the transport activity of GLUT1. Biochimie 125:179–185
96. Balan G, Bubalo N, Bubalo V, Zhminko P, Nedopytanska N, Babich V (2017) Family of nuclear peroxisome proliferator—activated receptors (PPARs): biological role in metabolic adaptation. Part III. PPARγ in energy homeostasis and formation of metabolic syndrome, hepatosteatosis, cardiovascular conditions and fibrosis (report 1). Ukranian J Mod Probl Toxicol (1-2):33–47
97. Mohammed F, Gurigis A, Abdel-Mageed W, Nassr A (2018) Improvement of insulin sensitivity and maintenance of glucose homeostasis in insulin-sensitive tissues via PPAR-γ and through activation of PI3K/p-Akt signaling pathway by resveratrol in type 2 diabetic rats. J Mol Immunol 3(119):2
98. Wilson C (2011) Diabetes: T2DM-PPARγ ligands without the adverse effects? Nat Rev Endocrinol 7(11):630. https://doi.org/10.1038/nrendo.2011.167
99. Aboonabi A, Rose'Meyer R, Singh I, Aboonabi A (2020) Anthocyanins reduce inflammation and improve glucose and lipid metabolism associated with inhibiting nuclear factor-kappaB activation and increasing PPAR-γ gene expression in metabolic syndrome subjects. Free Radic Biol Med 150:30–39
100. Naim MJ, Alam O, Alam MJ, Shaquiquzzaman M, Alam MM, Naidu VGM (2018) Synthesis, docking, in vitro and in vivo antidiabetic activity of pyrazole-based 2, 4-thiazolidinedione derivatives as PPAR-γ modulators. Arch Pharm 351(3–4):e1700223. https://doi.org/10.1002/ardp.201700223
101. Blanquicett C, Kang B-Y, Ritzenthaler JD, Jones DP, Hart CM (2010) Oxidative stress modulates PPARγ in vascular endothelial cells. Free Radic Biol Med 48(12):1618–1625
102. Rinwa P, Kaur B, Jaggi AS, Singh N (2010) Involvement of PPAR-gamma in curcumin-mediated beneficial effects in experimental dementia. Naunyn Schmiedeberg's Arch Pharmacol 381(6):529–539
103. Narala VR, Smith MR, Adapala RK, Ranga R, Panati K, Moore BB et al (2009) Curcumin is not a ligand for peroxisome proliferator-activated receptor-γ. Gene Ther Mol Biol 13(1):20–25
104. Luo J, Qu J, Yang R, Ge M-X, Mei Y, Zhou B-T et al (2016) Phytochemicals mediate the expression and activity of OCTN2 as activators of the PPARγ/RXRα pathway. Front Pharmacol 7:189
105. Jayakumar V, Ahmed SS, Ebenezar KK (2016) Multivariate analysis and molecular interaction of curcumin with PPARγ in high fructose diet induced insulin resistance in rats. Springerplus 5(1):1732. https://doi.org/10.1186/s40064-016-3364-1
106. El-Naggar ME, Al-Joufi F, Anwar M, Attia MF, El-Bana MA (2019) Curcumin-loaded PLA-PEG copolymer nanoparticles for treatment of liver inflammation in streptozotocin-induced diabetic rats. Colloids Surf B Biointerfaces 177:389–398
107. Singh S, Usman K, Banerjee M (2016) Pharmacogenetic studies update in type 2 diabetes mellitus. World J Diabetes 7(15):302–315
108. Zarvandi M, Rakhshandeh H, Abazari M, Shafiee-Nick R, Ghorbani A (2017) Safety and efficacy of a polyherbal formulation for the management of dyslipidemia and hyperglycemia in patients with advanced-stage of type-2 diabetes. Biomed Pharmacother 89:69–75
109. Weisberg S, Leibel R, Tortoriello D (2016) Proteasome inhibitors, including curcumin, improve pancreatic β-cell function and insulin sensitivity in diabetic mice. Nutr Diabetes 4:e205–e205
110. Soleimani V, Sahebkar A, Hosseinzadeh H. Turmeric (Curcuma longa) and its major constituent (curcumin) as nontoxic and safe substances: Review. Phytother Res 32(6):985–995. https://doi.org/10.1002/ptr.6054. Epub 2018 Feb 26. PMID: 29480523

The Effect of Curcumin in Improving Lipid Profile in Patients with Cardiovascular Risk Factors: A Systematic Review of Clinical Trials

10

Sahar Rafiee, Mohammad Bagherniya,
Gholamreza Askari, Thozhukat Sathyapalan,
Tannaz Jamialahmadi, and Amirhossein Sahebkar

Abstract

Cardiovascular disease (CVD) is a major cause of death worldwide. Lipid abnormalities are one of the major risk factors for CVD. Curcumin is a natural polyphenol with lipid-lowering properties. Therefore, we carried out a systematic review to summarize the randomized controlled trials (RCTs) investigating the effect of curcumin on lipid profile in patients at risk of CVD. A comprehensive systematic search was conducted in PubMed, Scopus, Web of Science, and Google Scholar up to March 1, 2020, to identify controlled clinical trials assessing the effects of curcumin on lipid profile in patients at risk of CVD. From 1051 initially identified studies, 22 met the eligibility criteria. Curcumin supplementation significantly reduced at least one of the lipid profile indices (triglycerides, total cholesterol, low-density lipoprotein-cholesterol, and high-density lipoprotein-cholesterol) in 15 studies and improved more than one index in five studies. However, curcumin had no effect on any of lipid profile indices in seven studies. Overall, studies using a bioavailable formulation of curcumin had a better impact on the lipid profile. The findings of this systematic review showed that curcumin supplementation significantly reduced at least one of the lipid profile indices in more than two-thirds of the included studies. Curcumin might be used as an accessible,

S. Rafiee · M. Bagherniya · G. Askari
Food Security Research Center, Department of Community Nutrition, School of Nutrition and Food Science, Isfahan University of Medical Sciences, Isfahan, Iran

T. Sathyapalan
Department of Academic Diabetes, Endocrinology and Metabolism, Hull York Medical School, University of Hull, Hull, UK

T. Jamialahmadi
Department of Food Science and Technology, Quchan Branch, Islamic Azad University, Quchan, Iran

Department of Nutrition, Faculty of Medicine, Mashhad University of Medical Sciences, Mashhad, Iran

A. Sahebkar (✉)
Biotechnology Research Center, Pharmaceutical Technology Institute, Mashhad University of Medical Sciences, Mashhad, Iran

Neurogenic Inflammation Research Center, Mashhad University of Medical Sciences, Mashhad, Iran

Polish Mother's Memorial Hospital Research Institute (PMMHRI), Lodz, Poland

Halal Research Center of IRI, FDA, Tehran, Iran
e-mail: sahebkara@mums.ac.ir

P. C. Guest (ed.), *Studies on Biomarkers and New Targets in Aging Research in Iran*, Advances in Experimental Medicine and Biology 1291, https://doi.org/10.1007/978-3-030-56153-6_10

inexpensive, and safe agent to reduce risk of CVD. More randomized, clinical controlled trials are needed to verify these results.

Keywords

Cardiovascular diseases · Curcumin · Lipid · Lipid profile · Systematic review

10.1 Introduction

One of the major causes of death worldwide is cardiovascular disease (CVD) [1, 2]. Overall, CVD accounts for 31% of global mortality [3]. It is estimated that by the year 2030, more than 22.2 million people will die of CVD each year [2]. CVD is not only an important cause of mortality but also a cause of major disability over many years of life [4]. Risk factors for this disease include age, sex, family history, unhealthy diet, sedentary lifestyle, alcohol and tobacco use, overweight, obesity, hyperglycemia, hyperlipidemia, hypertension, metabolic syndrome, and diseases such as nonalcoholic fatty liver disease and polycystic ovary syndrome [5–8]. The World Health Organization (WHO) has estimated that more than 75% of premature CVD can be prevented. In addition, minimizing risk factors can reduce the growing burden of CVD for both individuals and health care centers [9]. A study found that dyslipidemia resulted in residual macrovascular and microvascular risk in patients at high risk for CVD [10]. Indeed, lipid abnormalities in patients with the risk of CVD such as type 2 diabetes, metabolic syndrome, and hypertension are considered major risk factors of CVD [11–13].

Currently, various non-pharmacological options such as natural herbal bioactive compounds have attracted significant attention to prevent and treat CVD as an alternative or complementary medicine [14, 15]. Curcumin is the main natural polyphenol found in rhizomes of Curcuma longa (turmeric) [16]. The characteristics of this compound include its antioxidant, antidiabetic, anti-inflammatory, antimicrobial, gastroprotective, hepatoprotective, cardioprotective, anti-obesity, and antihypertensive properties [17–22]. In addition, some studies have investigated the lipid-lowering effects of curcumin [23–26]. One meta-analysis showed that curcumin can protect patients at risk for cardiovascular disease through lipid-lowering effects [27, 28].

Although many randomized clinical trials (RCTs) have investigated the effects of curcumin on lipid biomarker profiles in patients at risk of CVD, the results have conflicted. Only in some studies, positive effects of curcumin supplementation on lipid profiles were observed [23, 24, 29, 30], while in other studies no effect was found [25, 31–33]. Therefore, we carried out a systematic review to summarize the data from RCTs to investigate the effect of curcumin on lipid biomarker profiles in patients at risk of CVD.

10.2 Material and Methods

This systematic review was designed and reported based on the preferred reporting items for systematic reviews and meta-analyses (PRISMA) guidelines [34].

10.2.1 Search Strategy

To find related articles, we searched databases including ISI web of science, Scopus, PubMed, and Google Scholar until March 1, 2020. We systematically searched for articles that studied the efficacy of curcumin supplementation on lipid profiles in patients at risk of CVD. Our search did not limit the publication period and language. We used ("curcumin" [Mesh] OR "curcumin" [tiab] OR "curcuminoids" [tiab] OR "turmeric" [tiab] OR "Curcuma longa" [tiab] OR "C. longa" [tiab] OR "Curcuma domestica" [tiab]) AND ("hyperlipidemias" [Mesh] OR "hyperlipidemias" [tiab] OR "dislipidemic" [MeSH] OR "dislipidemic" [tiab] OR "hyperlipidemic" [tiab] OR "dyslipidemia" [tiab] OR "hyperlipidemia" [tiab] OR "hypercholesterolemia" [tiab] OR "hypercholesterolemic" [tiab] OR "hypocholesterolemic" [tiab] OR "total cholesterol" [tiab] OR TC [tiab] OR "low-density lipoprotein" [tiab] OR LDL [tiab] OR "LDL-C" [tiab] OR "LDL-cholesterol"

[tiab] OR "high-density lipoprotein" [tiab] OR HDL [tiab] OR "HDL-C" [tiab] OR "HDL-cholesterol" [tiab] OR "cholesterol" [tiab] OR "triglycerides" [tiab] OR "hypertriglyceridemia" [tiab] OR "hypotriglyceridemic" [tiab] OR "triacylglycerol" [tiab] OR "triacylglyceride" [tiab] OR TG [tiab]) AND (obesity [MeSH] OR obesity [tiab] OR obese [tiab] OR overweight [MesH] OR overweight [tiab] OR "Non-alcoholic Fatty Liver Disease" [MeSH] OR NAFLD [tiab] OR "fatty liver" [MeSH] OR "fatty liver" [tiab] OR "hepatic adipose infiltration" [tiab] OR "metabolic syndrome" [MeSH] OR "metabolic syndrome" [tiab] OR "coronary artery disease" [MeSH] OR "coronary artery disease" [tiab] OR angina [tiab] OR "heart failure" [MeSH] OR "heart failure" [tiab] OR hypertension [MeSH] OR hypertension [tiab] OR diabetes [tiab] OR "diabetes mellitus" [MeSH] OR hyperglycemia [MeSH] OR hyperglycemia [tiab] OR "polycystic ovary syndrome" [MeSH] OR "polycystic ovary syndrome" [tiab] OR PCOS [tiab]) as search terms. Additionally, the references of included studies and related reviews were screened to define more relevant studies. After removing duplicates, the remaining documents were reviewed based on title, abstract, or full text by two authors (SR and MB) separately. During the study selection process, discrepancies between the researchers were resolved by face-to-face discussion.

10.2.2 Inclusion Criteria

Original studies with following criteria were considered in our systematic review: (1) having a clinical trial design; (2) intervention with curcumin supplement; (3) testing interventions on participants at risk of CVD; and (4) considering triglycerides, total cholesterol, high-density lipoprotein cholesterol (HDL-C), and low-density lipoprotein cholesterol (LDL-C) as the primary or secondary outcomes.

10.2.3 Exclusion Criteria

Studies were excluded if they (1) had no control group; (2) reported duplicate data; (3) used turmeric or other curcumin-containing ingredients instead of curcumin supplements; (4) were reviews, letters, editorial articles, study protocol, or case reports; and (5) did not report significance levels between intervention and control groups.

10.2.4 Data Extraction

Articles were selected after reading titles, abstracts, and consideration of inclusion and exclusion criteria. The following information was extracted from articles and recorded in Table 10.1: publication information (first author's last name, publication date, and study location), participants' characteristics (total sample size, gender, mean age, mean body mass index (BMI), and health status), details of the clinical trial (study design, dose and type of interventions, and trial duration), and main results.

10.2.5 Quality Assessment

Researchers (SR and MB) independently assessed the quality of the studies via the Jadad scale, including three domains as follows: (1) randomization (0–2 points); (2) blinding (0–2 points); and (3) dropouts and withdrawals (0–1 points) [50]. According to these domains, the overall quality of each study was considered either low (≤2 scores) or high quality (≥3 scores) [51].

10.3 Results

10.3.1 Search Results and Study Selection

In total, we identified 770 non-duplicated records in searching the electronic databases. Most records ($n = 717$) were excluded after the title and abstract screening stage because they were either multiple publications ($n = 92$) or unrelated studies ($n = 625$). Then, the full text of 53 potentially related articles were assessed. Finally, 22 trials [23–25, 29–33, 35–41, 43–49] met all criteria and were included in the systematic review (Fig. 10.1).

Features of all involved studies were summarized in Table 10.1. The publication dates ranged from 2013 to 2020. All trials were of parallel design and enrolled in 1545 subjects aged 16–70 years. Apart from four studies which were conducted in one gender [30–32, 36], other studies enrolled both genders. More than half of the studies were conducted in Iran [23–25, 29–33, 35, 40–44, 46, 47], and remainder were conducted in Thailand, Kentucky (USA), Italy, Japan, China, Australia, or Taiwan [36–39, 45, 48, 49]. Six trials were conducted on patients with type 2 diabetes [24, 35, 37, 45, 47, 48], two on polycystic ovary syndrome (PCOS) [30, 31], six on NAFLD [23, 25, 33, 40–42, 46], one on coronary artery disease [43], one on patients with dyslipidemia [44], two on metabolic syndrome [29, 49], two on obese or overweight participants [32, 36], and two on patients with impaired glucose tolerance [38, 39]. The intervention periods ranged from 4 to 24 weeks. The dose of curcumin supplementation varied from 80 to 2000 mg/day.

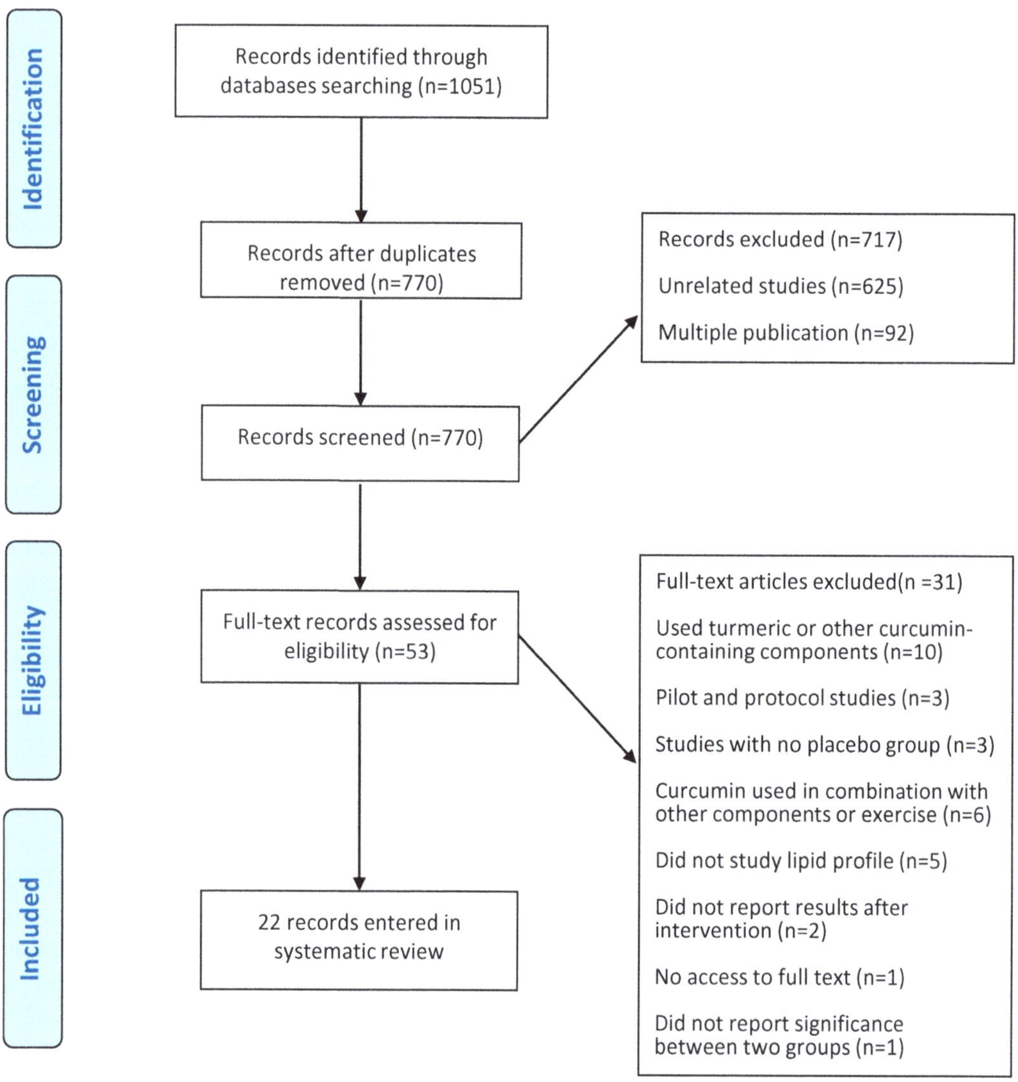

Fig. 10.1 Flowchart of the study selection process

Table 10.1 The characteristics of randomized controlled trials (RCTs) included in this systematic review (arranged alphabetically by first author's last name)

First author (publication year)	Country	N (Male/Female)	Target population	Mean age (or range)	Mean BMI	RCT design (blinding)	Duration	Intervention	Control	TC	TG	LDL	HDL	Jadad score
Adibian (2019) [35]	Iran	44 (22/22)	Patients with type II diabetes	40–70	ND	Parallel (yes)	10 weeks	Curcumin (1500 mg/day—three times)	Placebo (rice flour)	NS	NS	NS	NS	5
Campbell (2019) [36]	Kentucky	22 (male)	Obese subjects	18–35	$\leq$30.0 kg/m^2	Parallel (no)	12 weeks	Enhanced bioavailable curcumin (500 mg/day)	Placebo (soluble fibre)	ND	ND	NS	↑	3
Chuengsamarn (2014) [37]	Thailand	213 (97/116)	Patients with type II diabetes	59.36 ± 10.85	26.96 ± 4.87	Parallel (yes)	6 months	Curcumin (1500 mg/day—two times)	Placebo (starch)	ND	↓	ND	ND	4
Cicero (2019) [38]	Italy	80 (37/43)	Overweight subjects with suboptimal values of FPG	53.5 ± 4.12	27 ± 1.84	Parallel (yes)	8 weeks	Phytosomal curcumin (Curserin®) (1600 mg/day—two times)	Placebo (ND)	NS	↓	NS	NS	4
Funamoto (2019) [39]	Japan	33 (23/10)	Patients with impaired glucose tolerance or non-insulin-dependent diabetes mellitus	69.45 ± 6.48	24.95 ± 3.58	Parallel (yes)	6 months	Highly absorbable curcumin (Theracurmin®) (180 mg/day in two times)	Placebo (ND)	ND	NS	NS	NS	4
Jamalian (2020) [30]	Iran	50 (female)	Patients with PCOS	26.88 ± 3.8	27.87 ± 4.09	Parallel (yes)	12 weeks	Curcumin 500 mg/day	Placebo (starch)	↓	NS	↓	↑	5

(continued)

Table 10.1 (continued)

First author (publication year)	Country	N (Male/Female)	Target population	Mean age (or range)	Mean BMI	RCT design (blinding)	Duration	Intervention	Control	TC	TG	LDL	HDL	Jadad score
Jazayeri-Tehrani (2019) [40]	Iran	84 (46/38)	Overweight and obese patients with NAFLD	42.15 ± 5.88	30.71 ± 2.23	Parallel (yes)	3 months	Nano-curcumin (80 mg/day)	Placebo (ND)	NS	NS	NS	↑	5
Mirhafez (2019) [41, 42]	Iran	59 (37/22)	Patients with NAFLD	43.04 ± 11.51	29.05 ± 5.91	Parallel (yes)	8 weeks	Phytosomal curcumin (capsule composed of 250 mg/day equivalent to 50 mg/day pure curcumin)	Placebo (ND)	NS	NS	NS	↑	5
Mirzabeigi (2015) [43]	Iran	33 (24/9)	Patients with coronary artery disease	62.85 ± 8.55	28.21 ± 3.60	Parallel (yes)	8 weeks	Curcumin (2000 mg/day—four times)	Placebo (ND)	NS	NS	NS	NS	3
Mohammadi (2013) [44]	Iran	30 (83% female)	Dyslipi-demic obese patients	38.43 ± 10.84	32.60 ± 3.58	Cross-over	4 weeks	Curcuminoids (1000 mg/day + 5 mg bioperine)	Piperine (ND)	NS	↓	NS	NS	4
Na (2012) [45]	China	100 (49/51)	Patients with type II diabetes	55.07 ± 7.4	27.27 ± 2.6	Parallel (yes)	12 weeks	Curcuminoids 300 mg/day	Starch	NS	↓	NS	NS	5
Panahi (2014) [29]	Iran	100 (50/50)	Patients with metabolic syndrome	44.13 ± 9.17	24.13 ± 4.36	Parallel (yes)	8 weeks	(Curcuminoids-piperine) 1000 mg/day	Lactose +5 mg Bioperine	↓	↓	↓	NS	4
Panahi (2016) [46]	Iran	87 (51/36)	Patients with NAFLD	46.08 ± 11.5	29.02 ± 3.43	Parallel (yes)	8 weeks	Curcumin (Meriva®) (phytosomal form; 1000 mg/day in 2 divided doses)	Lactose	↓	↓	↓	NS	2

Panahi (2017) [24]	Iran	100 (51/49)	Patients with type II diabetes	42 ± 7.54	26.93 ± 2.01	Parallel (yes)	12 weeks	Curcuminoids (1000 mg/day coadministered with piperine 10 mg/day)	Micro-crystalline cellulose	↓	NS	NS	↑	4
Panahi (2019) [25]	Iran	70 (39/31)	Patients with NAFLD	47.07 ± 2.36	–	Parallel (yes)	12 weeks	500 mg curcuminoids coadministered with 5 mg piperine	Placebo (ND)	NS	NS	NS	NS	4
Rahimi (2016) [47]	Iran	70 (31/39)	Patients with type II diabetes	58.64 ± 11.13	27.09 ± 3.16	Parallel (yes)	12 weeks	Curcumin nanomicelle 80 mg/day	Placebo (ND)	NS	NS	↓	NS	4
Rahmani) 2016) [23]	Iran	80 (38/42)	Patients with NAFLD	47.71 ± 10.68	31.10 ± 5.09	Parallel (yes)	8 weeks	Curcumin formulation (500 mg/day equivalent to 70-mg curcumin)	Placebo (ND)	↓	↓	↓	NS	4
Saadati (2019) [33]	Iran	50 (27/23)	Patients with NAFLD	45.70 ± 11.13	32.33 ± 4.72	Parallel (yes)	12 weeks	Curcumin (1500 mg/day in three times)	Maltod-exterin	NS	NS	NS	NS	5
Saraf (2019) [32]	Iran	60 (female)	Overweight and obese adolescents	16.00 ± 1.63	30.85 ± 2.87	Parallel (yes)	10 weeks	Standardized 95% turmeric extract (500 mg curcumin)	Placebo (ND)	NS	NS	NS	NS	5
Sohaei (2019) [31]	Iran	51 (female)	Patients with PCOS	29.48 ± 5.12	30.45 ± 4.20	Parallel (yes)	6 weeks	Curcumin (1000 mg/day) in two times	Placebo (ND)	NS	NS	NS	NS	5
Thota (2019) [48]	Australia	64 (26/38)	Patients with type II diabetes	55 ± 1.3	31.1 ± 0.7	Parallel (yes)	12 weeks	Curcumin (Meriva®) (1000 mg/day in two times)	Placebo (ND)	NS	↓	NS	NS	5

(continued)

Table 10.1 (continued)

First author (publication year)	Country	N (Male/Female)	Target population	Mean age (or range)	Mean BMI	RCT design (blinding)	Duration	Intervention	Control	TC	TG	LDL	HDL	Jadad score
Yang (2014) [49]	Taiwan	65 (29/36)	Patients with metabolic syndrome	59.3 ± 12.0	30.06 ± 4.12	Parallel (yes)	12 weeks	Curcumin extract capsule (1890 mg/day in three times)	Placebo (ND)	NS	NS	↓	NS	5

M male, *F* female, *ND* no data, *NAFLD* nonalcoholic fatty liver disease, *BMI* body mass index, *TC* total c**holes**terol, *TG* triglyceride, *HDL* high-density lipoprotein, *LDL* low-density lipoprotein, *FPG* fasting plasma glucose, *PCOS* polycystic ovary syndrome, *NS* not significant

Two studies used nano-curcumin [40, 47], two studies used phytosomal curcumin [38, 46], three studies used curcumin-piperine [24, 25, 29], one study used phospholipidated curcumin [41, 42], one study used highly absorbable curcumin [39], and one study used enhanced bioavailable curcumin [36] for supplementation.

10.3.2 Quality Assessment

Apart from one study that had a low-quality score [46], all of the remaining studies ($n = 21$) were categorized as high quality using the Jadad scale.

10.3.3 Effect of Curcumin Supplementation on Triglycerides

Curcumin supplementation significantly reduced triglyceride levels in 8 out of 21 studies [23, 24, 37, 38, 44–46, 48]. In the remaining studies, the results were not significant. In one study, triglycerides were not evaluated [36].

10.3.4 Effect of Curcumin Supplementation on Total Cholesterol

Of the 22 studies included in the present review, 21 assessed total cholesterol, and only 5 of these studies showed a significant reduction after curcumin supplementation [23, 24, 29, 30, 46]. No significant effects were observed in other studies. In one study, total cholesterol was not assessed [37].

10.3.5 Effect of Curcumin Supplementation on LDL-C

In 6 out of 21 studies which assessed LDL-C, a significant reduction was observed after intervention with curcumin supplementation [23, 29, 30, 46, 47, 49]. The other studies did not report any significant results, and LDL-C was not assessed in one study [37].

10.3.6 The Effect of Curcumin Supplementation on HDL-C

In five of the included studies, supplementation with curcumin significantly increased HDL-C levels [24, 30, 36, 40–42]. However, other studies did not report significant results. In one study, HDL-C was not assessed [37].

10.4 Discussion

Evidence shows that hyperlipidemia is one of the significant risk factors for cardiovascular disease [52]. Recent studies have also shown that supplementation with curcumin can have beneficial effects on lipid profiles [26], although the results of some studies were contradictory [23–25, 31]. Therefore, this systematic review of existing clinical studies has summarized the results to help determine what conclusions can be drawn. For this, we analyzed 22 clinical studies which examined triglycerides, cholesterol, LDL-C, and HDL-C biomarker indices. In general, the present study showed that supplementation with curcumin improved at least one of the lipid profile indices in 15 out of 22 studies and improved more than one index in 5 studies.

Hypercholesterolemia is one of the main cardiovascular risk factors and plays a key role in the development of CVD [53]. However, only five studies have reported positive effects with curcumin supplementation, with most of these using other compounds in addition to curcumin to improve its bioavailability. One of the suggested mechanisms for the cholesterol-lowering effects of curcumin is via reduced absorption of cholesterol from the intestine to the circulatory system [54].

Increasing evidence has shown that hypertriglyceridemia increases the risk of atherosclerotic CVD [55]. Eight trials showed that curcumin supplementation reduced triglyceride levels, with most of the studies using high bioavailability curcumin supplements. Also, most of these studies used curcumin at a dose greater than or equal to 1000 mg/day. Therefore, these factors may have contributed to the efficacy of curcumin supplementation in this case. The triglyceride-reducing effects of curcumin could be due to effects on various pathways involved in homeostasis of triglycerides and free fatty acids. These pathways could include those regulated by peroxisome proliferator-activated receptors alpha and gamma, cholesterol ester transfer protein (CETP), and lipoprotein lipase [49].

A review of studies involving the LDL-C index showed that six clinical trials found beneficial lowering effects of curcumin supplementation, and in most of these studies, the duration of the intervention was 8 weeks or more. Half of these studies used high bioavailability supplements which might have had an impact on the observed effect. It has been proposed that the mechanism underlying the effect of curcumin supplementation on LDL-C levels may be through upregulation of the LDL receptor in hepatocytes [56]. Moreover, curcumin can decrease the levels of apoB100 which is the major apolipoprotein of LDL-C [57, 58].

Evidence suggests an inverse relationship between HDL and CVD risk [59]. In five studies, supplementation with curcumin had an additive effect on HDL levels. It should be noted that in most of the studies, the duration of supplementation was 8 weeks or more. Moreover, most of the studies used curcumin supplements with high bioavailability, which could have contributed to the observed beneficial effects. However, there are some inconsistencies of these findings with those of other studies [23, 35, 38, 39, 47]. The finding of little or no effect in these other studies may be due to possible upregulation of the scavenger receptor class B type 1 (SRB1) in the liver. This would lead to increased absorption of cholesterol from HDL into hepatocytes [42].

The main limitations of this systematic review include the points that the analyzed studies used varying dosages and formulations of curcumin, as well as different intervention periods, which made direct comparisons difficult. Secondly, more than half of the studies were from Iran, which restricts the ability to extend the results to other populations. Thus, further studies should be conducted in different locations using standardized conditions.

10.5 Conclusions

The results of this systematic review showed that in more than two-thirds of the included studies, curcumin supplementation significantly reduced at least one of the lipid profile indices and in five studies improved more than one index in patients with the risk of CVD. Overall, studies using the formulations of curcumin with higher bioavailability had a better impact on the lipid profile. Considering the fact that curcumin is a natural, accessible, safe, and inexpensive plant extract [60], the findings of this review highlight the potential of curcumin supplementation to improve lipid profiles in patients at risk of CVDs. However, in seven included studies, curcumin had no effect on any of lipid profile indices. Future studies are still needed to explore the lipid-modifying effects of curcuminoids in dyslipidemic populations at a high risk of CVD and in the setting of large-scale randomized placebo-controlled trials. Such studies are encouraged to look at the long-term effects on lipid indices and assess if the putative improvement in lipid profile by curcuminoids is associated with a reduction in cardiovascular events.

References

1. Deng J (2011) Journal of Cardiovascular Disease Research celebrates its anniversary. J Cardiovasc Dis Res 2(1):1–2
2. Ruan Y, Guo Y, Zheng Y, Huang Z, Sun S, Kowal P et al (2018) Cardiovascular disease (CVD) and associated risk factors among older adults in six low-and middle-income countries: results from SAGE Wave 1. BMC Public Health 18:778. https://doi.org/10.1186/s12889-018-5653-9
3. World Health Organisation (2016) Cardiovascular diseases (CVDs) (2017) Fact Sheet. https://www.who.int/news-room/fact-sheets/detail/cardiovascular-diseases-(cvds)
4. Perk J, De Backer G, Gohlke H, Graham I, Reiner Z, European Association for Cardiovascular Prevention & Rehabilitation (EACPR), ESC Committee for Practice Guidelines (CPG) et al (2012) European Guidelines on cardiovascular disease prevention in clinical practice (version 2012). The Fifth Joint Task Force of the European Society of Cardiology and Other Societies on Cardiovascular Disease Prevention in Clinical Practice (constituted by representatives

of nine societies and by invited experts). Eur Heart J 33(13):1635–1701

5. World Health Organization (2011) Global atlas on cardiovascular disease prevention and control. https://www.who.int/cardiovascular_diseases/publications/atlas_cvd/en/
6. Tana C, Ballestri S, Ricci F, Di Vincenzo A, Ticinesi A, Gallina S et al (2019) Cardiovascular risk in non-alcoholic fatty liver disease: mechanisms and therapeutic implications. Int J Environ Res Public Health 16(17):3104. https://doi.org/10.3390/ijerph16173104
7. Papadakis G, Kandaraki E, Papalou O, Vryonidou A, Diamanti-Kandarakis E (2017) Is cardiovascular risk in women with PCOS a real risk? Current insights. Minerva Endocrinol 42(4):340–355
8. Buttar HS, Li T, Ravi N (2005) Prevention of cardiovascular diseases: role of exercise, dietary interventions, obesity and smoking cessation. Exp Clin Cardiol 10(4):229–249
9. Stewart J, Manmathan G, Wilkinson P (2017) Primary prevention of cardiovascular disease: a review of contemporary guidance and literature. JRSM Cardiovasc Dis 6:2048004016687211. https://doi.org/10.1177/2048004016687211
10. Fruchart JC, Sacks F, Hermans MP, Assmann G, Brown WV, Ceska R et al (2008) The residual risk reduction initiative: a call to action to reduce residual vascular risk in patients with dyslipidemia. Am J Cardiol 102(10):1K–34K
11. Battisti WP, Palmisano J, Keane WF (2003) Dyslipidemia in patients with type 2 diabetes. Relationships between lipids, kidney disease and cardiovascular disease. Clin Chem Lab Med 41(9):1174–1181
12. Johnson ML, Pietz K, Battleman DS, Beyth RJ (2004) Prevalence of comorbid hypertension and dyslipidemia and associated cardiovascular disease. Am J Manag Care 10:926–932
13. Srikanth S, Deedwania P (2016) Management of dyslipidemia in patients with hypertension, diabetes, and metabolic syndrome. Curr Hypertens Rep 18(10):76. https://doi.org/10.1186/1758-5996-2-55
14. Huffman MD, Bhatnagar D (2012) Novel treatments for cardiovascular disease prevention. Cardiovasc Ther 30(5):257–263
15. Adegbola P, Aderibigbe I, Hammed W, Omotayo T (2017) Antioxidant and anti-inflammatory medicinal plants have potential role in the treatment of cardiovascular disease: a review. Am J Cardiovasc Dis 7(2):19–32
16. Hewlings SJ, Kalman DS (2017) Curcumin: a review of its effects on human health. Foods 6(10):92. https://doi.org/10.3390/foods6100092
17. Rahmani AH, Alsahli MA, Aly SM, Khan MA, Aldebasi YH (2018) Role of curcumin in disease prevention and treatment. Adv Biomed Res 7:38. https://doi.org/10.4103/abr.abr_147_16
18. Bagherniya M, Nobili V, Blesso CN, Sahebkar A (2018) Medicinal plants and bioactive natural compounds in the treatment of non-alcoholic fatty liver disease: a clinical review. Pharmacol Res 130:213–240
19. Teymouri M, Pirro M, Johnston TP, Sahebkar A (2017) Curcumin as a multifaceted compound against human papilloma virus infection and cervical cancers: A review of chemistry, cellular, molecular, and preclinical features. BioFactors 43(3):331–346.
20. Momtazi AA, Derosa G, Maffioli P, Banach M, Sahebkar A (2016) Role of microRNAs in the Therapeutic Effects of Curcumin in Non-Cancer Diseases. Mol Diagn Ther 20(4):335–345.
21. Iranshahi M, Sahebkar A, Hosseini ST, Takasaki M, Konoshima T, Tokuda H (2010) Cancer chemopreventive activity of diversin from Ferula diversivittata in vitro and in vivo. Phytomedicine 17(3-4):269–273
22. Mollazadeh H, Cicero AFG, Blesso CN, Pirro M, Majeed M, Sahebkar A (2019) Immune modulation by curcumin: The role of interleukin-10 (2019) Critical Reviews in Food Science and Nutrition, 59(1):89–101.
23. Rahmani S, Asgary S, Askari G, Keshvari M, Hatamipour M, Feizi A et al (2016) Treatment of non-alcoholic fatty liver disease with curcumin: a randomized placebo-controlled trial. Phytother Res 30(9):1540–1548
24. Panahi Y, Khalili N, Sahebi E, Namazi S, Reiner Ž, Majeed M et al (2017) Curcuminoids modify lipid profile in type 2 diabetes mellitus: a randomized controlled trial. Complement Ther Med 33:1–5
25. Panahi Y, Valizadegan G, Ahamdi N, Ganjali S, Majeed M, Sahebkar A (2019) Curcuminoids plus piperine improve nonalcoholic fatty liver disease: a clinical trial. J Cell Biochem 120(9):15989–15996
26. Panahi Y, Ahmadi Y, Teymouri M, Johnston TP, Sahebkar A (2018) Curcumin as a potential candidate for treating hyperlipidemia: a review of cellular and metabolic mechanisms. J Cell Physiol 233(1):141–152
27. Qin S, Huang L, Gong J, Shen S, Huang J, Ren H, Hu H (2017) Efficacy and safety of turmeric and curcumin in lowering blood lipid levels in patients with cardiovascular risk factors: a meta-analysis of randomized controlled trials. Nutr J 16(1):68. https://doi.org/10.1186/s12937-017-0293-y
28. Simental-Mendía LE, Pirro M, Gotto AM, Jr., Banach M, Atkin SL, Majeed M, Sahebkar A. Lipid-modifying activity of curcuminoids: A systematic review and meta-analysis of randomized controlled trials (2019). Critical Reviews in Food Science and Nutrition, 59(7):1178–1187
29. Panahi Y, Khalili N, Hosseini MS, Abbasinazari M, Sahebkar A (2014) Lipid-modifying effects of adjunctive therapy with curcuminoids–piperine combination in patients with metabolic syndrome: results of a randomized controlled trial. Complement Ther Med 22(5):851–857
30. Jamilian M, Foroozanfard F, Kavossian E, Aghadavod E, Shafabakhsh R, Hoseini A et al (2020) Effects of curcumin on body weight, glycemic control and serum lipids in women with polycystic ovary syndrome: a

randomized, double-blind, placebo-controlled trial. Clin Nutr ESPEN 36:128–133
31. Sohaei S, Amani R, Tarrahi MJ, Ghasemi-Tehrani H (2019) The effects of curcumin supplementation on glycemic status, lipid profile and hs-CRP levels in overweight/obese women with polycystic ovary syndrome: a randomized, double-blind, placebo-controlled clinical trial. Complement Ther Med 47:102201. https://doi.org/10.1016/j.ctim.2019.102201
32. Saraf-Bank S, Ahmadi A, Paknahad Z, Maracy M, Nourian M (2019) Effects of curcumin on cardiovascular risk factors in obese and overweight adolescent girls: a randomized clinical trial. Sao Paulo Med J 137(5):414–422
33. Saadati S, Hatami B, Yari Z, Shahrbaf MA, Eghtesad S, Mansour A et al (2019) The effects of curcumin supplementation on liver enzymes, lipid profile, glucose homeostasis, and hepatic steatosis and fibrosis in patients with non-alcoholic fatty liver disease. Eur J Clin Nutr 73(3):441–449
34. Moher D, Liberati A, Tetzlaff J, Altman DG (2009) Preferred reporting items for systematic reviews and meta-analyses: the PRISMA statement. Ann Intern Med 151(4):264–269
35. Adibian M, Hodaei H, Nikpayam O, Sohrab G, Hekmatdoost A, Hedayati M (2019) The effects of curcumin supplementation on high-sensitivity C-reactive protein, serum adiponectin, and lipid profile in patients with type 2 diabetes: a randomized, double-blind, placebo-controlled trial. Phytother Res 33(5):1374–1383
36. Campbell MS, Ouyang A, Krishnakumar I, Charnigo RJ, Westgate PM, Fleenor BS (2019) Influence of enhanced bioavailable curcumin on obesity-associated cardiovascular disease risk factors and arterial function: a double-blinded, randomized, controlled trial. Nutrition 62:135–139
37. Chuengsamarn S, Rattanamongkolgul S, Phonrat B, Tungtrongchitr R, Jirawatnotai S (2014) Reduction of atherogenic risk in patients with type 2 diabetes by curcuminoid extract: a randomized controlled trial. J Nutr Biochem 25(2):144–150
38. Cicero AF, Sahebkar A, Fogacci F, Bove M, Giovannini M, Borghi C (2020) Effects of phytosomal curcumin on anthropometric parameters, insulin resistance, cortisolemia and non-alcoholic fatty liver disease indices: a double-blind, placebo-controlled clinical trial. Eur J Nutr 59(2):477–483
39. Funamoto M, Shimizu K, Sunagawa Y, Katanasaka Y, Miyazaki Y, Kakeya H et al (2019) Effects of highly absorbable curcumin in patients with impaired glucose tolerance and non-insulin-dependent diabetes mellitus. J Diabetes Res 2019:8208237. https://doi.org/10.1155/2019/8208237
40. Jazayeri-Tehrani SA, Rezayat SM, Mansouri S, Qorbani M, Alavian SM, Daneshi-Maskooni M et al (2019) Nano-curcumin improves glucose indices, lipids, inflammation, and Nesfatin in overweight and obese patients with non-alcoholic fatty liver disease (NAFLD): a double-blind randomized placebo-controlled clinical trial. Nutr Metab (Lond) 16:8. https://doi.org/10.1186/s12986-019-0331-1
41. Mirhafez SR, Farimani AR, Dehhabe M, Bidkhori M, Hariri M, Ghouchani BFNM et al (2019) Effect of phytosomal curcumin on circulating levels of adiponectin and leptin in patients with non-alcoholic fatty liver disease: a randomized, double-blind, placebo-controlled clinical trial. J Gastrointestin Liver Dis 28:183–189
42. Mirzabeigi P, Mohammadpour AH, Salarifar M, Gholami K, Mojtahedzadeh M, Javadi MR (2015) The effect of curcumin on some of traditional and non-traditional cardiovascular risk factors: a pilot ramized, double-blind, placebo-controlled trial. Iran J Pharm Res 14(2):479–486
43. Mohammadi A, Sahebkar A, Iranshahi M, Amini M, Khojasteh R, Ghayour-Mobarhan M et al (2013) Effects of supplementation with curcuminoids on dyslipidemia in obese patients: a randomized crossover trial. Phytother Res 27(3):374–379
44. Na LX, Li Y, Pan HZ, Zhou XL, Sun DJ, Meng M et al (2013) Curcuminoids exert glucose-lowering effect in type 2 diabetes by decreasing serum free fatty acids: a double-blind, placebo-controlled trial. Mol Nutr Food Res 57(9):1569–1577
45. Panahi Y, Kianpour P, Mohtashami R, Jafari R, Simental-Mendía LE, Sahebkar A (2016) Curcumin lowers serum lipids and uric acid in subjects with non-alcoholic fatty liver disease: a randomized controlled trial. J Cardiovasc Pharmacol 68(3):223–229
46. Rahimi HR, Mohammadpour AH, Dastani M, Jaafari MR, Abnous K, Mobarhan MG et al (2016) The effect of nano-curcumin on HbA1c, fasting blood glucose, and lipid profile in diabetic subjects: a randomized clinical trial. Avicenna J Phytomed 6(5):567–577
47. Thota RN, Acharya SH, Garg ML (2019) Curcumin and/or omega-3 polyunsaturated fatty acids supplementation reduces insulin resistance and blood lipids in individuals with high risk of type 2 diabetes: a randomised controlled trial. Lipids Health Dis 18(1):31. https://doi.org/10.1186/s12944-019-0967-x
48. Yang YS, Su YF, Yang HW, Lee YH, Chou JI, Ueng KC (2014) Lipid-lowering effects of curcumin in patients with metabolic syndrome: a randomized, double-blind, placebo-controlled trial. Phytother Res 28(12):1770–1777
49. Nelson RH (2013) Hyperlipidemia as a risk factor for cardiovascular disease. Prim Care 40(1):195–211
50. Jadad AR, Moore RA, Carroll D, Jenkinson C, Reynolds DJM, Gavaghan DJ et al (1996) Assessing the quality of reports of randomized clinical trials: is blinding necessary? Control Clin Trials 17(1):1–12
51. Moher D, Cook D, Jadad A, Tugwell P, Moher M, Jones A et al (1999) Assessing the quality of reports of randomised trials: implications for the conduct of meta-analyses. Health Technol Assess 3(12):i–iv. 1–98
52. Cimminiello C, Zambon A, Polo HF (2016) Hypercholesterolemia and cardiovascular risk: advantages and limitations of current treatment options. G

Ital Cardiol (Rome) 17(4 Suppl 1):6S–13S. https://doi.org/10.1714/2254.24276
53. Baum L, Cheung SK, Mok VC, Lam LC, Leung VP, Hui E et al (2007) Curcumin effects on blood lipid profile in a 6-month human study. Pharmacol Res 56(6):509–514
54. Peng J, Luo F, Ruan G, Peng R, Li X (2017) Hypertriglyceridemia and atherosclerosis. Lipids Health Dis 16(1):233–233
55. Feng D, Ohlsson L, Duan R-D (2010) Curcumin inhibits cholesterol uptake in Caco-2 cells by down-regulation of NPC1L1 expression. Lipids Health Dis 9(1):40. https://doi.org/10.1186/1476-511X-9-40
56. Shin SK, Ha TY, McGregor RA, Choi MS (2011) Long-term curcumin administration protects against atherosclerosis via hepatic regulation of lipoprotein cholesterol metabolism. Mol Nutr Food Res 55(12):1829–1840
57. Jang EM, Choi MS, Jung UJ, Kim MJ, Kim HJ, Jeon SM et al (2008) Beneficial effects of curcumin on hyperlipidemia and insulin resistance in high-fat–fed hamsters. Metabolism 57(11):1576–1583
58. Bartlett J, Predazzi IM, Williams SM, Bush WS, Kim Y, Havas S et al (2016) Is isolated low HDL-C a CVD risk factor? New insights from the framingham offspring study. Circ Cardiovasc Qual Outcomes 9(3):206–212
59. Cuchel M, Rader DJ (2006) Macrophage reverse cholesterol transport: key to the regression of atherosclerosis? Circulation 113(21):2548–2555
60. Soleimani V, Sahebkar A, Hosseinzadeh H. Turmeric (Curcuma longa) and its major constituent (curcumin) as nontoxic and safe substances: Review (2018) Phytotherapy Research 32(6):985–995

11 The Effects of Curcumin in the Treatment of Gingivitis: A Systematic Review of Clinical Trials

Farinaz Shirban, Fateme Gharibpour, Ailin Ehteshami, Mohammad Bagherniya, Thozhukat Sathyapalan, and Amirhossein Sahebkar

Abstract

Different modalities of treatments are available for management of gingival disease but most have adverse effects. Curcumin has anti-inflammatory properties and can be used for management of various inflammatory processes. This systematic review evaluates the effects of curcumin as an adjuvant to oral hygiene on plaque index (PI), gingival index (GI), gingival bleeding index (GBI), and inflammation in patients with gingivitis. A comprehensive search was conducted using PubMed/MEDLINE, Cochrane, SCOPUS, and Google Scholar. Based on the Population, Intervention, Control, and Outcome (PICO) model, clinical trials which tested the effects of curcumin as an adjunctive product or alone in control of gingival inflammation up until 21 February 2020 with language restrictions were selected. From the 422 papers found, 14 met the eligibility criteria. In most of these studies, curcumin treatment achieved significant reductions in PI, GI, GBI, and microbial colony count and was as effective as chlorhexidine mouthwash, with no serious adverse effects. We conclude that treatment with curcumin for gingivitis is safe as a natural herbal compound and is as effective as chlorhexidine mouthwash.

F. Shirban · F. Gharibpour · A. Ehteshami
Department of Orthodontics, Dental Research Center, Dental Research Institute, School of Dentistry, Isfahan University of Medical Sciences, Isfahan, Iran

M. Bagherniya
Food Security Research Center, Department of Community Nutrition, School of Nutrition and Food Science, Isfahan University of Medical Sciences, Isfahan, Iran

T. Sathyapalan
Department of Academic Diabetes, Endocrinology and Metabolism, Hull York Medical School, University of Hull, Hull, UK

A. Sahebkar (✉)
Biotechnology Research Center, Pharmaceutical Technology Institute, Mashhad University of Medical Sciences, Mashhad, Iran

Neurogenic Inflammation Research Center, Mashhad University of Medical Sciences, Mashhad, Iran

Halal Research Center of IRI, FDA, Tehran, Iran

Polish Mother's Memorial Hospital Research Institute (PMMHRI), Lodz, Poland
e-mail: sahebkara@mums.ac.ir

Keywords

Curcumin · Gingivitis · Plaque index · Gingival index · Gingival bleeding index · Systematic review

P. C. Guest (ed.), *Studies on Biomarkers and New Targets in Aging Research in Iran*, Advances in Experimental Medicine and Biology 1291, https://doi.org/10.1007/978-3-030-56153-6_11

11.1 Introduction

Colonization of various microbiota starts in the oral cavity by the time of birth. Some microbiota can form dental plaque by adhering to the host, resulting in various pathologies such as gingivitis and periodontal disease [1, 2]. Gingivitis is an inflammatory disease that is prevalent in both children and adults and can lead to periodontitis unless treated [3, 4]. Several etiological factors have been attributed to gingivitis but a key contributor is biofilm accumulation (bacterial plaque) [3, 4]. Mechanical and chemical plaque reduction is effective and measures such as scaling, root planing, brushing, and floss are the first step in the management of gingivitis [4, 5]. However, the time-consuming nature of these approaches could lead to some patients neglecting their oral hygiene. This makes it challenging to accomplish satisfactory oral hygiene by mechanical methods alone [6, 7]. Hence, chemical therapies are often needed for controlling plaque formation and may be beneficial as an adjunctive agent in addition to mechanical therapy [8, 9].

The gold standard for treatment and prevention of gingivitis is chlorhexidine (CHX) because of its antibacterial and antiseptic properties [10, 11]. However, the side effects of CHX include brownish discoloration of teeth, taste alteration, erosion, and bitter taste [2, 6]. Recently, herbal medicines have been assessed as potential agents in oral and dental health, since these are natural, accessible, and inexpensive approaches without serious adverse effects [12–14]. The therapeutic use of natural compounds has a long history [15]. There has been an increase in interest in the use of herbal products against human diseases such as cardiovascular disease, cancer, diabetes, and inflammatory disease [16, 17]. They have also become the subject of several types of researches in the field of dentistry [13, 14, 18, 19]. The cost benefits and fewer side effects of herbal products in comparison with conventional drugs make them an economical alternative medicine [20].

Turmeric (*Curcuma longa* L.) is a spice that has been used in Asian countries for thousands of years [6]. The principal component of turmeric is curcumin (CUR), a yellow-colored chemical with a low molecular weight. This polyphenol has been shown to have analgesic, antibacterial, anti-inflammatory, antitumor, antiallergic, and antioxidant effects [21–27]. Also, it is known that massaging CUR can eliminate pain and swelling of the gingivae [28]. In gingivitis, pathological microorganisms can produce proinflammatory vasoactive mediators such as prostaglandin E2 (PGE2) and interleukin 1 (IL-1) by monocyte and macrophage activation. CUR can be effective against this progression due to its anti-inflammatory properties [29]. The anti-inflammatory properties of CUR are responsible for blockade of the arachidonic acid pathway, with effects on cyclooxygenase, lipoxygenase, tumor necrosis factor-alpha (TNF-α), nuclear factor kappa B (NF-κB), and various cytokines [30–43].

Wide-ranging research has been conducted to assess the efficacy of CUR against gingivitis. Thus, this systematic review was aimed to summarize and categorize the evidence of the efficacy of CUR for the management of gingivitis.

11.2 Materials and Methods

This systematic review was conducted in accordance with the Preferred Reporting Items for Systematic Reviews (PRISMA) Statement Guidelines [44]. We aimed to assess the null hypothesis that "there is no effect of consumption of CUR as an adjunctive product or alone in controlling gingival inflammation." The focused question according to the Population, Intervention, Control, and Outcome (PICO) format was: in patients with gingivitis, does CUR alone or in addition to conventional mechanical oral hygiene provide improvement in clinical outcome? In this case, the population was patients with gingival inflammation, intervention was treatment with CUR products, control was treatment with scaling and root planning or usage of CHX, and outcome was plaque, gingival and gingival bleeding index (GBI), probing depth (PD), and microbial colony count.

11.2.1 Search Strategy

An electronic search was performed up to 21 February 2020 to acquire potentially eligible studies with no time restrictions in the following electronic bibliographic databases: PubMed/MEDLINE, Cochrane, and SCOPUS. The first hundred records of Google Scholar were also screened. The search strategies of each database are presented in Table 11.1. The reference part of the retrieved full-text articles (cross-referencing) was also searched for further papers. Papers not written in English were excluded.

11.2.2 Eligibility Criteria

The following selection criteria were applied for this systematic review:

1. Inclusion criteria: randomized experimental trials, both parallel-group and split-mouth were considered for articles providing data regarding effects of CUR on gingivitis.
2. Exclusion criteria:
   (a) In vitro histological and animal studies.
   (b) Review articles, case reports, letters to editors.
   (c) Pilot studies without a control group.
   (d) Studies about the effects of CUR on periodontitis or patients with attachment loss or bone loss non-English language articles.

11.2.3 Study Selection

The titles and abstracts of the explored studies were screened by FS and FG. Studies were excluded if they were not either appropriate to the present study, duplicates, and because the focused question was not addressed. Any discrepancies were resolved through discussion. Publications were included for full text assessment if they met the inclusion criteria in the first analysis or if pieces of information were insufficient in the title and abstract to enable a decision to be made. The review authors were not blinded to the author(s), institution, or site of publication of studies.

11.2.4 Data Extraction

The following data were extracted:

1. Data from the study.
   (a) Authors, year of the publication, and country.
   (b) Design (randomized controlled trial (RCT) or clinical study, and duration of intervention and blindness.
2. Intervention and control.
   (a) Scientific or common name.
3. Protocol.
   (a) Dose and route of administration.
   (b) Instruction of herbal use.
4. Participants
   (a) Sample size.
   (b) Age.
   (c) Sex.
5. Results
   (a) Outcome measurements.
   (b) Adverse effect.
   (c) Main outcome.
   (d) Measure at baseline and end of the study.
   (e) *P*-value.

11.2.5 Quality Assessment in Included Studies

Two trained reviewers judged the quality of the articles in the current review (FS and FG). The Jadad scale of blinding, randomization, and account of all patients was used to assess the quality of studies. This scale consists of the following items: randomization (0–2 points); blinding (0–2 points), and dropouts and withdrawals (0–1 point). According to this scale, the overall score of a study ranges between 0 and 5, with higher scores indicative of better quality [45]. Studies with Jadad scores of ≤ 2 and ≥ 3 were considered as low- and high-quality, respectively [46].

Table 11.1 Databases, applied search strategy, and numbers of retrieved studies

Database of published trials	Search strategy used
MEDLINE searched via PubMed, searched on 21 February 2020, via www.ncbi.nlm.nih.gov/sites	((curcuma"[MeSH Terms] OR ("curcuma"[MeSH Terms] OR "curcuma"[All Fields]) OR ("curcuma"[MeSH Terms] OR "curcuma"[All Fields] OR "curcumas"[All Fields]) OR ("curcumin"[MeSH Terms] OR "curcumin"[All Fields]) OR curcuminoid[All Fields] OR ("diarylheptanoids"[MeSH Terms] OR "diarylheptanoids"[All Fields] OR "curcuminoids"[All Fields]) OR ("curcumin"[MeSH Terms] OR "curcumin"[All Fields] OR "tumeric"[All Fields] OR "curcuma"[MeSH Terms] OR "curcuma"[All Fields]) OR ("curcuma"[MeSH Terms] OR "curcuma"[All Fields] OR "curcumin"[All Fields]) OR "curcuma longa"[All Fields] OR "curcuma zedoaria"[All Fields] OR ("zingiberaceae"[MeSH Terms] OR "zingiberaceae"[All Fields])) AND ((((((((("gingivitis"[MeSH Terms] OR "gingivitis"[All Fields]) OR ("gingival diseases"[MeSH Terms] OR ("gingival"[All Fields] AND "diseases"[All Fields]) OR "gingival diseases"[All Fields] OR ("gingival"[All Fields] AND "disease"[All Fields]) OR "gingival disease"[All Fields])) OR ("gingival diseases"[MeSH Terms] OR ("gingival"[All Fields] AND "diseases"[All Fields]) OR "gingival diseases"[All Fields] OR ("gum"[All Fields] AND "disease"[All Fields]) OR "gum disease"[All Fields])) OR ("gingivitis"[MeSH Terms] OR "gingivitis"[All Fields] OR ("gum"[All Fields] AND "inflammation"[All Fields]) OR "gum inflammation"[All Fields])) OR ("gingivitis"[MeSH Terms] OR "gingivitis"[All Fields] OR ("gingival"[All Fields] AND "inflammation"[All Fields]) OR "gingival inflammation"[All Fields])) OR ("dental plaque"[MeSH Terms] OR ("dental"[All Fields] AND "plaque"[All Fields]) OR "dental plaque"[All Fields])) OR ("dental plaque index"[MeSH Terms] OR ("dental"[All Fields] AND "plaque"[All Fields] AND "index"[All Fields]) OR "dental plaque index"[All Fields])) OR ("gingival haemorrhage"[All Fields] OR "gingival hemorrhage"[MeSH Terms] OR ("gingival"[All Fields] AND "hemorrhage"[All Fields]) OR "gingival hemorrhage"[All Fields])) OR (("hemorrhage"[MeSH Terms] OR "hemorrhage"[All Fields] OR "bleeding"[All Fields]) AND probing[All Fields
Scopus searched via Scopus on 21 February 2020, via https://www.scopus.com	(*curcuma* **OR** *curcumas* **OR** *curcumin* **OR** *curcuminoid* **OR** *curcuminoids* **OR** *tumeric* **OR** *curcumin* **OR** *longa* **OR** *zedoaria* **OR** *zingiberaceae*) **AND** (*gingivitis* **OR** *gingival* **OR** *gum* **OR** *hemorrhage* **OR** *bleeding* **OR** *probing*) **SUBJAREA** (*dent*)
Cochrane Central Register of Controlled Trials searched via the Cochrane Library Searched on 21 February 2020 via www.thecochranelibrary.com	(curcuma OR curcuma OR curcuma OR curcumin OR curcuminoids OR curcuminoids OR tumeric OR curcumin OR curcuma OR longa OR curcuma OR zedoaria OR zingiberaceae) AND (gingivitis OR gingival OR gum OR hemorrhage OR bleeding OR probing) in Title Abstract Keyword - (word variations have been searched)

11.3 Results

11.3.1 Search and Selection Results

The search of PubMed, Cochrane, and Scopus databases returned 422 papers (Fig. 11.1). The first one hundred google scholar records were also screened and two additional articles were found to be appropriate to include in the review. Forty-three records were identified to be relative in the initial screening, and 381 articles were excluded due to the irrelevance of the title, not being in the English language, and not being a publication or a clinical trial publication. A full assessment of the 43 publications revealed that 14 publications were appropriate. Thus, data extraction was done on these 14 articles. An overview of each selected paper is shown in Table 11.2.

11.3.2 Assessment of the Study Quality

In Table 11.3, the methodological quality assessment is summarized. Seven studies were categorized as low quality due to having a Jadad scale of 2 or less [47–53]. The remaining studies were relatively high quality (Jadad scale ≥) [1, 54–59].

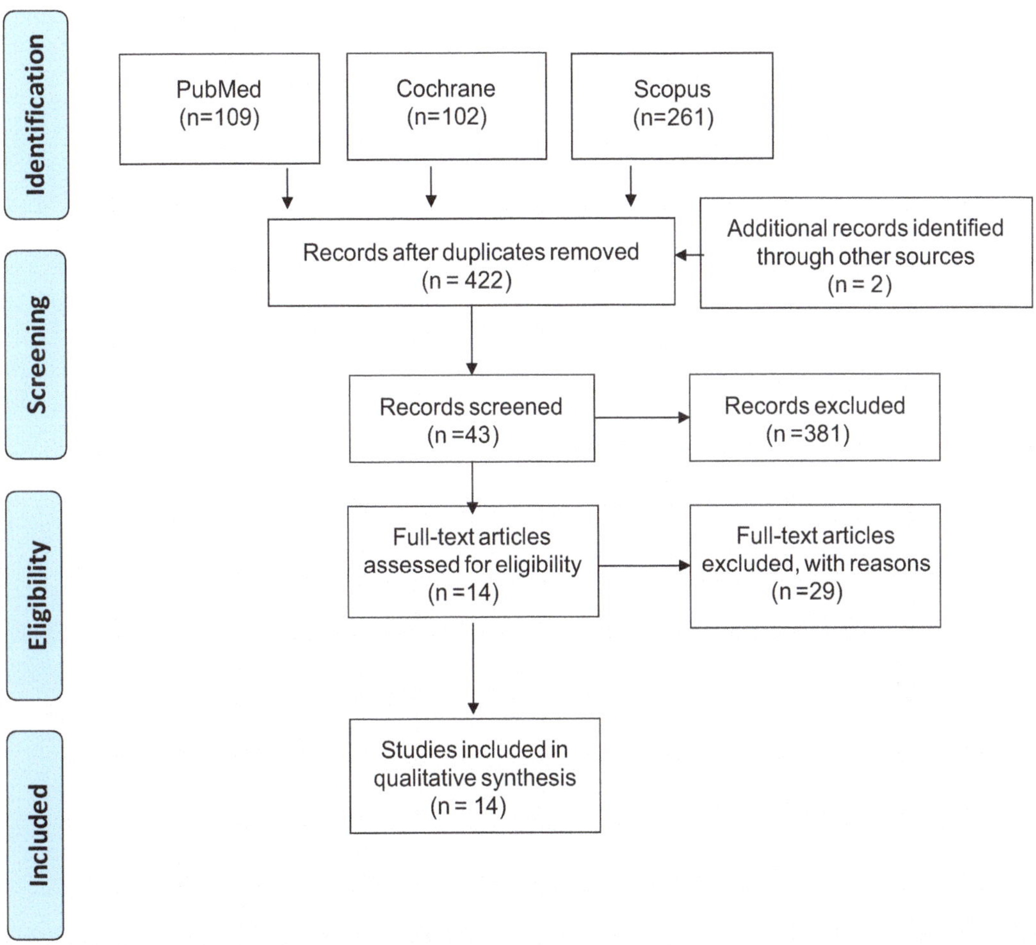

Fig. 11.1 Flow diagram of the study protocol

Table 11.2 General characteristic of the included studies

Author Year location	Type of article/ Duration of intervention	Intervention and control group/dose	Route of administration	Number of subjects/ gender	Age range	Outcome measurement	The adverse effect of patient	Main outcome	Instruction of curcumin used
Kandwal et al. (2015), India [48]	RCT 21 days	1: CUR (10 mg) gel 2: CHX 2% gel	Topical application 5 min/twice daily and 30 min after brushing in 3 weeks	?/60 Male:? Female:?	Range > 18	PI GI	CUR: NR CHX: minor tooth staining	Both CUR and CHX gel have comparable antiplaque and antigingivitis effect	Commercially available "CURENEXT" gel
Singh et al. (2015), India [49]	RCT 21 days	1: CHX 0.2% gel 2: CUR (?) gel	Twice daily after brushing	?/40 Male:? Female:?	20–35	PI GI SBI	NM	Both groups are effective, but CHX is better than CUR	An appropriate amount of crude extract and glycerin was mixed: glycerine 75:25
Muglikar et al. (2013), India [57]	RCT Blinded 21 days	1: SRP + 0.2% CHX MW 2:SRP + 20% CUR MW 3:SRP	Twice daily for 3 weeks/ 10 mL of MW for 1 min, 10 min after brushing	30/30 Male:? Female:?	20–40	PI GI	NM	CHX 0.2% and CUR 20% are equally effective in treatment of gingivitis	One teaspoon (5.139 g CUR) was mixed in 50 mL distilled water. Boiled down to 25 mL and filtered
Chunawala et al. (2018), India [50]	Clinical study 60 days	1: tablet AHPL/ AYTAB/1514	2 tablets 4 times a day, orally, after food eating	47/54 Male:28 Female:19	18–54	1. GI 2. PI 3. Bad breath analyzer 4. Organoleptic assessment 5. Sore throat and cough	Burning sensation in one patient	The tablet is safe and effective in alleviating halitosis, gingivitis and plaque control	Composition of tablet: 50 mg Curcuma longa 2.5 mg menthe arvensis 2.5 mg piper longum

Mali et al. (2012), India [51]	Clinical study 21 days	1: CHX 0.2% MW 2: CUR 0.1% MW	10 mL MW with equal dilution of water for 1 min, twice daily after 30 min brushing	60/100 Male:? Female:?	Range > 15 years	PI GI Microbial colony count	CUR: temporary tongue staining CHX: bitter taste and tongue staining	CUR MW showed antiplaque properties and effect on gingival inflammation was comparable to CHX MW	MW composition: CUR extract 0.1% Eugenol 0.01% Water Q.S
Paschoal et al. (2015), Brazil [54]	RCT Single-blind 4 weeks	1: CHX 2% varnish 2: Placebo varnish 3: C-pact	Once a week for 4 weeks 5 mL CUR gargle for 1 min and illuminated for 10 min	30/45 Male:? Female:?	13–18	PI GBI	No side effect	Both treatments controlled gingivitis after 1 month with no clinical evidence of decreased PI in patients	CUR solution dissolved in deionized water (final concentration 1.5 mg/mL) on same day of application
Arunachalam et al. (2017), India [55]	RCT Single-blind 4 weeks	1: Control (saline) 2: CHX MW 0.2% 3: CUR MW 0.1%	10 mL MW swish for 1 min, twice daily after brushing for 4 weeks	30/30 Male:12 Female:18	25–60	PI GI ROM	No side effect	CUR MW was comparable to CHX in plaque control and better (insignificant) in reducing gingival inflammation. Also, CUR was better (significant) in comparison with other groups in ROM value	Dissolved 10 mg CUR extract in 100 mL distilled water using 0.005% of flavoring agent Ph: 4
Waghmare et al. (2011) India [56]	RCT Double-blind 21 days	1: CHX MW 0.2% 2: CUR MW 0.1%	10 ml of MW, twice daily after brushing for 3 weeks	?/100 Male:? Female:?	25–35	PI GI Microbial colony count	CUR: NR CHX: NM	CHX, and CUR MW, can be effectively used as an adjunctive to mechanical plaque control in prevention of plaque and gingivitis	10 mg CUR extract dissolved in 100 mL distilled water and flavoring agent 0.005% peppermint oil Ph:4

(continued)

Table 11.2 (continued)

Author Year location	Type of article/ Duration of intervention	Intervention and control group/dose	Route of administration	Number of subjects/ gender	Age range	Outcome measurement	The adverse effect of patient	Main outcome	Instruction of curcumin used
Kaur et al. (2017), India [47]	RCT 21 days	1: SRP 2: SRP + CUR	Systemically used 350 dosages of CUR, twice daily for 21 days	?/60 Male:? Female:?	18–55	PI GI PD	NM	Both groups reduced scores and test group was more effective than control group	300 mg CUR 5 mg piperine 10 mg lycopene
Vinod et al. (2018), India [58]	RCT Double-blind 15 days	1: CHX 0.12% MW 2: herbal MW 20%	15 mL MW, twice daily for 30 s for 15 days with regular tooth brushing	200/200 Male:100 Female:100	18–22	Microbial colony count	Herbal MW: bitter taste and unattractive in color	CHX MW was better(insignificant) on seventh day in reducing microbial colony count while prolonged usage for 2 weeks showed the herbal preparation performed better than CHX	200 g powder of herbal (Tulsi, CUR, Triphala, haritak, bibitaki, neem) suspended in polyethylene glycol and 800 mL distilled water and allowed to evaporate to the final concentration (20%)
Chatterjee et al. (2017), India [1]	RCT Double-blind 28 days	1: CUR MW 0.1% 2: Placebo (saline) MW 3:CHX MW 0.2%	10 mL of MW for 1 min twice a day after 30 min brushing for 28 days	150/150 Male:? Female:?	20–40	PI GI SBI	CUR: NR CHX: Mild tooth staining	Antiplaque and antigingivitis efficacy of CUR MW was proved and comparable to CHX MW	0.3-5.4% raw CUR and volatile oil including tumerone, atlantone, zingiberone plus sugars, protein, and resin were the composition of MW

Pulikkotil et al. (2015), India [59]	RCT Double-blind 60 days	1: CUR gel (10 mg) 2: CHX gel 1% 3: CHX (0.25%) + MTZ (10 mg)	1 g gel in the tooth shield applied and left for 10 min, twice daily for 29 days without any mechanical removal of plaque in selected area. In 30 to 60 days experimental gel was used as adjunct with mechanical removal of plaque	56/60 Male:? Female:?	18–35	PI GI BOP PD	Cur gel caused a mild yellowish discoloration of teeth that disappeared after brushing	CUR and CHX + MTZ was superior to CHX in affecting the development of gingival inflammation and CUR has potential in the treatment of inflammatory periodontitis disease as an adjunctive to mechanical control	Commercially available CUR: "CURENEXT" gel CHX: Hexigel CHX + MTZ: Metrogyl DG gel
Sharma et al. (2016), India [52]	Split mouth Double-blind 2 weeks	1: SRP 2: SRP + CUR gel (10 mg) 3: CUR gel (10 mg)	Apply small pea size gel over gingivitis 1 h after oral hygiene practice with no mouth rinse for at least 10 min	?/20 Male:13 Female:7	25–30	PI GI SBI	NM	CUR effective against periodontal disease with or without SRP but more effective with SRP	Commercially available "CURENEXT" gel

(continued)

Table 11.2 (continued)

Author Year location	Type of article/ Duration of intervention	Intervention and control group/dose	Route of administration	Number of subjects/ gender	Age range	Outcome measurement	The adverse effect of patient	Main outcome	Instruction of curcumin used
Roopa et al. (2016), India [53]	RCT 3 weeks	1:SRP 1: SRP+ CUR gel1%	Twice daily for 3 weeks. Apply gel for at least 10 min	30/30 Male:? Female:?	16–35	PI GI SBI	NR	CUR was effective in treating gingivitis and group 2 was more effective than group 1	Commercially available "CURENEXT" gel

CHX chlorhexidine, *MW* mouthwash, *CUR* curcumin, *PI* plaque index, *GI* gingival index, *GBI* gingival bleeding index, *PD* probing depth, *SBI* sulcular bleeding index, *ROM* reactive oxygen metabolite, *NR* not reported, *NM* not mentioned, *C-pact* CUR-photodynamic antimicrobial chemotherapy, *MTZ* metronidazole

Table 11.3 Quality assessment of the included studies using Jadad Scale

	Paschoal et al. (2015)	Arunachalam et al. (2017)	Waghmare et al. (2011)	Kaur et al. (2017)	Kandwal et al. (2015)	Singh et al. (2015)	Muglikar et al. (2013)	Vinod et al. (2018)	Chunawala et al. (2018)	Mali et al. (2012)	Chatterjee et al. (2017)	Pulikkotil et al. (2015)	Sharma et al. (2016)	Roopa et al. (2016)
Blinded?	√	√	√	–	–	–	√	√	–	–	√	√	√	–
Description of blindness method appropriate?	√	–	–	–	–	–	√	–	–	–	–	√	–	–
Randomized?	√	√	√	√	√	√	√	√	–	–	√	√	√	√
Description of randomization appropriate?	–	–	√	–	√	–	–	–	–	–	√	√	–	–
Outcome of all patients mentioned?	√	√	–	–	–	–	√	√	√	√	√	√	–	√
Jadad scale	4	3	3	1	2	1	4	3	1	1	4	5	2	2

11.3.3 Baseline Prophylaxis, Polishing, and Process of Scaling and Root Planing

In six studies, scaling and root planing (SRP) was carried out at baseline [1, 48, 49, 51, 55, 59]. Also, in the study of Paschoal et al. [54], tooth brushing was done as a prophylaxis before varnish therapy. In five studies, a motivation protocol such as telephone calls or use of a reminder sheet or daily record was performed [50, 54–56, 58].

11.3.4 Intervention and Regimens

Various formulas of the herbal compounds were used in selected studies. Only four studies used a commercially available oral gel product named "Curenext" [48, 52, 53, 59]. Other studies used experimental products mentioned in Table 11.2.

11.3.5 Clinical Parameter

11.3.5.1 Plaque Index

Thirteen studies assessed the plaque index (PI) and are summarized in Table 11.4. Four studies used the Loe and Silness 1964 [47, 54, 55, 59], two used Loe and Silness 1963 [1, 49], and one used Loe and Silness 1967 [48] indices. One study did not mention the year of Loe and Silness PI [57]. One study used Tureskey modification of the Quigley Hein index 2006 [50] and one study used PI modified Quigley Hein and Elliat index 1987 [52]. Two studies used Tureskey-Gilmore-Glickman-modification of Quigley Hein 1970 [51, 56]. One study did not mention the type of index used [53].

11.3.5.2 Gingival Index

Twelve studies assessed the gingival index (GI) summarized in Table 11.5. Four studies used Loe and Silness 1963 [1, 47, 49, 56], two studies used Loe and Silness 1967 [48, 55], and one study used Loe and Silness 2011 [50] indices. In addition, two studies used Loe and Silness without mentioning the year of the index [51, 57]. Two studies used a modification of Loe and Silness (Lobene, 1986) [52, 59], and one study did not mention the type of index [53].

11.3.5.3 GBI

Six studies assessed GBI summarized in Table 11.6. Two studies used Loe and Silness 1964 [54, 59]. Two studies used the sulcular bleeding index (SBI) without mentioning the year of the index [49, 53]. One study used the Muhlemann 1977 SBI index [1], and another study used the modified Mombelli 1987 SBI index [52].

11.3.5.4 Other Clinical Parameters

Two studies assessed PD. Kaur et al. [47] used the Williams probe to measure PD, and Pulikkotil et al. [59] used the UNC-15 periodontal probe to measure pocket depth. The details are summarized in Table 11.7. In addition, three studies measured the microbial colony count (Table 11.8) [51, 56, 58].

11.3.6 Study Outcome

11.3.6.1 Intragroup Comparison

All groups showed a significant reduction in PI from baseline to the end of the trial except in the cases of the Chatterjee et al. study, in which the placebo group did not show any significant differences, and the Paschoal et al. study, which did not show any clinical evidence of a reduction in PI after 1 month [1, 54].

For GI, the groups in all 12 studies showed a significant decrease from baseline to the end of the trial except for the Arunachalam et al. and Chatterjee et al. studies, in which the placebo groups did not show any significant reduction [1, 55].

For GBI, four studies showed significant differences in all groups [49, 52, 53, 59]. The Chatterjee et al. and Paschoal et al. studies showed significant differences in the test group but not in the placebo group [1, 54].

For PD measurement, Kaur et al. [47] showed a significant reduction from baseline to the end of the study, and Pulikkotil et al. [59] showed that all values remained similar except in the case of the CHX gel group at 29 days.

For microbial colony count, all three studies showed a significant reduction in the total bacterial count [51, 56, 58].

11.3.6.2 Intergroup Comparison

For PI, several groups were compared. In all studies, CUR was compared to CHX or placebo in combination with SRP or alone. The results were in the following order:

Two studies mentioned that CHX was better than CUR but agents were significantly effective against plaque accumulation [49, 56]. In contrast, Pulikkotil et al. found that CUR was better than CHX [59]. In two studies, the differences between CUR and CHX were insignificant [48, 51]. Two other studies revealed that CUR and CHX were more effective than the control group, but there were no significant differences between CUR and CHX [1, 55]. Some studies that used CUR or CHX adjunct to SRP concluded that SRP plus CUR was significantly more effective than SRP alone [47, 53]. Two other studies found insignificant differences between SRP plus CHX/CUR or SRP alone [52, 57]. One study of plaque accumulation after a month's follow-up found insignificant differences between the groups but CUR-photodynamic antimicrobial chemotherapy (C-PACT) showed more PI in comparison with CHX or placebo varnish after 3 months' follow-up [54]. The Chunawal et al. study did not compare CUR with other groups [50].

For GI, Singh et al. mentioned that CHX was more effective than CUR despite the effectiveness of CUR against gingival inflammation [49]. Four studies demonstrated that CHX was as useful as CUR [1, 48, 51, 56]. Two studies stated that CUR was better than CHX in controlling inflammation [55, 59]. Other studies that used a chemical agent adjunct to SRP concluded that the combination of a mechanical and chemical procedure was significantly more effective than mechanical therapy alone [47, 53, 57]. Also, Sharma et al. stated that CUR in combination with SRP was more effective than SRP alone, and these two groups were more effective than CUR alone [52]. In Chunawala et al., no other groups were used as a comparison [50].

For GBI, Singh et al. mentioned that CHX was better than CUR for control of bleeding [49]. Pulikkotil et al. demonstrated that CUR was better than CHX/CHX+ metronidazole (MTZ) [59]. Two studies mentioned that CHX and CUR showed no significant differences [1, 54]. By combining SRP and chemical therapy, Sharma et al. found that there were no significant differences between SRP or SRP plus CUR, and these two groups were significantly more effective than CUR alone [52]. Roopa et al. stated that SRP plus CUR was more effective than SRP alone [53].

For microbial colony count, Vinod et al. pointed out that herbal mouthwash (MW) was more effective than CHX MW in the reduction of total bacterial count on day 14 [58]. Two other studies mentioned that there were no significant differences between CHX and CUR MW [51, 56].

For PD, Kaur et al. stated that SRP plus CUR was more effective than SRP alone, and Pulikkotil et al. found insignificant differences between the groups at 29 days, and significant differences were observed on day 60 [47, 59].

11.3.7 Adverse Effects

Four studies reported no side effects of either CHX or CUR [53–56]. Five studies did not mention anything about the side effects [47, 49, 52, 57]. In other studies, tooth and tongue staining and bitter taste were shown for CHX MW use [1, 48, 51]. Temporary tongue staining, mild yellowish tooth coloration that was cleared after brushing, and a burning sensation in one patient were observed for CUR products [50, 51, 59]. Also, the herbal MW group found the color unattractive and bitter taste was reported [58].

11.4 Discussion

The main finding of this systematic review is that CUR has a promising role in the management of gingivitis. It substantially reduced PI, GI, GBI, and microbial colony count without any unwanted adverse effect. To the best of the authors' knowl-

Table 11.4 Study protocols and main findings of the included studies regarding plaque index (PI)

Author Year	Intervention and control group/dose	Plaque index	Baseline	End			Comparison
Kandwal et al. (2015) [48]	1: CUR (10 mg) gel 2: CHX 2% gel	Silness and Loe (1967)	1: 2.337 ± 0.192 2:2.32 ± 0.169	14 days 1: 1.307 ± 0.222 2: 1.27 ± 0.168	21 days 1: 0.913 ± 0.138 2: 0.923 ± 0.147		There were significant differences in plaque index reduction from baseline to 14 and 21 days. Intergroup differences were insignificant
Singh et al. (2015) [49]	1: CHX 0.2% gel 2: CUR (?) gel	Silness and Loe (1963)	1: 1.58 ± 0.59 2: 1.64 ± 0.5	14 days 1: 0.47 ± 0.32 2: 0.76 ± 0.41	21 days 1: 0.38 ± 0.29 2: 0.71 ± 0.38		The mean PI in both groups decreased after treatment significantly from baseline to 2 and 3 weeks. Decreased rate in group 1 was significantly higher than group 2
Muglikar et al. (2013) [57]	1: SRP + 0.2% CHX MW 2: SRP + 20% CUR MW 3:SRP	Silness and Loe (?)	1: 2.33 ± 0.24 2: 2.32 ± 0.25 3: 2.27 ± 0.22	7 days 1: 1.39 ± 0.12 2: 1.36 ± 0.13 3: 1.31 ± 0.12	14 days 1: 1.47 ± 0.12 2: 1.43 ± 0.12 3: 1.43 ± 0.12	21 days 1: 1.57 ± 0.18 2: 1.54 ± 0.11 3: 1.57 ± 0.13	PI decreased significantly from baseline to 21 days in all groups. There were insignificant differences between the groups in all periods

Chunawala et al. (2018) [50]	Tablet AHPL/ AYTAB/1514 1: no plaque 2: isolated flecks of plaque 3:continuous band of plaque	Turesky modification of Quigley Hein index 2006	1: 2.1% 2: 91.5% 3: 6.4%	Days 15	Days 30	Days 45	Days 60	Days 75	In days 45 to 75, there were significant differences in the no plaque group. Other groups showed insignificant differences
				1: 2.1% 2:91.5% 3: 6.4%	1:8.5% 2: 87.2% 3: 4.3%	1: 19.2% 2: 78.7% 3: 2.1%	1: 17% 2: 76.7% 3: 6.4%	1: 23.4% 2: 72.3% 3: 4.3%	
Mali et al. (2012) [51]	1: CHX 0.2% MW 2: CUR 0.1% MW	Turesky-Gilmore-Glickman modification of Quigley-Hein 1970	1: 3.303 ± 0.443 2: 3.309 ± 0.399	14 days		21 days			PI was reduced during the study. No significant differences were found between the groups
				1: 1.873 ± 0.415 2: 1.855 ± 0.475		1: 1.147 ± 0.25 2: 0.99 ± 0.36			
Arunachalam et al. (2017) [55]	1: Control (saline) 2: CHX MW 0.2% 3: CUR MW 0.1%	Silness and Loe (1964)	1; 1.35 ± 0.141 2: 1.319 ± 0.136 3: 1.232 ± 0.224	4 weeks					PI showed significant differences in all groups from baseline to 4 weeks. Statistically significant differences for PI between control and CHX as well as control and CUR were found, but no significant differences were found between CHX and CUR
				1: 1.020 ± 0.042 2: 0.442 ± 0.160 3: 0.428 ± 0.291					

(continued)

Table 11.4 (continued)

Author Year	Intervention and control group/dose	Plaque index	Baseline	End			Comparison
Waghmare et al. (2011) [56]	1: CHX MW 0.2% 2: CUR MW 0.1%	Tureskey-Gilmore-Glickman modification of Quigley-Hein 1970	1: 3.31 ± 0.36 2: 3.27 ± 0.47	14 days 1: 1.72 ± 0.38 2: 2 ± 0.46	21 days 1: 0.83 ± 0.27 2: 1.22 ± 0.13		The reduction in PI from baseline to day 14 and 21 was significant in both group
Kaur et al. 2017 [47]	1: SRP 2: SRP + CUR	Silness and Loe (1964)	1: 1.52 ± 0.2 2: 1.38 ± 0.27	21 days 1: 0.93 ± 0.21 2: 0.74 ± 0.17			Both groups showed a significant PI reduction with greater reduction in the test group
Chatterjee et al. (2017) [1]	1: CUR MW 0.1% 2: Placebo (saline) MW 3: CHX MW 0.2%	Silness and Loe (1963)	1: 1.88 ± 0.19 2: 1.91 ± 0.31 3: 1.93 ± 0.22	7 days 1: 1.49 ± 0.2 2: 1.83 ± 0.12 3: 1.53 ± 0.19	14 days 1: 1.1 ± 0.19 2: 1.84 ± 0.15 3: 1.14 ± 0.2	28 days 1: 0.55 ± 0.08 2: 1.2 ± 0.43 3: 0.57 ± 0.11	In groups 1 and 3 the PI reduction from baseline to 28 days was statistically significant

Pulikkotil et al. (2015) [59]	1: CUR gel (10 mg) 2: CHX gel 1% 3: CHX(0.25%) + MTZ (10 mg)	Silness and Loe (1964)	1: 0.11 ± 0.11 2: 0.13 ± 0.13 3: 0.15 ± 0.11	29 days 1: 0.52 ± 0.14 2: 0.98 ± 0.29 3: 0.53 ± 0.17	60 days 1: 0.23 ± 0.23 2: 0.14 ± 0.17 3: 0.18 ± 0.13	There was an increase in a mean PI difference for all three groups from baseline to 29 days. Differences were significant after 60 days. There was a gradual decrease in all groups and no significant differences from baseline to 60 days. CUR and CHX-MTZ were similar in PI scores, and significantly lower than the CHX group at 29 days

(continued)

Table 11.4 (continued)

Author Year	Intervention and control group/dose	Plaque index	Baseline	End		Comparison
Sharma et al. (2016) [52]	1: SRP 2: SRP + CUR gel (10 mg) 3:CUR gel (10 mg)	PI modified Quigley Hein and Elliot index 1987	1: 2.78 ± 0.38 2: 2.69 ± 0.29 3: 2.58 ± 0.28	1 week 1: 1.74 ± 0.14 2: 1.48 ± 0.13 3: 2.14 ± 0.47	2 week 1: 1.22 ± 0.11 2: 1.16 ± 0.15 3: 1.88 ± 0.32	All groups showed significant differences from baseline to 1 and 2 weeks. The difference in a reduction in PI was not significant between groups 1 and 2 and was higher in group 2. The differences were significant between group 1 and 2 in comparison to group 3
Roopa et al. (2016) [53]	1: SRP 2: SRP+ CUR gel1%	(?)	1: 2.75 ± 0.28 2: 2.77 ± 0.26	7 days 1: 1.43 ± 0.33 2: 1.81 ± 0.21	21 days 1: 0.94 ± 0.51 2: 1.52 ± 0.25	Both groups showed decreased PI and improvement was evidently greater in the test group (group 2) than control group

Paschoal et al. (2015) [54]	1: CHX 2% varnish 2: Placebo varnish 3:c-pact	Silness and Loe (1964)	1: 0.62 ± 0.38 2: 0.60 ± 0.35 3: 0.8 ± 0.49	1 month	3 months	PI in 3-month follow-up was significantly higher relative to 1 month. Group 3 showed significant difference compare to other groups at the 3-month follow-up
				1: 0.57 ± 0.43 2: 0.47 ± 0.41 3: 0.46 ± 0.39	1: 0.91 ± 0.75 2: 1.03 ± 0.51 3: 1.52 ± 0.51	

CHX chlorhexidine, *MW* mouthwash, *CUR* curcumin, *SRP* scaling and root planning, *C-pact* CUR-photodynamic antimicrobial chemotherapy, *MTZ* metronidazole

Table 11.5 Study protocols and main findings of the included studies regarding gingivitis index (GI)

<table>
<tr><th rowspan="2">Author
Year</th><th rowspan="2">Intervention and control group/dose</th><th rowspan="2">Gingival index</th><th rowspan="2">Baseline</th><th colspan="4">End</th><th rowspan="2">Comparison</th></tr>
<tr><th></th><th></th><th></th><th></th></tr>
<tr><td rowspan="2">Kandwal et al. (2015) [48]</td><td rowspan="2">1: CUR (10 mg) gel
2: CHX 2% gel</td><td rowspan="2">Loe and Silness (1967)</td><td rowspan="2">1: 1.953 ± 0.281
2: 1.99 ± 0.283</td><td colspan="2">14 days</td><td colspan="2">21 days</td><td rowspan="2">There were significant differences in GI from baseline to 14 and 21 days. Intergroup differences were insignificant</td></tr>
<tr><td colspan="2">1: 1.11 ± 0.158
2:1.10 ± 0.146</td><td colspan="2">1: 0.55 ± 0.145
2:0.573 ± 0.17</td></tr>
<tr><td rowspan="2">Singh et al. (2015) [49]</td><td rowspan="2">1: CHX 0.2% gel
2: CUR (?) gel</td><td rowspan="2">Loe and Silness (1963)</td><td rowspan="2">1: 1.80 ± 0.54
2: 1.83 ± 0.57</td><td colspan="2">14 days</td><td colspan="2">21 days</td><td rowspan="2">The mean GI in both groups decreased after treatment significantly from baseline to 2 and 3 weeks. Decreasing rate in group 1 was significantly higher than group 2</td></tr>
<tr><td colspan="2">1: 0.52 ± 0.34
2: 0.84 ± 0.43</td><td colspan="2">1: 0.45 ± 0.34
2:0.82 ± 0.44</td></tr>
<tr><td rowspan="2">Muglikar et al. (2013) [57]</td><td rowspan="2">1: SRP + 0.2% CHX MW
2: SRP + 20% CUR MW
3: SRP</td><td rowspan="2">Loe and Silness (?)</td><td rowspan="2">1:2.29 ± 0.12
2: 2.37 ± 0.19
3: 1.61 ± 0.17</td><td>7 days</td><td colspan="2">14 days</td><td>21 days</td><td rowspan="2">GI decreased significantly from baseline to 21 days in all groups. In day 14 and 21, groups 1 and 2 showed significant differences compared to group 3 but was insignificant between group 1 & 2</td></tr>
<tr><td>1: 1.58 ± 0.23
2:1.62 ± 0.25
3: 1.61 ± 0.17</td><td colspan="2">1: 1.26 ± 0.15
2: 1.36 ± 0.18
3: 1.64 ± 0.13</td><td>1: 1.09 ± 0.13
2: 1.19 ± 0.17
3: 1.65 ± 0.12</td></tr>
</table>

Chunawala et al. (2018) [50]	Tablet AHPL/AYTAB/1514 1: Normal gingiva 2: Mild gingivitis 3: Moderate gingivitis	Loe and Silness (2011)	1: 0% 2: 57.6% 3: 42.6%	days 15 1: 0% 2: 66% 3: 34%	days 30 1: 4.3% 2: 80.8% 3: 14.9%	days 45 1: 4.3% 2: 91.4% 3: 4.3%	days 60 1: 2.1% 2: 93.6% 3: 4.3%	days 75 1: 4.3% 2: 93.6% 3: 2.1%	In days 30–75, there were significant differences in mild inflammation and decreased moderate inflammation significantly. Patients shifted from normal to mild gingivitis but this was insignificant
Mali et al. 2012 [51]	1: CHX 0.2% MW 2: CUR 0.1% MW	Loe and Silness (?)	1: 1.802 ± 0.123 2: 1.774 ± 0.186	14 days 1: 0.857 ± 0.198 2: 0.833 ± 0.199		21 days 1: 0.700 ± 0.153 2: 0.662 ± 0.240			GI was reduced during the study but no significant differences were observed between groups
Arunachalam et al. (2017) [55]	1: Control (saline) 2: CHX MW 0.2% 3: CUR MW 0.1%	Loe and Silness (1967)	1: 1.261 ± 0.195 2: 1.317 ± 0.120 3: 1.149 ± 0.318	4 weeks 1: 1.240 ± 0.139 2: 0.344 ± 0.154 3: 0.190 ± 0.063					GI showed significant differences in groups 2 and 3 from baseline to 4 weeks. For GI it was significant among all groups and CUR was the best

(continued)

Table 11.5 (continued)

Author Year	Intervention and control group/dose	Gingival index	Baseline	End			Comparison
Waghmare et al. (2011) [56]	1: CHX MW 0.2% 2: CUR MW 0.1%	Loe and Silness (1963)	1: 1.77 ± 0.19 2: 1.81 ± 0.13	14 days 1: 0.87 ± 0.12 2: 0.91 ± 0.09	21 days 1: 0.73 ± 052 2: 0.71 ± 0.12		Reduction in GI from baseline to days 14 and 21 was significant in both groups
Kaur et al. (2017) [47]	1: SRP 2: SRP + CUR	Loe and Silness (1963)	1: 1.71 ± 0.25 2: 1.73 ± 0.28	21 days 1: 0.95 ± 0.2 2: 0.77 ± 0.23			Both groups showed significant reduction in GI with a greater reduction in the test group
Chatterjee et al. (2017) [1]	1: CUR MW 0.1% 2: Placebo (saline) MW 3: CHX MW 0.2%	Loe and Silness (1963)	1: 1.83 ± 0.37 2: 1.85 ± 0.2 3: 1.89 ± 0.39	7 days 1: 1.61 ± 0.25 2: 1.8 ± 0.1 3: 1.61 ± 0.27	14 days 1: 1.2 ± 0.2 2: 1.76 ± 0.4 3: 1.21 ± 0.22	28 days 1: 0.65 ± 0.11 2: 1.09 ± 0.5 3: 0.66 ± 0.11	In groups 1 and 3, GI reduction from baseline to 28 days was statistically significant

Pulikkotil et al. 2015 [59]	1: CUR gel (10 mg) 2: CHX gel 1% 3: CHX(0.25%) + MTZ (10 mg)	MGI: lobene (1986)	1: 0.41 ± 0.2 2: 0.31 ± 0.13 3: 0.29 ± 0.18	29 days 1: 0.86 ± 0.27 2: 1.59 ± 0.49 3: 1.01 ± 0.31	60 days 1: 0.36 ± 0.34 2: 0.39 ± 0.18 3: 0.26 ± 0.24	There was an increase in mean difference of MGI for all three groups from baseline to 29 days, and differences were significant after 60 days. There was a gradual decrease in all groups and no significant differences from baseline to 60 days. CUR and CHX-MTZ were similar in MGI scores, and significantly lower than the CHX group at 29 days
Sharma et al. (2016) [52]	1: SRP 2: SRP + CUR gel (10 mg) 3: CUR gel (10 mg)	MGI: lobene (1986)	1: 1.82 ± 0.25 2: 1.92 ± 0.3 3: 1.8 ± 0.42	1 week 1: 1.43 ± 0.18 2: 1.36 ± 0.21 3: 1.59 ± 0.28	2 week 1: 1.38 ± 0.21 2: 1.32 ± 0.22 3: 1.45 ± 0.31	All groups showed significant differences from baseline to 1 and 2 weeks. Group 2 showed a greater reduction in GI in comparison to group 1 and 3

(continued)

Table 11.5 (continued)

Author Year	Intervention and control group/dose	Gingival index	Baseline	End		Comparison
Roopa et al. (2016) [53]	1: SRP 1: SRP+ CUR gel 1%	(?)	1: 2.67 ± 0.29 2: 2.67 ± 0.29	7 days	21 days	Both groups showed decreased GI and improvement was greater in the test than control group
				1: 2.15 ± 0.3 2 2.17 ± 0.26	1: 1.3 ± 0.13 2: 1.47 ± 0.1	

CHX Chlorhexidine, *MW* Mouthwash, *CUR* Curcumin, *SRP* Scaling and root planning, *MTZ* Metronidazole

Table 11.6 Study protocols and main findings of the included studies regarding gingival bleeding index (GBI)

Author Year	Intervention and control group/dose	Gingival bleeding index	baseline	End			*P* value Comparison
Singh et al. (2015) [49]	1: CHX 0.2% gel 2: CUR (?) gel	SBI (?)	1: 2.02 ± 0.94 2: 2.00 ± 0.83	14 days 1: 0.32 ± 0.29 2: 0.68 ± 0.4	21 days 1: 0.29 ± 0.29 2: 0.66 ± 0.38		The mean SBI in both groups decreased after treatment significantly from baseline to 2 and 3 weeks. Decreasing rate in group 1 was significantly higher than group 2
Chatterjee et al. (2017) [1]	1: CUR MW 0.1% 2: Placebo(saline) MW 3: CHX MW 0.2%	SBI Muhlemann (1977)	1: 2.13 ± 0.3 2: 2.08 ± 0.28 3: 2.16 ± 0.34	7 days 1: 1.75 ± 0.32 2:1.95 ± 0.33 3:1.74 ± 0.39	14 days 1: 1.2 ± 0.3 2: 1.8 ± 0.26 3: 1.19 ± 0.35	28 days 1: 0.65 ± 0.23 2: 1.13 ± 0.27 3:0.26 ± 0.23	In group 1 and 3, the PI reduction from baseline to 28 days was statistically significant
Pulikkotil et al. (2015) [59]	1: CUR gel(10 mg) 2: CHX gel 1% 3: CHX (0.25%) + MTZ (10 mg)	BOP Silness (1964)	1: 0 2: 0 3: 0	29 days 1: 0.26 ± 0.19 2: 0.68 ± 0.15 3: 0.5 ± 0.13	60 days 1: 0 2: 0 3: 0		There was an increase in a mean difference of BOP for all three groups from baseline to 29 days, and differences were significant after 60 days. There was a gradual decrease in all groups. BOP was significantly less in group 1 in comparison to other groups

(continued)

Table 11.6 (continued)

Author Year	Intervention and control group/dose	Gingival bleeding index	baseline	End		P value Comparison
Sharma et al. (2016) [52]	1: SRP 2: SRP + CUR gel (10 mg) 3: CUR gel (10 mg)	Modified SBI Mombelli 1987	1: 2.04 ± 0.43 2: 2.18 ± 0.51 3: 1.95 ± 0.42	1 week 1: 1.41 ± 0.26 2: 1.54 ± 0.15 3: 1.63 ± 0.18	2 week 1: 1.31 ± 0.19 2: 1.32 ± 0.17 3: 1.5 ± 0.23	All groups showed significant differences from baseline to 1 and 2 weeks. The difference in reduction of SBI was higher in group 2 compared to groups 1 and 3. However, the difference between groups 2 and 3 was significant and was insignificant between groups 2 and 1
Roopa et al. (2016) [53]	1: SRP 1: SRP+ CUR gel 1%	SBI(?)	1: 3.47 ± 0.38 2: 3.87 ± 0.34	7 days 1: 1.32 ± 0.45 2: 1.82 ± 0.48	21 days 1: 1.19 ± 0.5 2: 2.59 ± 0.53	Both groups showed decreased SBI and improvement was greater in the test compared to the control group

Paschoal et al. (2015) [54]	1: CHX 2% varnish 2: Placebo varnish 3: C-pact	GBI Silness and Loe (1964)	1: 0.69 ± 0.17 2: 0.71 ± 0.34 3: 0.81 ± 0.32	1 month 1: 0.19 ± 0.5 2: 0.69 ± 0.42 3: 0.21 ± 0.52	3 months 1: 0.77 ± 0.39 2: 0.73 ± 0.37 3: 0.79 v 0.42	There was a significant reduction in groups 1 and 3 at 1-month follow-up in comparison to 3 months' follow-up. Group 2 showed a significant difference in comparison to groups 1 and 3 at 1-month follow-up

CHX chlorhexidine, *MW* mouthwash, *CUR* curcumin, *SRP* scaling and root planning, *SBI* sulcular bleeding index, *C-pact* CUR-photodynamic antimicrobial chemotherapy, *BOP* bleeding on probing, *MTZ* metronidazole

Table 11.7 Study protocols and main findings of the included studies regarding probing depth (PD)

Author Year	Intervention and control group/dose	Probing measurement	baseline	End		Comparison
Kaur et al. (2017) [47]	1: SRP 2: SRP + CUR	Williams probe	1: 1.84 ± 0.22 2: 1.74 ± 0.22	21 days 1: 1.50 ± 0.15 2: 1.25 ± 0.15		Both groups showed significant reduction of PD with a greater reduction in the test group
Pulikkotil et al. (2015) [59]	1: CUR gel(10 mg) 2: CHX gel 1% 3: CHX(0.25%) + MTZ (10 mg)	UNC-15 periodontal probe	1: 2.09 ± 0.48 2: 2.09 ± 0.39 3: 2.26 ± 0.57	29 days 1: 2.38 ± 0.53 2: 2.31 ± 0.49 3: 2.42 ± 0.54	60 days 1: 2.23 ± 0.52 2: 2.03 ± 0.31 3: 2.42 ± 0.54	All values remained similar except group 2 at 29 days. Intergroup comparison was similar at baseline and was insignificant. However, it was significant at 60 days

CHX chlorhexidine, *MW* mouthwash, *CUR* curcumin, *SRP* scaling and root planning, *MTZ* metronidazole

edge, this is the first study that systematically reviewed the effects of CUR in the treatment of gingivitis based on clinical trials. It seems that the beneficial effects of CUR on gingivitis are due to its role in the reduction of gingival inflammation.

The anti-inflammatory potential of CUR can be attributed to some of the features including inhibition of NF-κB and downregulation of the pro-inflammatory enzyme cyclooxygenase-2 via inhibition of cytokine gene expression [41–43]. CUR inhibits the inflammatory molecules prostaglandin E2 and thromboxane selectively, without affecting the synthesis of prostacyclin [60]. CUR increases the number of fibroblasts, leading to a reduction of vascularization, which could explain the reduction in PD along with bleeding on probing (BOP) [61]. CUR increases fibronectin and transforming growth factor B transcription that causes wound healing [41, 62]. It also promotes wound reduction and enhancement of cellular proliferation by collagen, which acts as a supportive matrix, and incorporation of CUR in this process has been demonstrated in the Gopinath et al. study [63]. Expression of inflammatory mediators such as IL1β, IL2, IL6, IL8, IL12, TNF-α, matrix metalloproteinase-2 (MMP2), and MMP9 is reduced by CUR. In addition, transcription factors such as NF-κB, activator protein 1 (AP-1), and mitogen-activated protein (MAP) kinases will be repressed [64–66]. CUR can prevent plaque formation and could potentially remove plaque. This effect is due to the *Lekhaniya* and *anti-Kapa* properties of CUR that would lead to plaque biofilm detachment and prevention of coaggregation. In addition, the growth of *Prevotella intermedia* as a major periodontal pathogen has been shown to be suppressed by CUR [67].

11.4.1 Summary of Evidence

We have compared CUR as an herbal agent and CHX as the gold standard in these articles. The intragroup comparisons in all studies that evaluated PI and used CHX or CUR as interventions showed a significant reduction in PI but Paschoal et al. observed no clinical evidence of a reduction in plaque accumulation [54]. They explained this result by the organization of biofilm structures. The stability and structural integrity of the bio-

Table 11.8 Study protocols and main findings of the included studies regarding microbial colony count

Author Year	Intervention and control group/dose	Microbial colony count	Baseline	End		Comparison
Mali et al. (2012) [51]	1: CHX 0.2% MW 2: CUR 0.1% MW	Microbial colony count BAPNA value	1: 25.484 ± 8.682 2: 30.334 ± 10.143	14 days 1: 18.572 ± 7.931 2: 21.605 ± 8.915	21 days 1: 14.647 ± 7.550 2: 15.300 ± 6.765	The mean value of both groups from baseline to days 14 and 21 was significant, but no significant difference between groups
Waghmare et al. (2011) [56]	1: CHX MW 0.2% 2: CUR MW 0.1%	Microbial colony count	1: 139.15 ± 51.92 2: 203.02 ± 34.03	21 days 1: 12.28 ± 2.78 2: 24.34 ± 11.84		The difference between mean reduction of total bacterial count was insignificant
Vinod et al. (2018) [58]	1: CHX 0.12% MW 2: Herbal MW 20%	Microbial colony count	1: 149.5 ± 58.911 2: 138.8 ± 58.47	7 days 1: 106.23 ± 53.42 2: 109.72 ± 53.77	14 days 1: 88.74 ± 45.24 2: 54.53 ± 22.69	From baseline to day 7, a nonsignificant reduction was found in both MW groups. On day 14, microbial count was reduced in both groups but this was more significant in the herbal group(composition of MW mentioned in Table 11.2)

CHX chlorhexidine, *MW* mouthwash, *CUR* curcumin, *SRP* scaling and root planning, *BAPNA* N-benzoyl-l-arginine-*p*-nitroanilide assay

film matrix would limit the diffusion of substances. Also, it would protect bacteria from the action of antimicrobial agents. Furthermore, increasing biofilm complexity seems to be more resistant to the C-PACT process. Additionally, more viscosity in the matrix would protect bacteria from the antimicrobial agent. Another reason is reactive oxygen species (ROS) and its interaction with bacteria. To overcome this, they proposed that time of light illumination should be 10 min for lower production of ROS, and excess of dye in the solution would prevent the light from reducing the bacteria [54].

In intergroup comparisons for PI, two studies showed that CHX was more effective than CUR. They attributed this result to the action of CHX at different levels of plaque formation. Also, the dilution factor of CUR may be another reason [49, 56]. Pulikkotil et al. stated that CUR was more effective than CHX. They pointed out that CUR resolves inflammation in an earlier stage, reflected by lower cytokine levels. Thus, the reduction of clinical inflammation may be more evident in CUR than CHX [59]. Four studies observed no significant differences between CHX and CUR [1, 48, 51, 55]. Also, Chatterjee et al. pointed out that on day 7, CHX was better than CUR, and on day 14 CUR was better than CHX, although the differences were not significant at either time [1]. Two studies that evaluated the effectiveness of chemical therapy in combination with SRP and compared to SRP alone showed that the treatment with SRP plus CUR was significantly better than SRP alone in controlling plaque accumulation. They related this result to the anti-inflammatory effects of CUR and its anti-inflammatory properties that make mechanical therapy more effective [47, 53]. Two studies showed that SRP or SRP plus chemical therapy in plaque control showed no significant differences despite better effectiveness of SRP plus chemical therapy. The use of CUR appears to make SRP easier for clinicians and patients to reduce bleeding and inflammation [52, 57]. Also, Sharma et al. stated that CUR alone (in this case, an oral formulation) could be useful. This implies that due to the anti-inflammatory and antioxidant effects of CUR, it may be beneficial to alleviate plaque-induced gingivitis. Therefore, CUR may be a treatment option in patients that need to be made more comfortable in cases where oral prophylaxis and mechanical therapy are delayed or postponed [52].

For the GI and GBI in the intragroup comparison, all studies showed improvement of inflammation in CHX or CUR groups. For GI assessment, Singh et al. showed that CHX was better than CUR in controlling GI with a similar finding for PI [49]. Four studies showed insignificant differences between CHX and CUR [1, 48, 51, 56]. Two studies demonstrated that despite the effectiveness of both CHX and CUR against inflammation, CUR was significantly better than CHX [55, 59]. A similar effect was seen for plaque control in the latter study, and Arunachalam et al. showed that the antiplaque efficacy of CUR and CHX was similar, but CUR had better anti-inflammatory effects than CHX at 4 weeks. This finding contrasted with that of Singh et al., which may be due to the fact that formulation used in the Arunachalam et al. study was gel compared with mouthwash [55, 59]. Three studies indicated that SRP plus the chemical agent was more effective than SRP alone [47, 53, 57]. Sharma et al. stated that SRP plus CUR showed the highest reduction in GI scores, followed by SRP and then by CUR alone [52].

For microbial colony count, Vinod et al. observed that herbal MW was more effective when assessed at 2 weeks than CHX MW due to the nature of herbal ingredients [58]. They explained that CUR has a lipophilic nature that can permeate the cell membrane rapidly and induce apoptosis, thereby destroying the integrity of bacterial cells. Also, other herbal compounds such as neem, tulsi, triphala, haritak, and bibitaki possess antibacterial properties. Other studies did not find any significant differences between CHX and CUR [51, 56].

The food and Drug Administration have acknowledged CUR as generally safe in line with the existing evidence [68]. However, longer term evaluations would provide more reliable results, particularly at high doses and with bioavailability-enhanced preparations. Some adverse effects have been observed in animal studies such as an increase in liver weight, discoloration of the face, and mucosal epithelium hyperplasia in cecum and colon, following administration of large doses of CUR [1]. Nevertheless, the maximum dose of CUR included in the review articles was 20%, which showed no considerable side effects. Furthermore, a larger sample size is required to evaluate the effect of CUR on gingivitis clinically.

11.4.2 Strengths and Limitations

A sensitive search strategy was used for this systematic review and efforts were made to identify all relevant studies. Study selection, confirmation of eligibility, and assessment of the quality of evidence were performed in duplicate, and any disagreement was resolved by discussion or consultation with the third co-author until a final consensus was achieved.

Quality assessment in most of the included studies was insufficient. The nature and characteristics of included studies could have an impact on the outcome. Therefore, the results should be considered by taking this fact into account. Due to the heterogeneous nature of the included studies in terms of interventional approaches, the use of different forms of CUR with different dosages and durations and evaluating the effects of curcumin on gingivitis compared with different modalities of treatments, prevented us from drawing strong evidence-based conclusions regarding the effects of curcumin on gingivitis. Finally, 13 out of 14 studies were conducted in India and one of them in Brazil, which makes it difficult to generalize these results to other populations.

11.5 Conclusions

Under the limitation of our review, curcumin was shown to have antibacterial effects. It reduced plaque, gingival and gingival bleeding indices in almost all of the reviewed studies. Therefore, in may be useful in the management of gingivitis. The evidence suggested that it is comparable to chlorhexidine mouthwash in this regard and was relatively safe with no serious adverse effects reported. However, tongue staining and bitter taste were shown after long-term use of chlorhexidine mouthwash. We conclude that curcumin appears to be a safe natural herbal compound and is as effective as chlorhexidine mouthwash as a treatment for gingivitis. Thus, further studies are warranted using longer term evaluations in multiple locations and countries to demonstrate the generalizability of the findings.

References

1. Chatterjee A, Debnath K, Rao NKH (2017) A comparative evaluation of the efficacy of curcumin and chlorhexidine mouthrinses on clinical inflammatory parameters of gingivitis: a double-blinded randomized controlled clinical study. J Indian Soc Periodontol 21(2):132–137
2. Eid Abdelmagyd HA, Ram Shetty DS, Musa Musleh Al-Ahmari DM (2019) Herbal medicine as adjunct in periodontal therapies: a review of clinical trials in past decade. J Oral Biol Craniofac Res 9(3):212–217
3. Safiaghdam H, Oveissi V, Bahramsoltani R, Farzaei MH, Rahimi R (2018) Medicinal plants for gingivitis: a review of clinical trials. Iran J Basic Med Sci 21(10):978–991
4. Farjana HN, Chandrasekaran SC, Gita B (2014) Effect of oral curcuma gel in gingivitis management—a pilot study. J Clin Diagn Res 8(12):ZC08–ZC10
5. Figuero E, Nobrega DF, Garcia-Gargallo M, Tenuta LM, Herrera D, Carvalho JC (2017) Mechanical and chemical plaque control in the simultaneous management of gingivitis and caries: a systematic review. J Clin Periodontol 44(Suppl 18):S116–s134
6. Stoyell KA, Mappus JL, Gandhi MA (2016) Clinical efficacy of turmeric use in gingivitis: a comprehensive review. Complement Ther Clin Pract 25:13–17
7. Asadoorian J (2017) Therapeutic oral rinsing with non-commercially available products: position paper and statement from the Canadian Dental Hygienists Association, Part 2. Can J Dent Hyg 50(3):126–139
8. Barnett ML (2003) The role of therapeutic antimicrobial mouthrinses in clinical practice: control of supragingival plaque and gingivitis. J Am Dent Assoc 134(6):699–704
9. Barnett ML (2006) The rationale for the daily use of an antimicrobial mouthrinse. J Am Dent Assoc 137(Suppl):16s–21s
10. Fine DH (1995) Chemical agents to prevent and regulate plaque development. Periodontol 8:87–107
11. Sekino S, Ramberg P, Uzel NG, Socransky S, Lindhe J (2004) The effect of a chlorhexidine regimen on de novo plaque formation. J Clin Periodontol 31(8):609–614

12. Bagherniya M, Nobili V, Blesso CN, Sahebkar A (2018) Medicinal plants and bioactive natural compounds in the treatment of non-alcoholic fatty liver disease: a clinical review. Pharmacol Res 130:213–240
13. Taheri JB, Azimi S, Rafieian N, Akhavan Zanjani H (2011) Herbs in dentistry. Int Dent J 61(6):287–296
14. Kumar G, Jalaluddin M, Rout P, Mohanty R, Dileep C (2013) Emerging trends of herbal care in dentistry. J Clin Diagn Res 7(8):1827–1829
15. Chen Y, Wong RW, McGrath C, Hagg U, Seneviratne CJ (2014) Natural compounds containing mouthrinses in the management of dental plaque and gingivitis: a systematic review. Clin Oral Investig 18(1):1–16
16. Yuan G, Wahlqvist ML, He G, Yang M, Li D (2006) Natural products and anti-inflammatory activity. Asia Pac J Clin Nutr 15(2):143–152
17. Tilak JC, Banerjee M, Mohan H, Devasagayam TP (2004) Antioxidant availability of turmeric in relation to its medicinal and culinary uses. Phytother Res 18(10):798–804
18. Little JW (2004) Complementary and alternative medicine: impact on dentistry. Oral Surg Oral Med Oral Pathol Oral Radiol Endod 98(2):137–145
19. Jain N, Rajwar YC, Batra M, Singh HP, Bhandari R, Agarwal P (2014) Dentistry: turning towards herbal alternatives: a review. Sch J Appl Med Sci 2(1C):253–257
20. Santi SS, Casarin M, Grellmann AP, Chambrone L, Zanatta FB (2019) Effect of herbal mouthrinses on dental plaque formation and gingival inflammation: a systematic review. Oral Dis Dec 14. https://doi.org/10.1111/odi.13254. [Epub ahead of print]
21. Zorofchian Moghadamtousi S, Abdul Kadir H, Hassandarvish P, Tajik H, Abubakar S, Zandi K (2014) A review on antibacterial, antiviral, and antifungal activity of curcumin. Biomed Res Int 2014:186864. https://doi.org/10.1155/2014/186864
22. Abdollahi E, Momtazi AA, Johnston TP, Sahebkar A (2018) Therapeutic effects of curcumin in inflammatory and immune-mediated diseases: a nature-made jack-of-all-trades? J Cell Physiol 233(2):830–848
23. Mollazadeh H, Cicero AF, Blesso CN, Pirro M, Majeed M, Sahebkar A (2019) Immune modulation by curcumin: the role of interleukin-10. Crit Rev Food Sci Nutr 59(1):89–101
24. Teymouri M, Pirro M, Johnston TP, Sahebkar A (2017) Curcumin as a multifaceted compound against human papilloma virus infection and cervical cancers: A review of chemistry, cellular, molecular, and preclinical features. BioFactors 43(3):331–346.
25. Hasanzadeh S, Read MI, Bland AR, Majeed M, Jamialahmadi T, Sahebkar A (2020) Curcumin: an inflammasome silencer. Pharmacol Res 159:104921. https://doi.org/10.1016/j.phrs.2020.104921. Epub 2020 May 25. PMID: 32464325.
26. Iranshahi M, Sahebkar A, Hosseini ST, Takasaki M, Konoshima T, Tokuda H (2010) Cancer chemopreventive activity of diversin from Ferula diversivittata in vitro and in vivo. Phytomedicine 17(3–4):269–273
27. Momtazi AA, Derosa G, Maffioli P, Banach M, Sahebkar A (2016) Role of microRNAs in the Therapeutic Effects of Curcumin in Non-Cancer Diseases. Mol Diagn Ther 20(4):335–345.
28. Chaturvedi TP (2009) Uses of turmeric in dentistry: an update. Indian J Dent Res 20(1):107–109
29. Mahyari S, Mahyari B, Emami SA, Malaekeh-Nikouei B, Jahanbakhsh SP, Sahebkar A et al (2016) Evaluation of the efficacy of a polyherbal mouthwash containing Zingiber officinale, Rosmarinus officinalis and Calendula officinalis extracts in patients with gingivitis: a randomized double-blind placebo-controlled trial. Complement Ther Clin Pract 22:93–98
30. Srivastava KC, Bordia A, Verma SK (1995) Curcumin, a major component of food spice turmeric (Curcuma longa) inhibits aggregation and alters eicosanoid metabolism in human blood platelets. Prostaglandins Leukot Essent Fatty Acids 52(4):223–227
31. Ammon HP, Safayhi H, Mack T, Sabieraj J (1993) Mechanism of antiinflammatory actions of curcumine and boswellic acids. J Ethnopharmacol 38(2–3):113–119
32. Jian Y, Wang J, Mai G, Zhang Y, Lai Z (2005) Curcumin regulates cyclooxygenase-2 activity in trinitrobenzene sulfonic acid-induced colitis. J Fourth Mil Med Univ 26:521–524
33. Skrzypczak-Jankun E, McCabe NP, Selman SH, Jankun J (2000) Curcumin inhibits lipoxygenase by binding to its central cavity: theoretical and X-ray evidence. Int J Mol Med 6(5):521–526
34. Kobayashi T, Hashimoto S, Horie T (1997) Curcumin inhibition of dermatophagoides farinea-induced interleukin-5 (IL-5) and granulocyte macrophage-colony stimulating factor (GM-CSF) production by lymphocytes from bronchial asthmatics. Biochem Pharmacol 54(7):819–824
35. Kang BY, Chung SW, Chung W, Im S, Hwang SY, Kim TS (1999) Inhibition of interleukin-12 production in lipopolysaccharide-activated macrophages by curcumin. Eur J Pharmacol 384(2–3):191–195
36. Bremner P, Heinrich M (2005) Natural products and their role as inhibitors of the pro-inflammatory transcription factor NF-κB. Phytochem Rev 4(1):27–37
37. Chan MM, Huang HI, Fenton MR, Fong D (1998) In vivo inhibition of nitric oxide synthase gene expression by curcumin, a cancer preventive natural product with anti-inflammatory properties. Biochem Pharmacol 55(12):1955–1962
38. Chainani-Wu N (2003) Safety and anti-inflammatory activity of curcumin: a component of tumeric (Curcuma longa). J Altern Complement Med 9(1):161–168
39. Singh R, Chandra R, Bose M, Luthra PM (2002) Antibacterial activity of Curcuma longa rhizome extract on pathogenic bacteria. Curr Sci 83(6):737–740
40. Osawa T, Sugiyama Y, Inayoshi M, Kawakishi S (1995) Antioxidative activity of tetrahydrocurcuminoids. Biosci Biotechnol Biochem 59(9):1609–1612
41. Sidhu GS, Singh AK, Thaloor D, Banaudha KK, Patnaik GK, Srimal RC et al (1998) Enhancement of wound healing by curcumin in animals. Wound Repair Regen 6(2):167–177

42. Baldwin AS (2001) Control of oncogenesis and cancer therapy resistance by the transcription factor NF-kappaB. J Clin Invest 107(3):241–246
43. Plummer SM, Holloway KA, Manson MM, Munks RJ, Kaptein A, Farrow S et al (1999) Inhibition of cyclo-oxygenase 2 expression in colon cells by the chemopreventive agent curcumin involves inhibition of NF-kappaB activation via the NIK/IKK signalling complex. Oncogene 18(44):6013–6020
44. Liberati A, Altman DG, Tetzlaff J, Mulrow C, Gotzsche PC, Ioannidis JP et al (2009) The PRISMA statement for reporting systematic reviews and meta-analyses of studies that evaluate healthcare interventions: explanation and elaboration. BMJ 339:b2700. https://doi.org/10.1136/bmj.b2700
45. Jadad AR, Moore RA, Carroll D, Jenkinson C, Reynolds DJM, Gavaghan DJ et al (1996) Assessing the quality of reports of randomized clinical trials: is blinding necessary? Control Clin Trials 17(1):1–12
46. Moher D, Cook D, Jadad A, Tugwell P, Moher M, Jones A et al (1999) Assessing the quality of reports of randomised trials: implications for the conduct of meta-analyses
47. Kaur S, Sharma R, Sarangal V, Kaur N, Prashar P (2017) Evaluation of anti-inflammatory effects of systemically administered curcumin, lycopene and piperine as an adjunct to scaling and root planing: A clinical study. Ayu 38(3–4):117–121
48. Kandwal A, Mamgain RK, Mamgain P (2015) Comparative evaluation of turmeric gel with 2% chlorhexidine gluconate gel for treatment of plaque induced gingivitis: a randomized controlled clinical trial. Ayu 36(2):145–150
49. Singh V, Pathak AK, Pal M, Sareen S, Goel K (2015) Comparative evaluation of topical application of turmeric gel and 0.2% chlorhexidine gluconate gel in prevention of gingivitis. Natl J Maxillofac Surg 6(1):67–71
50. Chunawala YK, Nipanikar S, Morawala AK, Talathi RD, Shaikh S, Upasani SA et al (2018) An open-label, interventional, single-center, prospective clinical study to evaluate the efficacy and safety of "AHPL/AYTAB/1514" in patients suffering from halitosis. Indian J Dent Res 29(6):732–738
51. Mali AM, Behal R, Gilda SS (2012) Comparative evaluation of 0.1% turmeric mouthwash with 0.2% chlorhexidine gluconate in prevention of plaque and gingivitis: a clinical and microbiological study. J Indian Soc Periodontol 16(3):386–391
52. Sharma V, Kalsi DS (2016) Effects of topical application of Curcuma longa extract in the treatment of early periodontal diseases. Indian J Dent Sci 8(3):118. https://doi.org/10.4103/0976-4003.191725
53. Roopa D, Singh S, Gupta S, Pandey Y, Goswami A, Johari S (2016) Curcumin: a herbal approach in the management of gingivitis. Rama Univ J Dent Sci 3(1):1–5
54. Paschoal MA, Moura CM, Jeremias F, Souza JF, Bagnato VS, Giusti JS et al (2015) Longitudinal effect of curcumin-photodynamic antimicrobial chemotherapy in adolescents during fixed orthodontic treatment: a single-blind randomized clinical trial study. Lasers Med Sci 30(8):2059–2065
55. Arunachalam LT, Sudhakar U, Vasanth J, Khumukchum S, Selvam VV (2017) Comparison of anti-plaque and anti-gingivitis effect of curcumin and chlorhexidine mouth rinse in the treatment of gingivitis: a clinical and biochemical study. J Indian Soc Periodontol 21(6):478–483
56. Waghmare PF, Chaudhari AU, Karhadkar VM, Jamkhande AS (2011) Comparative evaluation of turmeric and chlorhexidine gluconate mouthwash in prevention of plaque formation and gingivitis: a clinical and microbiological study. J Contemp Dent Pract 12(4):221–224
57. Muglikar S, Patil KC, Shivswami S, Hegde R (2013) Efficacy of curcumin in the treatment of chronic gingivitis: a pilot study. Oral Health Prev Dent 11(1):81–86
58. Vinod KS, Sunil KS, Sethi P, Bandla RC, Singh S, Patel D (2018) A novel herbal formulation versus chlorhexidine mouthwash in efficacy against oral microflora. J Int Soc Prev Commun Dent 8(2):184–190
59. Pulikkotil SJ, Nath S (2015) Effects of curcumin on crevicular levels of IL-1beta and CCL28 in experimental gingivitis. Aust Dent J 60(3):317–327
60. Suhag A, Dixit J, Dhan P (2007) Role of curcumin as a subgingival irrigant: a pilot study. Periodontal Pract Today 4(2):115–121
61. Sajithlal GB, Chithra P, Chandrakasan G (1998) Effect of curcumin on the advanced glycation and cross-linking of collagen in diabetic rats. Biochem Pharmacol 56(12):1607–1614
62. Shukla PK, Khanna VK, Ali MM, Khan MY, Srimal RC (2008) Anti-ischemic effect of curcumin in rat brain. Neurochem Res 33(6):1036–1043
63. Gopinath D, Ahmed MR, Gomathi K, Chitra K, Sehgal PK, Jayakumar R (2004) Dermal wound healing processes with curcumin incorporated collagen films. Biomaterials 25(10):1911–1917
64. Guimaraes MR, Coimbra LS, de Aquino SG, Spolidorio LC, Kirkwood KL, Rossa CJ (2011) Potent anti-inflammatory effects of systemically administered curcumin modulate periodontal disease in vivo. J Periodontal Res 46(2):269–279
65. Kohli K, Ali J, Ansari M, Raheman Z (2005) Curcumin: a natural antiinflammatory agent. Indian J Pharmacol 37(3):141
66. Jurenka JS (2009) Anti-inflammatory properties of curcumin, a major constituent of Curcuma longa: a review of preclinical and clinical research. Altern Med Rev 14(2):141–153
67. Kim SJ (2011) Curcumin suppresses the production of interleukin-6 in Prevotella intermedia lipopolysaccharide-activated RAW 264.7 cells. J Periodontal Implant Sci 41(3):157–163
68. Soleimani V, Sahebkar A, Hosseinzadeh H (2018) Turmeric (Curcuma longa) and its major constituent (curcumin) as nontoxic and safe substances: Review. Phytotherapy Research 32(6):985–995

12 Does Curcumin Have an Anticaries Effect? A Systematic Review of In Vitro Studies

Ailin Ehteshami, Farinaz Shirban, Fateme Gharibpour, Mohammad Bagherniya, Thozhukat Sathyapalan, and Amirhossein Sahebkar

Abstract

Background: Dental caries is one of the most important oral health problems and a common infectious microbial disease. *Streptococcus mutans* (*S. mutans*) has been regarded as the primary etiologic factor in the formation of dental caries. Curcumin (CUR) has an antibacterial action and could be used in the eradication of *S. mutans* to control dental caries. This systematic review was undertaken with the aim of evaluating the anticaries effect of CUR.

Methods: A comprehensive search was conducted in the PubMed/Medline, Cochrane - CENTRAL, and Scopus databases. Based on the PICO model, studies which evaluated the anticaries effects of CUR up until 24 February 2020 with language restrictions were selected for this systematic review.

Results: From 753 papers found, 13 met the eligibility criteria and were included. In 12 out of 13 included studies, CUR had significant antibacterial and anticaries effects. CUR had inhibitory effects on *S. mutans* growth, acid production, ATPase and sortase A activity, biomass, viability and metabolism reduction of biofilm, reduced exopolysaccharide production of biofilms, changes in biofilm structure, and had anti-adhesion effects against *S. mutans*.

Conclusion: This systematic review suggests promising antibacterial and anticaries effects of CUR including inhibition of *S. mutans* growth, acid production, ATPase and sortase A activity. This review provides unique information regarding the potential role of CUR in the prevention and treatment of dental carries as a natural, accessible, safe, and inexpensive agent to increase oral and dental

A. Ehteshami · F. Shirban · F. Gharibpour
Department of Orthodontics, Dental Research Center, Dental Research Institute, School of Dentistry, Isfahan University of Medical Sciences, Isfahan, Iran

M. Bagherniya
Food Security Research Center, Department of Community Nutrition, School of Nutrition and Food Science, Isfahan University of Medical Sciences, Isfahan, Iran

T. Sathyapalan
Department of Academic Diabetes, Endocrinology and Metabolism, Hull York Medical School, University of Hull, Hull, UK

A. Sahebkar (✉)
Biotechnology Research Center, Pharmaceutical Technology Institute, Mashhad University of Medical Sciences, Mashhad, Iran

Neurogenic Inflammation Research Center, Mashhad University of Medical Sciences, Mashhad, Iran

Polish Mother's Memorial Hospital Research Institute (PMMHRI), Lodz, Poland

Halal Research Center of IRI, FDA, Tehran, Iran
e-mail: sahebkara@mums.ac.ir

P. C. Guest (ed.), *Studies on Biomarkers and New Targets in Aging Research in Iran*, Advances in Experimental Medicine and Biology 1291, https://doi.org/10.1007/978-3-030-56153-6_12

health. However, clinical randomized controlled trials are needed to confirm these results.

Keywords

Curcumin · Dental caries · Oral health · *Streptococcus mutans* · Systematic review

12.1 Introduction

Oral diseases continue to be a major health problem worldwide and are integrally associated with general well-being and quality of life [1]. Dental caries is one of the most important global oral health problems and is the most prevalent chronic disease in the dental field [2–7]. This disease is multifactorial, contagious, and site-specific [8] caused by interactions of multiple factors including diet constituents, cariogenic bacteria, and the composition of the tooth enamel [9, 10]. Dental caries result mainly by bacterial acid production diffusing into the enamel and dentine as consequences of ecologically driven imbalances of oral microbial biofilms [1, 11].

Streptococcus mutans (*S. mutans*) is a gram-positive endogenous bacterium which has been regarded as the primary etiologic factor in the formation of dental plaques and dental caries [5, 12–18]. It causes dental caries by fermenting dietary carbohydrates into lactic acid [1, 19]. Lactic acid reduces the pH of the biofilm matrix resulting in the dissolution of tooth surfaces, which initiates dental caries [20]. Therefore, eradication of dental biofilm and *S. mutans* is an effective method for controlling dental caries [3, 10].

Different types of antibacterial agents, including antibiotics, are available against oral bacteria. Short-term antibiotic therapy results in a significant reduction in the level of *S. mutans*. However, antibiotics have side effects including nausea, vomiting, diarrhea, and changes in oral microflora. Other antibacterial agents such as cetylpyridinium chloride, chlorhexidine, and amine fluorides have been reported to have various toxicities, teeth staining, or in the case of ethanol (commonly found in mouthwashes), they have been related to oral cancer [21–36]. Most antibacterial agents can also develop resistant bacterial strains [37, 38].

There is a need for alternative prevention and treatment options for oral diseases that are safe, effective, and economical [1, 11, 39, 40]. Naturally derived chemotherapeutic agents are an appealing option [41–44]. In recent years, the usage of medicinal plants for primary health care has increased worldwide [7, 45, 46]. There have been numerous reports of the use of plant products for the treatment of oral diseases [1, 20, 47–49]. Although several plants have shown antibacterial activity, to date, their antibacterial activity against bacteria causing dental caries is still under research [11].

One of these natural products is curcumin (CUR), a dietary natural product extracted from the root of turmeric, which is widely used as a flavoring and coloring agent. It also has many pharmacological activities including anti-inflammatory and antitumor properties [20, 41, 50–58]. In addition, CUR has been shown to have antibacterial, blood pressure lowering, whitening, and antioxidative activities [59, 60].

Since some studies have shown the effects of CUR on oral bacteria associated with dental caries, the purpose of this investigation was to conduct a systematic review of in vitro studies about the effects of CUR on dental caries and cariogenic bacteria.

12.2 Methods and Materials

12.2.1 Protocol

This systematic review was conducted according to the Preferred Reporting Items for Systematic Reviews and Meta-Analysis (PRISMA) Statement Guidelines [61].

12.2.2 Focus Question

The purpose of the present systematic review was to assess the null hypothesis that "there would be no difference in the number of dental caries by the administration of CUR." The focused question according to the PICO format (i.e., population, intervention, control group, and outcome) was are anticaries effects by administration of CUR compared to control group in in vitro studies significant (population: samples with caries or cariogenic bacteria, interventions: CUR administration and control intervention, control group: samples without CUR administration, outcome: anticaries effects)?

12.2.3 Search Strategy

An electronic search was performed until February 24, 2020, to acquire potentially eligible studies, with no time restrictions in the following electronic bibliographic databases: PubMed/MEDLINE, Cochrane – CENTRAL, and SCOPUS. Search strategies of each database are presented in Table 12.1. The references of retrieved full-text articles were also searched for further papers. Papers that were not in English were excluded.

12.2.4 Eligibility Criteria

The following selection criteria were applied for this systematic review:

1. Inclusion criteria: (a) in vitro studies; and (b) articles providing data regarding the effects of CUR on dental caries.
2. Exclusion criteria: (a) review articles, (b) case reports, (c) letters to editors, (d) none-English-language papers.

12.2.5 Study Selection

The titles and abstracts of the searched studies were screened by two authors independently (AE and FS). Papers were excluded if they were not relevant to the current study or if they were duplicates. Controversies were resolved through discussion. Publications were included for full-text evaluation if they met the inclusion criteria in the first analysis or if insufficient information was provided in the title and abstract to enable a decision to be made.

12.2.6 Data Extraction

All eligible studies underwent data extraction. Data were extracted by at least two review authors independently. Any disagreement was discussed and a third review author consulted as necessary. Finally, the following data were extracted from the eligible studies: study design; sample size; type of species; type and dosage of CUR; method of CUR administration; the outcome measurement method; and the outcome measured (effect on dental caries samples and bacteria).

12.3 Results

12.3.1 Search and Selection Results

The search of PubMed/MEDLINE, Cochrane – CENTRAL, SCOPUS databases returned 753 papers (Fig. 12.1). Thirty-one records were identified to be relative in the initial screening. After initial screening, 19 articles were selected for full assessment. A full assessment of 19 articles revealed that 13 articles were appropriate. Thus, data extraction was done on 13 articles. An overview of each selected paper is shown in Table 12.2.

12.3.2 Target Tissue/Cell

Seven studies analyzed *S. mutans* biofilms [3, 6, 10, 62–65]. Three studies analyzed *S. mutans* and *Lactobacillus acidophilus* [11], *S. mutans* and *Candida albicans* [41], and *S. mutans*, *Actinomyces naeslundii* and *Streptococcus oralis* [20]. Three studies worked on the bacterial biofilm on extracted teeth [8, 19, 66].

12.3.3 Type of Intervention

In nine studies, CUR alone at different concentrations was used [3, 6, 8, 19, 20, 41, 62–64]. The other four studies used other plants and herbal compounds in addition to CUR [10, 1, 65, 66].

Table 12.1 Databases, applied search strategy, and numbers of retrieved studies

Database of published trials	Search strategy used
MEDLINE searched via PubMed, searched on 24st February 2020, via www.ncbi.nlm.nih.gov/sites	("curcuma"[MeSH Terms] OR ("curcuma"[MeSH Terms] OR "curcuma"[All Fields]) OR ("curcuma"[MeSH Terms] OR "curcuma"[All Fields] OR "curcumas"[All Fields]) OR ("curcumin"[MeSH Terms] OR "curcumin"[All Fields]) OR curcuminoid[All Fields] OR "curcuminoids"[All Fields]) OR ("curcumin"[MeSH Terms] OR "curcumin"[All Fields] OR "tumeric"[All Fields] OR "curcuma"[MeSH Terms] OR "curcuma"[All Fields]) OR ("curcuma"[MeSH Terms] OR "curcuma"[All Fields] OR "turmeric"[All Fields]) OR "curcuma longa"[All Fields] OR "curcuma zedoaria"[All Fields])("zingiberaceae"[MeSH Terms] OR "zingiberaceae"[All Fields]) AND (("dental caries"[MeSH Terms] OR ("dental"[All Fields] AND "caries"[All Fields]) OR "dental caries"[All Fields] OR ("dental"[All Fields] AND "decay"[All Fields]) OR "dental decay"[All Fields]) OR ("dental caries"[MeSH Terms] OR ("dental"[All Fields] AND "caries"[All Fields]) OR "dental caries"[All Fields] OR "caries"[All Fields]) OR ("dental caries"[MeSH Terms] OR ("dental"[All Fields] AND "caries"[All Fields]) OR "dental caries"[All Fields] OR ("dental"[All Fields] AND "decay"[All Fields]) OR "dental decay"[All Fields]) OR "dental"[All Fields]) AND Carious[All Fields]) OR (("dentin"[MeSH Terms] OR "dentin"[All Fields]) AND Carious[All Fields]) OR ("dentin"[MeSH Terms] OR "dentin"[All Fields] OR "dentins"[All Fields]) OR ("dentin"[MeSH Terms] OR "dentin"[All Fields]) OR ("dental caries"[MeSH Terms] OR ("dental"[All Fields] AND "caries"[All Fields]) OR "dental caries"[All Fields] OR ("carious"[All Fields] AND "dentins"[All Fields])) OR Carious[All Fields] OR ("dental caries"[MeSH Terms] OR ("dental"[All Fields] AND "caries"[All Fields]) OR "dental caries"[All Fields] OR ("dental"[All Fields] AND "white"[All Fields] AND "spot"[All Fields]) OR "dental white spot"[All Fields]) OR ("dental caries"[MeSH Terms] OR ("dental"[All Fields] AND "caries"[All Fields]) OR "dental caries"[All Fields] OR ("white"[All Fields] AND "spots"[All Fields]) OR "white spots"[All Fields]) OR ("dental caries"[MeSH Terms] OR ("dental"[All Fields] AND "caries"[All Fields]) OR "dental caries"[All Fields] OR ("dental"[All Fields] AND "white"[All Fields] AND "spots"[All Fields]) OR "dental white spots"[All Fields]) OR Spot[All Fields] OR ("dental caries"[MeSH Terms] OR ("dental"[All Fields] AND "caries"[All Fields]) OR "dental caries"[All Fields] OR ("white"[All Fields] AND "spots"[All Fields]) OR "white spots"[All Fields]) OR "white"[All Fields]) AND ("dental caries"[MeSH Terms] OR ("dental"[All Fields] AND "caries"[All Fields]) OR "dental caries"[All Fields] OR ("white"[All Fields] AND "spot"[All Fields]) OR "white spot"[All Fields])) OR ("dental caries"[MeSH Terms] OR ("dental"[All Fields] AND "caries"[All Fields]) OR "dental caries"[All Fields] OR ("dental"[All Fields] AND "white"[All Fields] AND "spots"[All Fields]) OR "dental white spots"[All Fields])
Scopus searched via Scopus on February 24, 2020, via https://www.scopus.com	(curcuma OR curcumas OR curcumin OR curcuminoid OR curcuminoids OR turmeric OR curcumin OR longa OR zedoaria OR zingiberaceae) AND (dental OR decay OR caries OR carious OR dentin OR carious OR dentins OR white OR spot OR spots) SUBJAREA (*dent*)
Cochrane Central Register of Controlled Trials searched via the Cochrane Library Searched on February 24, 2020 via www.thecochranelibrary.com	(curcuma OR curcumas OR curcumin OR curcuminoid OR c urcuminoids OR tumeric OR turmeric OR curcuma OR longa OR zedoaria ORzingiberaceae) AND (caries OR Dental OR decay OR dental OR Caries OR Carious OR dentin OR Dentins OR white OR spot OR spots)) in Title Abstract Keyword – (Word variations have been searched)

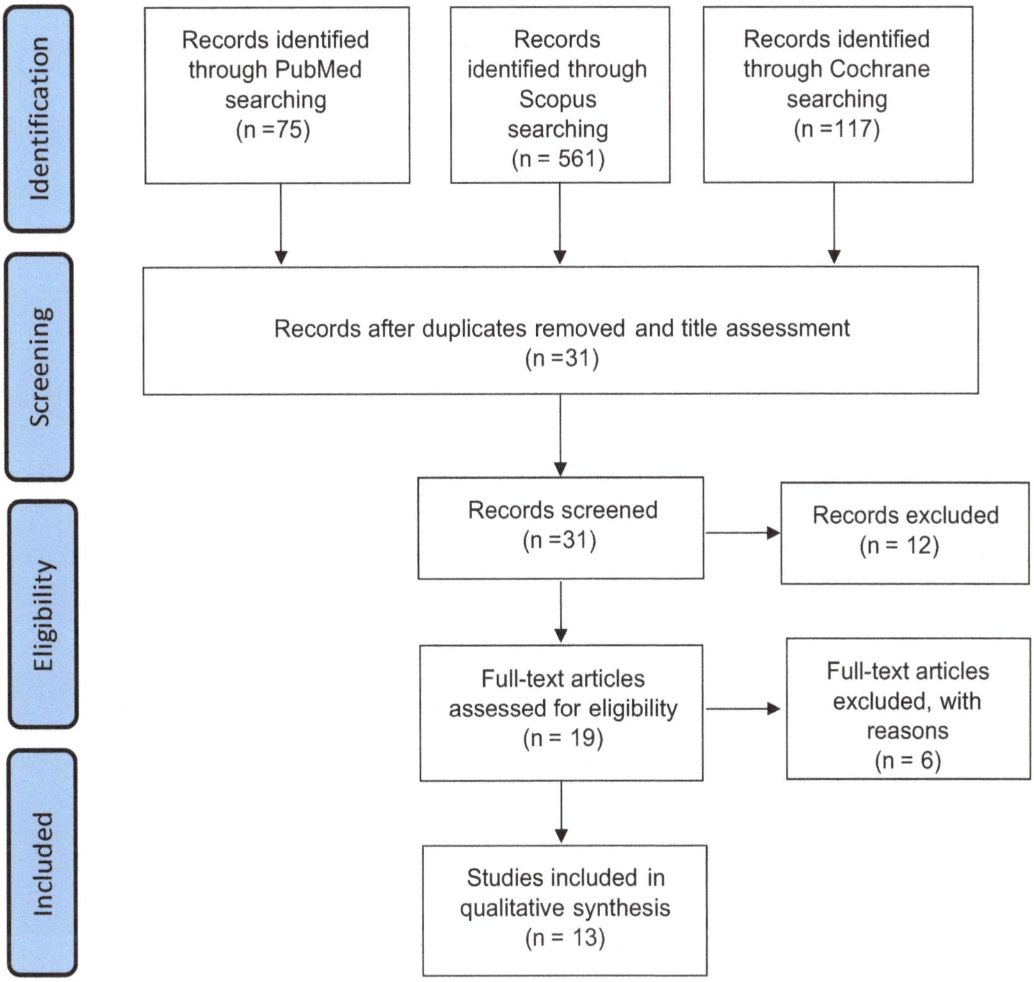

Fig. 12.1 Flowchart showing the process of the study selection

12.3.4 Mode of CUR Administration

In ten studies, CUR was dissolved in dimethyl sulfoxide (DMSO) [3, 6, 8, 10, 19, 41, 62–65]. Two studies dissolved CUR in methanol or used methanol fractions [11, 20], and one study used distilled water as the solvent [65].

12.3.5 Methods of Evaluation

Different methods of evaluation were used in these 13 studies. The most common method used was the minimum inhibitory concentration (MIC) [3, 10, 11, 19, 20, 62, 63]. The methods used were disc diffusion [6, 10, 11, 66], viability and biomass of biofilm [3, 41, 64, 65], scanning electron microscopy (SEM) [3, 6, 41, 64], pH assessment [6, 8, 20, 64], polymerase chain reaction (PCR) [3, 41, 63], confocal laser scanning microscopy (CLSM) [3, 19, 41], bacterial adhesion [6, 19, 20], minimum bactericidal concentration (MBC) [11, 62], and quantifying increased fluorescent intensity [62, 63]. Other sporadic methods are listed in Table 12.2.

Table 12.2 General characteristics of the included studies

Author/year/location	Study design	Study subjects/total number	Target tissue/cell	Study groups (number/amount of test/control)	Mode of curcumin administration/dosage	Primary methods of evaluation
Elgamily et al. (2019) Egypt [11]	In-vitro study	10/60	*S. mutans* and *Lactobacillus acidophilus*	5 test groups (cinnamon, turmeric, ginger, clove, black seed), 1 control (CHX) ($n = 10$)	Fifty grams plant powder with an organic solvent 90% methanol	Disc diffusion method (measuring the diameter of the inhibition zone (mm)), MIC, and MBC
Li et al. (2019) China [41]	In-vitro study	96-well flat-bottom plates	Biofilms formed by *S. mutans* and *Candida albicans*	3 test groups with CUR (*S. mutans* biofilm, *C. albicans* biofilm, and *S. mutans-C. albicans* biofilm) and 1 control group (DMSO)	0.5 mM CUR in DMSO for liquid storage	MBEC by the colony counting method, quantitative biofilm biomass and viability, live–dead PCR, CLSM, SEM, detection of gene expression
Li et al. (2018) China [3]	In-vitro study	96-well flat-bottom plates	Biofilms formed by *S. mutans*	7 test group with different CUR concentration (2000, 1000, 500, 250, 125, 62.5, and 31.25 μM) and 1 control group (DMSO) for MIC 2 test groups (500 μM CUR, 0.12% CHX) and 1 control group (1% BHIS) for CLSM & SEM	250 mM CUR in DMSO	MIC, SMIC, assays of the viability and biomass of biofilm, analysis of Live and dead bacteria and EPS by CLSM, SEM, assay of gene expression by real-time PCR
Kanth et al. (2016) India [66]	In-vitro study	Eleven circular disk	Aerobic culture from caries material of extracted teeth	10 test groups (clove oil, ginger and garlic paste, Tea tree oil, neem, ginger, garlic, cinnamon oil, green tea, turmeric, eucalyptus oil) and 1 control group (ciprofloacxin)	10 g CUR with 100 mL distilled water	Vernier's caliper for measurement the zone of inhibition
Hu et al. (2013) China [62]	In-vitro study	NR	*S. mutans* UA159 and its genome	1 test group (CUR) 2 control groups (pHMB as a positive control 1% DMSO as a negative control)	CUR dissolved in 1% DMSO	Sortase A activity inhibition assay by quantifying increased fluorescent intensity, MIC and MBC

Author/year/location	Study design	Study subjects/total number	Target tissue/cell	Study groups (number/amount of test/control)	Mode of curcumin administration/dosage	Primary methods of evaluation
Hu et al. (2013) China [63]	In-vitro study	96-well flat-bottom plates	*S. mutans* UA159 and its genome	1 test group(CUR) 2 control groups (pHMB as a positive control 0.5% DMSO as a negative control)	Different concentrations of CUR (1 μM, 2 μM, 4 μM, 8 μM, 16 μM, 32 μM, 64 μM, 128 μM) dissolved in 1% DMSO 15 μmol/L CUR for PCR	Sortase A activity inhibition assay by quantifying increased fluorescent intensity, MIC, real-time PCR assays, SDS–PAGE and Western blot analysis, biofilm formation assay
Joycharat et al. (2012) Thailand [10]	In-vitro study	96-well sterile micro titer plates	*S. mutans*	35 different herbal formulas test groups 1 control group (1%DMSO and penicillin G (10 U/disc))	200 g/ethanol extracts of CUR dissolved DMSO to give a concentration of 250 mg/mL	MIC, agar disc diffusion method
Lee et al. (2011) South Korea [6]	In-vitro study	24-well plates	*S. mutans*	5 test groups with different concentration of curcuma longa (0.25, 0.5, 1, 2, 4 mg/mL) and 1 control group (0.1% of DMSO)	0.25–4 mg/mL essential oil of curcuma longa dissolved in DMSO	Inhibition of bacterial growth assay ([CFU]/mL), acid production, bacterial adherence assay, biofilm assay by CFU]/mL, SEM and GC and GC-MS analysis
Pandit et al. (2011) Republic of Korea [20]	In-vitro study	24 well plate	*S. mutans*, *Actinomyces naeslundii*, and *Streptococcus oralis*	5 test groups (methanol extract, *n*-hexane fraction, chloroform fraction, ethyl acetate fraction, 70% aqueous methanol fraction) One control group (4% DMSO)	Turmeric extracted with methanol and its fractions	MIC, glycolytic pH-drop assay, biofilm formation, bacterial adherence by counting number of colony forming units (CFUs), acid killing and F-ATPase activity assay

(continued)

Table 12.2 (continued)

Author/year/location	Study design	Study subjects/total number	Target tissue/cell	Study groups (number/amount of test/control)	Mode of curcumin administration/dosage	Primary methods of evaluation
Kim et al. (2008) Republic of Korea [64]	In-vitro study	NR	S. mutans	1 test group (curcuma xanthorrhiza) 3 control groups (2 mg/mL CHX as the positive control Saline and 1% DMSO as negative controls)	0.1 mg/mL curcuma xanthorrhiza in 1% DMSO 3 times (5, 30, and 60 min)	Counting the number of colonies for bacterial viability, measuring the pH of biofilm, SEM
Basir et al. (2018) Iran [8]	In-vitro study	8/40	Enamel blocks with *S. mutans* biofilm	3 test groups (different concentration of curcuma longa (5, 10, 20 mg/mL)) 2 control groups (sucrose 10% + NACL 0.9% as the positive and NACL 0.9% as a negative controls)	Solution from CUR 5, 10, 20 mg/mL in 100% DMSO	Superficial hardness of enamel based on VICKER'S scale, pH measurement of biofilm
Song et al. (2012) Korea [19]	In-vitro study	96-well plate	Tooth with *S. mutans*	9 test groups with different concentration of curcuma longa (1–256 μg/mL) 1 control group (BHI medium supplemented with 1% sucrose)	Solution of CUR prepared at 5 mg/mL in 100% DMSO	MIC, CLSM, photographic presentation and quantified by measuring the CV-stained cells with a densitometer, fibronectin and collagen immobilization
Mizuki Sekiya et al. (2019) Japan [65]	In-vitro study	NR	*S. mutans* cells, F-ATPases of *E. coli and S. mutans*	4 test groups (CUR, demethoxyCUR, piceatannol, and citreoviridin) 1 control group (0.5% DMSO)	30 μM CUR and 30 μM demethoxyCUR dissolved in DMSO	Enzyme assays, counting the number of colonies for bacterial viability (CFUs)

CHX chlorhexidine, *CUR* curcumin, *MIC* minimum inhibitory concentration, *MBC* minimum bactericidal concentration, *MBEC* minimum biofilm eradication concentration, *CLSM* confocal laser scanning microscopy, *SEM* scanning electron microscopy, *SMIC* sessile minimum inhibitory concentration, *DMSO* dimethyl sulfoxide, *S. mutans Streptococcus mutans*, *Phmb* ρ-hydroxymercuribenzoic acid, *NR* not reported

12.3.6 Study Outcomes

Twelve studies showed a significant antibacterial and anticaries effect of CUR [3, 6, 8, 10, 11, 19, 20, 41, 62–65]. These effects were reduction of biofilm biomass and viability in total/live cells, reduction of exopolysaccharide (EPS) production of both mono- and dual-species biofilms, lower mRNA levels of all genes tested, changes in biofilm structure, inhibition of *S. mutans* growth and acid production, reduction in production and metabolism of biofilm, anti-adhesion effect against *S. mutans* on extracellular matrices and tooth surfaces, inhibition of ATPase and sortase A activity of *S. mutans*, and reduction in enamel surface hardness loss. The details of the study outcomes are listed in Table 12.3.

One study found that there was no zone of inhibition for CUR [66]. In three studies, a dose-dependent antibacterial effect for CUR was seen [6, 62, 63]. Three studies compared CUR to chlorhexidine (CHX). Two of them revealed greater antibacterial activity for CHX [6, 11]. One showed more bactericidal activity for Curcuma xanthorrhiza than CHX [64].

12.4 Discussion

The use of natural products in the treatment of oral disease is increasing. As CUR has a wide range of effects in different fields, we decided to review its effects on one of the most common oral diseases, dental caries. In this study, we performed a systematic review of in vitro studies to investigate whether or not CUR has anticaries effects.

Overall, based on the information provided from these in vitro studies, CUR had anticaries effects and antibacterial properties against the microorganisms associated with dental caries. These effects occurred in a variety of ways. In addition to the bacteriostatic effects and reductions in biofilm and bacterial growth, other specific effects are discussed in the following paragraphs.

EPS production by *S. mutans* augments the local aggregation and clustering of microbes. In addition, during the development of biofilm, EPS synthesis forms a complex, three-dimensional matrix architecture and makes a safe environment within the biofilm that can facilitate the pathogenesis of dental caries [67, 68]. CUR has been shown to destroy the structure of EPS in the short term [3]. *S. mutans* extracellular glucosyltransferases (Gtfs), encoded by the gtfBCD genes, are inevitable for glucose utilization in EPS synthesis since they play a role in the formation of biofilm and caries development [3, 69]. GtfB and GtfC synthesize water-insoluble polysaccharides [13, 41, 70], which comprise the main part of the EPS and relates to the structural scaffold of the biofilms [13]. The inhibition of gtfB and gtfC by CUR might be related to the reduction of the EPS biomass [3, 41].

The initial step in dental plaque formation is the adhesion of *S. mutans* to tooth surfaces [71]. Inhibition of *S. mutans* adherence to the tooth surface is essential for prevention of plaque formation [6]. Three studies showed that addition of CUR significantly reduced the adherence of *S. mutans* to tooth surfaces and inhibited dental plaque formation and progression of dental caries [6, 19, 20]. Hence, CUR may have anti-adhesion-mediated properties for controlling dental caries. Furthermore, surface proteins are major factors in the adherence ability of *S. mutans* to the oral mucosa and tooth surfaces [63]. Sortase A is a membrane-anchoring protein and can control *S. mutans* adherence ability and biofilm formation, and it has been mentioned as one of the caries virulence factors [62, 63]. According to studies of Hu et al. and Song et al., CUR inhibits sortase A activity in vitro [19, 62, 63].

Acidogenicity and aciduricity are other important virulence properties of *S. mutans* biofilms [20]. Acidic environment in the mouth can decrease dental plaque pH, demineralize tooth enamel, and cause dental caries [72]. Three studies showed an inhibitory effect of CUR on acid production of *S. mutans* [6, 20, 64]. Furthermore, the proton-pumping F-ATPase enzyme has been identified in bacterial membranes and is related to bacterial acid tolerance [65, 73] and aciduricity of *S. mutans* [74]. Two studies reported a reduction of F-ATPase activity by CUR [20, 65].

Table 12.3 Characteristics of bacteria/biofilm and results in included studies

Author	Type of biofilm/cell	Amount of bacteria/ biofilm	Type of culture	Results
Elgamily et al. (2019) [11]	*S. mutans* Serotype c. Carious dentin ATCC 25175 type strain and Lactobacillus acidophilus CH-2	100 µL from each bacterial strain in Muller Hinton agar plates	*S. mutans* on TSA *Lactobacillus acidophilus* on MRS agar	Antibacterial activity of turmeric against *Lactobacillus acidophilus* with inhibition zones diameters of 9.33 mm
Li et al. (2019) [41]	*S. mutans* UA159 (ATCC 700610) and *Candida albicans* SC5314 (ATCC MYA-2876)	For dual-species biofilms, 1.5 mL *S. mutans* and 1.5 mL *C. albicans*. For mono-species biofilms, 3 mL *S. mutans* and 3 mL *C. albicans* separately	*Candida albicans* in Sabouraud's dextrose broth. *S. mutans* in BHI	In CUR-treated groups: reduction in biomasses, decrease in total/live cells, production of EPS of both mono- and dual-species biofilms and mRNA levels of all genes tested, changes in biofilm structure
Li et al. (2018) [3]	*S. mutans* strain UA159 (ATCC700610)	100 µL overnight culture of *S. mutans*	BHI	CUR inhibited growth of planktonic *S. mutans* with an MIC of 125 µM, decreased biofilm viability by 50%, at 500 µM, decreased biofilm metabolism at 5 min and 24 h, reduced *S. mutans* biomass at 24 h, reduced live bacteria, reduction in overall quantity of *S. mutans* EPS, CUR had both short- and long-term antibacterial effect
Kanth et al. (2016) [66]	Aerobic culture from caries material of extracted teeth	500 µL of suspension	Hartley's broth	Turmeric showed no zone of inhibition
Hu et al. (2013) [62]	*S. mutans* UA159 and its genome	200 µL standardized cell suspension (10^7 CFU/mL)	Luria–Bertani agar plates, BHI	Inhibition sortase activity, dose-dependent antibacterial activity
Hu et al. (2013) [63]	*S. mutans* UA159 and its genome	200 µL standardized cell suspension (10^7 CFU/mL)	Luria–Bertani agar plates, BHI	Inhibition sortase A activity, dose-dependent antibacterial activity, reduced biofilm formation (15 µmol/L CUR)

(continued)

Table 12.3 (continued)

Author	Type of biofilm/cell	Amount of bacteria/biofilm	Type of culture	Results
Joycharat et al. (2012) [10]	*S. mutans* ATCC 25175 and GS-5	Three to five colonies on Brain Heart Infusion (BHI) Agar transferred to TSB	TSB	Zone of inhibition: 7 mm for GS-5 and ≤ 6.0 for ATCC 25175
Lee et al. (2011) [6]	*S. mutans* ATCC 25175	5×10^5 [CFU]/mL Optical density (OD) of cells measured at 550 nm	BHI, MSA, and phenol red broth	Inhibition growth of *S. mutans* in a dose-dependent manner at concentrations higher than 0.5 mg/mL, inhibition acid production and Inhibition of bacterial adherence with 0.5–4 mg/mL extract, decreased biofilm formation at concentration of (2–4 mg/mL)
Pandit et al. (2011) [20]	*S. mutans* UA159, Actinomyces naeslundii12104, *Streptococcus oralis* 35,037	100 μL suspensions of each bacterial strain	UTE broth containing 1% glucose or 1% sucrose	Mild bacteriostatic effects against *S. mutans*, inhibitory effects on glycolytic pH-drop, anti-biofilm activity, 40% inhibition of bacterial adherence, inhibitory effect of on the acidogenicity and aciduricity of *S. mutans* biofilms
Kim et al. (2008) [64]	*S. mutans* ATCC 25175	The *S. mutans* biofilms on each slide contained 2×10^7 [CFU]/mL	Tryptone yeast extract broth containing 1% (w/v) sucrose	Bactericidal effects against *S. mutans* biofilms Inhibition of acid production, alteration of microstructure of *S. mutans*
Basir et al. (2018) [8]	*S. mutans* ATCC 35668	1.8×10^8/mL *S. mutans*	BHI broth containing sucrose 1%	1. Decrease in surface hardness loss by inhibiting bacterial activity 2. CUR had the most anticaries effect and most PH at the lowest concentration (5 mg/mL)

(continued)

Table 12.3 (continued)

Author	Type of biofilm/cell	Amount of bacteria/ biofilm	Type of culture	Results
Song et al. (2012) [19]	*S. mutans* ATCC25175	5×10^6 cfu/mL	BHI medium with 1% sucrose	strong antimicrobial activity against the organism at 128 µg/mL, CUR at a concentration of 8 µg/mL decreased bacterial adhesion to the glass surface, attachment of *S. mutans* grown on either collagen- or fibronectin-immobilized surfaces was significantly reduced, anti-adhesive activity of CUR to tooth surface
Mizuki Sekiya et al. (2019) [65]	*S. mutans* ATCC 25175 cells, F-ATPases of *E. coli* and *S. mutans*	–	TSB	Inhibition ATPase activity of *S. mutans* (the effect of demethoxy analogue of CUR, was more potent than that of CUR), reduced growth of *S. mutans* at pH 5.3, reduced the number of CFUs at pH 4.3, inhibition bacterial growth and survival under acidic conditions

CUR curcumin, *S. mutans Streptococcus mutans*, *TSA* tryptic soy agar, *BHI* brain-heart infusion, *MSA* mitis-salivarius agar, *UTE* tryptone; 4-yeast extract, *TSB* tryptic soy broth

In four studies, SEM was used to assess the bacterial cell morphology and biofilm formation with CUR administration [3, 6, 41, 64]. These studies demonstrated that biofilm structure and formation seemed more disordered in the CUR groups [3, 6, 41] with an irregular cell wall structure, accommodated material of debris and fibers, and fewer cells in chains [64]. Basir et al. [8] applied VICKER'S micro-hardness tester to assess the surface hardness loss that happens during the dental caries process and concluded that CUR has a relatively high anticaries activity at low concentrations.

In three studies, CUR was compared to CHX [6, 11, 64]. The antibacterial characteristics of both substances were similar. Two studies showed greater antibacterial activity for CHX [6, 11]. One of the studies, which used Curcuma xanthorrhiza, showed more bactericidal activity of CUR than CHX [64]. Specifically, this used an ethyl-acetate fraction of the methanol extract of Javanese turmeric. The authors reported that the superior antibacterial effect of Curcuma xanthorrhiza might be the result of bacterial peptidoglycan layer damage [64]. Also, Mizuki Sekiya et al. [65] compared CUR with demethoxyCUR and concluded that an antibacterial effect of the demethoxy analogue of CUR was more potent than that of CUR. Pandit et al. [20] assessed different fractions of CUR including a methanol extract, and *n*-hexane, chloroform, ethyl acetate, and 70% aqueous methanol fractions. The study showed that the anti-acidogenic activity of the methanol extract and the fractions could be related to the concentration of curcuminoids. Hence, the chloroform fraction, which showed a strong anti-acidogenic effect, contained the highest content of curcuminoids [20]. It should be mentioned that the effects of different fractions and combinations of CUR should be assessed in future studies. Only one study showed that there was no zone of inhibition for CUR [10], most likely due to the small sample size of that study.

Finally, none of the studies reported any adverse effects of CUR, which is consistent with the putative safety of this phytochemical [75]. However, it should be used with caution until clinical trials have determined the range of side effects and safe dosage. The strength of the present systematic review was the use of a robust methodology. Efforts were made for reduction of bias, and screening, verification of eligibility, and the abstraction of information were performed in duplicate, and any disagreement was resolved by discussion or consultation with the third coauthor until a final consensus was achieved. There are also some limitations to the present review, arising mainly from the nature and the characteristics of the included studies. All studies were in vitro studies and not necessarily translatable to human studies. In addition, the results were derived after administration of CUR for short periods of time. CUR was administrated in experimental dosages, which may be different from those used in clinical studies. Therefore, well-designed clinical trial studies are needed to assess the effects of CUR and its precise clinical dosage on dental caries.

12.5 Conclusion

Within the limitations of this review, it can be concluded that CUR has anticaries and antibacterial effects on caries microorganisms in in vitro testing. These effects consisted of inhibition of *S. mutans* and biofilm growth, acid production, ATPase and aortase A activity, and a decrease in biomass, viability and biofilm metabolism, EPS production of biofilms and mRNA levels, and changes in biofilm structure and an anti-adhesion effect against *S. mutans*. For determining the safe dosage and mode of usage of CUR, clinical studies are warranted.

References

1. Ramakrishna Y, Goda H, Baliga MS, Munshi AK (2011) Decreasing cariogenic bacteria with a natural, alternative prevention therapy utilizing phytochemistry (plant extracts). J Clin Pediatr Dent 36(1):55–63
2. Petersen PE (2003) The World Oral Health Report 2003: continuous improvement of oral health in the 21st century—the approach of the WHO Global Oral Health Programme. Community Dent Oral Epidemiol 31:3–24
3. Li B, Li X, Lin H, Zhou Y (2018) Curcumin as a promising antibacterial agent: effects on metabolism and biofilm formation in S. mutans. Biomed Res Int 2018:4508709. https://doi.org/10.1155/2018/4508709
4. Caufield PW, Griffen AL (2000) Dental caries: an infectious and transmissible disease. Pediatr Clin N Am 47(5):1001–1019
5. Loesche WJ (1986) Role of Streptococcus mutans in human dental decay. Microbiol Rev 50(4): 353–380
6. Lee KH, Kim BS, Keum KS, Yu HH, Kim YH, Chang BS et al (2011) Essential oil of Curcuma longa inhibits Streptococcus mutans biofilm formation. J Food Sci 76(9):H226–H230
7. Shetty RN, Shetty SB, Janardhanan S, Shetty S, Shetty S, Raj K (2017) Comparative evaluation of effect of use of toothbrush with paste and munident on levels of Streptococcus mutans and gingival health in children: an in vivo study. J Indian Soc Pedod Prev Dent 35(2):162–166
8. Basir L, Kalhori S, Zare Javid A, Khaneh Masjedi M (2018) Anticaries activity of curcumin on decay process in human tooth enamel samples (in vitro study). J Natl Med Assoc 110(5):486–490
9. Samaranayake L (2002) Microbiology of dental caries. Essential microbiology for dentistry
10. Joycharat N, Limsuwan S, Subhadhirasakul S, Voravuthikunchai SP, Pratumwan S, Madahin I et al (2012) Anti-Streptococcus mutans efficacy of Thai herbal formula used as a remedy for dental caries. Pharm Biol 50(8):941–947
11. Elgamily H, Safy R, Makharita R (2019) Influence of medicinal plant extracts on the growth of oral pathogens Streptococcus mutans and Lactobacillus acidophilus: an in-vitro study. Open Access Maced J Med Sci 7(14):2328–2334
12. Miller WD (1902) The presence of bacterial plaques on the surface of teeth and their significance. Dent Cosmos 44:425–446
13. Bowen W, Koo H (2011) Biology of Streptococcus mutans-derived glucosyltransferases: role in extracellular matrix formation of cariogenic biofilms. Caries Res 45(1):69–86
14. Smith E, Spatafora G (2012) Gene regulation in S. mutans: complex control in a complex environment. J Dent Res 91(2):133–141

15. Rawashdeh RY, Malkawi HI, Al-Hiyasat AS, Hammad MM (2008) A fast and sensitive molecular detection of Streptococcus mutans and Actinomyces viscosus from dental plaques. Jordan J Biol Sci 1(3)
16. Wiater A, Choma A, Szczodrak J (1999) Insoluble glucans synthesized by cariogenic streptococci: a structural study. J Basic Microbiol 39(4):265–273
17. Selwitz RH, Ismail AI, Pitts NB (2007) Dental caries. Lancet 369(9555):51–59
18. Lemos JA, Quivey RG Jr, Koo H, Abranches J (2013) Streptococcus mutans: a new Gram-positive paradigm? Microbiology 159.(Pt 3:436–445
19. Song J, Choi B, Jin EJ, Yoon Y, Choi KH (2012) Curcumin suppresses Streptococcus mutans adherence to human tooth surfaces and extracellular matrix proteins. Eur J Clin Microbiol Infect Dis 31(7):1347–1352
20. Pandit S, Kim HJ, Kim JE, Jeon JG (2011) Separation of an effective fraction from turmeric against Streptococcus mutans biofilms by the comparison of curcuminoid content and anti-acidogenic activity. Food Chem 126(4):1565–1570
21. Sheen S, Addy M (2003) An in vitro evaluation of the availability of cetylpyridinium chloride and chlorhexidine in some commercially available mouthrinse products. Br Dent J 194(4):207–210
22. Neturi RS, Srinivas R, Vikram Simha B, Sandhya Sree Y, Chandra Shekar T, Kumar S (2014) Effects of green tea on Streptococcus mutans counts-a randomised control trail. J Clin Diagn Res 8(11):ZC128–ZC130
23. Van der Weijden G, Timmer C, Timmerman M, Reijerse E, Mantel M, Van der Velden U (1998) The effect of herbal extracts in an experimental mouthrinse on established plaque and gingivitis. J Clin Periodontol 25(5):399–403
24. Jeng JH, Hsieh C, Lan W, Chang M, Lin S, Hahn L, Kuo M (1998) Cytotoxicity of sodium fluoride on human oral mucosal fibroblasts and its mechanisms. Cell Biol Toxicol 14(6):383–389
25. Ciancio SG (2007) Improving our patients' oral health: the role of a triclosan/copolymer/fluoride dentifrice. Compend Contin Educ Dent 28(4):178–180. 182–173
26. Phan TN, Marquis RE (2006) Triclosan inhibition of membrane enzymes and glycolysis of Streptococcus mutans in suspensions and biofilms. Can J Microbiol 52(10):977–983
27. Goodson J, Tanner A (1992) Antibiotic resistance of the subgingival microbiota following local tetracycline therapy. Oral Microbiol Immunol 7(2):113–117
28. Davies J (1994) Inactivation of antibiotics and the dissemination of resistance genes. Science 264(5157):375–382
29. Preus HR, Lassen J, Aass AM, Ciancio SG (1995) Bacterial resistance following subgingival and systemic administration of minocycline. J Clin Periodontol 22(5):380–384
30. Gilbert P, Das J, Foley I (1997) Biofilm susceptibility to antimicrobials. Adv Dent Res 11(1):160–167
31. Tenover FC (2001) Development and spread of bacterial resistance to antimicrobial agents: an overview. Clin Infect Dis 33(Suppl 3):S108–S115
32. Roberts MC (2002) Antibiotic toxicity, interactions and resistance development. Periodontol 28(1):280–297
33. Knoll-Köhler E, Stiebel J (2002) Amine fluoride gel affects the viability and the generation of superoxide anions in human polymorphonuclear leukocytes: an in vitro study. Eur J Oral Sci 110(4):296–301
34. Lachenmeier DW (2008) Safety evaluation of topical applications of ethanol on the skin and inside the oral cavity. J Occup Med Toxicol 3(1):26. https://doi.org/10.1186/1745-6673-3-26
35. McCullough M, Farah C (2008) The role of alcohol in oral carcinogenesis with particular reference to alcohol-containing mouthwashes. Aust Dent J 53(4):302–305
36. Neumegen RA, Fernández-Alba AR, Chisti Y (2005) Toxicities of triclosan, phenol, and copper sulfate in activated sludge. Environ Toxicol: Int J 20(2):160–164
37. Chung J, Choo J, Lee M, Hwang JK (2006) Anticariogenic activity of macelignan isolated from Myristica fragrans (nutmeg) against Streptococcus mutans. Phytomedicine 13(4):261–266
38. Marc G, Araniciu C, Oniga SD, Vlase L, Pîrnău A, Duma M et al (2018) New N-(oxazolylmethyl)-thiazolidinedione active against Candida albicans biofilm: potential Als proteins inhibitors. Molecules 23(10):E2522. https://doi.org/10.3390/molecules23102522
39. Tichy J, Novak J (1998) Extraction, assay, and analysis of antimicrobials from plants with activity against dental pathogens (Streptococcus sp.). J Altern Complement Med 4(1):39–45
40. Badria FA, Zidan OA (2004) Natural products for dental caries prevention. J Med Food 7(3):381–384
41. Li X, Yin L, Ramage G, Li B, Tao Y, Zhi Q, Lin H, Zhou Y (2019) Assessing the impact of curcumin on dual-species biofilms formed by Streptococcus mutans and Candida albicans. Microbiology 8(12):e937. https://doi.org/10.1002/mbo3.937
42. Iranshahi M, Sahebkar A, Hosseini ST, Takasaki M, Konoshima T, Tokuda H (2010) Cancer chemopreventive activity of diversin from Ferula diversivittata in vitro and in vivo. Phytomedicine 17(3–4):269–273
43. Teymouri M, Pirro M, Johnston TP, Sahebkar A (2017) Curcumin as a multifaceted compound against human papilloma virus infection and cervical cancers: A review of chemistry, cellular, molecular, and preclinical features. BioFactors 43(3):331-346.
44. Shahabipour F, Caraglia M, Majeed M, Derosa G, Maffioli P, Sahebkar A (2017) Naturally occurring anti-cancer agents targeting EZH2. Cancer Lett 400:325–335. https://doi.org/10.1016/j.canlet.2017.03.020. Epub 2017 Mar 18. PMID: 28323035
45. Dubey N, Kumar R, Tripathi P (2004) Global promotion of herbal medicine: India's opportunity. Curr Sci 86(1):37–41
46. Momtazi AA, Derosa G, Maffioli P, Banach M, Sahebkar A (2016) Role of microRNAs in the thera-

peutic effects of curcumin in non-cancer diseases. Mol Diagn Ther 20(4):335–345
47. Cowan MM (1999) Plant products as antimicrobial agents. Clin Microbiol Rev 12(4):564–582
48. Kalemba D, Kunicka A (2003) Antibacterial and antifungal properties of essential oils. Curr Med Chem 10(10):813–829
49. Kim HS (2005) Do not put too much value on conventional medicines. J Ethnopharmacol 100(1–2):37–39
50. Esatbeyoglu T, Huebbe P, Ernst IM, Chin D, Wagner AE, Rimbach G (2012) Curcumin-from molecule to biological function. Angew Chem Int Ed 51(22):5308–5332
51. Liu Q, Loo WT, Sze S, Tong Y (2009) Curcumin inhibits cell proliferation of MDA-MB-231 and BT-483 breast cancer cells mediated by down-regulation of NFκB, cyclinD and MMP-1 transcription. Phytomedicine 16(10):916–922
52. Singla A, Kaur I (2003) Inhibitory effect of dibenzoylmethane on mutagenicity of food-derived heterocyclic amine mutagens. Phytomedicine 10(6–7):575–582
53. Ruby AJ, Kuttan G, Babu KD, Rajasekharan K, Kuttan R (1995) Anti-tumour and antioxidant activity of natural curcuminoids. Cancer Lett 94(1):79–83
54. Bagherniya M, Nobili V, Blesso CN, Sahebkar A (2018) Medicinal plants and bioactive natural compounds in the treatment of non-alcoholic fatty liver disease: a clinical review. Pharmacol Res 130:213–240
55. Abdollahi E, Momtazi AA, Johnston TP, Sahebkar A (2018) Therapeutic effects of curcumin in inflammatory and immune-mediated diseases: a nature-made jack-of-all-trades? J Cell Physiol 233(2):830–848
56. Mollazadeh H, Cicero AFG, Blesso CN, Pirro M, Majeed M, Sahebkar A (2019) Immune modulation by curcumin: the role of interleukin-10. Crit Rev Food Sci Nutr 59(1):89–101
57. Panahi Y, Ahmadi Y, Teymouri M, Johnston TP, Sahebkar A. Curcumin as a potential candidate for treating hyperlipidemia: A review of cellular and metabolic mechanisms. J Cell Physiol. 2018 Jan;233(1):141-152. https://doi.org/10.1002/jcp.25756. Epub 2017 Jun 6. PMID: 28012169.
58. Bagheri H, Ghasemi F, Barreto GE, Sathyapalan T, Jamialahmadi T, Sahebkar A. The effects of statins on microglial cells to protect against neurodegenerative disorders: A mechanistic review. Biofactors. 2020 May;46(3):309-325. https://doi.org/10.1002/biof.1597. Epub 2019 Dec 17. PMID: 31846136.
59. An BJ, Lee JY, Park TS, Pyeon JR, Bae HJ, Song M et al (2006) Antioxidant activity and whitening effect of extraction conditions in curcuma longa L. Korean J Med Crop Sci 14(3):168–172
60. Sun XY, Zheng YP, Lin DH, Zhang H, Zhao F, Yuan CS (2009) Potential anti-cancer activities of furanodiene, a sesquiterpene from Curcuma wenyujin. Am J Chinese Med 37(03):589–596
61. Liberati A, Altman DG, Tetzlaff J, Mulrow C, Gøtzsche PC, Ioannidis JP et al (2009) The PRISMA statement for reporting systematic reviews and meta-analyses of studies that evaluate health care interventions: explanation and elaboration. Ann Intern Med 151(4):W-65–W-94
62. Hu P, Huang P, Chen WM (2013) Curcumin inhibits the Sortase A activity of the Streptococcus mutans UA159. Appl Biochem Biotechnol 171(2):396–402
63. Hu P, Huang P, Chen MW (2013) Curcumin reduces Streptococcus mutans biofilm formation by inhibiting sortase A activity. Arch Oral Biol 58(10):1343–1348
64. Kim JE, Kim HE, Hwang JK, Lee HJ, Kwon HK, Kim BI (2008) Antibacterial characteristics of Curcuma xanthorrhiza extract on Streptococcus mutans biofilm. J Microbiol 46(2):228–232
65. Sekiya M, Izumisawa S, Iwamoto-Kihara A, Fan Y, Shimoyama Y, Sasaki M, Nakanishi-Matsui M (2019) Proton-pumping F-ATPase plays an important role in Streptococcus mutans under acidic conditions. Arch Biochem Biophys 666:46–51
66. Kanth MR, Prakash AR, Sreenath G, Reddy VS, Huldah S (2016) Efficacy of specific plant products on microorganisms causing dental caries. J Clin Diagn Res 10(12):ZM01–ZM03
67. Klein MI, Hwang G, Santos PH, Campanella OH, Koo H (2015) Streptococcus mutans-derived extracellular matrix in cariogenic oral biofilms. Front Cell Infect Microbiol 5:10. https://doi.org/10.3389/fcimb.2015.00010
68. Koo H, Falsetta M, Klein M (2013) The exopolysaccharide matrix: a virulence determinant of cariogenic biofilm. J Dent Res 92(12):1065–1073
69. Liu S, Qiu W, Zhang K, Zhou X, Ren B, He J et al (2017) Nicotine enhances interspecies relationship between Streptococcus mutans and Candida albicans. Biomed Res Int 2017:7953920. https://doi.org/10.1155/2017/7953920
70. Aoki H, Shiroza T, Hayakawa M, Sato S, Kuramitsu H (1986) Cloning of a Streptococcus mutans glucosyltransferase gene coding for insoluble glucan synthesis. Infect Immun 53(3):587–594
71. Matsumoto M, Minami T, Sasaki H, Sobue S, Hamada S, Ooshima T (1999) Inhibitory effects of oolong tea extract on caries–inducing properties of Mutans streptococci. Caries Res 33(6):441–445
72. Köhler B, Birkhed D, Olsson S (1995) Acid production by human strains of Streptococcus mutans and Streptococcus sobrinus. Caries Res 29(5):402–406
73. Bender GR, Sutton SV, Marquis RE (1986) Acid tolerance, proton permeabilities, and membrane ATPases of oral streptococci. Infect Immun 53(2):331–338
74. Lemos JA, Abranches J, Burne RA (2005) Responses of cariogenic streptococci to environmental stresses. Curr Issues Mol Biol 7(1):95–108
75. Soleimani V, Sahebkar A, Hosseinzadeh H (2018) Turmeric (Curcuma longa) and its major constituent (curcumin) as nontoxic and safe substances: Review. Phytother Res 32(6):985–995. https://doi.org/10.1002/ptr.6054. Epub 2018 Feb 26. PMID: 29480523

The Clinical Use of Curcumin for the Treatment of Recurrent Aphthous Stomatitis: A Systematic Review of Clinical Trials

13

Fateme Gharibpour, Omid Fakheran, Alireza Parvaneh, Farinaz Shirban, Mohammad Bagherniya, Thozhukat Sathyapalan, and Amirhossein Sahebkar

Abstract

Recurrent aphthous stomatitis (RAS) lesions are inflammatory painful oral ulcers with uncertain etiology. Curcumin acts as an effective anti-inflammatory and antibacterial agent in the treatment of various oral diseases. This systematic review aimed to assess the effects of curcumin on RAS. A systematic search of the medical databases, PubMed, Scopus, ISI, Science Direct, and Google Scholar was performed up to March 30, 2020, to identify clinical trials assessing the effect of curcumin on aphthous ulcers. Nine studies comprising of 469 participants met all criteria and were analyzed. Treatment with curcumin significantly reduced aphthous ulcer size (seven studies), pain intensity (eight studies), number of aphthous ulcers (three studies), erythematous halo (one study), and erythema and exudate of the aphthous (one study). In four studies, the effect

F. Gharibpour
Department of Orthodontics, School of Dentistry, Isfahan University of Medical Sciences, Isfahan, Iran

O. Fakheran
Dental Research Center, Department of Periodontics, Dental Research Institute, Isfahan University of Medical Sciences, Isfahan, Iran

A. Parvaneh
Department of Gastroenterology and Hepatology, Faculty of Medicine, Mashhad University of Medical Sciences, Mashhad, Iran

F. Shirban
Department of Orthodontics, Dental Research Center, Dental Research Institute, School of Dentistry, Isfahan University of Medical Sciences, Isfahan, Iran

M. Bagherniya
Food Security Research Center, Department of Community Nutrition, School of Nutrition and Food Science, Isfahan University of Medical Sciences, Isfahan, Iran

T. Sathyapalan
Department of Academic Diabetes, Endocrinology and Metabolism, Hull York Medical School, University of Hull, Hull, UK

A. Sahebkar (✉)
Biotechnology Research Center, Pharmaceutical Technology Institute, Mashhad University of Medical Sciences, Mashhad, Iran

Neurogenic Inflammation Research Center, Mashhad University of Medical Sciences, Mashhad, Poland

Polish Mother's Memorial Hospital Research Institute (PMMHRI), Lodz, Poland

Halal Research Center of IRI, FDA, Tehran, Iran
e-mail: sahebkara@mums.ac.ir

P. C. Guest (ed.), *Studies on Biomarkers and New Targets in Aging Research in Iran*, Advances in Experimental Medicine and Biology 1291, https://doi.org/10.1007/978-3-030-56153-6_13

of curcumin on aphthous ulcer was assessed in comparison to the effects of the standard medication, triamcinolone. In all of these studies, curcumin had similar beneficial effects on the aphthous ulcer as measured by ulcer size, number, and pain. Only three studies were categorized as high quality using the Jadad scale. Within the limitations of this review, it can be concluded that curcumin may have a beneficial role in the treatment of recurrent aphthous ulcers. However, more randomized clinical trials are needed to validate these findings.

Keywords

Curcumin · Turmeric · Recurrent aphthous stomatitis · Aphthous ulcer · Systematic review

13.1 Introduction

Recurrent aphthous stomatitis (RAS) is one of the most common oral mucosa lesions worldwide, affecting 5–25% of the population [1, 2]. It can occur in both men and women of all ages, as well as in all races and geographic regions [3, 4]. RAS is characterized by multiple recurrent small, round ulcers with erythematous haloes with a chronic and self-limiting course in most cases [2, 5]. Aphthous ulcers are usually painful, especially in the first 4–5 days and this impacts on eating, speaking, and quality of life [6]. Recurrent aphthous ulcers (RAUs) are classified in three subtypes as minor, major, and herpetiform, based on the clinical features and course of the lesions. Minor recurrent aphthous ulcer is the most common form of the disease, affecting around 80% of patients suffering from RAU [5].

The etiology of RAS has not been clearly defined. However, various factors have been reported to be associated with the occurrence of RAS. The most known predisposing factors are stress and anxiety, chemical products, microbial agents, endocrine alterations, food allergens, and genetic susceptibility [7, 8]. As the cause of the disease is not well defined, many therapeutic strategies have been assessed in an attempt to palliate the symptoms. The main targets in almost all of these treatments involve controlling the inflammation and reducing the pain.

Cucumin is a dietary polyphenol and the major identified bioactive ingredient of turmeric [9–13]. Furthermore, the efficacy of curcumin as a therapeutic agent in the management of various pathological conditions has been shown in modern dentistry and medicine [14, 15]. In the scientific literature, a wide spectrum of biological actions has been reported for curcumin [16–18]. These include anti-inflammatory [19–23], anti-allergic [24], neuroprotective [25, 26], anti-microbial [27], and antioxidant [28] activities.

In this systematic review, we have analyzed the efficacy of curcumin as a treatment for RAS, based on a systematic search of the current evidence. Moreover, we attempted to determine whether or not this evidence is applicable to clinical practice and suggest future research foci in this field.

13.2 Methods

We used the guidelines of the preferred reporting items for systematic reviews and meta-analyses (PRISMA) [29] to investigate the effects of curcumin therapy on aphthous ulcers in clinical trials.

13.2.1 Search Strategy

We conducted a systematic and comprehensive search of the online medical databases, PubMed, Scopus, ISI, Science Direct, and Google Scholar up to March 30, 2020, using the following search terms in titles and abstracts to identify the potential interest articles: "Stomatitis" OR "mouth ulcer" OR "aphthous" OR "aphthous stomatitis" OR "recurrent ulcerative stomatitis" OR "recurrent oral ulcer" AND "curcumin" OR "turmeric." The language of articles and time constraints were not considered in the search of these databases. Whenever possible, Medical Subject Headings (MESH) terms were used. Finally, we performed a manual search of the reference lists of the included articles to identify other relevant works for inclusion in the current review.

13.2.2 Study Selection

FG and MB independently reviewed the titles and abstracts of all papers found in the first stage of the search to assess these according to the set inclusion and exclusion criteria (see below). Any disagreements were discussed and a solution agreed upon.

13.2.3 Inclusion and Exclusion Criteria

Relevant articles were included if they:

(a) Examined the effects of curcumin or turmeric alone or in combination with the other herbs.
(b) Used a clinical trial design.
(c) Provided sufficient information in terms of the effects of curcumin or turmeric on the aphthous ulcer.

Articles were excluded if they were:

(a) Experimental studies.
(b) Duplicated records.
(c) Reviews, letters, editorial articles, or case reports.

13.2.4 Data Extraction

FG and MB extracted the following required data:

(a) First author's name.
(b) Publication year.
(c) Country of investigation.
(d) Study design.
(e) Sample size.
(f) Gender.
(g) Mean or range of age of subjects.
(h) Treatment (dosage and type of curcumin/turmeric).
(i) Duration of intervention.
(j) Results.
(k) Jadad score.

13.2.5 Quality Assessment

We used the Jadad scale to assess the quality of the studies. This scale provides an independent assessment of the methodological quality of a clinical trial according to the following items: randomization (0–2 points), blinding (0–2 points), and dropouts and withdrawals (0–1 point). Therefore, the overall score of a study assessed by this scale can range between 0 and 5, with higher scores indicating higher quality [30]. Studies with Jadad scores less than or equal to 2 are considered as low quality and those greater than or equal to 3 as high quality [31]. FG and MB independently assessed the quality of the studies using this scale.

13.3 Results

13.3.1 Search Results and Study Selection

The literature search yielded 948 published records, after exclusion of duplicated studies 938 remained. Due to the obvious irrelevant titles or abstracts, 928 references were excluded. This led to identification of ten articles which passed all criteria, although one of these was omitted due to inaccessible full text. This resulted in a final total of nine articles that were eligible for analysis (Fig. 13.1).

13.3.2 Characteristics of the Included Studies

The main characteristics of the trials included in this systematic review are summarized in Table 13.1. All of the included studies were conducted in Asian countries, namely India (four

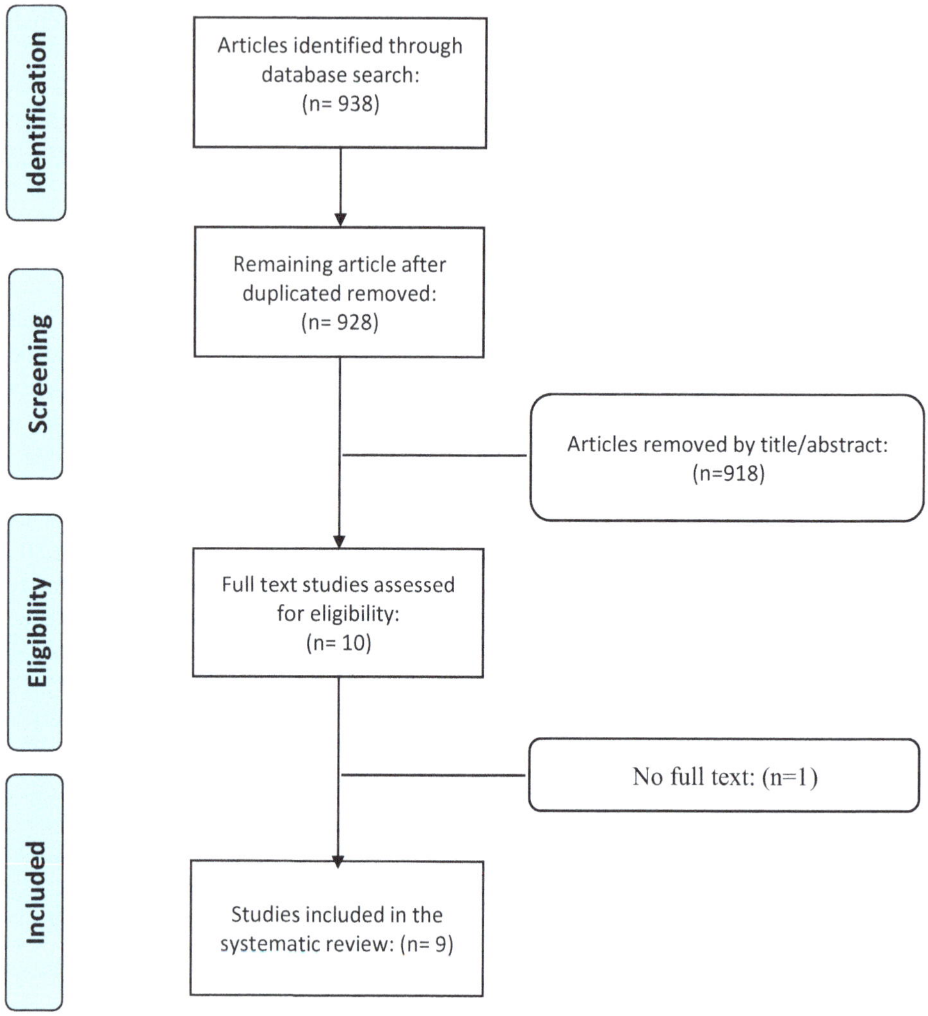

Fig. 13.1 Schematic showing the study selection procedure

articles) [32–35], Iran (two articles) [36, 37], Indonesia (one article) [38], Iraq (one article) [39], and Malaysia (one article) [40]. All of these studies were published between 2006 and 2020. A total of 469 participants were enrolled in the included studies with a sample size ranging from 16 to 105 participants. In four studies, the effect of curcumin on aphthous ulcer was assessed in comparison to the effects of triamcinolone [32, 35, 36, 40], two studies had no control group [33, 38], one study used placebo gel as the control [37], and two studies had more than two groups, using either two intervention (curcumin) and two control groups [39] or curcumin, honey, and Orabase (benzocaine, a topical anesthetic) groups [34]. All studies were conducted in both male and females except for one study, which did not include information on the gender of participants [40]. Six studies used curcumin (2–50%) [32, 34–37, 39], two applied turmeric [38, 40], and one used Haridra, which contained (*Curcuma longa*), Nimba Patra (*Azadirachta indica*), Yastimadhu (*Glycyrrhiza glabra*), and Neelkamal (*Nelumbo nucifera*) in equal quantities with sesame oil [33]. The duration of treatment varied between 3 and 14 days.

Table 13.1 The effects of curcumin on recurrent aphthous stomatitis

First author, Year, Country	Study design	Sample size	Male/female	Age (years; mean or range)	Treatment (dosage and type of curcumin/turmeric)	Study duration	Results	Adverse effects	Jadad Score
Nurdiana (2016), Indonesia [38]	Trial	16 patients with minor RAS	3/13	18–29	Turmeric extract gel (2%; (three times a day))	3 days	Significant reduction in erythematous halo, ulcer size, and pain intensity	Not reported	0
Al-Saffar (2006), Iraq [39]	Controlled trial	83 patients with minor RAS	25/58	28.41 ± 9.67	(A) Viscous solution of curcumin 10% (B) Viscous solution of curcumin 50% (C) Control (negative) (D) Glycerin treatment (control positive)	11 days	In comparison to both control groups (positive and negative controls), a significant % difference of complete healing (no pain and no inflammation) was found in both curcumin groups No difference was observed between two curcumin groups	Not reported	1
Lahankar (2013), India [33]	Trial	30 patients with mild-to-severe RAS	16/14	35.21 ± 14.97	Haridradi Tail "Gandoosh," every 4 h, oral administration, 10 mL twice a day Haridra: (*Curcuma longa*), Nimba patra (*Azadirachta indica*), Yastimadhu (*Glycyrrhiza glabra*), and Neelkamal (*Nelumbo nucifera*) in equal quantity and sesame oil	10 days	The trial preparation produced significant symptomatic relief and marked improvement in ulceration, present in buccal mucosal layer, burning sensation of palate, redness and erosion of oral cavity, difficulty in swallowing & chewing pungent things, enlargement of lymph nodes, etc. Ulcer size and numbers both significantly reduced	Not reported	0
Pandharipande, India (2019) [34]	RCT	105 patients with minor RAS	41/64	18–65	(A) Curcumin (B) Honey (C) Orabase (three times daily)	7 days	Size of ulcer, VAS score for pain, erythema, and exudate of aphthous ulcer significantly reduced in curcumin group compared with honey and Orabase groups	Not reported	2
Singh, India (2019) [35]	Clinical trial	40 RAS patients	20/20	Not mentioned	(A) Curcumin gel (B) Triamcinolone (three times daily)	7 days	VAS score for pain and number of ulcers reduced in both groups without significant differences between groups	Not reported	0

(continued)

Table 13.1 (continued)

First author, Year, Country	Study design	Sample size	Male/female	Age (years; mean or range)	Treatment (dosage and type of curcumin/turmeric)	Study duration	Results	Adverse effects	Jadad Score
Deshmukh, India (2014) [32]	RCT-double blind	60 RAS patients	31/29	32.51 ± 11.797	(A) Curcumin gel (1 g gel = 10 mg curcuma longa extract) (B) Triamcinolone 0.1% (three times daily)	7 days	Size, pain (VAS score), number, and duration of ulcers significantly reduced in both groups without significant differences between groups	Not reported	3
Halim, Malaysia (2013) [40]	Randomized clinical trial	20 RAS patients	Not mentioned	≥18 years	(A) Turmeric powder (B) Triamcinolone acetonide (0.1%), (5 min twice per day)	5 days	VAS score for pain and ulcer size reduced in both groups without significant differences between groups	Not reported	2
Manifar Iran, (2012) [37]	RCT—double blind, placebo	57 patients with minor aphthous	32/25	34.67 ± 9.57	(A) Curcumin gel (2% curcumin) (B) Placebo gel (Twice per day)	14 days	Curcumin gel significantly reduced pain intensity and size of aphthous ulcer compared to placebo	None	3
Kia, Iran (2020) [36]	Double-blind randomized clinical trial	58 patients with aphthous Stomatitis	36/22	44.38 ± 9.21	(A) Curcumin orabase 5% (B) Triamcinolone acetonide orabase 0.1% (three times daily)	10 days	(1) Average size of lesions and VAS score for pain dramatically decreased in both groups without significant differences between groups (2) No burning sensation, the abnormal change in the mucosa, or taste sense malfunction observed in the group which received 5% of curcumin	None	4

RAS Recurrent aphthous stomatitis, *VAS* Visual analogue scale, *RCT* randomized controlled trial

13.3.3 Quality Assessment

Five trials used randomization [32, 34, 36, 37, 40], but only one of these described the randomization method [36]. A blinded approach was used in four papers [32, 36, 37, 40], although none of the studies reported the blinding method. Five references reported the number and cause of participants who dropped out of the study [32, 34, 36, 37, 39]. Overall, three studies were categorized as high quality [32, 36, 37], and six studies had low methodological quality [33–35, 38–40].

13.3.4 Effects of Curcumin/Turmeric on RAS: Intragroup Comparison

The effects of curcumin/turmeric on aphthous ulcer size were assessed in seven studies. In all of these trials, treatment with curcumin or turmeric significantly reduced ulcer size [32–34, 36–38, 40]. All eight studies that evaluated the intensity of pain in patients found significant pain reduction in response to the use of curcumin/turmeric therapy [32, 34–40]. In three studies, the number of ulcers were measured and curcumin significantly reduced the number aphthous ulcers in all of these [32, 33, 35]. The use of turmeric extract gel significantly decreased erythematous halo in one study [38] and curcumin significantly reduced erythema, and exudate of aphthous in another trial [34]. In one study, it was shown that treatment with Haridra was associated with a relief of redness and erosion of the oral cavity, difficulty in swallowing, and in chewing pungent things and enlargement of lymph nodes [33]. Two studies reported measures of adverse effects, but none were found in any of these studies [36, 37].

13.3.5 Effects of Curcumin/Turmeric on RAS: Intergroup Comparison

In three studies, the effect of curcumin [32, 35, 36] and in one study the effect of turmeric [40] on aphthous ulcer was assessed in comparison to the effect of triamcinolone. In all of these studies, curcumin/turmeric had beneficial effects on the aphthous ulcer as measured by ulcer size, number, and pain compared with triamcinolone. In one study, the use of a curcumin gel led to a significant reduction in pain intensity and size of aphthous ulcers compared to a placebo gel [37]. In another study, two viscous solutions of curcumin (10% and 50%) were used in comparison with two control groups consisting of no treatment (negative control) and glycerin treatment (positive control). The results of this study indicated that both curcumin groups significantly reduced the percentage of complete healing with no pain or inflammation, compared to the control groups, and no significant differences were found between the two curcumin groups [39]. Finally, one study showed that curcumin had a greater beneficial effect than honey and benzocaine on ulcer size, pain, erythema, and exudate of aphthous ulcer [34].

13.4 Discussion

To the best of our knowledge, this is the first systematic review assessing the effects of curcumin or turmeric on RAS. The main findings suggest that these natural substances may offer a potential treatment strategy for this inflammatory condition. Due to the reduced quality of life and significant morbidity, an effective therapeutic strategy for managing RAS is necessary. Although RAS is a self-limiting disease, the high frequency of recurrence has been a bothersome factor for patients [19]. The treatment goal is the relief of symptoms, including the pain and ulcerations [4, 41, 42]. The pain results from the tissue distortion and is induced by inflammatory mediators such as bradykinin, serotonin, and prostaglandins [38]. Tumor necrosis factor alpha (TNF-α) as an inflammatory agent, contributes to the activation and recruitment of leukocyte in the RAS lesions [43]. Also, increased blood flow caused by vasodilation results in erythema in the injured area [38]. It has been shown that superinfection can increase healing time and worsen the pain in RAS lesions [44]. In addition, oxidative stress has been found to be associated with RAS [45].

In view of these findings, different anti-inflammatory drugs have traditionally been used for the treatment of RAS [36]. The corticosteroid triamcinolone is considered as the gold standard treatment, although this is contraindicated in pregnant and breastfeeding women, children, and older adults or patients with significant renal and liver pathologies [46]. In addition, long-term administration of corticosteroids is associated with adrenal suppression and hyperglycemia [47]. In this systematic review, some studies showed that curcumin had comparable efficacy to triamcinolone in the treatment of RAS [32, 35, 36, 40], and it had a greater beneficial effect than placebo [37], honey, and benzocaine [34].

The anti-inflammatory, antitumor, antibacterial, analgesic, and antioxidant properties of curcumin [48–51] are likely to be the causative factors in reducing healing time and alleviating ulcer-related pain in RAS. The anti-inflammatory activity of curcumin is mediated by effects on multiple pathways including downregulation of cytokines such as various interleukin proteins, TNF-α, and monocyte chemoattractant protein-1 (MCP-1) [52]. The review demonstrated that curcumin significantly reduced RAS ulcer size [32–34, 36–38, 40] and the number aphthous ulcers [32, 33, 35]. Curcumin may also accelerate the healing process by attenuation of matrix metalloproteinase (MMP)-9 activity and improvement of MMP-2 activity [53]. This mechanism has also been shown to involve increased cellular proliferation, collagen type 3 synthesis, and contraction and cross-linking of collagen in wound areas [40, 54]. Furthermore, the antibacterial effects of curcumin could inhibit bacterial growth around the ulcer and thereby prevent superinfection [32].

The results of this systematic review suggest that curcumin is a useful alternative in the treatment of RAS. However, several limitations should be acknowledged. First, the review only included a small number of studies and low overall numbers of participants. Second, the low quality of the included studies prevented us from drawing strong evidence-based conclusions. Third, the included studies were heterogeneous regarding the interventional approaches. Finally, all of the studies were conducted in Asian countries, making it difficult to generalize the findings to other populations. Considering the above limitations, more randomized clinical controlled trials with larger sample sizes and more standardized treatment regimes are needed to clarify the effects of curcumin and related compounds on aphthous stomatitis.

13.5 Conclusions

Within the limitations of this review, it can be concluded that curcumin and/or turmeric may have a beneficial role in the treatment of recurrent aphthous stomatitis. In all of the included studies, these natural compounds were found to be effective in the treatment of aphthous ulcers even in comparison to the gold standard treatment, triamcinolone. This was particularly apparent with respect to effects on ulcer size, number, and pain intensity. However, more carefully designed randomized clinical trials are needed to clarify these results.

References

1. Chavan M, Jain H, Diwan N, Khedkar S, Shete A, Durkar S (2012) Recurrent aphthous stomatitis: a review. J Oral Pathol Med 41(8):577–583
2. Preeti L, Magesh K, Rajkumar K, Karthik R (2011) Recurrent aphthous stomatitis. J Oral Maxillofac Pathol 15(3):252–256
3. Barrons RW (2001) Treatment strategies for recurrent oral aphthous ulcers. Am J Health Syst Pharm 58(1):41–50. quiz 51–43
4. Akintoye SO, Greenberg MS (2005) Recurrent aphthous stomatitis. Dent Clin N Am 49(1):31–47. vii–viii
5. Giannetti L, Murri Dello Diago A, Lo Muzio L (2018) Recurrent aphtous stomatitis. Minerva Stomatol 67(3):125–128
6. Wiriyakijja P, Fedele S (2017) Patient-reported outcome measures in recurrent aphthous stomatitis: a critical assessment of quality properties. Oral Dis 23(8):1168–1179
7. Saikaly SK, Saikaly TS, Saikaly LE (2018) Recurrent aphthous ulceration: a review of potential causes and novel treatments. J Dermatolog Treat 29(6):542–552

8. Bilodeau EA, Lalla RV (2019) Recurrent oral ulceration: etiology, classification, management, and diagnostic algorithm. Periodontol 80(1):49–60
9. Aggarwal BB, Sundaram C, Malani N, Ichikawa H (2007) Curcumin: the Indian solid gold. Adv Exp Med Biol 595:1–75
10. Abdollahi E, Momtazi AA, Johnston TP, Sahebkar A (2018) Therapeutic effects of curcumin in inflammatory and immune-mediated diseases: a nature-made jack-of-all-trades? J Cell Physiol 233(2):830–848
11. Momtazi AA, Derosa G, Maffioli P, Banach M, Sahebkar A (2016) Role of microRNAs in the therapeutic effects of curcumin in non-cancer diseases. Mol Diagn Ther 20(4):335–345
12. Momtazi AA, Sahebkar A (2016) Difluorinated curcumin: a promising curcumin analogue with improved anti-tumor activity and pharmacokinetic profile. Curr Pharm Des 22(28):4386–4397
13. Teymouri M, Pirro M, Johnston TP, Sahebkar A (2017) Curcumin as a multifaceted compound against human papilloma virus infection and cervical cancers: A review of chemistry, cellular, molecular, and preclinical features. BioFactors 43(3):331–346.
14. Lestari ML, Indrayanto G (2014) Curcumin. Profiles Drug Subst Excip Relat Methodol 39:113–204
15. Yeung AWK, Horbanczuk M, Tzvetkov NT (2019) Curcumin: total-scale analysis of the scientific literature. Molecules 24(7):E1393. https://doi.org/10.3390/molecules24071393
16. Pulido-Moran M, Moreno-Fernandez J, Ramirez-Tortosa C, Ramirez-Tortosa M (2016) Curcumin and health. Molecules 21(3):264. https://doi.org/10.3390/molecules21030264
17. Iranshahi M, Sahebkar A, Takasaki M, Konoshima T, Tokuda H (2009) Cancer chemopreventive activity of the prenylated coumarin, umbelliprenin, in vivo. Eur J Cancer Prev 18(5):412–415
18. Mohajeri M, Bianconi V, Ávila-Rodriguez MF, Barreto GE, Jamialahmadi T, Pirro M, Sahebkar A (2020) Curcumin: a phytochemical modulator of estrogens and androgens in tumors of the reproductive system. Pharmacol Res 156:104765. https://doi.org/10.1016/j.phrs.2020.104765. Epub 2020 Mar 23. PMID: 32217147.
19. Chainani-Wu N (2003) Safety and anti-inflammatory activity of curcumin: a component of tumeric (Curcuma longa). J Altern Complement Med 9(1):161–168
20. Kocaadam B, Sanlier N (2017) Curcumin, an active component of turmeric (Curcuma longa), and its effects on health. Crit Rev Food Sci Nutr 57(13):2889–2895
21. Bagherniya M, Nobili V, Blesso CN, Sahebkar A (2018) Medicinal plants and bioactive natural compounds in the treatment of non-alcoholic fatty liver disease: a clinical review. Pharmacol Res 130:213–240
22. Mollazadeh H, Cicero AFG, Blesso CN, Pirro M, Majeed M, Sahebkar A (2019) Immune modulation by curcumin: the role of interleukin-10. Crit Rev Food Sci Nutr 59(1):89–101
23. Sahebkar A (2010) Molecular mechanisms for curcumin benefits against ischemic injury. Fertil Steril 94(5):e75–e76
24. Altintoprak N, Kar M, Acar M, Berkoz M, Muluk NB, Cingi C (2016) Antioxidant activities of curcumin in allergic rhinitis. Eur Arch Otorhinolaryngol 273(11):3765–3773
25. Bagheri H, Ghasemi F, Barreto GE, Sathyapalan T, Jamialahmadi T, Sahebkar A (2020) The effects of statins on microglial cells to protect against neurodegenerative disorders: A mechanistic review. Biofactors 46(3):309–325. https://doi.org/10.1002/biof.1597. Epub 2019 Dec 17. PMID: 31846136
26. Bagheri H, Ghasemi F, Barreto GE, Rafiee R, Sathyapalan T, Sahebkar A (2020) Effects of curcumin on mitochondria in neurodegenerative diseases. Biofactors 46(1):5–20. https://doi.org/10.1002/biof.1566. Epub 2019 Oct 3. PMID: 31580521
27. Sharifi S, Fathi N (2020) Anti-microbial activity of curcumin nanoformulations: new trends and future perspectives. Phytother Res Mar 13. https://doi.org/10.1002/ptr.6658. [Epub ahead of print]
28. Llano S, Gomez S, Londono J, Restrepo A (2019) Antioxidant activity of curcuminoids. Phys Chem Chem Phys 21(7):3752–3760
29. Moher D, Liberati A, Tetzlaff J, Altman DG (2009) Preferred reporting items for systematic reviews and meta-analyses: the PRISMA statement. Ann Intern Med 151(4):264–269
30. Jadad AR, Moore RA, Carroll D, Jenkinson C, Reynolds DJM, Gavaghan DJ, McQuay HJ (1996) Assessing the quality of reports of randomized clinical trials: is blinding necessary? Control Clin Trials 17(1):1–12
31. Moher D, Cook D, Jadad A, Tugwell P, Moher M, Jones A et al (1999) Assessing the quality of reports of randomised trials: implications for the conduct of meta-analyses. Health Technol Assess 3(12):i–iv, 1–98
32. Deshmukh RA, Bagewadi AS (2014) Comparison of effectiveness of curcumin with triamcinolone acetonide in the gel form in treatment of minor recurrent aphthous stomatitis: a randomized clinical trial. Int J Pharm Investig 4(3):138–141
33. Lahankar M, Magar SV, Dwivedi AP (2013) Management of 'Mukhapaka' by 'Haridradi Tail'wsr to recurrent Aphthous ulcer. Int J Adv Ayurveda, Yoga, Unani, Siddha, Homeopathy 2(1):119–124
34. Pandharipande R, Chandak R, Sathawane R, Lanjekar A, Gaikwad R, Khandelwal V, Kurawar K (2019) To evaluate efficiency of curcumin and honey in patients with recurrent aphthous stomatitis: a randomized clinical controlled trial. Int J Res Rev 6(12):449–455
35. Singh H, Singh S, Singh N, Singh P, Sharma K (2018) Comparative analysis of therapeutic efficacy of curcumin & triamcinolone acetonide in recurrent aphthous stomatitis—a clinical study. J Adv Med Dent Sci Res 6(4):136–138
36. Kia SJ, Mansourian A, Basirat M, Akhavan M, Mohtasham-Amiri Z, Moosavi M-S (2020) New

concentration of curcumin orabase in recurrent aphthous stomatitis: a randomized, controlled clinical trial. J Herb Med:100336. https://doi.org/10.1016/j.hermed.2020.100336
37. Manifar S, Obwaller A, Gharehgozloo A, Boorboor Shirazi Kordi HR, Akhondzadeh S (2012) Curcumin gel in the treatment of minor aphthous ulcer: a randomized, placebo-controlled trial. J Med Plants 11(41):40–45
38. Nurdiana N, Krishnasamy S (2016) Effect of two percent turmeric extract gel on minor recurrent aphthous stomatitis. Padjadjaran J Dent 28(1). https://doi.org/10.24198/pjd.vol28no1.13503
39. Al–Saffar MT (2006) The therapeutic effect of viscous solution of curcumine in the treatment of recurrent aphthous stomatitis (RAS). Al-Rafidain Dent J 6(7):48–52
40. Halim DS, Khalik N, Taib H, Pohchi A, Hassan A, Alam MK (2013) Novel material in the treatment of minor oral recurrent aphthous stomatitis. Int Med J 20(3):392–394
41. Ship JA (1996) Recurrent aphthous stomatitis. An update. Oral Surg Oral Med Oral Pathol Oral Radiol Endod 81(2):141–147
42. Shashy RG, Ridley MB (2000) Aphthous ulcers: a difficult clinical entity. Am J Otolaryngol 21(6):389–393
43. Natah SS, Hayrinen-Immonen R, Hietanen J, Malmstrom M, Konttinen YT (2000) Immunolocalization of tumor necrosis factor-alpha expressing cells in recurrent aphthous ulcer lesions (RAU). J Oral Pathol Med 29(1):19–25
44. Lee W, Lee DG (2014) An antifungal mechanism of curcumin lies in membrane-targeted action within Candida albicans. IUBMB Life 66(11):780–785
45. Tugrul S, Kocyigit A, Dogan R, Eren SB, Senturk E, Ozturan O et al (2016) Total antioxidant status and oxidative stress in recurrent aphthous stomatitis. Int J Dermatol 55(3):e130–e135
46. Xiong C, Li Q, Lin M, Li X, Meng W, Wu Y et al (2009) The efficacy of topical intralesional BCG-PSN injection in the treatment of erosive oral lichen planus: a randomized controlled trial. J Oral Pathol Med 38(7):551–558
47. Lourenco SV, Hussein TP, Bologna SB, Sipahi AM, Nico MM (2010) Oral manifestations of inflammatory bowel disease: a review based on the observation of six cases. J Eur Acad Dermatol Venereol 24(2):204–207
48. Zheng LW, Hua H, Cheung LK (2011) Traditional Chinese medicine and oral diseases: today and tomorrow. Oral Dis 17(1):7–12
49. Cheng AL, Hsu CH, Lin JK, Hsu MM, Ho YF, Shen TS et al (2001) Phase I clinical trial of curcumin, a chemopreventive agent, in patients with high-risk or pre-malignant lesions. Anticancer Res 21(4B):2895–2900
50. Guimaraes MR, Coimbra LS, de Aquino SG, Spolidorio LC, Kirkwood KL, Rossa CJ (2011) Potent anti-inflammatory effects of systemically administered curcumin modulate periodontal disease in vivo. J Periodontal Res 46(2):269–279
51. Haukvik T, Bruzell E, Kristensen S, Tonnesen HH (2011) A screening of curcumin derivatives for antibacterial phototoxic effects studies on curcumin and curcuminoids. XLIII. Pharmazie 66(1):69–74
52. Aggarwal BB, Harikumar KB (2009) Potential therapeutic effects of curcumin, the anti-inflammatory agent, against neurodegenerative, cardiovascular, pulmonary, metabolic, autoimmune and neoplastic diseases. Int J Biochem Cell Biol 41(1):40–59
53. Swarnakar S, Ganguly K, Kundu P, Banerjee A, Maity P, Sharma AV (2005) Curcumin regulates expression and activity of matrix metalloproteinases 9 and 2 during prevention and healing of indomethacin-induced gastric ulcer. J Biol Chem 280(10):9409–9415
54. Panchatcharam M, Miriyala S, Gayathri VS, Suguna L (2006) Curcumin improves wound healing by modulating collagen and decreasing reactive oxygen species. Mol Cell Biochem 290(1–2):87–96

Antibacterial Activity of Curcumin Against Periodontal Pathogens: A Systematic Review

14

Omid Fakheran, Abbasali Khademi,
Mohammad Bagherniya,
Mandana Dehghannejad, Thozhukat Sathyapalan,
and Amirhossein Sahebkar

Abstract

Periodontitis is a chronic inflammatory disease characterized by destruction of the supporting structures of teeth caused by development of dental plaques and accumulation of microorganism around the gingival tissue. Curcumin has been shown to improve clinical parameters in periodontal diseases. However, the efficacy of curcumin in the elimination of periodontal pathogens is not clearly defined. The purpose of this study was to carry out a systematic review of the antibacterial activity of curcumin against periodontal pathogens. An electronic literature search in Medline, Scopus, Science Direct, Web of Science, Cochrane library, and Google scholar was performed up to February 29, 2020, to identify studies assessing the antibacterial activity of curcumin against periodontal pathogens. From 1238 publications, three clinical trials and five in vitro studies met the eligibility criteria. All three clinical studies reported improvement in restoring gingival health in clinical and microbiological parameters, following adjunctive use of curcumin for treatment of periodontitis. All five in vitro studies showed that curcumin could inhibit the growth of bacterial strains. Three of the five in vitro studies evaluated the effect of curcumin on mixed biofilm of periopathogens, which showed a significant inhibitory effect of

O. Fakheran
Dental Research Center, Department of Periodontics, Dental Research Institute, Isfahan University of Medical Sciences, Isfahan, Iran

A. Khademi
Dental Research Center, Department of Endodontics, Dental Research Institute, Isfahan University of Medical Sciences, Isfahan, Iran

M. Bagherniya
Food Security Research Center, Department of Community Nutrition, School of Nutrition and Food Science, Isfahan University of Medical Sciences, Isfahan, Iran

M. Dehghannejad
Dental Research Center, Dental Research Institute, Isfahan University of Medical Sciences, Isfahan, Iran

T. Sathyapalan
Department of Academic Diabetes, Endocrinology and Metabolism, Hull York Medical School, University of Hull, Hull, UK

A. Sahebkar (✉)
Biotechnology Research Center, Pharmaceutical Technology Institute, Mashhad University of Medical Sciences, Mashhad, Iran

Neurogenic Inflammation Research Center, Mashhad University of Medical Sciences, Mashhad, Iran

Polish Mother's Memorial Hospital Research Institute (PMMHRI), Lodz, Poland

Halal Research Center of IRI, FDA, Tehran, Iran
e-mail: sahebkara@mums.ac.ir

P. C. Guest (ed.), *Studies on Biomarkers and New Targets in Aging Research in Iran*, Advances in Experimental Medicine and Biology 1291, https://doi.org/10.1007/978-3-030-56153-6_14

curcumin on periodontal biofilms. This systematic review found that curcumin has antibacterial activity against periopathogens. The anti-biofilm activity of curcumin is reported as one of the mechanisms for this phenomenon. Curcumin could improve the clinical parameters of periodontal tissue not only by inhibition of the pathogens but also by modulating the host response.

Keywords

Curcumin · Periodontitis · Periopathogens · Systematic review

14.1 Introduction

Periodontitis is a chronic inflammatory disease characterized by the destruction of supporting structures of the teeth. The primary etiology of periodontitis is the development of dental plaques and accumulation of microorganism around the gingival tissue. The purpose of non-surgical periodontal therapy is to eliminate bacterial plaques and remove the factors that contribute to plaque accumulation [1]. The first line of management in periodontal therapy is scaling and root planing (SRP). SRP involves mechanical removal of supragingival and subgingival plaques and prepares the periodontal environment for tissue healing. However, removal of calculus and other pathogenic factors from deep periodontal pockets is not possible through SRP alone. Thus, deep pockets require adjunctive anti-microbial regimens to aid the mechanical debridement of tooth surfaces. In this regard, herbal plants are increasingly used as an alternative to systemic anti-microbial agents. Because of the relatively safe nature of herbal extracts, many such products and their components have been used in periodontics as topical agents. It has been shown that these herbal agents can significantly improve gingival health when used as an adjunct therapy in periodontitis [2].

Turmeric is a popular spice, derived from the rhizomes of the *Curcuma Longa* plant, and is frequently used in Indian foods and curries. Turmeric has been used in Indian systems of medicine as a "detoxifier of the body," for perhaps thousands of years [3, 4]. The application of turmeric for the treatment of various diseases has been mentioned in traditional Indian and Chinese historical medical documents. The most active component of turmeric and the one responsible for its vibrant yellow color is curcumin, which makes up 2–5% of the spice [5, 6]. In recent years, both pre-clinical and clinical studies have reported that curcumin is safe and has pharmacological activitiles incudig anti-oxidant, anti-microbial, and anti-inflammatory effects, which suggest that it could be used as a natural remedy in multiple disease states and conditions [7–14].

Previous investigations have demonstrated the anti-microbial activities of curcumin, although some in vivo studies reported less efficacious effects of this compound [15]. Curcumin has shown to be effective against a wide array of microbial pathogens ranging from *Candida albicans* to *Trichomonas vaginalis* and *Helicobacter pylori* [16–18]. Moreover, curcumin has been shown to have an inhibitory effect on biofilm formation [19].

The efficacy of curcumin in the treatment of periodontal disease has been investigated previously [20]. Almost all of the studies in this regard reported a significant beneficial effect of curcumin in improving various clinical parameters of this condition [21, 22]. However, the efficacy profile of curcumin in the elimination of periodontal pathogens has not been clearly defined. The purpose of this study was to systematically review the current evidence regarding the antibacterial activity of curcumin against various periodontal pathogens.

14.2 Methods

This systematic review was conducted following the PRISMA recommendations for the transparent reporting of systematic reviews and meta-analyses.

14.2.1 Focused Question

The aim of the current systematic review was to answer the specific question: does curcumin have antibacterial activity against periodontal pathogens?

14.2.2 Search Strategy

A search strategy was developed to collect all scientific papers. MEDLINE (PubMed), Scopus, The Cochrane Central Register of Controlled Trials (CENTRAL), Science Direct, Web of Science, and Google Scholar were systematically searched up to the January 31, 2020, without any restrictions on language or date of publication. The structured search strategy used key words as follows: (curcumin OR turmeric) AND (antimicrobial OR antibacterial OR antiplaque OR oral-pathogens OR periodontal pathogens) AND (periodontal OR gingival OR periodontal-disease OR gingivitis OR periodontitis OR periodontal-therapy OR periodontal-treatment OR periodontal-care OR plaque-control OR oral-hygiene OR periodontal surgery). The electronic search was complemented by a manual search in the reference lists of relevant papers.

14.2.3 Screening and Identification of Eligible Studies

A systematic selection of the articles was primarily carried out by two independent examiners based on the titles and abstracts. Full-text versions of selected papers were collected for further assessments. In this step, the reports that satisfied all of the inclusion criteria were used for data extraction.

14.2.4 Data Extraction

One investigator (OF) extracted the data, and two investigators (MD and MB) checked the data extraction independently for completeness and accuracy. Disagreements were resolved by discussion. The extracted data were classified into two tables based on the design of studies.

14.2.5 Quality Assessment of Studies

The Jadad scale was used to assess the quality of the included randomized controlled clinical studies. This scale has the following items: randomization (0–2 points), blinding (0–2 points), and dropouts and withdrawals (0–1 point). Therefore, the overall score of a study can range from 0 to 5, with higher scores indicative of better quality [23]. Studies with a Jadad score of ≤ 2 and ≥ 3 were considered as low- and high quality, respectively [24].

14.3 Results

14.3.1 Search

A total of 1238 publications were found in the search results of the six databases. By screening the titles and abstracts and removing duplicates, 19 papers were retrieved, for which full-text versions were obtained for detailed assessment. Manual examination of the reference lists of these 19 retrieved papers did not find any additional relevant papers. Finally, eight eligible articles were included in the current systematic review. More details of the data search are described in the flowchart (Fig. 14.1).

14.3.2 Study Characteristics

Among the eight included studies, three were randomized clinical studies, and the other five reports were based on in vitro investigations. The three clinical studies were conducted in India and exhibited substantial heterogeneity in terms of treatment protocol, follow up duration, and clinical dose of curcumin used for the treatment of periodontitis [25–27]. However, the clinical parameters for diagnosing periodontitis were similar in all three clinical studies and included plaque index (PI), sulcus bleeding index (SBI), probing pocket depth (PPD), clinical attachment

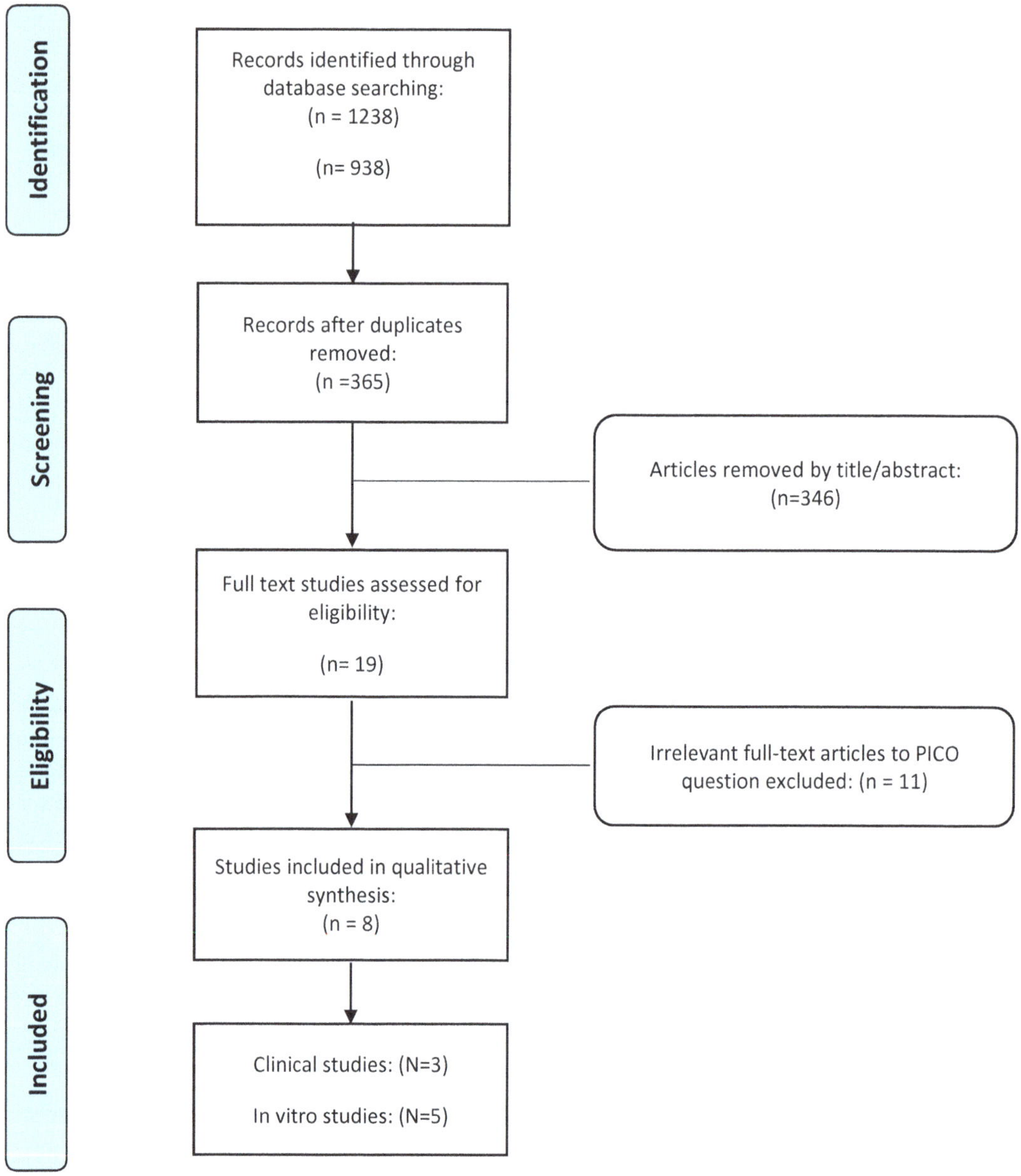

Fig. 14.1 Flowchart showing the study selection process

level (CAL), and gingival index (GI). The included clinical studies tested a variety of periodontal pathogens (Table 14.1). *Porphyromonas gingivalis* was the only common periopathogen in all the three studies. Based on the Jadad scale, one of the included articles was of high quality [25], and two were judged to be of low quality [26, 27].

The five in vitro studies also had substantial heterogeneity regarding the methodologies and targeted outcomes (Table 14.2). These studies used a variety of laboratory techniques including spectrofluorophotometric assay, disk diffusion method, and standardized broth microdilution assays.

Table 14.1 General characteristics of the clinical trial studies included (arranged alphabetically by first author's last name)

First author, year, country	Design	Sample	Treatment modalities	Randomization	Clinical Parameters	Microbiological parameters	Follow up	Jadad score
Bhatia (2014), India [26]	Split mouth clinical study	– 25 patients with chronic periodontitis – Female: 10 Male: 15 – Age range: 21–45 years	G1: SRP alone G2: SRP+ 1% Curcumin gel inserted deep into periodontal pockets on day 0	Yes	PI, SBI, PPD, CAL	*P. gingivalis, P. intermedia, Capnocytophaga, Fusobacterium nucleatum*	6 months	1
Nagasri (2015) [27]	Split mouth clinical study	– 30 patients with chronic periodontitis – Female: 18 Male: 12 – Age range: 35–60 years Mean age: 38 ± 4	G1: SRP alone G2: SRP + Curcumin gel inserted deep into the periodontal pockets on day 0	Yes	PI, GI, PPD, CAL	*P. gingivalis, T. forsythia, T. denticola*	4 weeks	2
Sreedhar (2015), India [25]	Split mouth clinical study	– 60 sites in 15 untreated chronic periodontitis patients – Female: 8 Male: 7 – Age range: 35–55 years	G1: SRP alone G2: SRP + Curcumin application for 5 min G3: SRP + curcumin + PDT (blue light emitting diode of wavelength 470 nm for 5 min) on 0 day G4: SRP + curcumin + PDT on day 0, 7 and 21	Yes	PI, SBI, PPD, CAL	Aa, *P. gingivalis, P. intermedia*	21 days	3

G treatment group, *SRP* scaling and root planing, *PDT* photodynamic therapy, *PI* plaque index, *SBI* sulcus bleeding index, *PPD* probing pocket depth, *CAL* clinical attachment level, *GI* gingival index, *Aa Aggregatibacter actinomycetemcomitans, P. gingivalis: Porphyromonas gingivalis, P. intermedia: Prevotella intermedia, T. forsythia: Tanerella forsythia, T. denticola: Treponema denticola*

Table 14.2 Summary of the in vitro studies included in the systematic review (arranged alphabetically by first author's last name)

Study	Method	Results	Conclusion
Izui (2016), Japan [28]	Inhibitory effect of curcumin on periodontopathic bacteria assessed using spectrofluorophotometric assay Analysis of biofilm formation was conducted using CLSM	**MIC (mg/mL) of curcumin against tested periodontopathic bacteria:** *A. actinomycetemcomitans* > 100 *F. nucleatum* = 10 *P. gingivalis* Type I FimA = 15 *P. gingivalis* Type II FimA = 10 *P. intermedia* = 10 *T. denticola* = 5 Curcumin inhibited RGP and KGP activities in dose-dependent manner ($P < 0.001$) in both *P. gingivalis* (type I and II) strains Curcumin inhibited *P. gingivalis* homotypic and heterotypic (with *Streptococcus gordonii*) biofilm formation in dose-dependent manner ($P < 0.001$)	Curcumin significantly inhibited growth of periodontopathic bacteria (*P. gingivalis, F. nucleatum, P. intermedia and T. denticola*) with exception of *A. actinomycetemcomitans* Curcumin suppressed *P. gingivalis* homotypic and *P. gingivalis—S. gordonii* biofilm formations Curcumin may be considered as a potent agent for prevention of periodontal diseases
Rudhra (2019), India [29]	1 g Curenext® 2% oral gel (Curcuma longa) dissolved in 100 mL water used for antibacterial assay Disk diffusion method with tetracycline as a positive control used for antibacterial assay	MIC and MBC of curcumin against *P. gingivalis* and *P. intermedia* was 0.5% and 1%, respectively Results statistically significant compared to positive control tetracycline ($P < 0.001$) when compared to negative control analyzed by one-way ANOVA	Curenext® 2% oral gel (Curcuma longa) can be used as a prophylactic and adjunctive natural agent for management of periodontal disease
Sha (2019), Iraq [30]	*P. gingivalis* clinically isolated strain cultured MIC and MBC determined by incubation of two-fold serial dilution broth media containing curcumin (100–0.05 μg/mL) for 48 h at 37 °C	The MIC and MBC values of curcumin against clinically isolated *P. gingivalis* were both 12.5 μg/mL	This study showed that clinically isolated *P. gingivalis* from chronic periodontitis patients is highly sensitive to curcumin at a low MIC and MBC (12.5 μg/mL)
Shahzad (2014), UK [31]	Effect of selected (poly) phenols (PPs), ubiquitous in plant foods, against *P. gingivalis*, *S. mitis*, *F. nucleatum* and Aa were assessed Antibacterial potential each compound evaluated by PMIC and PMBC using standardized broth microdilution assays	Curcumin was the most effective PPs for inhibition of planktonic growth of bacterial strains (PMIC range = 7.8–62.5 μg/mL) Among all PPs, only curcumin-treated biofilms displayed a significantly reduced metabolic activity The potential of curcumin to adsorb onto hydroxyapatite (HA), a key constituent of tooth enamel, was tested (after 1 h incubation, a significant adsorption was observed for curcumin ($P < 0.001$) compared with the control	Curcumin can affect metabolic activity and architecture of mature, multi-species, pathogenic oral biofilms. Curcumin also possess high adsorptive affinity for HA that can be employed in increasing the local (oral) availability of material

(continued)

Table 14.2 (continued)

Study	Method	Results	Conclusion
Singh (2018), India [32]	The study evaluated a new quantum curcumin (CurQDs) as an anti-microbial, anti-gingipains and anti-biofilm agent against *P. gingivalis* and select reference strains The study was conducted on 11 isolated *P. gingivalis* strains from patients with chronic periodontitis MIC was determined against 11 clinical isolates of *P. gingivalis* by micro-broth dilution	The MIC and MBC of curcumin against *P. gingivalis* was 1.114 μM Based on growth curves, CurQDs inhibited growth of *P. gingivalis* at a minimal concentration of 0.557 μM. At concentrations >8.913 μM, complete inhibition of the bacterial growth was manifested CurQDs inhibited both RGP and KGP of *P. gingivalis* in dose-dependent manner ($P < 0.001$) CurQDs dampened RGP and KGP by 58.2% and 51.4% respectively at 0.557 μM while at the 17.826 μM concentration 98.7% and 89.4% inhibition was noted Based on the analysis by CLSM, CurQDs corroded the mixed biofilm (*P. gingivalis, S. mutans, and A. viscosus*) in a dose-dependent manner ($P < 0.001$)	The results showed strong antibacterial activity of CurQDs against *P. gingivalis* even at significantly low concentrations and potent antibiofilm activities against mixed biofilms formed by *P. gingivalis, A. viscosus,* and *S. Mutans*

CLSM confocal laser-scanning microscopy, *MIC* minimum inhibitory concentration, *MBC* minimum bactericidal concentration, *RGP P. gingivalis* Arg-specific proteinase, *KGP P. gingivalis* Lys-specific proteinase, *S. mitis Streptococcus mitis*, *PMIC* planktonic minimum inhibitory concentration, *PMBC* planktonic minimum bactericidal concentration

14.3.3 Antibacterial Effects

All three clinical studies reported a marked improvement in restoring gingival health, as assessed by the various clinical and microbiological parameters, following adjunctive use of curcumin for treatment of periodontitis. In the study of Sreedhar et al., the effect of curcumin on growth inhibition of *Aggregatibacter actinomycetemcomitans*, *Porphyromonas gingivalis (P. gingivalis)*, and *Prevotella intermedia (P. intermedia)* was reported [25]. The result obtained from the investigation of Bhatia et al. showed a significant reduction in *P. gingivalis, P. intermedia, Capnocytophaga* spp., and *Fusobacterium nucleatum* [26]. The clinical study of Nagasri et al. showed a significant reduction in all microbiological parameters, including *P. gingivalis, Tanerella forsythia*, and *Treponema denticola* [27] (Table 14.3).

As mentioned above, the included in vitro investigations used a variety of methods to assess the antibacterial effects of curcumin against periodontal pathogens. All five in vitro studies showed a potential inhibitory effect of curcumin on the planktonic growth of bacterial strains and all reported the minimum inhibitory concentration (MIC) to achieve this effect [28–32]. Three of the in vitro studies evaluated the effect of curcumin on mixed biofilms of periopathogens and all showed a significant inhibitory effect [28, 31, 32]. Furthermore, the studies of Izui et al. and Singh et al. reported that curcumin inhibited both gingivalis arginine-specific proteinase (RGP) and gingivalis lysine-specific proteinase (KGP) activities in a dose-dependent manner ($P < 0.001$) in P. gingivalis strains [28, 32]. Moreover, Shahzad et al. demonstrated the potential of curcumin to adsorb onto hydroxyapatite (HA), which could be potentially beneficial for controlling periodontal pathogens in the management of periodontitis [31].

14.4 Discussion

Curcumin has stood out as a promising anti-inflammatory natural agent with potential application in the management of various periodontal diseases [33]. The data presented in this review demonstrated the anti-microbial efficacy of curcumin against major pathogenic bacteria which

Table 14.3 Main outcomes of the clinical trial studies included (arranged alphabetically by first author's last name)

First author, Year, country	Main outcomes	Conclusions
Bhatia (2014), India [26]	Highly significant reduction at 1, 3, and 6 months between G1 and G2 regarding all microbiological parameters ($P < 0.001$) including *P. gingivalis, P. intermedia, Capnocytophaga, and Fusobacterium nucleatum*	Curcumin gel used as local drug delivery as an adjunct to SRP showed marked improvement in restoring gingival health by improvement in clinical and microbiological parameters
Nagasri (2015), India [27]	Pre- and post-operative comparison in control group showed significant reduction in *P. gingivalis* ($P = 0.001$) and no reduction in T. forsythia ($P = 0.813$) and *T. denticola* ($P = 0.070$) Pre- and post-operative comparison in test group showed significant reduction in all microbiological parameters including *P. gingivalis* ($P = 0.001$), *T. forsythia* ($P = 0.043$) and *T. denticola* ($P = 0.001$)	The microbiological inhibitory effects of curcumin support its utility as a prophylactic and therapeutic agent for periodontitis
Sreedhar (2015), India [25]	Significant reduction in PI, SBI, PPD, and CAL in all treatment modalities ($P < 0.001$) Significant reduction in Aa, *P. gingivalis and P. intermedia* in all groups, except in G1 from baseline to 1-month postoperatively ($P < 0.001$). Significant reduction in Aa, *P. gingivalis and P. intermedia* at 3-month postoperative interval between G1 and G2. All other microbiological parameters were not significant ($P > 0.05$)	Adjunctive use of curcumin gel as a local sub-gingival drug with SRP has antibacterial action on periodontal pathogens such as Aa, *P. gingivalis*, and *P. intermedia*

have been implicated in the development of periodontitis. Three clinical studies and five in vitro investigations included in this review together described significant antibacterial effects of curcumin against different periopathogens such as *P. gingivalis, P. intermedia, A. actinomycetemcomitans, Fusobacterium nucleatum, T. denticola, T. forsythia,* and *Capnocytophaga.*

Efficacy of curcumin in the treatment of periodontal diseases has been shown in many studies and review articles. However, this is the first review article focusing on the antibacterial effect of curcumin in the management of periodontitis. In a recent study, researchers demonstrated the protective effect of curcumin in an experimental rat model of periodontitis [34]. They revealed that curcumin attenuated the production of interleukin (IL)-1β and tumor necrosis factor (TNF)-α in rat gingival fibroblasts stimulated by lipopolysaccharides (LPS), and inhibited the LPS-induced decrease in the osteoprotogerin/receptor activator of nuclear factor (NF-κB) receptor activator of NF-κB ligand [osteoprotegerin/soluble receptor activator of nuclear factor-κB (OPG/sRANKL)] ratio. Curcumin significantly reduced gingival inflammation and modulated collagen fiber and alveolar bone loss in vivo.

The therapeutic effect of curcumin has also been shown in some human clinical studies. Application of curcumin as an adjunct to nonsurgical periodontal therapy showed substantial improvement in clinical parameters and inflammatory criteria [35–37]. It is notable that the antimicrobial effect of curcumin against non-periodontal bacteria was demonstrated in various studies. For examples, the efficacy of curcumin in the inhibition of Streptococcus mutans has been shown in some experimental and clinical studies [38–40]. Furthermore, the antibacterial activity of curcumin against other microorganisms such as viruses, fungi, and parasites has been described in another review article [41].

The studies included in this review proposed a variety of parameters for the antibacterial activity of curcumin against periopathogens. The minimum inhibitory concentration (MIC) showed the capacity of curcumin to inhibit the growth of bacteria in planktonic forms. This antibacterial capacity was reported in all five of the in vitro studies included in this review. The main pathogenic activities of periopathogens in periodontal tissue occur in the form of mixed microbial biofilms [42]. We anticipated that curcumin would be effective in inhibiting such biofilms and such an effect was reported in the three in vitro studies of this review. The antibiofilm activity of curcumin in different colonies has also been reported in other studies [38, 43, 44].

Curcumin not only inhibits the growth of periopathogens directly but could also eliminate their pathogenesis by controlling their harmful products. This capacity has been shown in two of the included studies which reported an inhibitory activity of curcumin against the RGP and KGP enzymes [28, 32]. RGP and KGP are significant virulence factors of *P. gingivalis*, and therefore the inhibitory effects of curcumin on their activities may be an important means of controlling the disease. Also, *P. gingivalis* is one of the main pathogenic bacteria in developing periodontitis and has been named as "keystone pathogen" in this regard [45].

Although this is the first systematic review assessing the antibacterial activity of curcumin against periodontal pathogens, several limitations should be addressed. First, the small number of included studies, especially in the number of clinical trials, prevented us from drawing firm conclusions. Second, the heterogeneous nature of included studies in terms of interventional approaches, using different forms of curcumin with dissimilar doses and durations, made comparisons across studies difficult. Moreover, all of the clinical trial studies were conducted in one country, India, which makes it difficult to generalize these results to other populations. Therefore, more randomized controlled trials taking these limitations into account will be needed to clarify the effects of curcumin against periodontal pathogens.

14.5 Conclusion

The results of this review showed the extensive antibacterial activity of curcumin against periopathogens. The anti-biofilm activity and anti-gingipains activity of curcumin are some of the reported mechanisms for this phenomenon. Curcumin could improve the clinical parameters of periodontal tissue not only by inhibition of the pathogens but also by modulating the host response. The low numbers of studies included in this systematic review showed the importance of further research in this regard. Conducting more controlled clinical studies evaluating the long-term efficacy of curcumin and possible side effects are necessary.

References

1. Smiley CJ, Tracy SL, Abt E, Michalowicz BS, John MT, Gunsolley J et al (2015) Evidence-based clinical practice guideline on the nonsurgical treatment of chronic periodontitis by means of scaling and root planing with or without adjuncts. J Am Dent Assoc 146(7):525–535
2. Abdelmagyd HAE, Shetty SR, Al-Ahmari MMM (2019) Herbal medicine as adjunct in periodontal therapies: a review of clinical trials in past decade. J Oral Biol Craniofac Res 9(3):212–217
3. Aggarwal BB, Sundaram C, Malani N, Ichikawa H (2007) Curcumin: the Indian solid gold. Adv Exp Med Biol 595:1–75
4. Teymouri M, Pirro M, Johnston TP, Sahebkar A (2017) Curcumin as a multifaceted compound against human papilloma virus infection and cervical cancers: A review of chemistry, cellular, molecular, and preclinical features. Biofactors 43(3):331–346. https://doi.org/10.1002/biof.1344. Epub 2016 Nov 29. PMID: 27896883.
5. Abdollahi E, Momtazi AA, Johnston TP, Sahebkar A (2018) Therapeutic effects of curcumin in inflammatory and immune-mediated diseases: a nature-made jack-of-all-trades? J Cell Physiol 233(2):830–848
6. Mollazadeh H, Cicero AFG, Blesso CN, Pirro M, Majeed M, Sahebkar A (2019) Immune modulation by curcumin: the role of interleukin-10. Crit Rev Food Sci Nutr 59(1):89–101
7. Zorofchian Moghadamtousi S, Abdul Kadir H, Hassandarvish P, Tajik H, Abubakar S, Zandi K (2014) A review on antibacterial, antiviral, and antifungal activity of curcumin. Biomed Res Int 2014:186864. https://doi.org/10.1155/2014/186864
8. Gunes H, Gulen D, Mutlu R, Gumus A, Tas T, Topkaya AE (2016) Antibacterial effects of curcumin: an in vitro minimum inhibitory concentration study. Toxicol Ind Health 32(2):246–250
9. Bagherniya M, Nobili V, Blesso CN, Sahebkar A (2018) Medicinal plants and bioactive natural compounds in the treatment of non-alcoholic fatty liver disease: a clinical review. Pharmacol Res 130:213–240

10. Momtazi AA, Sahebkar A (2016) Difluorinated curcumin: a promising curcumin analogue with improved anti-tumor activity and pharmacokinetic profile. Curr Pharm Des 22(28):4386–4397
11. Soleimani V, Sahebkar A, Hosseinzadeh H (2018) Turmeric (Curcuma longa) and its major constituent (curcumin) as nontoxic and safe substances: Review. Phytother Res 32(6):985–995. https://doi.org/10.1002/ptr.6054. Epub 2018 Feb 26. PMID: 29480523
12. Hasanzadeh S, Read MI, Bland AR, Majeed M, Jamialahmadi T, Sahebkar A (2020) Curcumin: an inflammasome silencer. Pharmacol Res 159:104921. https://doi.org/10.1016/j.phrs.2020.104921. Epub 2020 May 25. PMID: 32464325
13. Bagheri H, Ghasemi F, Barreto GE, Rafiee R, Sathyapalan T, Sahebkar A (2020) Effects of curcumin on mitochondria in neurodegenerative diseases. Biofactors 46(1):5–20. https://doi.org/10.1002/biof.1566. Epub 2019 Oct 3. PMID: 31580521
14. Momtazi AA, Derosa G, Maffioli P, Banach M, Sahebkar A (2016) Role of microRNAs in the Therapeutic Effects of Curcumin in Non-Cancer Diseases. Mol Diagn Ther 20(4):335–45. https://doi.org/10.1007/s40291-016-0202-7. PMID: 27241179
15. Praditya D, Kirchhoff L, Brüning J, Rachmawati H, Steinmann J, Steinmann E (2019) Anti-infective properties of the golden spice curcumin. Front Microbiol 10:912. https://doi.org/10.3389/fmicb.2019.00912
16. Marini E, Di Giulio M, Magi G, Di Lodovico S, Cimarelli ME, Brenciani A et al (2018) Curcumin, an antibiotic resistance breaker against a multiresistant clinical isolate of *Mycobacterium abscessus*. Phytother Res 32(3):488–495
17. Jianhua W, Hai W (2009) Antifungal susceptibility analysis of berberine, baicalin, eugenol and curcumin on Candida albicans. J Med Colleges PLA 24(3):142–147
18. De R, Kundu P, Swarnakar S, Ramamurthy T, Chowdhury A, Nair GB et al (2009) Antimicrobial activity of curcumin against *Helicobacter pylori* isolates from India and during infections in mice. Antimicrob Agents Chemother 53(4):1592–1597
19. Singh AK, Prakash P, Singh R, Nandy N, Firdaus Z, Bansal M et al (2017) Curcumin quantum dots mediated degradation of bacterial biofilms. Front Microbiol 8:1517. https://doi.org/10.3389/fmicb.2017.01517
20. Harshan P, Fathima L, Manipal S (2019) Systematic review on effectiveness of curcumin in treatment of periodontitis. J Pharm Sci Rev Res 11(9):3174–3180
21. Bhuvaneswarri J, Valiathan M (2019) Clinical efficacy and anti-inflammatory property of curcumin in periodontal disease: a systematic review. Indian J Public Health Res Dev 10(12):1163–1166
22. Asteriou E, Gkoutzourelas A, Mavropoulos A, Katsiari C, Sakkas LI, Bogdanos DP (2018) Curcumin for the management of periodontitis and early ACPA-positive rheumatoid arthritis: killing two birds with one stone. Nutrients 10.(7). pii: E908. https://doi.org/10.3390/nu10070908
23. Jadad AR, Moore RA, Carroll D, Jenkinson C, Reynolds DJM, Gavaghan DJ et al (1996) Assessing the quality of reports of randomized clinical trials: is blinding necessary? Control Clin Trials 17(1):1–12
24. Moher D, Cook D, Jadad A, Tugwell P, Moher M, Jones A et al (1999) Assessing the quality of reports of randomised trials: implications for the conduct of meta-analyses. Health Technol Assess 3(12):i–iv., 1–98
25. Sreedhar A, Sarkar I, Rajan P, Pai J, Malagi S, Kamath V et al (2015) Comparative evaluation of the efficacy of curcumin gel with and without photo activation as an adjunct to scaling and root planing in the treatment of chronic periodontitis: a split mouth clinical and microbiological study. J Nat Sci Biol Med 6(Suppl 1):S102–S109
26. Bhatia M, Urolagin SS, Pentyala KB, Urolagin SB, KB M, Bhoi S (2014) Novel therapeutic approach for the treatment of periodontitis by curcumin. J Clin Diagn Res 8(12):ZC65–ZC69. https://doi.org/10.7860/JCDR/2014/8231.5343
27. Nagasri M, Madhulatha M, Musalaiah S, Kumar PA, Krishna CM, Kumar PM (2015) Efficacy of curcumin as an adjunct to scaling and root planning in chronic periodontitis patients: a clinical and microbiological study. J Pharm Bioallied Sci 7(Suppl 2):S554–S558
28. Izui S, Sekine S, Maeda K, Kuboniwa M, Takada A, Amano A et al (2016) Antibacterial activity of curcumin against periodontopathic bacteria. J Periodontol 87(1):83–90
29. Rudhra K, Ilango P, Chandran V, Abirami T, Vummidi AV, Mahalingam A (2020) An in vitro evaluation of curcumin oral gel antimicrobial efficacy against *Porphyromonas gingivalis* and *Prevotella intermedia*. Drug Invent Today 13(1):97–101
30. Sha AM, Garib BT (2019) Antibacterial effect of curcumin against clinically isolated *Porphyromonas gingivalis* and connective tissue reactions to curcumin gel in the subcutaneous tissue of rats. Biomed Res Int 2019:6810936. https://doi.org/10.1155/2019/6810936
31. Shahzad M, Millhouse E, Culshaw S, Edwards CA, Ramage G, Combet E (2015) Selected dietary (poly) phenols inhibit periodontal pathogen growth and biofilm formation. Food Funct 6(3):719–729
32. Singh AK, Yadav S, Sharma K, Firdaus Z, Aditi P, Neogi K et al (2018) Quantum curcumin mediated inhibition of gingipains and mixed-biofilm of *Porphyromonas gingivalis* causing chronic periodontitis. RSC Adv 8(70):40426–40445
33. Asteriou E, Gkoutzourelas A, Mavropoulos A, Katsiari C, Sakkas LI, Bogdanos DP (2018) Curcumin for the management of periodontitis and early ACPA-positive rheumatoid arthritis: killing two birds with one stone. Nutrients 10(7):E908. https://doi.org/10.3390/nu10070908

34. Livada R, Shiloah J, Tipton DA, Dabbous MK (2017) The potential role of curcumin in periodontal therapy: a review of the literature. J Int Acad Periodontol 19(3):70–79
35. Anusha D, Chaly PE, Junaid M, Nijesh JE, Shivashankar K, Sivasamy S (2019) Efficacy of a mouthwash containing essential oils and curcumin as an adjunct to nonsurgical periodontal therapy among rheumatoid arthritis patients with chronic periodontitis: a randomized controlled trial. Indian J Dent Res 30(4):506–511
36. Raghava KV, Sistla KP, Narayan SJ, Yadalam U, Bose A, Mitra K (2019) Efficacy of curcumin as an adjunct to scaling and root planing in chronic periodontitis patients: a randomized controlled clinical trial. J Contemp Dent Pract 20(7):842–846
37. Kaur H, Grover V, Malhotra R, Gupta M (2019) Evaluation of curcumin gel as adjunct to scaling & root planing in management of periodontitis- randomized clinical & biochemical investigation. Infect Disord Drug Targets 19(2):171–178
38. Li B, Li X, Lin H (2018) Curcumin as a promising antibacterial agent: effects on metabolism and biofilm formation in *S. mutans*. Biomed Res Int 2018:4508709. https://doi.org/10.1155/2018/4508709
39. Li X, Yin L, Ramage G, Li B, Tao Y, Zhi Q et al (2019) Assessing the impact of curcumin on dual-species biofilms formed by *Streptococcus mutans* and *Candida albicans*. Microbiol Open 8(12):e937. https://doi.org/10.1002/mbo3.937
40. Maghsoudi A, Yazdian F, Shahmoradi S, Ghaderi L, Hemati M, Amoabediny G (2017) Curcumin-loaded polysaccharide nanoparticles: optimization and anticariogenic activity against *Streptococcus mutans*. Mater Sci Eng C Mater Biol Appl 75:1259–1267
41. Moghadamtousi SZ, Kadir HA, Hassandarvish P, Tajik H, Abubakar S, Zandi K (2014) A review on antibacterial, antiviral, and antifungal activity of curcumin. Biomed Res Int 2014:186864. https://doi.org/10.1155/2014/186864
42. Larsen T, Fiehn NE (2017) Dental biofilm infections—an update. APMIS 125(4):376–384
43. Khezri A, Karimi A, Yazdian F, Jokar M, Mofradnia SR, Rashedi H et al (2018) Molecular dynamic of curcumin/chitosan interaction using a computational molecular approach: Emphasis on biofilm reduction. Int J Biol Macromol 114:972–978
44. Kali A, Bhuvaneshwar D, Charles PM, Seetha KS (2016) Antibacterial synergy of curcumin with antibiotics against biofilm producing clinical bacterial isolates. J Basic Clin Pharm 7(3):93–96
45. Hajishengallis G, Darveau RP, Curtis MA (2012) The keystone-pathogen hypothesis. Nat Rev Microbiol 10(10):717–725

The Clinical Use of Curcumin for the Treatment of Rheumatoid Arthritis: A Systematic Review of Clinical Trials

15

Mohammad Bagherniya, Mina Darand, Gholamreza Askari, Paul C. Guest, Thozhukat Sathyapalan, and Amirhossein Sahebkar

Abstract

Rheumatoid arthritis (RA) is a chronic inflammatory disease of the joints, which is prevalent in about 0.5–1.0% of the world population. Newer therapies for RA have only minimal efficacy in some cases and some adverse effects. Curcumin with anti-antioxidant, anti-inflammatory, and immunomodulatory properties might have beneficial effects on RA. We have carried out a systematic review with the main aim of estimating the effect of curcumin supplementation on RA. A systematic search of the medical databases, PubMed, Scopus, ISI, and Google Scholar was performed up to March 21, 2020 to identify clinical trials assessing the effect of turmeric or curcumin on RA. Six studies, comprising 259 patients with RA of 6–12 weeks duration, were included. Disease activity was assessed using 28 joints (DAS-28), visual analog scale (VAS), and American College of Rheumatology (ACR-20) scores. Treatment with curcumin significantly reduced DAS-28 scores in four out of five studies and VAS scores for pain in all three studies and significantly increased ACR-20

M. Bagherniya · G. Askari
Department of Community Nutrition, School of Nutrition and Food Science, Food Security Research Center, Isfahan University of Medical Sciences, Isfahan, Iran

M. Darand
Students' Research Committee, Department of Community Nutrition, School of Nutrition and Food Science, Isfahan University of Medical Sciences, Isfahan, Iran

P. C. Guest
Laboratory of Neuroproteomics, Department of Biochemistry and Tissue Biology, Institute of Biology, University of Campinas (UNICAMP), Campinas, Brazil

T. Sathyapalan
Department of Academic Diabetes, Endocrinology and Metabolism, Hull York Medical School, University of Hull, Hull, UK

A. Sahebkar (✉)
Biotechnology Research Center, Pharmaceutical Technology Institute, Mashhad University of Medical Sciences, Mashhad, Iran

Neurogenic Inflammation Research Center, Mashhad University of Medical Sciences, Mashhad, Iran

Polish Mother's Memorial Hospital Research Institute (PMMHRI), Lodz, Poland

Halal Research Center of IRI, FDA, Tehran, Iran
e-mail: sahebkara@mums.ac.ir

P. C. Guest (ed.), *Studies on Biomarkers and New Targets in Aging Research in Iran*, Advances in Experimental Medicine and Biology 1291, https://doi.org/10.1007/978-3-030-56153-6_15

scores in all three studies in which it was measured. Erythrocyte sedimentation rate (ESR) and circulating C-reactive protein (CRP) were assessed in six and five studies, respectively, out of which four studies reported significant reductions in these parameters in response to curcumin treatment. Rheumatoid factor (RF) was significantly reduced after consumption of curcumin in all three relevant studies. None of the studies reported serious adverse effects with curcumin consumption. The present systematic review suggests that curcumin could be used as a safe agent to treat RA. Thus, further validation is justified.

Keywords

Curcumin · Arthritis · Rheumatoid arthritis · DAS-28 · Systematic review

15.1 Introduction

Rheumatoid arthritis (RA) is a chronic inflammatory joints disease that involves the fingers, wrists, and feet. The most important clinical manifestations of RA include joint pain and rigidity [1]. The global prevalence of RA is estimated to be about 0.5–1.0% and varies based on age, sex, race, and the population studied [2, 3]. It is estimated that medications used to treat RA cost about $20 billion worldwide and this has been projected to grow to $36 billion by the year 2021 [4].

The increasing prevalence of RA imposes enormous economic pressure on the healthcare systems and decreases the quality of life and ability of the affected individuals to do effective social activities [5, 6]. The pathogenesis of RA is complex and includes both genetic and environmental triggers, such as infection, smoking, alcohol consumption, and hormonal imbalances [7]. Multiple therapies to relieve pain and suppress the immune system in patients with RA include the use of various medications, joint injections, physiotherapy, rest, and surgery [8].

Unfortunately, some of these therapies, such as the use of immunosuppressants, have significant side effects on patients [9]. Hence, the use of various complementary and alternative medicine (CAM) therapies such as dietary approaches has become more favorable in the treatment of RA [10]. Medicinal plants are an important source of natural bioactive compounds, which can often deliver some therapeutic efficacy without serious side effects [11, 12].

Curcumin is one of the most studied hydrophobic polyphenols, derived mainly from *Curcuma longa* L. (turmeric). Curcumin has been widely and safely used in the treatment and prevention of various diseases [13, 14–17, 18]. The potential protective mechanisms of curcumin have been claimed to be related to antioxidant [19], anti-inflammatory [20, 21], anti-microbial [22], neuroprotective [23, 24], anti-tumor [25, 26, 27], and immunomodulatory [28] properties.

Since inflammation is one of the major pathogenic factors of RA, curcumin may be able to improve RA by suppressing pro-inflammatory signaling pathways such as those involving NF-kB pathways and pro-inflammatory cytokines [29]. Previous clinical trials have demonstrated that curcumin supplementation can alleviate the clinical manifestations of RA [30–32], although another study found that curcumin had no beneficial effect on RA patients [33].

To the best of our knowledge, there is no systematic review to assess the effects of curcumin on RA. Considering the numerous benefits of curcumin and the inconclusive evidence in this area, we conducted a systematic review of all the available randomized clinical trials to investigate this question.

15.2 Methods

This systematic review was designed and reported based on the guidelines of the preferred reporting items for systematic reviews and meta-analyses (PRISMA) [34].

15.2.1 Search Strategy

A systematic and comprehensive search of the online medical databases, PubMed, Scopus, ISI, and Google Scholar, was performed up to March 21, 2020, without any limitations for language and date. The following search terms were used to identify potential articles of interest in titles and abstracts: "curcumin" OR "curcuminoids" OR "Curcuma domestica" OR "Curcuma Longa" OR "Curcuma" OR "turmeric" AND "rheumatoid arthritis" OR "rheumatoid." Whenever possible, Medical Subject Headings (MESH) terms were used. Finally, a manual search of the reference lists of the included original papers and review articles was carried out to identify other relevant works for inclusion.

15.2.2 Study Selection

After the initial electronic search, the results were exported to End-Note software (v. X8; Thomson Reuters) and duplicate publications were removed. A careful, independent screening of the titles and abstracts was then conducted and irrelevant articles were removed. When required, the full-text version of remaining publications was screened by two independent researchers (MB and MD) and any disagreements were discussed and resolved. Finally, all human clinical trials that examined the effects of turmeric or curcumin supplementation on RA were included.

15.2.3 Inclusion and Exclusion Criteria

Relevant articles were included if they examined the effects of turmeric/curcumin alone or in combination with the other herbs, applied a clinical trial design, and provided sufficient information in terms of effects on RA. Articles were excluded if they were experimental studies, duplicate records, reviews, letters, editorial articles, or case reports.

15.2.4 Data Extraction

The following data were extracted from the full text of included studies using a predesigned abstraction form as follows:

(a) First author's last name, publication year, location of the study.
(b) Sample size.
(c) Inclusion criteria.
(d) Age.
(e) Study design.
(f) Intervention.
(g) Control.
(h) Study duration.
(i) Main results.
(j) Side effects.

Data extraction was conducted by MB.

15.2.5 Quality Assessment of Studies

We used the Jadad scale to assess the quality of the included studies. This scale accounts for the following items:

1. Randomization (0–2 points).
2. Blinding (0–2 points).
3. Dropouts and withdrawals (0–1 point).

According to this scale, the overall score of a study can range between 0 and 5, with higher scores indicating better quality [35]. Studies with scores less than or equal to 2 and greater than or equal to 3 were considered to be low and high quality, respectively [36].

15.3 Results

15.3.1 Search Results and Study Selection

After a systematic search of databases, a total of 727 references were identified. After omitting duplicated studies, 500 references

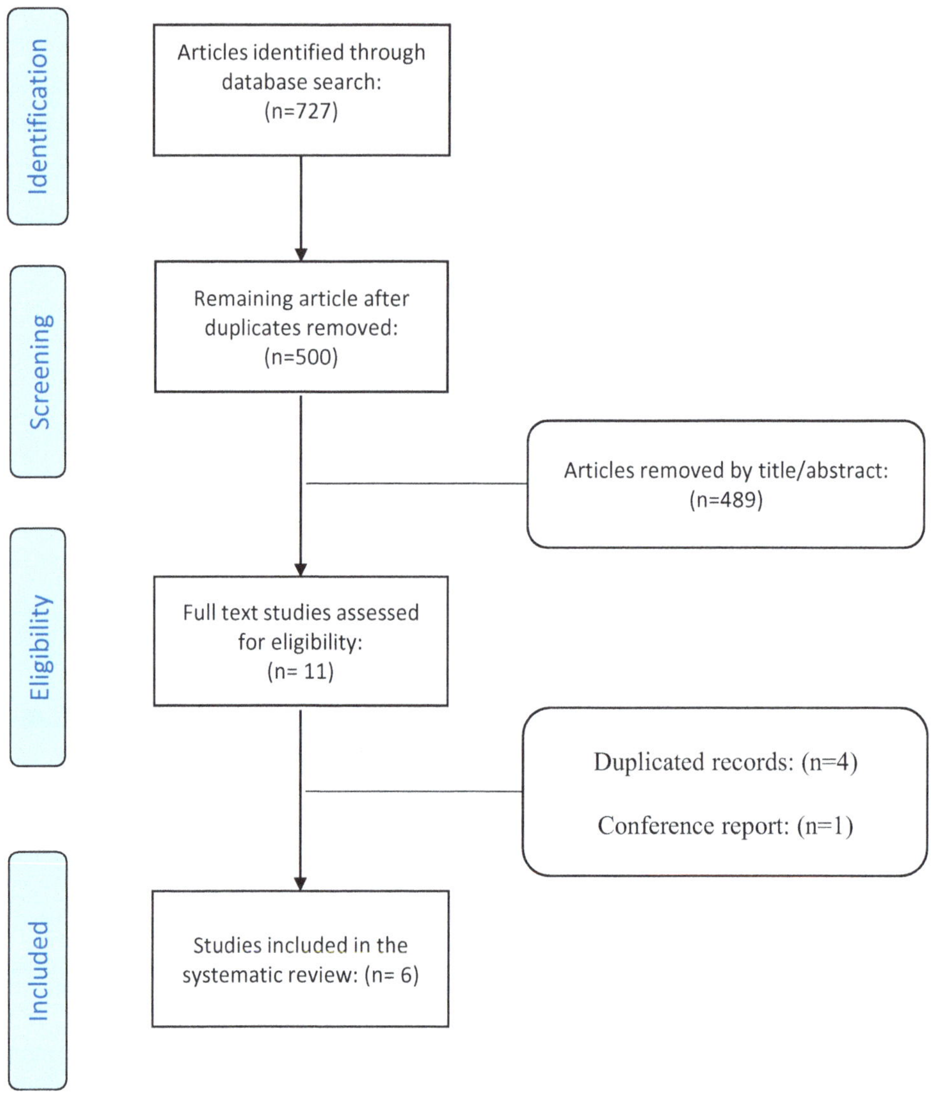

Fig. 15.1 Flowchart showing the study selection process

remained. After assessing titles and abstracts, 489 papers were deleted and leaving 11 articles. Full texts of these 11 articles were reviewed and five references were omitted as four contained duplicated data and one was a conference report (Fig. 15.1). This left six articles for a systematic review.

15.3.2 Characteristics of the Included Studies

The main characteristics of the included studies are summarized in Table 15.1. All of the studies were published between 2012 and 2019. Four studies were performed in India [29–31, 37] and two were conducted in Iran [32, 33]. The total

Table 15.1 Summary of the studies included in the systematic review

First author, year	Sample size (male/female)	Inclusion criteria	Age range (years) or mean	Study design	Intervention	Control	Duration	Main results	Side effects
Javadi (2019), Iran [33]	49 (5/44)	(1) Patients with RA according to ACR criteria (2) 20–80 years old	55.0 ± 2.9	RCT	Curcumin nanomicelle (40 mg) 3/day	Placebo (500 mg, wheat flour)	12 weeks	(1) No significant differences regarding DAS-28, TJC, and SJC in curcumin nanomicelle groups compared with placebo (2) No significant changes in ESR levels between the two groups of RA patients	None
Jacob (2019), India [30]	24 (12/12)	(1) 18–65 years old (2) Diagnosed with RA according to ACR criteria (with RA functional class or II)	18–65	RCT	Hydrogenated bioavailable curcumin CuroWhite: (A) low dose (250 mg) (B) high dose (500 mg)	Placebo	3 months	(1) Patients in both treatment groups (low and high doses) reported significant changes in their clinical symptoms compared with placebo (2) Significant changes in DAS-28, VAS, RF values, ACR response, TJC, and SJC for CuroWhite groups compared with placebo (3) ESR and CRP significantly decreased in CuroWhite groups compared with placebo	None
Anusha (2019), India [37]	45 (female)	(1) RA patients with chronic periodontitis (2) Diagnosed with RA according to ACR criteria	42.9 ± 6.9	RCT	Group A: scaling and root planing (SRP) with 0.2% chlorhexidine mouthwash as an adjunct Group B: SRP with mouthwash containing essential oils and curcumin (MEC) as an adjunct	Group C: SRP alone	6 weeks	Highest % mean reduction in plaque index and RA parameters (ESR, RF, CRP) from baseline to 6 weeks in group B followed by groups A and C	Not reported

(continued)

First author, year	Sample size (male/female)	Inclusion criteria	Age range (years) or mean	Study design	Intervention	Control	Duration	Main results	Side effects
Amalraj (2017), India [29]	36 (21/15)	(1) RA according to ACR criteria (with RA functional class of II) (2) DAS > 5.1 (3) Sum of SJC and TJC > 8	38.2 ± 8.5	RCT	(A) Low-dose curcumin (2 capsules 250 mg curcumin/day) (B) High-dose curcumin (2 capsules/day 500 mg)	Placebo (500 mg food grade starch)	3 months	(1) A significant reduction in DAS-28 and VAS scores in both curcumin groups compared with baseline and placebo (2) ACR-20 significantly increased and TJC and SJC significantly reduced in both curcumin groups, but no significant improvement in the placebo group (3) CRP, ESR, and RF values significantly reduced in both curcumin groups compared to the baseline and placebo	None
Hemmati (2016), Iran [32]	60 (no data reported on males and females)	RA according to ACR criteria	≥18	RCT	Curcumex (ginger, curcumin, and black pepper)	Placebo	8 weeks	(1) TJC, SJC, DAS-28, and ESR significantly reduced in the Curcumex group compared with placebo (2) CRP levels did not significantly change	Low

First author, year	Sample size (male/female)	Inclusion criteria	Age range (years) or mean	Study design	Intervention	Control	Duration	Main results	Side effects
Chandran, India, 2012 [31]	45 (7/38)	RA according to ACR and DAS > 5.1	47.8 ± 12.0	Pilot RCT	(A) Curcumin (500 mg) (B) Diclofenac sodium (50 mg) (C) Curcumin (500 mg) + diclofenac sodium (50 mg)	–	8 weeks	(1) All three treatments showed significant improvement in DAS-28 scores (no significant differences between groups) (2) VAS score significantly reduced in all three groups. Curcumin group showed highest reduction in VAS score from baseline (59.9%) (3) Although difference between treatment groups in the proportion of ACR20 responders was not significant, there was a marked trend toward ACR20 response rate in the curcumin group compared with the other two groups (4) TJC and SJC significantly decreased in all three groups (5) CRP showed a significant change only in the curcumin group (6) ESR did not significantly change in all three groups	Adverse events reported in the curcumin group (mild fever and throat infection)

RA rheumatoid arthritis, *ACR* American College of Rheumatology, *RCT* randomized control trial, *DAS-28* disease activity score of 28 joints, *TJC* tender joint count, *SJC* swollen joint count, *ESR* erythrocyte sedimentation rate, *VAS* visual analog scale, *RF* rheumatoid factor, *CRP* C-reactive protein

number of participants was 259 across all six studies. Participants were ≥ 18 years of age ranging from 38 to 55 years old. The duration of the studies was between 6 and 12 weeks. Five out of the six studies had a parallel design. One study used curcumin nanomicelle (120 mg/day) [33], one used bioavailable hydrogenated curcumin (250 and 500 mg/day) [30], two studies analyzed curcumin (500 and 1000 mg/day) [29, 31], one study used curcumin in a mouthwash (for treatment of patients with both RA and periodontal diseases) [37], and in the final study, curcumin was used in conjunction with other natural compounds [32]. One study was performed on females [37], in one study there were no data regarding the gender of participants [32], and the remainder were conducted on both genders [29–31, 33]. All of the studies used the American College of Rheumatology (ACR) criteria for the diagnosis of patients with RA. The six clinical papers exhibited substantial heterogeneity in terms of the treatment protocol, follow-up duration, the clinical dose of curcumin, and scales to assess the efficacy of treatment (Table 15.1).

15.3.3 Quality Assessment

All trials used randomization, and three articles described their randomization method [31, 33, 37]. Blindness was used in all studies, and two articles described the method of blinding [33, 37]. Four studies reported the number and cause of participants who dropped out of the studies and two did not address this issue [31, 32]. Five articles were categorized as being of high quality [29–31, 33, 37] and only one study was of low quality [32] (Table 15.2).

15.3.4 Effects of Curcumin on RA Disease Activity

15.3.4.1 Disease Activity Score of 28 Joints (DAS-28) Scale

The DAS-28 is a composite score derived from analysis of 28 joints for the following criteria: number of swollen joints; number of tender joints; erythrocyte sedimentation rate (ESR) or C reactive protein (CRP) levels in the blood; and the patient's own global assessment of health (from very good to very bad). In five studies, the DAS-28 scale was used as the main tool to assess the effects of curcumin in RA patients [29–33]. In all of these studies, treatment with curcumin was associated with a significant reduction in the score compared with baseline. Two of these studies used curcumin [29, 30] and one used Curcumex (consisting of ginger, curcumin, and black pepper) [32], both in comparison to placebo. One study tested three interventions (diclofenac, curcumin, and diclofenac + curcumin combination) which all had significant effects compared to baseline, but no placebo group was analyzed [31]. Finally, one study found significant effects of curcumin nanomicelle (120 mg/day) in comparison to baseline, although comparison with the placebo group did not show a significant difference [33].

15.3.4.2 Visual Analog Scale (VAS)

In the VAS, the patient is asked to place a mark on a 10 cm line indicating their level of pain, from 0 cm (no pain) to 10 cm (severe pain). Three of the six studies assessed VAS scores [29–31]. Two of these showed a significant decrease after curcumin administration, compared to the placebo [29, 30], and one study found a significant

Table 15.2 Assessment of the quality of the included studies based on the Jadad scale

First author, year, reference	Randomization	Blinding	An account of all patients	Total
Javadi (2019) [33]	2	2	1	5
Jacob (2019) [30]	1	1	1	3
Anusha (2019) [37]	2	2	1	5
Amalraj (2017) [29]	1	1	1	3
Hemmati (2016) [32]	1	1	0	2
Chandran (2012) [31]	2	1	0	3

decrease in the curcumin group compared to baseline (no placebo group) [31].

15.3.4.3 American College of Rheumatology (ACR-20) Scale

The ACR-20 uses a composite measure of 20% improvement in number of tender and number of swollen joints, and in three out of five other criteria: patient global assessment; physician global assessment; functional ability measure; VAS score; and ESR or CRP. Three studies assessed ACR-20 scores [29–31]. In two of these studies, the ACR-20 score was significantly increased in response to curcumin consumption compared with placebo [29, 30] and in the other study there was a trend toward an increase in the curcumin group compared with the other two groups (diclofenac and diclofenac + curcumin combination) although the difference between treatment groups was not significant and no placebo group was used [31]. However, the same study showed a significant reduction in total painful and swollen joints in all three intervention groups [31]. Two of the three studies with curcumin [29, 30] and the one study with Curcumex [32] also showed a significant reduction in total painful and swollen joints, compared to placebo, although one of the curcumin studies showed no effect compared to placebo [33]. In two studies, two different doses of bioavailable curcumin were used at low and high doses, although no significant differences were observed compared to placebo [29, 30].

15.3.4.4 ESR, CRP, and Rheumatoid Factor (RF)

All six studies assessed ESR as an important inflammatory factor related to RA. In two studies, there was a significant reduction in ESR after curcumin administration compared to placebo [29, 30]. Likewise, in the study which used Curcumex, ESR was significantly decreased compared to the placebo group [32]. One study, which used a mouthwash containing curcumin and essential oils, showed a greater reduction in ESR compared with the other treatment groups, although there was no placebo group [37]. In two studies, curcumin had no effect on ESR [31, 33]. CRP was assessed in five studies and in two of these, this biomarker was significantly reduced in the curcumin group compared to the placebo [29, 30]. In the study which used three intervention groups, CRP was significantly decreased only in the curcumin group [31]. In the study which used a mouthwash containing curcumin and essential oils, there was a greater reduction in CRP levels compared with the other treatment groups [37]. In the Curcumex study, CRP levels were not significantly changed [32]. Curcumin administration significantly reduced RF in the three studies which assessed this factor [29, 30, 37], although one of these studies had no placebo group [37].

15.3.5 Side Effects

In all studies, curcumin consumption was safe without any major side effects on RA patients. One study did not report any side effects of curcumin treatment on RA patients [37], three studies reported that consumption of curcumin had no significant adverse effects on patients [29, 30, 33], the Curcumex study reported minimum side effects [32], and one study reported mild fever and throat infection as a side effects of curcumin consumption [31].

15.4 Discussion

To the best of our knowledge, this is the first systematic review that assessed the effects of curcumin on RA patients. The analysis demonstrated that curcumin supplementation has a significant role in alleviating some symptoms of RA, as reflected by varying improvements in the assessed efficacy measures. These included positive effects on ACR-20, DAS-28, and VAS assessments, and on biomarker readouts such as RF, ESR, and CRP levels, with few or no side effects.

It has been shown that patients with RA have a high prevalence and incidence of cardiovascular risk factors including hypertension, diabetes mellitus, hyperlipidemia, and obesity [38]. In all of these diseases, systemic inflammation plays a significant role, which results in accelerated atherosclerosis. In addition, the rate of psychological problems such as depression and anxiety is high in RA patients [39, 40]. Combined with the pain and disease activity, this leads to a considerable reduction in the quality of life for these patients [41, 42]. Furthermore, there is no cure available to treat RA, and the aims of treatment protocols are to reduce pain and inflammation, to delay joint erosion, and to maintain some functionality [43]. Although new therapies have been introduced for RA such as disease-modifying anti-rheumatic drugs (DMARDs) and biological DMARDs, such therapies have had only minimal impact in some cases with some adverse effects [4, 44].

Recent reviews and meta-analyses of clinical trials have shown that curcumin has a beneficial effect on several chronic diseases such as non-alcoholic fatty liver diseases [45], metabolic syndrome [46], hyperlipidemia [47], hypertension [48], and diabetes mellitus [49]. Since systemic inflammation is involved in all of these chronic diseases, it has been suggested that the protective role of curcumin in preventing and treating these diseases could be explained by its anti-inflammatory and antioxidant properties [11]. Although the precise pathophysiological mechanisms of RA are still unclear, it is known that common pro-inflammatory cytokines such as interleukin (IL)-1, IL-6, and tumor necrosis factor-α (TNF-α) are produced within affected joints, where they play a substantial role in development and progression of the disease [4, 50, 51]. The rationale for testing curcumin as a potential treatment for RA includes the point that it is an effective inhibitor of such inflammatory factors [13, 52–55]. The results of the current study showed that treatment with curcumin had significant effects on both ESR and CRP in most of the studies. In one study, curcumin had a better effect on reducing CRP levels than the non-steroidal anti-inflammatory drug (NSAID), diclofenac [31]. Interestingly, in one study which found no significant effect of curcumin on ESR levels, no significant differences were observed regarding clinical symptoms of RA (as measured by DAS-28 score, and swollen/tender joint counts) between the intervention and placebo groups [33]. Together, these findings lend support to the case that the beneficial effects of curcumin on RA diseases are due to its anti-inflammatory properties. In all three studies in which it was assessed, curcumin treatment led to a significant reduction in RF [29, 30, 37]. This factor is an autoantibody found in the blood of patients with RA and other inflammatory diseases [56, 57]. The current finding that curcumin reduces the levels of RF is consistent with previous reports showing that it also has immunomodulatory properties [28, 58].

Results of the current review showed that curcumin has favorable effects on clinical symptoms in RA patients as measured by DAS-28, VAS, and ACR-20 scores. These results are in line with findings of the recent systematic reviews and meta-analyses, which showed the efficacy of curcumin on the treatment of osteoarthritis [12, 59]. The DAS-28 scale assesses 28 joints and determines the number of those which are tender or swollen, as well as ESR or CRP levels and general health [60]. Significantly reduced DAS-28 scores were found in four out of five studies which measured this parameter [29–32]. Likewise, VAS was significantly decreased in three studies in which this score for pain was assessed [29–31]. Moreover, in response to curcumin, ACR-20 score improved in all three studies which reported the use of this scale [29–31].

Inflammation plays a substantial role in both the initiation and persistence of pathological pain through activation of nociceptive sensory neurons [61, 62]. Thus, the anti-inflammatory effects of curcumin might also help to treat pain associated with RA. It has been proposed that curcumin might target various components of pain networks, as several preclinical studies have shown that it has favorable effects on chronic pain [63, 64]. Curcumin has even been found to attenuate neuropathic pain, which is one of the most difficult forms of pain to treat effectively [65–67].

One of the major drawbacks of curcumin is its poor absorption and bioavailability which limits its application in the treatment of various diseases [68]. To overcome this limitation, in the studies reviewed here, novel curcumin compounds, which have a higher solubility and bioavailability compared to curcumin, were used. One of these was BCM-95®, which is not only natural, but also has a six- to eightfold enhanced bioavailability [31]. The other curcumin-related compounds with higher bioavailability were a hydrogenated curcuminoid formulation (CuroWhite™) [30], curcuminoids in a turmeric matrix [29], and curcumin nanomicelles [33]. Therefore, the use of these might account for the efficacy of curcumin on the reduction of RA symptoms in these studies.

Although this is the first systematic review assessing the efficacy of curcumin on RA in clinical trials, several limitations should be considered in the interpretation of the results. First, the small number of included studies prevented us from drawing strong evidence-based conclusions. Second, the heterogeneous nature of the included studies in terms of the interventional approaches, the different forms of curcumin used at dissimilar dosages and durations, comparisons with and without placebo or with other therapeutic drugs made it difficult to compare findings across the different studies. Moreover, all of the studies were conducted in two countries, India and Iran, which makes it difficult to generalize the results to the other populations. Finally, the small number of total participants in these studies was also a negative point. Therefore, more randomized clinical controlled trials with larger sample sizes and standardized treatment groups will be needed to clarify the effects of curcumin on RA.

15.5 Conclusion

The results of this systematic review suggest that curcumin supplementation has promising effects on various clinical symptoms and inflammatory parameters of RA patients. Curcumin treatment improved DAS-28 score, VAS for pain, and ACR-20 scores and reduced ESR, CRP, and RF levels in patients with RA. Low numbers of participants reported side effects and there were no serious adverse effects following curcumin consumption. The limitations regarding small sample size, low participant numbers, and non-standardized protocols in this systematic review support the need for further investigations to clarify the effects of curcumin as an alternative or complementary medication to treat RA.

References

1. Firestein GS (2003) Evolving concepts of rheumatoid arthritis. Nature 423(6937):356–361
2. Symmons DP (2002) Epidemiology of rheumatoid arthritis: determinants of onset, persistence and outcome. Best Pract Res Clin Rheumatol 16(5):707–722
3. Symmons D, Mathers C, Pfleger B (2006) The global burden of rheumatoid arthritis in the year 2000. https://www.who.int/healthinfo/statistics/bod_rheumatoidarthritis.pdf
4. Oliveira RAD, Fierro IM (2018) New strategies for patenting biological medicines used in rheumatoid arthritis treatment. Expert Opin Ther Pat 28(8):635–646
5. Cooper N (2000) Economic burden of rheumatoid arthritis: a systematic review. Rheumatology (Oxford) 39(1):28–33
6. Haroon N, Aggarwal A, Lawrence A, Agarwal V, Misra R (2007) Impact of rheumatoid arthritis on quality of life. Mod Rheumatol 17(4):290–295
7. Paul B, Pariyapurath R (2018) Risk factor assessment of rheumatoid arthritis in North Kerala. Eur J Rheumatol 5(3):184–190
8. Bullock J, Rizvi SA, Saleh AM, Ahmed SS, Do DP, Ansari RA et al (2018) Rheumatoid arthritis: a brief overview of the treatment. Med Prin Pract 27(6):501–507
9. Quan L-D, Thiele GM, Tian J, Wang D (2008) The development of novel therapies for rheumatoid arthritis. Expert Opin Ther Pat 18(7):723–738
10. Efthimiou P, Kukar M (2010) Complementary and alternative medicine use in rheumatoid arthritis: proposed mechanism of action and efficacy of commonly used modalities. Rheumatol Int 30(5):571–586
11. Bagherniya M, Nobili V, Blesso CN, Sahebkar A (2018) Medicinal plants and bioactive natural compounds in the treatment of non-alcoholic fatty liver disease: a clinical review. Pharmacol Res 130:213–240
12. Fahira A, Soraya A, Achmadsyah A, Hakim RW (2019) Efficacy and safety of Curcuma longa extract in the treatment of osteoarthritis: a systematic review. Int J Appl Pharm 11:140–145
13. Abdollahi E, Momtazi AA, Johnston TP, Sahebkar A (2018) Therapeutic effects of curcumin in inflammatory and immune-mediated diseases: a nature-made jack-of-all-trades? J Cell Physiol 233(2):830–848
14. Pari L, Tewas D, Eckel J (2008) Role of curcumin in health and disease. Arch Physiol Biochem 114(2):127–149
15. Mollazadeh H, Cicero AFG, Blesso CN, Pirro M, Majeed M, Sahebkar A (2019) Immune modulation by curcumin: the role of interleukin-10. Crit Rev Food Sci Nutr 59(1):89–101
16. Teymouri M, Pirro M, Johnston TP, Sahebkar A (2017) Curcumin as a multifaceted compound against human papilloma virus infection and cervical cancers: A review of chemistry, cellular, molecular, and preclinical features. BioFactors 43(3):331–346

17. Momtazi AA, Derosa G, Maffioli P, Banach M, Sahebkar A (2016) Role of microRNAs in the therapeutic effects of curcumin in non-cancer diseases. Mol Diagn Ther 20(4):335–345
18. Soleimani V, Sahebkar A, Hosseinzadeh H (2018) Turmeric (Curcuma longa) and its major constituent (curcumin) as nontoxic and safe substances: Review. Phytother Res 32(6):985–995. https://doi.org/10.1002/ptr.6054. Epub 2018 Feb 26. PMID: 29480523
19. Sun YM, Wang RX, Yuan SL, Lin XJ, Liu CB (2004) Theoretical study on the antioxidant activity of curcumin. Chin J Chem 22(8):827–830
20. Srimal R, Dhawan B (1973) Pharmacology of diferuloyl methane (curcumin), a non-steroidal anti-inflammatory agent. J Pharm Pharmacol 25(6):447–452
21. Hasanzadeh S, Read MI, Bland AR, Majeed M, Jamialahmadi T, Sahebkar A (2020) Curcumin: an inflammasome silencer. Pharmacol Res 159:104921. https://doi.org/10.1016/j.phrs.2020.104921. Epub 2020 May 25. PMID: 32464325.
22. Fagbemi FJ, Ugoji E, Adenipekun T et al (2009) Evaluation of the antimicrobial properties of unripe banana (Musa sapientum L.), lemon grass (Cymbopogon citratus S.) and turmeric (Curcuma longa L.) on pathogens. Afr J Biotechnol 8(7):1176–1182
23. Forouzanfar F, Read MI, Barreto GE, Sahebkar A (2020) Neuroprotective effects of curcumin through autophagy modulation. IUBMB Life 72(4):652–664. https://doi.org/10.1002/iub.2209. Epub 2019 Dec 5. PMID: 31804772
24. Bagheri H, Ghasemi F, Barreto GE, Rafiee R, Sathyapalan T, Sahebkar A (2020) Effects of curcumin on mitochondria in neurodegenerative diseases. Biofactors 46(1):5–20. https://doi.org/10.1002/biof.1566. Epub 2019 Oct 3. PMID: 31580521
25. Iranshahi M, Sahebkar A, Hosseini ST, Takasaki M, Konoshima T, Tokuda H (2010) Cancer chemopreventive activity of diversin from Ferula diversivittata in vitro and in vivo. Phytomedicine 17(3–4):269–73. https://doi.org/10.1016/j.phymed.2009.05.020. Epub 2009 Jul 3. PMID: 19577457
26. Momtazi AA, Sahebkar A (2016) Difluorinated curcumin: a promising curcumin analogue with improved anti-tumor activity and pharmacokinetic profile. Curr Pharm Des 22(28):4386–4397
27. Mohajeri M, Bianconi V, Ávila-Rodriguez MF, Barreto GE, Jamialahmadi T, Pirro M, Sahebkar A (2020) Curcumin: a phytochemical modulator of estrogens and androgens in tumors of the reproductive system. Pharmacol Res 156:104765. https://doi.rg/10.1016/j.phrs.2020.104765. Epub 2020 Mar 23. PMID: 32217147
28. Srivastava RM, Singh S, Dubey SK, Misra K, Khar A (2011) Immunomodulatory and therapeutic activity of curcumin. Int Immunopharmacol 11(3):331–341
29. Amalraj A, Varma K, Jacob J, Divya C, Kunnumakkara AB, Stohs SJ et al (2017) A novel highly bioavailable curcumin formulation improves symptoms and diagnostic indicators in rheumatoid arthritis patients: a randomized, double-blind, placebo-controlled, two-dose, three-arm, and parallel-group study. J Med Food 20(10):1022–1030
30. Jacob J, Amalraj A, Raj KJ, Divya C, Kunnumakkara AB, Gopi S (2019) A novel bioavailable hydrogenated curcuminoids formulation (CuroWhite™) improves symptoms and diagnostic indicators in rheumatoid arthritis patients: a randomized, double blind and placebo controlled study. J Tradit Complement Med 9(4):346–352
31. Chandran B, Goel A (2012) A randomized, pilot study to assess the efficacy and safety of curcumin in patients with active rheumatoid arthritis. Phytother Res 26(11):1719–1725
32. Hemmati AA, Rajaee E, Houshmand G, Fakhroddin MA, Dargahi-MalAmir M, Hesam S et al (2016) Study the effects of anti-inflammatory curcumex capsules containing three plants (ginger, curcumin and black pepper) in patients with active rheumatoid arthritis. IIOAB J 7:389–392
33. Javadi M, Khadem Haghighian H, Goodarzy S, Abbasi M, Nassiri-Asl M (2019) Effect of curcumin nanomicelle on the clinical symptoms of patients with rheumatoid arthritis: a randomized, double-blind, controlled trial. Int J Rheum Dis 22(10):1857–1862
34. Moher D, Liberati A, Tetzlaff J, Altman DG (2009) Preferred reporting items for systematic reviews and meta-analyses: the PRISMA statement. Ann Intern Med 151(4):264–269
35. Jadad AR, Moore RA, Carroll D, Jenkinson C, Reynolds DJM, Gavaghan DJ et al (1996) Assessing the quality of reports of randomized clinical trials: is blinding necessary? Control Clin Trials 17(1):1–12
36. Moher D, Cook D, Jadad A, Tugwell P, Moher M, Jones A et al (1999) Assessing the quality of reports of randomised trials: implications for the conduct of meta-analyses. Health Technol Assess 3(12):i–iv. 1–98
37. Anusha D, Chaly P, Junaid M, Nijesh J, Shivashankar K, Sivasamy S (2019) Efficacy of a mouthwash containing essential oils and curcumin as an adjunct to nonsurgical periodontal therapy among rheumatoid arthritis patients with chronic periodontitis: a randomized controlled trial. Indian J Dent Res 30(4):506–511
38. Radner H, Lesperance T, Accortt NA, Solomon DH (2017) Incidence and prevalence of cardiovascular risk factors among patients with rheumatoid arthritis, psoriasis, or psoriatic arthritis. Arthritis Care Res (Hoboken) 69(10):1510–1518
39. Iaquinta M, McCrone S (2015) An integrative review of correlates and predictors of depression in patients with rheumatoid arthritis. Arch Psychiatr Nurs 29(5):265–278
40. Matcham F, Ali S, Irving K, Hotopf M, Chalder T (2016) Are depression and anxiety associated with disease activity in rheumatoid arthritis? A prospective study. BMC Musculoskelet Disord 17:155. https://doi.org/10.1186/s12891-016-1011-1
41. Rezaei F, Doost HTN, Molavi H, Abedi MR, Karimifar M (2014) Depression and pain in patients

with rheumatoid arthritis: mediating role of illness perception. Egypt Rheumatol 36(2):57–64
42. Tander B, Cengiz K, Alayli G, İlhanli İ, Canbaz S, Canturk F (2008) A comparative evaluation of health related quality of life and depression in patients with fibromyalgia syndrome and rheumatoid arthritis. Rheumatol Int 28(9):859–865
43. Deighton C, O'Mahony R, Tosh J, Turner C, Rudolf M (2009) Management of rheumatoid arthritis: summary of NICE guidance. BMJ 338:b702. https://doi.org/10.1136/bmj.b702
44. Guo Q, Wang Y, Xu D, Nossent J, Pavlos NJ, Xu J (2018) Rheumatoid arthritis: pathological mechanisms and modern pharmacologic therapies. Bone Res 6:1):1–1)14
45. Wei Z, Liu N, Tantai X, Xing X, Xiao C, Chen L et al (2019) The effects of curcumin on the metabolic parameters of non-alcoholic fatty liver disease: a meta-analysis of randomized controlled trials. Hepatol Int 13(3):302–313
46. Sahebkar A (2013) Why it is necessary to translate curcumin into clinical practice for the prevention and treatment of metabolic syndrome? BioFactors 39(2):197–208
47. Qin S, Huang L, Gong J, Shen S, Huang J, Ren H et al (2017) Efficacy and safety of turmeric and curcumin in lowering blood lipid levels in patients with cardiovascular risk factors: a meta-analysis of randomized controlled trials. Nutr J 16(1):68. https://doi.org/10.1186/s12937-017-0293-y
48. Hadi A, Pourmasoumi M, Ghaedi E, Sahebkar A (2019) The effect of Curcumin/Turmeric on blood pressure modulation: a systematic review and meta-analysis. Pharmacol Res 150:104505. https://doi.org/10.1016/j.phrs.2019.104505
49. Poolsup N, Suksomboon N, Kurnianta PDM, Deawjaroen K (2019) Effects of curcumin on glycemic control and lipid profile in prediabetes and type 2 diabetes mellitus: a systematic review and meta-analysis. PLoS One 14(4):e0215840. https://doi.org/10.1371/journal.pone.0215840
50. Libby P (2008) Role of inflammation in atherosclerosis associated with rheumatoid arthritis. Am J Med 121(10):S21–S31
51. McInnes IB, Schett G (2007) Cytokines in the pathogenesis of rheumatoid arthritis. Nat Rev Immunol 7(6):429–442
52. Sahebkar A, Cicero AF, Simental-Mendía LE, Aggarwal BB, Gupta SC (2016) Curcumin downregulates human tumor necrosis factor-α levels: a systematic review and meta-analysis of randomized controlled trials. Pharmacol Res 107:234–242
53. Sahebkar A (2014) Are curcuminoids effective C-reactive protein-lowering agents in clinical practice? Evidence from a meta-analysis. Phytother Res 28(5):633–642
54. Derosa G, Maffioli P, Simental-Mendia LE, Bo S, Sahebkar A (2016) Effect of curcumin on circulating interleukin-6 concentrations: a systematic review and meta-analysis of randomized controlled trials. Pharmacol Res 111:394–404
55. Ghandadi M, Sahebkar A (2017) Curcumin: an effective inhibitor of interleukin-6. Curr Pharm Des 23(6):921–931
56. Sokolove J, Johnson DS, Lahey LJ, Wagner CA, Cheng D, Thiele GM et al (2014) Rheumatoid factor as a potentiator of anti–citrullinated protein antibody–mediated inflammation in rheumatoid arthritis. Arthritis Rheum 66(4):813–821
57. Bovin LF, Rieneck K, Workman C, Nielsen H, Sørensen SF, Skjødt H et al (2004) Blood cell gene expression profiling in rheumatoid arthritis: discriminative genes and effect of rheumatoid factor. Immunol Lett 93(2–3):217–226
58. Yadav V, Mishra K, Singh D, Mehrotra S, Singh V (2005) Immunomodulatory effects of curcumin. Immunopharmacol Immunotoxicol 27(3):485–497
59. Daily JW, Yang M, Park S (2016) Efficacy of turmeric extracts and curcumin for alleviating the symptoms of joint arthritis: a systematic review and meta-analysis of randomized clinical trials. J Med Food 19(8):717–729
60. Fransen J, Stucki G, van Riel PL (2003) Rheumatoid arthritis measures: disease activity score (DAS), disease activity score-28 (DAS28), rapid assessment of disease activity in rheumatology (RADAR), and rheumatoid arthritis disease activity index (RADAI). Arthritis Care Res 49(S5):S214–S224
61. Zhang JM, An J (2007) Cytokines, inflammation, and pain. Int Anesthesiol Clin 45(2):27–37
62. Longoni M, Ferrarese C (2006) Inflammation and excitotoxicity: role in migraine pathogenesis. Neurol Sci 27(Suppl. 2):S107–S110
63. Mittal N, Joshi R, Hota D, Chakrabarti A (2009) Evaluation of antihyperalgesic effect of curcumin on formalin-induced orofacial pain in rat. Phytother Res 23(4):507–512
64. Banafshe HR, Hamidi GA, Noureddini M, Mirhashemi SM, Mokhtari R, Shoferpour M (2014) Effect of curcumin on diabetic peripheral neuropathic pain: possible involvement of opioid system. Eur J Pharmacol 723:202–206
65. Sharma S, Chopra K, Kulkarni SK (2007) Effect of insulin and its combination with resveratrol or curcumin in attenuation of diabetic neuropathic pain: participation of nitric oxide and TNF-alpha. Phytother Res 21(3):278–283
66. Sharma S, Kulkarni SK, Agrewala JN, Chopra K (2006) Curcumin attenuates thermal hyperalgesia in a diabetic mouse model of neuropathic pain. Eur J Pharmacol 536(3):256–261
67. Li Y, Zhang Y, Liu DB, Liu HY, Hou WG, Dong YS (2013) Curcumin attenuates diabetic neuropathic pain by downregulating TNF-α in a rat model. Int J Med Sci 10(4):377–381
68. Lopresti AL (2018) The problem of curcumin and its bioavailability: could its gastrointestinal influence contribute to its overall health-enhancing effects? Adv Nutr 9(1):41–50

A Systematic Review of the Clinical Use of Curcumin for the Treatment of Osteoarthritis

16

Nafiseh Shokri-Mashhadi, Mohammad Bagherniya, Gholamreza Askari, Thozhukat Sathyapalan, and Amirhossein Sahebkar

Abstract

Osteoarthritis is characterized by degeneration of joint structure over time, resulting in limitation of joint mobility. There is growing evidence that curcumin has anti-inflammatory properties and could be a potential therapeutic option for chronic inflammatory diseases. Hence, curcumin could potentially have a positive impact on osteoarthritis symptoms. This systematic review aimed to estimate the effects of curcumin on osteoarthritis. We systematically searched PubMed, ISI, Scopus, and Google Scholar up to March 4, 2020 to identify randomized controlled trials that evaluated the effects of consumption of all types of curcumin compounds in the treatment of osteoarthritis, especially in patients with knee osteoarthritis. Seventeen trials were identified. The duration of the included studies varied from 4 weeks to 8 months. Across all trials, 13 studies involved screening using Western Ontario and McMaster Universities (WOMAC) scores and 11 studies used visual analog scales (VAS) for recording pain from baseline to post-intervention. There was a significant improvement in VAS and overall WOMAC scores with oral administration of various types of curcumin formulations with no severe adverse effects. In conclusion, different types of curcumin compounds may be beneficial as an alternative or complementary agent for the management of osteoarthritis. Moreover, certain curcumin compounds with higher bioavailability tended to show more positive effects.

Keywords

Osteoarthritis · Arthritis · Curcumin · WOMAC · Systematic review

N. Shokri-Mashhadi
Department of Clinical Nutrition, School of Nutrition and Food Science, Food Security Research Center, Isfahan University of Medical Sciences, Isfahan, Iran

M. Bagherniya · G. Askari
Food Security Research Center, Department of Community Nutrition, School of Nutrition and Food Science, Isfahan University of Medical Sciences, Isfahan, Iran

T. Sathyapalan
Department of Academic Diabetes, Endocrinology and Metabolism, Hull York Medical School, University of Hull, Hull, UK

A. Sahebkar (✉)
Biotechnology Research Center, Pharmaceutical Technology Institute, Mashhad University of Medical Sciences, Mashhad, Iran

Neurogenic Inflammation Research Center, Mashhad University of Medical Sciences, Mashhad, Iran

Polish Mother's Memorial Hospital Research Institute (PMMHRI), Lodz, Poland

Halal Research Center of IRI, FDA, Tehran, Iran
e-mail: sahebkara@mums.ac.ir

P. C. Guest (ed.), *Studies on Biomarkers and New Targets in Aging Research in Iran*, Advances in Experimental Medicine and Biology 1291, https://doi.org/10.1007/978-3-030-56153-6_16

16.1 Introduction

Osteoarthritis is one of the most common types of arthritis and is characterized by degeneration of the joint structure [1]. This condition exacerbates over time and has been associated with knee pain and the progression of cartilage degeneration and leading to chronic disability [2]. The prevalence of osteoarthritis is increasing even in early age groups, resulting in considerable morbidity and a significant economic burden [3, 4]. Studies have shown that although the role of aging in the pathogenesis of osteoarthritis is significant, there are other underlying mechanisms that contribute to osteoarthritis [5, 6]. An increasing number of studies have shown that the release of inflammatory cytokines in OA activates several molecular pathways involved in the dysregulated metabolism of chondrocytes, the cells involved in cartilage formation and maintenance [7].

One of the main treatment strategies for pain relief and functional improvement in knee osteoarthritis is the use of anti-inflammatory drugs such as non-steroidal anti-inflammatory agents (NSAIDs), although these have some gastrointestinal side effects [8]. There is growing evidence that nutraceuticals as well as medicinal plants could potentially act as a complementary therapy for the management of many chronic inflammatory conditions to reduce the safety risks and side effects associated with NSAIDs [9–14]. Curcumin, commonly called diferuloylmethane, is a hydrophobic polyphenol derived from the rhizome of *Curcuma longa* (the culinary spice plant), which has a multitude of biological and pharmacological effects [15–20]. Furthermore, curcumin has been used as a health-promoting agent for many ailments due to its properties [21–25]. Biochemically, the curcuminoid compounds include curcumin, dimethoxy curcumin (DMC), bisdemethoxycurcumin (BDMC), and keto-enol tautomers [26]. In vivo and in vitro studies have suggested that curcuminoid formulations may have anti-apoptotic effects and inhibit or slow the catabolic actions of key inflammatory mediators in the early stages of the disease [27–31]. Also, it has been reported that they scavenge reactive oxygen and nitrogen species and may block inflammatory pathways that have been associated with the progression of knee osteoarthritis [29].

Several clinical trial studies have now been conducted to evaluate the effect of curcuminoids on osteoarthritis [32]. Although a recently conducted meta-analysis concluded that curcuminoid formulations could be potentially beneficial for OA patients, the sample size (5 studies) was not adequate and had significant heterogeneity. In addition, publication bias was not assessed. To our knowledge, there is no comprehensive review assessing the effectiveness of different curcumin compounds in the management of OA. Therefore, the aim of this study was to systematically review and summarize the anti-osteoarthritic effects of various curcumin compounds utilizing published randomized controlled trial (RCT) data in adults.

16.2 Materials and Methods

16.2.1 Eligibility Criteria

This study reviewed evidence from RCTs of children, adults, and older people that evaluated the effects of all types of curcumin compounds in the management of osteoarthritis, especially in knee osteoarthritis patients. As the main outcome, the therapeutic effect of curcumin was measured using the improvement of Western Ontario and McMaster Universities (WOMAC) scores and visual analog scales (VAS). As a secondary outcome, evaluation of changes in other pains and performance, as well as knee injury questionnaires has been considered. Additionally, any reports or data in selected studies on improvement in inflammatory biomarkers and adverse have been included.

16.2.2 Search Strategy

We conducted a systematic review of the literature according to the PRISMA guidelines in order to answer our objectives. A comprehensive search was performed on March 4, 2020, in which we searched the PubMed, ISI, Scopus, and Google Scholar databases, using keywords related to osteoarthritis and *Curcuma longa* without time restrictions. Only English articles were included. The following keywords were used in searches of all the databases: "curcumin" OR "curcuminoids" OR "Curcuma domestica" OR "Curcuma Longa" OR "curcuma" OR "turmeric" AND "osteoarthritis" OR "arthritis." Whenever possible, Medical Subject Headings (MESH) terms were used. Finally, a search was performed on the reference lists of the included original papers and review articles to identify other relevant works for inclusion.

16.2.3 Selection of Studies

RCTs were included that evaluated the effects of different types of curcumin and curcumin complexes in patients with osteoarthritis, in comparison to placebo or to patients who used standard analgesics treatments such as diclofenac, glucosamine, and ibuprofen. Studies with related titles were then collected and screened. Studies found in more than one database were excluded. After this, full paper manuscripts were studied, and irrelevant publications were removed.

16.2.4 Data Extraction

Relevant information was extracted from selected studies which included the following: (1) study design; (2) patient characteristics; (3) intervention regimens; (4) controls (placebo or standard treatment) regimens; (5) safety and adverse effects from treatment; and (6) changes in inflammatory marker levels if they reported. The main outcome of concern was the treatment efficacy of curcumin complex, classified as improvements in WOMAC or VAS scores, compared to the control groups.

16.2.5 Quality Assessment of Studies

The Jadad scale was used to assess the quality of the included studies. This scale considers the following items: randomization (0–2 points), blinding (0–2 points), and dropouts and withdrawals (0–1 point). According to this scale, the overall score of a study ranges between 0 and 5, with higher scores indicative of better quality [33]. Studies with Jadad scores of ≤ 2 and ≥ 3 were considered as low and high quality, respectively [34].

16.3 Results

16.3.1 Search Results

A total of 1251 studies were selected during the initial search, of which 817 were not duplicates. After reading the titles and abstracts, 784 references were omitted after which 33 articles remained. Full texts of these 33 articles were reviewed and 16 articles were omitted due to duplication of data ($n = 3$); curcumin was administrated in combination with other substances ($n = 9$); the source was a conference report ($n = 2$); the source was a study protocol ($n = 1$); or it was a non-English language article ($n = 1$). All included studies were RCTs. The results of the search are shown in Fig. 16.1.

16.3.2 Study Characteristics

The main characteristics of the trials included in this systematic review are summarized in Table 16.1. The total number of participants enrolled in the included studies was 1758, with sample sizes ranging from 25 to 367 participants and a mean age ranging from 38 to 80 years. Three studies reported different outcomes with the same population [35–37]. The most majority of studies were conducted on Asian populations and one study was from Belgium [38]. Follow-up analyses ranged from 4 weeks to 8 months. The treatment formulation assigned to the patients included lipid-based curcumin, curcuminoids, enhanced curcumin, and *Curcuma longa* extracts. The dosage of supplemental curcumin or lipid-based curcumin ranged from 40 mg to 200 mg per day [40–44]. Five studies used *Curcuma*

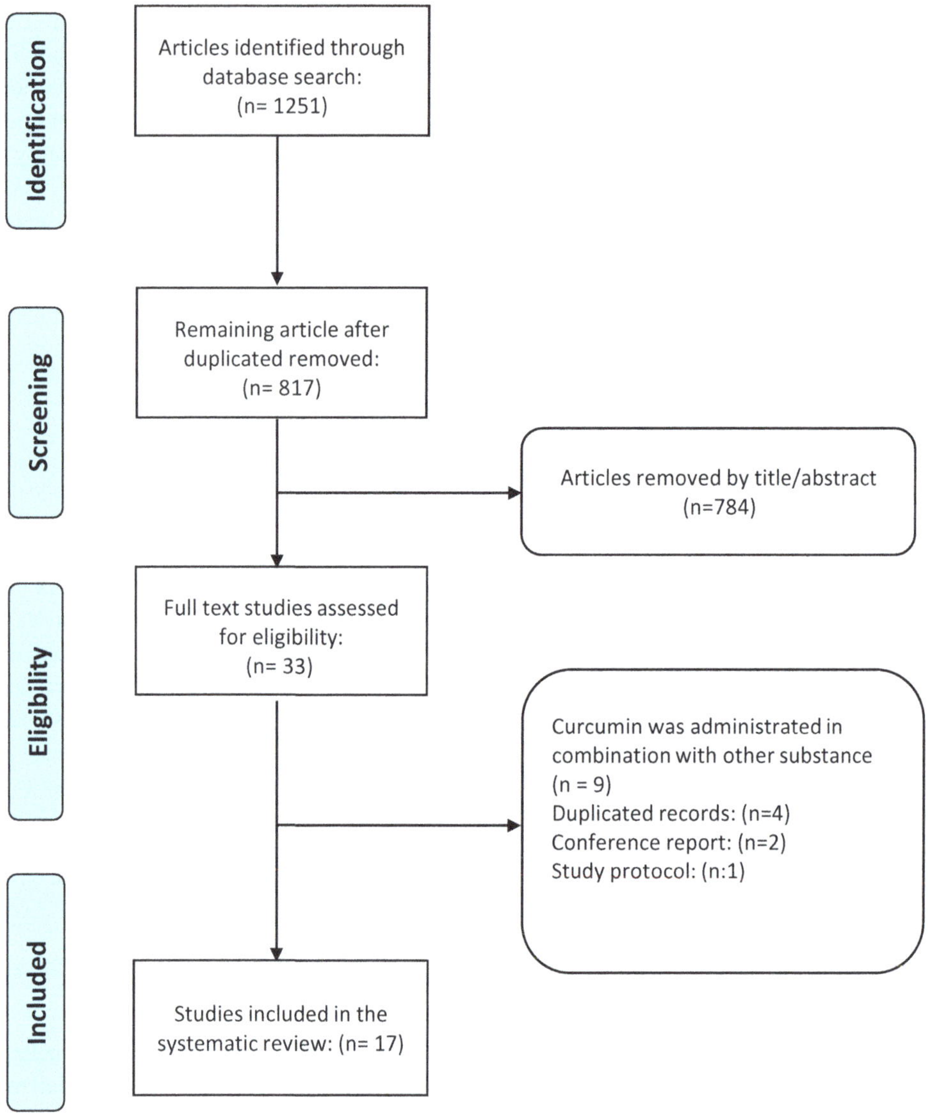

Fig. 16.1 Flow chart of the study selection process

longa extract supplements with a dose range of 0.5–1 g [38, 44–47]. Seven studies used pure curcumin at different dosages ranging from 0.5 to 2.5 g curcuminoids [35–37, 48–50]. Among the 17 randomized clinical trials, 12 compared the effectiveness of curcumin with placebo, two studies were a comparison with the usual treatments [39, 40], two compared curcumin with NSAIDs [44, 49], and one had no control group but had two intervention groups (curcumin, curcumin plus exercise) [51].v

16.3.3 WOMAC and VAS Outcomes

The main outcome for 11 out of 15 studies [35–37] was the WOMAC index [35, 39–42, 44–47, 50, 51], while 7 studies also measured components of the WOMAC assessment, including pain sensation, joint stiffness, and physical function [35, 39, 40, 44, 46, 47, 50]. In addition, the VAS tool was used by 10 studies to measure amounts of pain [35, 38, 41, 43, 45–49, 51]. Three studies also used the global Knee injury and Osteoarthritis

Table 16.1 Summary of the studies included in the systematic review (arranged alphabetically by first author's last name)

Author, year, country	Size	Inclusion criteria	Age Range (years) or mean	Study design	Intervention	Control	Duration	Main results	Jadad score
Belcaro et al. (2010), Italy [39]	100	Criteria for primary knee osteoarthritis (OA) (grade 1 or 2) according to American Rheumatism Association	In: 43.6 ± 5.5/ Co: 44.2 ± 6	Trial	Usual treatment+ Meriva (curcumin-phosphatidylcholine complex): Two 500-mg tablets daily, (1000 mg/day, corresponding to 200 mg curcumin/ day)	No placebo- only usual treatment	8 months	(1) Pain sensation, joint stiffness, and physical function improved significantly with Meriva, using Western Ontario and McMaster Universities (WOMAC) scores (pain, stiffness, and physical function were all positively affected by treatment) (2) Meriva significantly improved the Karnofsky Scale (from 73.3 at inclusion to 92.2 at study completion), with no significant improvements in controls (3) A significant improvement of mobility, evaluated by a treadmill was found in the intervention compared with the control group (4) sCD40L, IL-1β, IL-6, sVCAM-1, and ESR significantly decreased in the intervention group with no changes in the control group	1

(continued)

Table 16.1 (continued)

Author, year, country	Size	Inclusion criteria	Age Range (years) or mean	Study design	Intervention	Control	Duration	Main results	Jadad score
Belcaro et al. (2010), Italy [39]	50	Knee OA patients confirmed by X-ray	In: 44.4 ± 7.2/ Co: 45.3 ± 8.6	Trial	Usual treatment + Curcumin with soy phosphatidylcholine (Meriva®, Indena SpA) (200 mg curcumin)	No placebo- only usual treatment	3 months	In the treatment compared to the control group: (1) The global WOMAC score significantly decreased by 58% (pain, stiffness, and physical function were all positively affected by treatment) (2) Walking distance in the treadmill test was significantly prolonged (3) CRP levels decreased in the subpopulation with high CRP (4) Treatment costs (use of anti-inflammatory drugs, treatment, and hospitalization) were reduced significantly in the treatment group	0
Gupte et al. (2019), India [41]	50	Knee OA as confirmed by the American College of Rheumatology (ACR) criteria like knee pain, stiffness/ crepitus and osteophytes in X-ray	In: 57 ± 7.5/ Co: 54 ± 8	RCT	Solid lipid curcumin particles (SLCP) (400 mg twice daily delivering 80 mg of curcumin per capsule) or Ibuprofen with the placebo group (400 mg each once daily)	Capsule Ibuprofen (400 mg) + placebo capsule (dextrin)	90 days	(1) The WOMAC and visual analog scale (VAS) score significantly decreased in both groups compared to baseline. There was no significant difference between both the groups (2) No difference between the groups was found regarding inflammatory markers differences (3) Levels of urine CTX-II were maintained in both groups (4) There were no serious adverse events reported in both groups	3

Haroyan et al. (2018), Armenia [50]	210	Degenerative hypertrophic knee OA (M 17, according to International Statistical Classification of Diseases and Related Health Problems tenth Revision (ICD-10) version for 2014 of bone joints and verified by Radiography)	56.2	RCT	CuraMed® 500 mg capsules (333 mg curcuminoids) and Curamin® 500 mg capsules (350 mg curcuminoids and 150 mg boswellic acid) taken orally three times	Placebo	12 weeks	(1) A significant effect of Curamin® (curcuminoids and 150 mg boswellic acid) compared to placebo in physical performance tests and the WOMAC total and joint pain scores (2) Superior efficacy of CuraMed (curcuminoids) vs placebo was observed only in physical performance tests (3) No difference was found between groups regarding the WOMAC stiffness index (4) The effect size compared to placebo was comparable for both treatment groups but was superior in the Curamin® group (5) The treatments were well tolerated	5
Hashemzadeh et al. (2019), Iran [42]	71	Knee OA patients[a]	[a]	RCT	40 mg nanocurcumin capsule every 12 h	Placebo	6 weeks	A significant decrease was observed in the overall score, along with the scores of pain, stiffness, and physical activity subscales of the WOMAC questionnaire in patients of the nanocurcumin group compared with the placebo group	-[a]

(continued)

Table 16.1 (continued)

Author, year, country	Size	Inclusion criteria	Age Range (years) or mean	Study design	Intervention	Control	Duration	Main results	Jadad score
Henrotin et al. (2019), Belgium [38]	150	Femorotibial and/or femoropatellar symptomatic knee OA diagnosed according to the clinical and radiological criteria of ACR	(A) 60.9 ± 9.78 (B) 61.4 ± 7.49 (C) 63.3 ± 7.69	RCT	(A) Bio-optimized *Curcuma longa* extracts (BCL) low dosage 2 × 2 caps/day plus placebo 2 × 1 cap/day (B) BCL high dosage 2 × 3 caps/day (each capsule contained 46.67 mg of turmeric rhizome extract (*Curcuma longa* L.), polysorbate 80 [E433] as an emulsifier, and citric acid [E330] as an acidity regulator)	(C) Placebo 2 × 3 capsules /day Placebo (sunflower seed oil)	3 months	(1) Low and high doses of BCL showed a greater decrease of Patient Global Assessment of Disease Activity (PGADA) than placebo (2) Pain (VAS) reduction at day 90 in low- and high-dose BCL groups was significantly higher than that in the placebo (3) Global Knee injury and Osteoarthritis Outcome Score (KOOS) significantly decreased over time, but changes did not significant across treatment arms (4) The ratio of patients with adverse events related to the product was similar in the placebo and treatment groups	5
Kuptniratsaikul et al. (2014), Thailand [44]	367	Knee OA patients according to the American Rheumatism Association criteria who had a numerical rating scale of knee pain of ≥5 out of 10	Curcumin: 60.3 ± 6.8/ Ibuprofen: 60.9 ± 6.9	RCT	*Curcuma domestica* extracts 1500 mg/day	Ibuprofen 1200 mg/day	4 weeks	(1) The mean of all WOMAC scores (total and sub-total scores for pain, stiffness and physical function) showed significant improvement when compared with the baseline in both groups but there were no differences between groups (2) The side effect profile was similar but with fewer gastrointestinal adverse event reports in the *C. domestica* extract group	5

Madhu et al. (2013), India [45]	120	Clinical evidence confirming the diagnosis of knee OA, duration of pain was at least 6 months, radiological evidence of OA grades 2 and 3	Mean of all: 57.3 ± 9.4	RCT	(A) Polysaccharide-rich *Curcuma longa* extract (NR-INF-02) (500 mg twice daily) (B) Glucosamine sulfate (GS) (750 mg twice daily) (C) combination of NR-INF-02 and GS	Placebo (400 mg twice daily)	42 days	(1) Administration of NRINF-02 using VAS, WOMAC, and Global Impression Change (CGIC) at each clinical visit showed a significant decrease compared to placebo (2) NR-INF-02 treated group showed a significant decrease in the use of rescue medication, with clinical and subjective improvement compared to placebo (3) The tolerability and acceptability profile of NR-INF-02 was better during the trial period	4
Nakagawa et al. (2014), Japan [43]	50	Knee OA confirmed by radiographic	In: 71.9 ± 5.3/ Co: 66.1 ± 7.2	RCT	Theracurmin containing 180 mg/ day of curcumin (developed a surface-controlled water-dispersible curcumin, named Theracurmin®)	Placebo)starch, dextrin, and maltose(	8 weeks	(1) Knee pain VAS scores were significantly lower in the Theracurmin group than in the placebo group, except in patients with initial VAS scores of 0.15 or less (2) Celecoxib dependence significantly deceased in Theracurmin than placebo (3) No major side effects were observed with Theracurmin treatment	4

(continued)

Table 16.1 (continued)

Author, year, country	Size	Inclusion criteria	Age Range (years) or mean	Study design	Intervention	Control	Duration	Main results	Jadad score
Panahi et al. (2014, 2016), Rahimnia et al. (2015), Iran [35–37]	40	Knee OA was based on the clinical and radiological criteria defined by the ACR and personal report of pain with mild-to-moderate degree on active movement	In: 57.32 ± 8.78/ Co: 57.57 ± 9.05	RCT	Curcuminoids (1500 mg/day in 3 divided doses)	Placebo	6 weeks	(1) Treatment with curcuminoids was associated with significantly greater reductions in WOMAC, VAS, and Lequesne's pain functional index (LPFI) scores compared with placebo (2) With respect to WOMAC subscales, there were significant improvements in pain and physical function scores but not stiffness score (3) A significant elevation in serum superoxide dismutase (SOD) activities and significant reduction in malondialdehyde (MDA) concentrations observed in curcuminoid compared with the placebo group (4) No significant changes in serum levels of interleukins 4 (IL-4) and 6 (IL-6), tumor necrosis factor-α (TNF-α), transforming growth factor-β (TGF-β), high-sensitivity C-reactive protein (hs-CRP), and erythrocyte sedimentation rate (ESR) found between groups (5) No considerable adverse effect in both groups	4[b]

Panda et al. (2018), India [46]	50	Suffering from unilateral or bilateral OA of the knee for greater than 3 months ACR criteria	In: 55.2 ± 8.6/ Co: 53.1 ± 8.3	RCT	Curene 500 mg once daily (Curene is a bioavailable formulation of turmeric *Curcucma longa* extract)	Placebo (microcrystalline cellulose)	3 months	(1) Significant reduction in WOMAC (total and in the pain, physical function scores and stiffness score) and VAS scores in the intervention compared with the placebo group (2) It had a good safety profile	5
Pinsornsak et al. (2012), Thailand [48]	88	Knee osteoarthritis according to the ACR (by history and physical examination)	45–80	RCT	Diclofenac at 75 mg/day + curcumin (curcuminoid capsules) at 1000 mg/day (study group)	75 mg/day Diclofenac with placebo control	3 months	No significant differences found regarding VAS for pain and KOOS scales	3
Shep et al. (2019), India [49]	139	ACR criteria for knee OA (confirmed by X-ray)	Curcumin: 53.1 ± 4.2/ Diclofenac: 52.1 ± 3.8	RCT	Curcumin (BCM-95®) 500-mg capsule 3 times daily	Diclofenac 50-mg tablet two times daily	28 days	(1) Similar improvement in the severity of VAS (pain) and KOOS (pain, symptoms, function in daily living, function in sport and recreation, and quality of life) scales found in two groups (2) Curcumin had weight-lowering effect and anti-ulcer effect (3) None of the patients required H2 blockers in the curcumin group and 19 patients required H2 blockers in the diclofenac group (0% versus 28%, respectively; $P < 0.01$) (4) Adverse effects were significantly less in the curcumin group (5) Patient and physician global assessment of therapy was similar in the two treatment groups	3

(continued)

Table 16.1 (continued)

Author, year, country	Size	Inclusion criteria	Age Range (years) or mean	Study design	Intervention	Control	Duration	Main results	Jadad score
Shin et al. (2017), South Korea [51]	25	Radiological diagnostic criteria for knee OA proposed by the ACR	T: 44.8 ± 5.4/T + E: 40.6 ± 7.7	Trial	(A) Theracurmin intake (T) group (capsule of 700 mg, 3 times per day, (total 2100 mg)) (B) Theracurmin is combined with exercise (T + E) (Theracurmin: with activity of curcumin increased 27-fold)	–	4 weeks	(1) The VAS was significantly decreased after 4 weeks in both group but the difference between groups did not significant (2) WOMAC score was significantly decreased in pain, physical function difficulties and total score but the difference between groups was not significant	1
Srivastava et al. (2016), India [47]	160	Suffering from knee OA was according to the guidelines by the ACR	In: 50.2 ± 8.1/ Co: 50.3 ± 8.6	RCT	*Curcuma longa* L. extract (500 mg) along with Diclofenac twice a day	Placebo capsules along with Diclofenac twice a day	4 months	(1) The VAS and WOMAC (total and pain, physical function and stiffness) scores improved in the intervention compared with the placebo group (2) IL-1b, ROS, and MDA significantly improved in the curcumin compared with the placebo group (3) *Curcuma longa* had no side effects compared with placebo	5

[a]No full text available (Epub ahead of print)
[b]Jadad score calculated for the study of Panahi et al. (2014) [36]

Outcome Score (KOOS) for assessing views of the patients about their knee- and osteoarthritis-related problems [38, 48, 49].

In comparison to controls, various curcumin formulations significantly improved the overall WOMAC scores in nine studies [35, 39, 40, 42, 44–47, 50], whereas one study did not see a significant difference between the intervention and control groups [41]. In one of these two studies, WOMAC scores were significantly decreased in both intervention groups (curcumin, curcumin plus exercise), although the differences between the groups were not significant [51].

Regarding VAS assessments, a significant decrease was observed in seven intervention groups compared to the placebo groups [35, 38, 43, 45–47, 49]. In one study, VAS scores were significantly decreased in both intervention groups (curcumin, curcumin plus exercise), although the differences between the groups were not significant [51]. However, two studies reported no significant differences for pain and knee injury in the treatment group [41, 48] compared to the placebo control group.

In cases where KOOS was assessed, scores were decreased in one study after intervention [39], although no significant differences were found in two studies between the intervention groups compared with placebo [48] or NSAID [49] treatments.

16.3.4 Other Outcomes and Quality Assessment

Some studies also assessed changes in various inflammatory markers in the interventional groups compared to the control. The most investigated inflammatory markers at follow up were C-reactive protein (CRP) [36, 39, 41]; interleukin (IL)-1β [36, 40, 41, 47]; IL-6 and erythrocyte sedimentation rate (ESR) [36, 40, 41]; and tumor necrosis factor-alpha (TNF-α) [36, 41]. After intervention, curcumin compounds led to significantly decreased levels of CRP [39], IL-6 and ESR [40], and IL-1β [40, 47], and two studies did not observe changes in biomarkers [36, 41]. Two studies showed that biomarkers of oxidative stress malondialdehyde (MDA), reactive oxygen species (ROS), and superoxide dismutase (SOD) were significantly reduced in the intervention groups compared with control groups [37, 47].

There were no severe adverse effects reported in nine studies [35–38, 41, 43, 46, 47, 49, 50]. Adverse effects were mostly manifested in the gastrointestinal system. However, Kuptniratsaiku et al. reported that curcumin showed better safety profiles than ibuprofen in terms of abdominal pain and distension [44]. Furthermore, of the 17 RCTs evaluated, only three studies were of low quality [39, 40, 51], as assessed by the Jadad scoring system and one study had no full text [42].

16.4 Discussion

16.4.1 Summary of Evidence

The main objective of this systematic study was to identify improvements in the WOMAC and VAS scores from baseline to post-intervention in clinical trials of curcumin-related treatments. We found that oral administration of various curcumin formulations could improve VAS and overall WOMAC scores significantly with no severe adverse effects after a minimum of four weeks of treatment. However, some curcumin preparations were more effective in improving knee stiffness.

The WOMAC index is a global questionnaire that measures the health status of osteoarthritis patients using 24 items divided into three subscales. It is generally used in the evaluation of patients with osteoarthritis knee and hip to evaluate the effects of various interventions [52]. The subscales of the WOMAC assessment consist of various aspects of osteoarthritis, including pain, stiffness, and physical function [53]. Belcaro et al. showed that 200 mg of curcumin-phosphatidylcholine compound (known as Meriva®) per day improved overall WOMAC scores and also the various components including pain sensation, joint stiffness, and physical function after 3 months treatment of osteoarthritis patients [39]. The mobility assessed by walking performance (treadmill) also increased after treatment with curcumin-phosphatidyl-choline compared with a standard treatment. These results were confirmed by a subsequent study by the same research team which investigated the long-term efficacy and safety of the curcumin-phosphatidyl-choline compound over a longer time period (8 months) in 100 patients with osteoarthritis [40]. They showed that the global WOMAC score significantly decreased by 58% after treatment with the curcumin–phosphatidyl–choline compound. There was also a reduction in inflammation indicated by changes in CD40L, IL-1β, IL-6, and ESR biomarkers. Another study showed a significant improvement in global WOMAC scores after treatment with a 350 mg curcuminoid and 150 mg boswellic acid complex for 3 months compared to standard care [51]. However, the positive effects were observed only in the subscale of physical performance tests. In addition, studies conducted by Panahi et al. found a significant improvement in total WOMAC scores following treatment with a curcuminoid complex (1500 mg/day in 3 divided doses) for 6 weeks [35–37]. However, with respect to WOMAC components, the improvement was detected only in the pain and physical function scores but not in stiffness scores.

In studies of curcumin extracted from turmeric, Shin et al. investigated the effect of the physiologic activity of 90 mg/day curcumin in 25 patients with knee osteoarthritis for 4 weeks and showed that total WOMAC score was significantly improved by the intervention, although there was no control group in this study [51]. However, a dosage of 180 mg/day of curcumin in a study of 50 subjects for 8 weeks was found to reduce knee pain [43]. Additionally, the use of a *Curcuma longa* extract at a different dosage significantly reduced overall WOMAC scores [44–47] as well as WOMAC subscales [46]. It is known that curcumin compounds have antioxidant properties and can facilitate scavenging of free radicals such as reactive oxygen and nitrogen species [54, 55]. These natural dietary antioxidants can also modulate the activity of glutathione (GSH), catalase, and SOD enzymes and thereby inhibit the production of inflammation by chondrocytes in OA patients, especially with a treatment duration of more than 6 weeks [56, 57]. In spite of its reported benefits via inflammatory and antioxidant mechanisms, curcumin has poor bioavailability due to its poor absorption, rapid metabolism, and excretion [58, 59]. To overcome this problem, there are now several types of curcumin preparations with varying oral bioavailability and stability [60–62]. A number of studies have shown that curcumin nanoparticles have a significantly higher bioavailability compared to free curcumin crystals [63]. This supports the idea that nanocurcumin supplementation could manage knee osteoarthritis complications such as the pain and stiffness of joints better than other curcumin compounds [42, 64]. The results of another study also suggested that a curcumin formulation with a lipid component (Meriva) has more bioavailability than unformulated curcumin [65]. Therefore, variations in bioavailability may explain the observed differing effects of the various curcumin preparations on components of WOMAC, such as knee stiffness.

Despite the overall improvement as seen with WOMAC scores, there were no significant differences between the patients taking anti-inflammatory drugs (400 mg ibuprofen) alone and those who were treated with 400 mg solid lipid curcumin and anti-inflammatory medications for 3 months [41]. However, Srivastava et al. showed that diclofenac with supplementary curcumin treatment led to a significant improvement in ROS and MDA levels as well as VAS and WOMAC scores, and this was better than diclofenac treatment alone [47]. This shows that curcumin could be as effective as an anti-inflammatory medication on WOMAC global function scores in osteoarthritis. However, additional studies are needed to confirm these findings.

This systematic review represents the highest level of available evidence which shows that different curcumin compounds are generally able to reduce overall WOMAC scores and some WOMAC subscales such as pain and physical function. In addition, some products such as curcumin phosphatidylcholine (lipid-based curcumin) [39] and nanocurcumin [42] can also modulate the stiffness subscale compared to placebo, most likely due to their increased bioavailability [66, 67].

In addition, the results of the present study demonstrate the efficacy of different types of curcumin mixtures in alleviating VAS scores in the intervention groups compared to placebo. This instrument has been designed to measure pain intensity and has also been used in various populations, including those with chronic diseases such as osteoarthritis [68, 69]. In the study by Henrotin et al., both low (2 × 2 46.67 mg capsules/day) and high (2 × 3 capsules/day) dosages of bio-optimized *Curcuma longa* extracts for 3 months decreased the Patient Global Assessment of Disease Activity (PGADA) and VAS scores, compared to placebo, in 150 patients with osteoarthritis [38]. The adverse events related to the product were similar in the placebo and the treatment groups. Similarly, intake of 100 or 500 mg / day of the bio-optimized extract had similar results over treatment periods of 3 and 12 weeks, respectively [45, 46]. Moreover, Nakagawa et al. indicated improvement in the VAS score after 8 weeks of treatment with 180 mg/day of curcumin (Theracurmin) compared to its control [43]. Interestingly, dependence on Celecoxib was significantly reduced in the intervention group compared to the placebo. In addition, Shin et al. demonstrated that treatment with 90 mg/day curcumin (Theracurmin) for 4 weeks could also improve VAS scores without side effects [51]. Furthermore, in the other three studies [35–37], the administration of curcuminoids (1500 mg/day in 3 divided doses) was associated with significantly greater reductions in VAS scores and MDA concentrations. However, no significant changes were observed in the serum levels of IL-4, IL-6, TNF-α, transforming growth factor-β (TGF-β), CRP, and ESR between groups. Although osteoarthritis is a complex condition that affects different joints, many previous studies have indicated that progression is significantly related to oxidative stress and ROS. ROS-induced damage leads to an increase in the irritation of the nerves resulting in pain in the synovial, bone, and soft tissues [70]. The mechanism underlying the analgesic effect of curcumin may be based on the elevation of serum SOD and GSH levels and the reduction of MDA serum levels through its antioxidant activity [56, 71]. In turn, this would lead to an improved oxidant status, increased joint mobility, and reduced pain in patients with osteoarthritis [35–37, 47].

Nevertheless, it should be noted some types of curcumin preparations (solid lipid curcumin or curcumin) did not result in a significant difference in the VAS scores compared to the control group receiving anti-inflammatory agents, diclofenac or ibuprofen) [44, 49]. However, both groups showed clinical improvements after medication. It should also be noted that the effectiveness of curcumin as a supplement in combination with the anti-osteoarthritic medications in the treatment of primary knee osteoarthritis was more effective than anti-osteoarthritic drugs alone [47].

The results of the study which used curcumin to evaluate the changes in short- and long-term symptoms and function of patients with a knee injury, using KOOS assessment, showed significant positive changes over time compared to placebo [38]. However, these changes were not significant when the control group received anti-inflammatory drugs [49]. Based on these results, it can be concluded that curcumin compounds can be effective in reducing symptoms and improving function in osteoarthritis patients. These compounds are also safe and well tolerated.

16.4.2 Limitations

Although our study indicated the promising role of various curcumin compounds in the improvement of function and reduction of pain using three global questionnaires (WOMAC, VAS, and KOOS) in patients with knee osteoarthritis for several durations, there were some limitations. The scales used as the main outcome in the mentioned studies are based on the patient perspectives of their symptoms. Secondly, the measurement of circulating metabolites after consumption of different curcumin compounds in the above studies should be taken into account in the interpretation of the results. Another common limitation shown in these studies is the lack of sufficient data to compare the bioavailability of the different curcumin compounds used together and there are also no comprehensive studies comparing the efficacy of different curcumin combinations in the management of osteoarthritis. Higher quality RCTs should be designed specifically to examine the role of different types of curcumin preparations as a supplementary therapy from the perspective of their metabolites and the bioavailability of various curcumin compounds.

16.5 Conclusion

This review provides evidence that different types of curcumin complexes are beneficial as an alternative or complementary medication for osteoarthritis treatment. Curcumin and various curcumin compounds reduced VAS and WOMAC scores in all the reviewed studies. Curcumin compounds with more bioavailability have a more positive effect on knee stiffness.

References

1. Poole AR (2012) Osteoarthritis as a whole joint disease. HSS J 8(1):4–6
2. Loeser RF (2011) Aging and osteoarthritis. Curr Opin Rheumatol 23(5):492–496
3. Ma VY, Chan L, Carruthers KJ (2014) Incidence, prevalence, costs, and impact on disability of common conditions requiring rehabilitation in the United States: stroke, spinal cord injury, traumatic brain injury, multiple sclerosis, osteoarthritis, rheumatoid arthritis, limb loss, and back pain. Arch Phys Med Rehab 95(5):986–995.e981
4. Fransen M, Bridgett L, March L, Hoy D, Penserga E, Brooks P (2011) The epidemiology of osteoarthritis in Asia. Int J Rheum Dis 14(2):113–121
5. Loeser RF, Collins JA, Diekman BO (2016) Ageing and the pathogenesis of osteoarthritis. Nat Rev Rheumatol 12(7):412–420
6. Zhang R, Ma J, Yao J (2013) Molecular mechanisms of the cartilage-specific microRNA-140 in osteoarthritis. Inflamm Res 62(10):871–877
7. Lee AS, Ellman MB, Yan D, Kroin JS, Cole BJ, van Wijnen AJ et al (2013) A current review of molecular mechanisms regarding osteoarthritis and pain. Gene 527(2):440–447
8. Crawford DC, Miller LE, Block JE (2013) Conservative management of symptomatic knee osteoarthritis: a flawed strategy? Orthop Rev (Pavia) 5(1):e2. https://doi.org/10.4081/or.2013.e2
9. Liu X, McLachlan A, Hunter D (2019) Vitamin and mineral supplements: the role of supplements and complementary medicines in osteoarthritis: an evidence-based guide to practice. Aust J Pharm 100(1180):62–68
10. Shirali S, Mashhadi NS, Ashtary-Larky D, Safania T, Barari A (2016) Effects of silymarin supplementation on leptin, adiponectin and paraoxanase levels and body composition during exercise: a randomized double-blind placebo controlled clinical trial. Jundishapur J Nat Pharm Prod 11(4):e30044. https://doi.org/10.17795/jjnpp-30044
11. Mashhadi NS, Ghiasvand R (2019) The role of nutraceuticals in preeclampsia and eclampsia. In: Bhatt P, Miraghajani MS, Pathak S, Pathak Y (eds) Nutraceuticals for prenatal, maternal, and offspring's nutritional health. CRC Press, Cleveland, OH. 1138345822
12. Abdollahi E, Momtazi AA, Johnston TP, Sahebkar A (2018) Therapeutic effects of curcumin in inflammatory and immune-mediated diseases: a nature-made jack-of-all-trades? J Cell Physiol 233(2):830–848
13. Sahebkar A, Saboni N, Pirro M, Banach M (2017) Curcumin: An effective adjunct in patients with statin-associated muscle symptoms? J Cachexia Sarcopenia Muscle 8(1):19–24. https://doi.org/10.1002/jcsm.12140. Epub 2016 Sep 22. PMID: 27897416; PMCID: PMC5326825
14. Soleimani V, Sahebkar A, Hosseinzadeh H (2018) Turmeric (Curcuma longa) and its major constituent (curcumin) as nontoxic and safe substances: Review. Phytother Res 32(6):985–995. https://doi.org/10.1002/ptr.6054. Epub 2018 Feb 26. PMID: 29480523
15. Anand P, Thomas SG, Kunnumakkara AB, Sundaram C, Harikumar KB, Sung B et al (2008) Biological activities of curcumin and its analogues (Congeners) made by man and mother nature. Biochem Pharmacol 76(11):1590–1611
16. Bagherniya M, Nobili V, Blesso CN, Sahebkar A (2018) Medicinal plants and bioactive natural compounds in the treatment of non-alcoholic fatty liver disease: a clinical review. Pharmacol Res 130:213–240
17. Iranshahi M, Sahebkar A, Hosseini ST, Takasaki M, Konoshima T, Tokuda H (2020) Cancer chemopreventive activity of diversin from Ferula diversivittata in vitro and in vivo. Phytomedicine 17(3–4):269-–73. https://doi.org/10.1016/j.phymed.2009.05.020. Epub 2009 Jul 3. PMID: 19577457.
18. Momtazi AA, Sahebkar A (2016) Difluorinated curcumin: a promising curcumin analogue with improved anti-tumour activity and pharmacokinetic profile. Curr Pharm Des 22(28):4386–4397
19. Teymouri M, Pirro M, Johnston TP, Sahebkar A (2017) Curcumin as a multifaceted compound against human papilloma virus infection and cervical cancers: A review of chemistry, cellular, molecular, and preclinical features. BioFactors 43(3):331–346
20. Mollazadeh H, Cicero AFG, Blesso CN, Pirro M, Majeed M, Sahebkar A (2019) Immune modulation by curcumin: the role of interleukin-10. Crit Rev Food Sci Nutr 59(1):89–101
21. Shishodia S, Sethi G, Aggarwal BB (2005) Curcumin: getting back to the roots. Ann N Y Acad Sci 1056(1):206–217
22. Momtazi AA, Derosa G, Maffioli P, Banach M, Sahebkar A (2016) Role of microRNAs in the therapeutic effects of curcumin in non-cancer diseases. Mol Diag Ther 20(4):335–345
23. Bagheri H, Ghasemi F, Barreto GE, Rafiee R, Sathyapalan T, Sahebkar A (2020) Effects of curcumin on mitochondria in neurodegenerative diseases. Biofactors 46(1):5–20. https://doi.org/10.1002/biof.1566. Epub 2019 Oct 3. PMID: 31580521
24. Sadeghian M, Rahmani S, Jamialahmadi T, Johnston TP, Sahebkar A (2020) The effect of oral curcumin supplementation on healthrelated quality of life: A

systematic review and meta-analysis of randomized controlled trials. J Affect Disord 29;278:627–636. https://doi.org/10.1016/j.jad.2020.09.091. Epub ahead of print. PMID: 33038707
25. Mohajeri M, Bianconi V, Ávila-Rodriguez MF, Barreto GE, Jamialahmadi T, Pirro M, Sahebkar A (2020) Curcumin: a phytochemical modulator of estrogens and androgens in tumors of the reproductive system. Pharmacol Res 156:104765. https://doi.org/10.1016/j.phrs.2020.104765. Epub 2020 Mar 23. PMID: 32217147
26. Onakpoya IJ, Spencer EA, Perera R, Heneghan CJ (2017) Effectiveness of curcuminoids in the treatment of knee osteoarthritis: a systematic review and meta-analysis of randomized clinical trials. Int J Rheum Dis 20(4):420–433
27. Mathy-Hartert M, Jacquemond-Collet I, Priem F, Sanchez C, Lambert C, Henrotin Y (2009) Curcumin inhibits pro-inflammatory mediators and metalloproteinase-3 production by chondrocytes. Inflamm Res 58(12):899. https://doi.org/10.1007/s00011-009-0063-1
28. Yeh CC, Su YH, Lin YJ, Chen PJ, Shi CS, Chen CN et al (2015) Evaluation of the protective effects of curcuminoid (curcumin and bisdemethoxycurcumin)-loaded liposomes against bone turnover in a cell-based model of osteoarthritis. Drug Des Devel Ther 9:2285–2300
29. Zhang Z, Leong DJ, Xu L, He Z, Wang A, Navati M et al (2016) Curcumin slows osteoarthritis progression and relieves osteoarthritis-associated pain symptoms in a post-traumatic osteoarthritis mouse model. Arthritis Res Ther 18(1):128. https://doi.org/10.1186/s13075-016-1025-y
30. Chin KY (2016) The spice for joint inflammation: anti-inflammatory role of curcumin in treating osteoarthritis. Drug Des Devel Ther 10:3029–3042
31. Hasanzadeh S, Read MI, Bland AR, Majeed M, Jamialahmadi T, Sahebkar A (2020) Curcumin: an inflammasome silencer. Pharmacol Res 159:104921. https://doi.org/10.1016/j.phrs.2020.104921. Epub 2020 May 25. PMID: 32464325.
32. Mirzaei H, Shakeri A, Rashidi B, Jalili A, Banikazemi Z, Sahebkar A (2017) Phytosomal curcumin: a review of pharmacokinetic, experimental and clinical studies. Biomed Pharmacother 85:102–112
33. Jadad AR, Moore RA, Carroll D, Jenkinson C, Reynolds DJM, Gavaghan DJ et al (1996) Assessing the quality of reports of randomized clinical trials: is blinding necessary? Control Clin Trials 17(1):1–12
34. Moher D, Cook D, Jadad A, Tugwell P, Moher M, Jones A et al (1999) Assessing the quality of reports of randomised trials: implications for the conduct of meta-analyses. Health Technol Assess 3(12):i–iv. 1–98
35. Panahi Y, Rahimnia AR, Sharafi M, Alishiri G, Saburi A, Sahebkar A (2014) Curcuminoid treatment for knee osteoarthritis: a randomized double-blind placebo-controlled trial. Phytother Res 28(11):1625–1631
36. Rahimnia AR, Panahi Y, Alishiri G, Sharafi M, Sahebkar A (2015) Impact of supplementation with curcuminoids on systemic inflammation in patients with knee osteoarthritis: findings from a randomized double-blind placebo-controlled trial. Drug Res 65(10):521–525
37. Panahi Y, Alishiri GH, Parvin S, Sahebkar A (2016) Mitigation of systemic oxidative stress by curcuminoids in osteoarthritis: results of a randomized controlled trial. J Diet Suppl 13(2):209–220
38. Henrotin Y, Malaise M, Wittoek R, de Vlam K, Brasseur JP, Luyten F et al (2019) Bio-optimized Curcuma longa extract is efficient on knee osteoarthritis pain: a double-blind multicenter randomized placebo controlled three-arm study. Arthritis Res Ther 21(1):179. https://doi.org/10.1186/s13075-020-2109-2
39. Belcaro G, Cesarone M, Dugall M, Pellegrini L, Ledda A, Grossi M et al (2010) Product-evaluation registry of Meriva®, a curcumin-phosphatidylcholine complex, for the complementary management of osteoarthritis. Panminerva Med 52(2 Suppl 1):55–62
40. Belcaro G, Cesarone MR, Dugall M, Pellegrini L, Ledda A, Grossi MG et al (2010) Efficacy and safety of Meriva®, a curcumin-phosphatidylcholine complex, during extended administration in osteoarthritis patients. Altern Med Rev 15(4):337–344
41. Gupte PA, Giramkar SA, Harke SM, Kulkarni SK, Deshmukh AP, Hingorani LL et al (2019) Evaluation of the efficacy and safety of Capsule Longvida® Optimized Curcumin (solid lipid curcumin particles) in knee osteoarthritis: a pilot clinical study. J Inflamm Res 12:145–152
42. Hashemzadeh K, Davoudian N, Jaafari MR, Mirfeizi Z (2019) The effect of nanocurcumin on the improvement symptoms of knee osteoarthritis: a randomized clinical trial. Curr Rheumatol Rev. https://doi.org/10.2174/1874471013666191223152658. [Epub ahead of print]
43. Nakagawa Y, Mukai S, Yamada S, Matsuoka M, Tarumi E, Hashimoto T et al (2014) Short-term effects of highly-bioavailable curcumin for treating knee osteoarthritis: a randomized, double-blind, placebo-controlled prospective study. J Orthop Sci 19(6):933–939
44. Kuptniratsaikul V, Dajpratham P, Taechaarpornkul W, Buntragulpoontawee M, Lukkanapichonchut P, Chootip C et al (2014) Efficacy and safety of Curcuma domestica extracts compared with ibuprofen in patients with knee osteoarthritis: a multicenter study. Clin Interv Aging 9:451–458
45. Madhu K, Chanda K, Saji M (2013) Safety and efficacy of Curcuma longa extract in the treatment of painful knee osteoarthritis: a randomized placebo-controlled trial. Inflammopharmacology 21(2):129–136
46. Nirvanashetty S, Parachur VA, Mohanty N, Swain T (2018) A randomized, double blind, placebo controlled, parallel-group study to evaluate the safety and efficacy of Curene® versus placebo in reducing symptoms of knee OA. Biomed Res Int 2018:5291945. https://doi.org/10.1155/2018/5291945

47. Srivastava S, Saksena AK, Khattri S, Kumar S, Dagur RS (2016) Curcuma longa extract reduces inflammatory and oxidative stress biomarkers in osteoarthritis of knee: a four-month, double-blind, randomized, placebo-controlled trial. Inflammopharmacology 24(6):377–388
48. Pinsornsak P, Niempoog S (2012) The efficacy of Curcuma Longa L. extract as an adjuvant therapy in primary knee osteoarthritis: a randomized control trial. J Med Assoc Thail 95(Suppl 1):S51–S58
49. Shep D, Khanwelkar C, Gade P, Karad S (2019) Safety and efficacy of curcumin versus diclofenac in knee osteoarthritis: a randomized open-label parallel-arm study. Trials. https://doi.org/10.1186/s13063-019-3327-2
50. Haroyan A, Mukuchyan V, Mkrtchyan N, Minasyan N, Gasparyan S, Sargsyan A et al (2018) Efficacy and safety of curcumin and its combination with boswellic acid in osteoarthritis: a comparative, randomized, double-blind, placebo-controlled study. BMC Complement Altern Med 18:7. https://doi.org/10.1186/s12906-017-2062-z
51. Shin YA, Suk MH, Jang HS, Choi HJ (2017) Short-term effects of Theracurmin dose and exercise type on pain, walking ability, and muscle function in patients with knee osteoarthritis. J Exerc Rehabil 13(6):684–692
52. McConnell S, Kolopack P, Davis AM (2001) The Western Ontario and McMaster Universities Osteoarthritis Index (WOMAC): a review of its utility and measurement properties. Arthritis Care Res 45(5):453–461
53. Ebrahimzadeh MH, Makhmalbaf H, Birjandinejad A, Keshtan FG, Hoseini HA, Mazloumi SM (2014) The Western Ontario and McMaster Universities Osteoarthritis Index (WOMAC) in Persian speaking patients with knee osteoarthritis. J Bone Joint Surg 2(1):57–62
54. Hewlings SJ, Kalman DS (2017) Curcumin: a review of its' effects on human health. Foods 6(10):92. https://doi.org/10.3390/foods6100092
55. Menon VP, Sudheer AR (2007) Antioxidant and anti-inflammatory properties of curcumin. In: Aggarwal BB, Surh YJ, Shishodia S (eds) The molecular targets and therapeutic uses of curcumin in health and disease. Springer, New York, NY, pp 105–125. ISBN-10: 038746400X
56. Chainani-Wu N (2003) Safety and anti-inflammatory activity of curcumin: a component of turmeric (Curcuma longa). J Altern Complement Med 9(1):161–168
57. Sahebkar A, Serban MC, Ursoniu S, Banach M (2015) Effect of curcuminoids on oxidative stress: a systematic review and meta-analysis of randomized controlled trials. J Funct Foods 18:898–909
58. Anand P, Kunnumakkara AB, Newman RA, Aggarwal BB (2007) Bioavailability of curcumin: problems and promises. Mol Pharm 4(6):807–818
59. Shoba G, Joy D, Joseph T, Majeed M, Rajendran R, Srinivas P (1998) Influence of piperine on the pharmacokinetics of curcumin in animals and human volunteers. Planta Med 64(4):353–356
60. Zou L, Liu W, Liu C, Xiao H, McClements DJ (2015) Designing excipient emulsions to increase nutraceutical bioavailability: emulsifier type influences curcumin stability and bioaccessibility by altering gastrointestinal fate. Food Funct 6(8):2475–2486
61. Mondal S, Ghosh S, Moulik SP (2016) Stability of curcumin in different solvent and solution media: UV–visible and steady-state fluorescence spectral study. J Photochem Photobiol B 158:212–218
62. Zheng B, Zhang Z, Chen F, Luo X, McClements DJ (2017) Impact of delivery system type on curcumin stability: comparison of curcumin degradation in aqueous solutions, emulsions, and hydrogel beads. Food Hydrocoll 71:187–197
63. Peng S, Li Z, Zou L, Liu W, Liu C, McClements DJ (2018) Enhancement of curcumin bioavailability by encapsulation in sophorolipid-coated nanoparticles: an in vitro and in vivo study. J Agric Food Chem 66(6):1488–1497
64. Cheragh-Birjandi S, Moghbeli M, Haghighi F, Safdari MR, Baghernezhad M, Akhavan A et al (2020) Impact of resistance exercises and nano-curcumin on synovial levels of collagenase and nitric oxide in women with knee osteoarthritis. Transl Med Commun 5(1):1–6
65. Marczylo TH, Verschoyle RD, Cooke DN, Morazzoni P, Steward WP, Gescher AJ (2007) Comparison of systemic availability of curcumin with that of curcumin formulated with phosphatidylcholine. Cancer Chemother Pharmacol 60(2):171–177
66. Bhadoriya SS, Mangal A, Madoriya N, Dixit P (2011) Bioavailability and bioactivity enhancement of herbal drugs by "Nanotechnology": a review. J Curr Pharm Res 8:1–7
67. Gupta NK, Dixit VK (2011) Bioavailability enhancement of curcumin by complexation with phosphatidyl choline. J Pharm Sci 100(5):1987–1995
68. Klimek L, Bergmann KC, Biedermann T, Bousquet J, Hellings P, Jung K et al (2017) Visual analogue scales (VAS): Measuring instruments for the documentation of symptoms and therapy monitoring in cases of allergic rhinitis in everyday health care. Allergo J Int 26(1):16–24
69. Hawker GA, Mian S, Kendzerska T, French M (2011) Measures of adult pain: Visual analog scale for pain (vas pain), numeric rating scale for pain (nrs pain), mcgill pain questionnaire (mpq), short-form mcgill pain questionnaire (sf-mpq), chronic pain grade scale (cpgs), short form-36 bodily pain scale (sf-36 bps), and measure of intermittent and constant osteoarthritis pain (icoap). Arthritis Care Res 63(S11):S240–S252
70. Lepetsos P, Papavassiliou AG (2016) ROS/oxidative stress signaling in osteoarthritis. Biochim Biophys Acta 1862(4):576–591
71. Ozgocmen S, Ardicoglu O, Erdogan H, Fadillioglu E, Gudul H (2005) In vivo effect of celecoxib and tenoxicam on oxidant/anti-oxidant status of patients with knee osteoarthritis. Ann Clin Lab Sci 35(2):137–143

Effects of Curcumin on Depression and Anxiety: A Narrative Review of the Recent Clinical Data

17

Mohammad Amin Khodadadegan*,
Shakiba Azami*, Paul C. Guest,
Tannaz Jamialahmadi, and Amirhossein Sahebkar

Abstract

Depressive and anxiety disorders affect a significant proportion of the global population and constitute one of the highest disease burdens worldwide. Conventional pharmacological treatments are traditionally the first line of therapy for individuals affected by these conditions although these are only successful approximately half of the time and are often associated with undesirable side effects. This review describes the use of the natural substance curcumin as a potential alternative treatment of these mental disorders. With this in mind, we analyzed the effects of curcumin in eight clinical studies of depression and five studies of anxiety and assessed these using psychiatric symptom scores and molecular biomarker readouts.

Keywords

Curcumin · Therapy · Add-on treatment · Depression · MDD · Antidepressant · Anxiety · Axiolytic

17.1 Introduction

Mental disorders are established features of the distribution of cognitive ability and responses to emotional experiences in the presence of psychological or biological dysfunction [1]. A wide range of concepts has also been applied to define psychiatric disorders, such as disability, irrationality, and etiology [2]. Mental disorders are

*Mohammad Amin Khodadadegan and Shakiba Azami are equally contributed.

M. A. Khodadadegan · S. Azami
Faculty of Medicine, Mashhad University of Medical Sciences, Mashhad, Iran

P. C. Guest
Laboratory of Neuroproteomics, Department of Biochemistry and Tissue Biology, Institute of Biology, University of Campinas (UNICAMP), Campinas, Brazil

T. Jamialahmadi
Department of Food Science and Technology, Quchan Branch, Islamic Azad University, Quchan, Iran

Faculty of Medicine, Department of Nutrition, Mashhad University of Medical Sciences, Mashhad, Iran

A. Sahebkar (✉)
Biotechnology Research Center, Pharmaceutical Technology Institute, Mashhad University of Medical Sciences, Mashhad, Iran

Neurogenic Inflammation Research Center, Mashhad University of Medical Sciences, Mashhad, Iran

Polish Mother's Memorial Hospital Research Institute (PMMHRI), Lodz, Poland

Halal Research Center of IRI, FDA, Tehran, Iran
e-mail: sahebkara@mums.ac.ir

P. C. Guest (ed.), *Studies on Biomarkers and New Targets in Aging Research in Iran*, Advances in Experimental Medicine and Biology 1291, https://doi.org/10.1007/978-3-030-56153-6_17

major leading causes of disability and loss of productive life years across a wide range of countries [3]. Recent evidence suggests that one out of every five people suffers from a common psychiatric disorder [4]. In general, the prevalence of mental illnesses is higher than any other chronic condition [3, 5]. The most frequent mental conditions among the general population are major depressive disorder (MDD) anxiety and somatoform disorders [6]. The prevalence of depression increased by 18.4% from 2005 to 2015 [7]. Depression is defined as a psychological state characterized with low mood, in which patients experience hopelessness, difficulty in thinking and concentration, feelings of dejection, and sometimes suicidal thoughts [8]. More than half of patients with depression demonstrate anxiety symptoms [9].

Chronic psychiatric disorders necessitate consecutive decision-making involving adaptive treatment tactics and specific methodological approaches [10]. Likewise, depending on the severity, a combination of pharmacotherapy and psychotherapy is recommended for the management of depression [11]. Pharmacological treatments are efficient in numerous patients with mental disorders, while some patients fail to respond to these drugs [12]. Most psychiatric drugs are accompanied by neurological, cardiovascular, gastrointestinal, and sexual side effects [13–16], and low compliance can therefore occur [17]. Moreover, the lack of suitable animal models, the high rate of clinical trial failure, and absence of novel confirmed targets of drugs are challenges faced by the pharmaceutical industry in developing new drug from the concept to the marketing stage [18]. Hence, finding new suitable treatment options is of prime importance.

With this objective, herbal medicines are becoming increasingly more popular in a number of disease areas [19, 20]. Also, the toxicity and side effects of herbal medications seem to be lower than synthetic drugs. Curcumin is a non-flavonoid polyphenol compound derived from the rhizomes of turmeric (*Curcuma longa* L), a member of the ginger family, which has ben identified to be safe [21] and to have many biological activities [22–30]. Over the last few decades, there has been increasing attention paid to curcumin due to its therapeutic properties and multiple biological and pharmacological roles, including antioxidant [31, 33], cardioprotective [32], thrombo-suppressive [33], lipid modulating [34], antidiabetic [35], neuroprotective [36], and antidepressant [37] properties. Furthermore, multiple studies have reported that curcumin has little or no toxicities, even at high dosages [38].

The anti-inflammatory effects of curcumin have been demonstrated to affect pathways in multiple psychiatric disorders, including neurochemical imbalances, hypothalamus–pituitary–adrenal (HPA) axis dysfunction, inflammatory dysregulation, reduced oxidative and nitrosative cellular defense, and mitochondrial dysfunction [39, 40]. Accordingly, curcumin may have therapeutic potential in conditions such as Parkinson's disease [41], schizophrenia [42], Alzheimer's disease [43], and several neurological diseases [23, 44]. Since the pathophysiology of depression is strongly related to dysregulation of oxidative-reductive and inflammatory pathways, it has been hypothesized that curcumin might have beneficial effects in the treatment of MDD.

Remarkably, several experimental studies on the antidepressant and anxiolytic effects of curcumin have been conducted in animal models and human clinical studies [45–47]. Given the vast literature on MDD and secondary forms of depression, we decided to collect and critically review interventional clinical studies which have addressed the bioavailability and potential benefits of curcumin in studies of depression and anxiety. With this objective, MEDLINE and Scopus databases were meticulously searched using relevant keywords to identify well-designed clinical studies addressing these points.

17.2 Curcumin and MDD

We identified eight randomized controlled trials that addressed the role of curcumin supplementation on depression and anxiety in MDD patients (Table 17.1).

The effect of adjuvant curcumin consumption on MDD patients was assessed by Kanchanatawan

Table 17.1 Clinical studies assessing the effect of curcumin on depression and/or anxiety in MDD patients

Study type	Intervention period	Follow-up	Study population and treatment dose	Outcomes	*p*-value	Ref.
Randomized, double-blind, placebo-controlled	12 weeks	2, 4, 8, 12, and 16 weeks	65 MDD patients treated with 500–1500 mg/day adjunctive curcumin (n = 30) or placebo (n = 31)	From weeks 12–16, adjunctive curcumin improved MADRS scores more effectively than placebo Greater response in men Curcumin treatment was safe and well tolerated	<0.05	[48]
Randomized, double-blind, placebo-controlled	12 weeks	4, 8, and 12 weeks	123 MDD patients treated with 500 mg/day curcumin (n = 26), 1000 mg/day curcumin (n = 33), 500 mg/day curcumin +30 mg/day saffron (n = 24), or placebo (n = 31)	Significant reduction of IDS-SR30 and STAI scores in major and atypical depression identified in groups receiving active drugs	<0.05	[49]
Controlled trial	6 weeks	6 weeks	111 MDD patients treated with standard antidepressant plus 1000–10 mg/day curcuminoids–piperine combination (n = 61) or standard antidepressant alone (n = 50)	↓ HADS and ↓ BDI-II	<0.05	[37]
Randomized, double-blind, placebo-controlled	8 weeks	8 weeks	50 participants treated with 1000 mg/day curcumin	Increased urinary thromboxane B2 and substance P. Greater reductions in IDS-SR30 score related to higher baseline plasma endothelin-1 and leptin in curcumin-treated individuals Significant reduction of cortisol and aldosterone in placebo group	<0.05	[50]
Randomized, double-blind, placebo-controlled pilot study	6 weeks	6 weeks	108 male MDD patients treated with 2000 mg/day curcumin (n = 50) or placebo soybean powder (n = 50)	↑ ↓HDRS-17 and ↓MADRS scales. ↓ IL-1β, TNFα, and ↑ BDNF	<0 0.05	[51]
Randomized, observer masked, controlled trial	6 weeks	2, 4, and 6 weeks	60 MDD patients treated with 1000 mg/day curcumin (n = 16), 20 mg/day fluoxetine (n = 17), or 1000–20 mg/day curcumin plus fluoxetine (n = 18)	Curcumin-fluoxetine combination showed higher non-significant response in HAM-D17 score and was safe	NS	[12]

(continued)

Table 17.1 (continued)

Study type	Intervention period	Follow-up	Study population and treatment dose	Outcomes	*p*-value	Ref.
Randomized, double-blind, placebo-controlled	8 weeks	4 and 8 weeks	56 MDD patients treated with 1000 mg/day curcumin ($n = 28$) or placebo ($n = 28$)	Depressive symptoms improved significantly in group × time interaction for IDS-SR30 total score and IDS-SR30 mood score during the second 4 weeks. Atypical depression showed greater response. Only curcumin group showed improvement in STAI at weeks 4–8	<0 0.05	[52]
Randomized, double-blind, placebo-controlled pilot study	5 weeks	Each visit up to 6 weeks	40 patients with first-episode MDD treated with 500-mg/day curcumin ($n = 20$) or placebo ($n = 20$) together with antidepressants (escitalopram or venlafaxine)	No difference between curcumin and placebo based on MADRS, HDRS, and CGI-S	NS	[53]

MDD major depressive disorder, *MADRS* Montgomery-Asberg Depression Rating Scale, *STAI* Spielberger State-Trait Anxiety Inventory, *IDS-SR30* Symptomatology self-rated version, *HADS* Hospital Anxiety and Depression Scale, *BDI-II* Beck Depression Inventory, *HDRS-17* 17-item Hamilton Depression Rating Scale, *IL-1β* interleukin 1β, *TNF α* tumor necrosis factor α, *BDNF* brain-derived neurotrophic factor, *HAM-D17* Hamilton Depression Rating Scale, *HDRS* Hamilton Depression Rating Scale, *CGI-S* Clinical Global Impression-Severity Scale

et al. [48]. In this randomized, double-blind, placebo-controlled study, 33 out of 65 patients received curcumin (increasing doses ranging from 500 to 1500 mg/day) and care as usual for 12 weeks. Between 12 and 16 weeks after treatment initiation, Montgomery-Asberg Depression Rating Scale (MADRS) scores were significantly improved in the group receiving curcumin. In addition, the study confirmed a higher therapeutic efficacy of curcumin in males compared to females. Combining curcumin with usual antidepressant therapy was safe and well tolerable.

Lopresti et al. conducted a randomized, double-blind, placebo-controlled study to assess the therapeutic effects of different doses of curcumin and a curcumin-saffron combination on depression and anxiety symptoms [49]. MDD patients ($n = 123$) were randomized to one of four treatment regimes, consisting of two doses of curcumin extract (500 or 1000 mg/day), curcumin/saffron combination (30 mg/day), or placebo for a 12-week period. There were positive responses to all three doses of active drug treatment using Spielberger State-Trait Anxiety Inventory (STAI) and the Inventory of Depressive Symptomatology self-rated version (IDS-SR30) scores in comparison to the placebo group. No significant difference was reported between the three real drug treatment groups, although active all interventions had a higher response in the atypical depression subgroup compared to the remainder of patients. In respect of this study, it was speculated that curcumin could exert antidepressant and anti-anxiety effects in MDD and particularly in atypical depressive symptoms.

Panahi et al. aimed to evaluate whether curcuminoids as an add-on to convention antidepressants improved symptoms of MDD patients [37]. The trial was conducted with 111 subjects with MDD. Patients were divided into two groups that received either curcuminoids (Complex®; 1000 mg/day) plus 10 mg/day piperine (Bioperine®) and standard antidepressant treatment ($n = 61$) or standard care ($n = 50$) over 6 weeks. By the end of the study, the analysis of the Hospital Anxiety and Depression Scale and the Beck Depression

Inventory II data showed a substantial reduction of anxiety and depression severity in the curcuminoid group compared to the controls.

In a study by Lopresti et al., an analysis of previous findings was carried out to investigate the underlying antidepressant mechanisms of curcumin supplementation [50]. Urinary, salivary, and blood samples of 50 MDD patients were collected to evaluate peripheral biomarkers pre- and post-consumption of curcumin (500 mg twice/day) or placebo for 8 weeks. The primary assessment of depressive symptoms was analyzed using the Inventory of Depressive Symptomatology (IDS)-SR30 scale. Curcumin supplementation (BCM-95 s) was found to significantly elevate urinary thromboxane B2, and substance P levels, while the placebo treatment was associated with significant reductions in urinary cortisol and aldosterone. Higher baseline plasma leptin and endothelin-1 in the curcumin group was related to a further decrease in IDS-SR30 scores over 8 weeks of treatment. The findings highlighted the association between the antidepressant mechanism of curcumin and the alteration of both clinical and molecular biomarkers.

In a randomized controlled trial, Yu et al. studied the efficacy of standard antidepressant-treatment improvement with the addition of curcumin in MDD patients [51]. One hundred and eight male adults recruited into the study and divided into two groups: (1) curcumin capsules (2000 mg/day) or (2) placebo in addition to their current antidepressant for 6 weeks. The efficacy of chronic curcumin treatment was observed as a significant reduction in the Chinese version of the 17-item Hamilton Depression (HAM-D17) and MADRS scales. Furthermore, the levels of the cytokines interleukin 1β (IL-1β), tumor necrosis factor α (TNF-α), and salivary cortisol were decreased, whereas brain-derived neurotrophic factor (BDNF) was increased in the curcumin compared to the placebo group.

Sanmukhani et al. performed a randomized, observer-masked intervention-controlled trial to investigate the safety and efficacy of curcumin along with the standard antidepressant fluoxetine in people with MDD without concurrent suicidal ideation or psychosis [12]. Sixty subjects were randomized into three parallel treatment arms, including 1 g curcumin, 20 mg fluoxetine, and a combination of curcumin and fluoxetine. After 6 weeks, the reduction of scores in the HAM-D17 scale was greater, but nonsignificant in the combination group. The mean change in HAM-D17 scores was comparable in all three groups. On account of the good tolerability of curcumin, these findings supported its use a safe intervention in MDD.

In 2014, Lopresti et al. evaluated the response of individuals with MDD to treatment with curcumin through a randomized, double-blind, placebo-controlled study [52]. Fifty-six patients were asked to take a capsule containing either curcumin (500 mg twice/day) or placebo over an 8-week period. Exclusively, the IDS-SR30 total and mood score showed a significant drop in the curcumin group compared with control in the last 4 weeks of the study. Likewise, a substantial reduction in anxiety symptoms was found only in the curcumin group (particularly in the atypical depression sub-group) based on STAI scores. These findings suggested that curcumin may have both antidepressant and anxiolytic effects in MDD.

In a pilot clinical study by Bergman et al., curcumin was applied as an add-on treatment with existing antidepressant drugs to define its efficacy [53]. Forty patients with first episode MDD were randomized into two groups of 20. Participants were prescribed either Curcumin Forte (500 mg/day) or placebo with antidepressants (escitalopram or venlafaxine) for 5 weeks. According to HAMD-17, Clinical Global Impression-Severity (CGI-S) Scale, and MADRS scores, both groups showed significant improvements in depressive symptoms after 7 days of treatment, although no significant difference was detected between the curcumin and placebo groups. However, patients in the curcumin group showed a nonsignificant trend toward a more rapid recovery from depressive symptoms compared to those in the placebo.

17.3 Curcumin and Other Disorder Apart from MDD

Five clinical studies considered the impact of curcumin supplementation on depression and anxiety disorders among subjects without MDD (Table 17.2).

In a randomized, double-blind, placebo-controlled clinical study of diabetic patients with peripheral neuropathy, Asadi et al. assessed the efficacy of curcumin on anxiety, depression, and stress [54]. Eighty patients were randomly assigned to receive either 80 mg/day curcumin ($n = 40$) or placebo for 8 weeks. To improve the bioavailability, a nanocurcumin supplement was used. This revealed a significant drop in Depression, Anxiety, Stress Scale (DASS-21-items) scores in the patients.

The objectives of another randomized, placebo-controlled trial were to compare the effects of curcumin on depression, anxiety, and other cardiovascular risk factors among patients

Table 17.2 Clinical studies assessing the effect of curcumin on depression and anxiety in non-MDD patients

Study type	Intervention period	Follow-up	Study population and treatment dose	Outcomes	*P*-value	Ref.
Randomized, double-blind, placebo-controlled	8 weeks	8 weeks	80 diabetic patients treated with 80 mg/day curcumin ($n = 40$) or placebo ($n = 40$)	Curcumin improved anxiety and depression scores compared to placebo based on DASS-21-items	< 0.05	[54]
Randomized, double-blind, placebo-controlled	6 weeks	6 weeks	120 metabolic syndrome patients treated with 1 g/day phospholipidated curcumin (PC) ($n = 40$), unformulated curcumin (UC) ($n = 40$) or placebo ($n = 40$)	According to BAI and BDI, significant decrease in severity but no improvement in anxiety identified in PC and UC groups No significant improvement in depressive symptoms	< 0.05	[55]
Double-blind, placebo-controlled	4 weeks	4 weeks	14 systemic lupus erythematosus patients treated with 50 mg/day curcuminoids ($n = 10$) or placebo ($n = 4$)	Significant reduction of TNF α but non-significant reduction of BDI Positive correlation between TNF α and BDI	<0.001	[56]
Randomized, double-blind, placebo-controlled	30 days	30 days	60 patients with occupational stress and anxiety treated with 1000 mg/day CGM ($n = 20$), unformulated standard curcumin ($n = 20$) or placebo ($n = 20$)	Improvement in anxiety symptoms, safety, and tolerance among CGM-treated patients compared with unformulated curcumin based on PSS-14, BAI, and SF-36	<0 0.05	[44]
Randomized, double-blind, cross-over trial	30 days intervention and 2 weeks wash-out, followed by 30 days intervention	4, 6, and 10 weeks	30 obese individuals treated with 1 g/day curcumin ($n = 15$) or placebo ($n = 15$)	↓mean BAI scores No impact on BDI	0.03	[57]

CGM curcumagalactomannoside, *DASS-21-items* depression, anxiety, stress scale, *BDI* beck depression inventory, *BAI* beck anxiety inventory, *TNF α* tumor necrosis factor α, *PSS-14* perceived stress scale with 14 items

with metabolic syndrome [55]. For this study, 120 subjects were assigned to one of three treatment groups: (1) phospholipidated curcumin (PC) capsules (1 g/day); (2) unformulated curcumin (UC) capsules (1 g/day); or (3) placebo capsule for 6 weeks. There were no substantial differences between UC and PC groups in biochemical readouts, mean blood pressure, or complete blood count, and the PC complex was no more effective than UC in any of these parameters. According to the Beck Depression Inventory (BDI) and Beck Anxiety Inventory (BAI) scores, curcumin supplementation (PC and UC) led to a substantial decrease in the severity but not the improvement of anxiety. In addition, no significant improvement in severity of depression was reported.

Setiawati et al. investigated the effect of curcumin on the appearance of depression in outpatients with systemic lupus erythematosus (SLE) [56]. In this double-blind, placebo-controlled study, 14 SLE patients were randomly assigned to groups for treatment with either Curcuma xanthorrhiza Roxb. extract (50 mg/day) or placebo for 4 weeks. Measurement of TNF-α concentrations and BDI scores were employed to evaluate efficacy. The results showed a significantly lower level of TNF-α in the treatment group, but the BDI score reduction was not significant. A positive correlation was observed between these two components.

Sudheeran et al. compared the safety, antioxidant effectiveness, and bioavailability of CurQfen (curcumagalactomannoside [CGM]) with standard curcumin and placebo in a 30-day study [44]. In this randomized, double-blind, placebo-controlled trial, 60 individuals with occupational stress were enrolled to receive either 1000 mg/day CGM, unformulated curcumin, or placebo. Based on the variations of BAI, Perceived Stress Scale with 14 items, and SF-36 mean scores, a substantial reduction was found in stress and anxiety symptoms of CGM-treated patients in comparison to the standard curcumin group. Moreover, correlations were found between the anxiety improvement, endogenous antioxidant biomarkers, and lipid peroxidation levels. In addition, data from hematological and biochemical evaluation highlighted the safety, tolerance, and bioavailability of oral CGM.

Esmaily et al. investigated the improvement of depressive and anxiety symptoms in 30 obese individuals treated with curcuminoids in a randomized, double-blind, crossover trial [57]. Patients were allocated to receive either curcumin (1 g/day) or placebo in a 1:1 ratio for 30 days. After a 2-week wash-out period, participants were then crossed-over to the other arm for a further 30 days. BAI and BDI scores were measured at the beginning and at weeks 4, 6, and 10. The impact of curcumin supplementation on improving anxiety symptoms was significant. However, no substantial improvement of depression was observed.

17.4 Mechanism

Several investigations have clarified the relationship between curcumin supplementation and various biomarkers. However, an understanding of the detailed mechanisms involved in the antidepressant and anxiolytic properties of curcumin requires further investigation.

17.4.1 Depression

MDD is generally considered a condition associated with a range of abnormalities, including reduced neurogenesis and increased glucocorticoid resistance, inflammatory response, oxidative stress, and disturbances in the HPA axis and neurotransmitter release. Interestingly, all of these abnormalities could be modulated by curcumin [40]. The most common pro-inflammatory cytokines that have been evaluated in depression include IL-1β, IL-6, and TNF-α [58]. In line with the anti-inflammatory properties of the curcuminoids, these compounds inhibit lipid peroxidation and free radical scavenging and suppress key regulators of inflammatory pathways such as mitogen-activated protein kinase, cyclooxygenase-II, NF-κb, and inducible nitric oxide synthase [44, 59, 60]. Most likely linked to these effects, curcumin appears to ameliorate mitochondrial dysfunction which is often found in psychiatric disorders [37]. Furthermore, curcuminoids can mitigate other physiological and molecular characteristics of depression, such as the neurotoxicity mediated by corticosterone, HPA imbalances [61], and reduced BDNF levels in the hippocampus [62]. Curcuminoids may also increase dopamine and serotonin concentrations via inhibition of the monoamine oxidase enzyme in an animal model of depression [63].

In a randomized controlled trial, Lopresti et al. explored relevant biomarkers associated with the antidepressant mechanisms of curcumin. They found associations with urinary thromboxane B2, substance P, cortisol and aldosterone, as well as baseline plasma concentrations of endotellin and leptin [50].

Substance P has many functions but is known as a stress-related neuropeptide and is highly expressed in the cerebrospinal fluid of depressed patients [64, 65]. It is reported that stressors activate the substance P preferred receptor, neurokinin-1 (NK1) [66]. Moreover, NK1 antagonists have been shown to have antidepressant and anxiolytic properties in clinical trials and experimental studies [64]. It is interesting in this regard that serum substance P levels appear to be decreased following curcumin administration [67].

Thromboxane B2 is an inactive metabolite of thromboxane A2 expressed in activated thrombocytes of MDD patients [49, 68, 69]. High levels of this metabolite have been reported in MDD patients [69]. Curcumin treatment also increased urinary level thromboxane B2 in MDD patients [50] and inhibited its production in human thrombocytes [70].

Changes in cortisol and aldosterone levels in MDD patients have been reported previously [71, 72] and the randomized, double-blind, placebo-controlled trial described above found decreased urinary aldosterone and cortisol in the placebo treatment group [50].

In addition, several investigations have noted an impaired function of the endothelin system in MDD patients [73]. In the line of this finding, an elevated level of endothelin-1, a potent vasoactive peptide, has been observed in severe depression [74]. However, in another randomized controlled trial on MDD patients, endothelin-1was not altered by curcumin, but higher antidepressant efficacy was observed in curcumin-treated individuals with a higher baseline plasma level of this biomarker [50].

In addition to the clinical evidence, several studies have shown decreased levels of leptin in animal models of depression, although the findings regarding this adipokine hormone have been inconsistent in human studies [75, 76]. From investigations on leptin in depression, effects on HPA axis regulation, monoamine activity, immune responses, and neurotrophic factors have been reported [76]. In vitro and animal models of inflammation have also found that curcumin can lower serum leptin levels [77]. In a randomized trial, leptin levels of diabetic patients decreased following curcumin administration [78]. In addition, high baseline plasma concentration of leptins was related to a greater antidepressant effect of curcumin [50].

As the effects of curcumin appear to involve multiple physiological pathways, these findings suggest that evaluating changes in biomarkers collectively is a more effective approach to increasing out knowledge about its mechanisms of action, rather than observing changes in isolated biomarkers.

17.4.2 Anxiety Disorder

Several studies have noted that curcumin could attenuate anxiety-like behaviors in animal models. The most commonly described mediators of anxiety disorder pathogenicity are noradrenaline, dopamine, serotonin, and gamma-aminobutyric acid [79], and there is now considerable evidence that the anxiolytic properties of curcumin could be associated with an amelioration in the levels of these neurotransmitters in the brain [80].

A preclinical study in mice exposed to chronic social stress found that curcumin boosted resilience to the stress and diminished anxiety-like symptoms in these mice, and this was associated with reduced levels of corticosterone and IL-6 [47].

In line with the effects of curcumin on inhibition of nitric oxide, intraperitoneal injection of curcumin was found to reduce anxiety behavior and lower nitrite levels in rats [81]. Moreover, the involvement of only inducible nitric oxide synthase (NOS) and not neuronal NOS was associated with these anxiolytic effects [82]. According to the serotonin theory of psychiatric diseases, changes in the expression of serotonin 5HT-1A receptors can lead to anxiety and depression [83]. This is consistent with findings that 5HT-1A mRNA expression can be upregulated by curcumin in rats under chronic stress, in association with anxiolytic effects [62].

Finally, effects of curcumin in depression and anxiety disorders support the case that it might be of therapeutic value in a range of neurological disorders such as epilepsy, Parkinson's disease, ischemic stroke, Alzheimer's disease, and depression [84].

17.5 Conclusion

A number of clinical and preclinical studies have now shown encouraging results concerning the beneficial effects of curcumin supplementation as a treatment for mental disorders. A recent systenatic review and meta-analysis of clinical trials has also supported the positive impact of curcuminoids supplementation on health-related quality of life [85]. Notably, certain clinical studies have documented that the curcumin administration may exert beneficial effects by improving the effects of standard treatments in depression and anxiety. Moreover, curcumin consumption has been reported as a safe and effective co-treatment in MDD patients. Nevertheless, some clinical reports have shown no definitive evidence along these lines. It could be that curcumin offers the most benefit when used as an add-on with conventional treatments. In line with this, a meta-analysis by Fusar-Poli et al. found that curcumin could alleviate depressive and anxiety symptoms when added to conventional pharmaceutical regimens in patients with depression [86]. Taken together, these findings support the case for further studies on the potential effects of curcumin in the treatment of mental disorders in well-designed randomized controlled trials.

References

1. First MB, Wakefield JC (2013) Diagnostic criteria as dysfunction indicators: bridging the chasm between the definition of mental disorder and diagnostic criteria for specific disorders. Can J Psychiatry 58(12):663–669
2. Stein DJ, Phillips KA, Bolton D, Fulford K, Sadler JZ, Kendler KS (2010) What is a mental/psychiatric disorder? From DSM-IV to DSM-V. Psychol Med 40(11):1759–1765
3. Murray CJL, Lopez AD (1996) The global burden of disease: a comprehensive assessment of mortality and disability from diseases, injuries, and risk factors in 1990 and projected to 2020: summary, World Health Organization. Harvard University Press, Cambridge, MA. ISBN-10: 0965546608
4. Steel Z, Marnane C, Iranpour C, Chey T, Jackson JW, Patel V et al (2014) The global prevalence of common mental disorders: a systematic review and meta-analysis 1980–2013. Int J Epidemiol 43(2):476–493
5. Kessler RC, Greenberg PE, Mickelson KD, Meneades LM, Wang PS (2001) The effects of chronic medical conditions on work loss and work cutback. J Occup Environ Med 43(3):218–225
6. Herzig L, Muhlemann N, Burnand B, Favrat B, Haftgoli N, Verdon F et al (2012) Development of mental disorders one year after exposure to psychosocial stressors; a cohort study in primary care patients with a physical complaint. BMC Psychiatry 12:120. https://doi.org/10.1186/1471-244X-12-120
7. Vos T, Allen C, Arora M, Barber RM, Bhutta ZA, Brown A et al. (2016) Global, regional, and national incidence, prevalence, and years lived with disability for 310 diseases and injuries, 1990–2015: a systematic analysis for the Global Burden of Disease Study 2015. Lancet 388(10053):1545-1602.
8. de Zwart PL, Jeronimus BF, de Jonge P (2019) Empirical evidence for definitions of episode, remission, recovery, relapse and recurrence in depression: a systematic review. Epidemiol Psychiatr Sci 28(5):544–562
9. Weitz E, Kleiboer A, van Straten A, Cuijpers P (2018) The effects of psychotherapy for depression on anxiety symptoms: a meta-analysis. Psychol Med 48(13):2140–2152
10. Murphy SA, Oslin DW, Rush AJ, Zhu J, MCATS (2007) Methodological challenges in constructing effective treatment sequences for chronic psychiatric disorders. Neuropsychopharmacology 32(2):257–262
11. National Collaborating Centre for Mental Health (UK) (2010) Depression: the treatment and management of depression in adults (updated edition). British Psychological Society, Leicester, UK. PMID: 22132433 NBK63748
12. Sanmukhani J, Satodia V, Trivedi J, Patel T, Tiwari D, Panchal B et al (2014) Efficacy and safety of curcumin in major depressive disorder: a randomized controlled trial. Phytother Res 28(4):579–585

13. Stanilla JK, Simpson GM (1997) Clozapine withdrawal resulting in delirium with psychosis: a report of three cases. J Clin Psychiatry 58(6):252–255
14. Rothschild AJ (2000) Sexual side effects of antidepressants. J Clin Psychiatry 61(Suppl 11):28–36
15. Haddad PM, Dursun SM (2008) Neurological complications of psychiatric drugs: clinical features and management. Hum Psychopharmacol Clin 23(S1):S15–S26
16. Borison RL, Diamond B, Sinha D, Gupta R, Ajiboye P (1988) Clozapine withdrawal rebound psychosis. Psychopharmacol Bull 24(2):260–263
17. Goethe JW, Woolley SB, Cardoni AA, Woznicki BA, Piez DA (2007) Selective serotonin reuptake inhibitor discontinuation: side effects and other factors that influence medication adherence. J Clin Psychopharmacol 27(5):451–458
18. Insel TR (2012) Next-generation treatments for mental disorders. Sci Transl Med 4(155):155. ps19. https://doi.org/10.1126/scitranslmed.3004873
19. Akhondzadeh S, Abbasi SH (2006) Herbal medicine in the treatment of Alzheimer's disease. Am J Alzheimers Dis Other Demen 21(2):113–118
20. Banach M, Patti AM, Giglio RV, Cicero AFG, Atanasov AG, Bajraktari G, Bruckert E, Descamps O, Djuric DM, Ezhov M, Fras Z, von Haehling S, Katsiki N, Langlois M, Latkovskis G, Mancini GBJ, Mikhailidis DP, Mitchenko O, Moriarty PM, Muntner P, Nikolic D, Panagiotakos DB, Paragh G, Paulweber B, Pella D, Pitsavos C, Reiner Ž, Rosano GMC, Rosenson RS, Rysz J, Sahebkar A, Serban MC, Vinereanu D, Vrablík M, Watts GF, Wong ND, Rizzo M (2018) International Lipid Expert Panel (ILEP). The Role of Nutraceuticals in Statin Intolerant Patients. J Am Coll Cardiol 72(1):96–118. https://doi.org/10.1016/j.jacc.2018.04.040. PMID: 29957236
21. Soleimani V, Sahebkar A, Hosseinzadeh H (2018) Turmeric (Curcuma longa) and its major constituent (curcumin) as nontoxic and safe substances: Review. Phytotherapy Research 32(6):985–995
22. Bar-Sela G, Epelbaum R, Schaffer M (2010) Curcumin as an anti-cancer agent: review of the gap between basic and clinical applications. Curr Med Chem 17(3):190–197
23. Olivera A, Moore TW, Hu F, Brown AP, Sun A, Liotta DC et al (2012) Inhibition of the NF-κB signaling pathway by the curcumin analog, 3,5-Bis (2-pyridinylmethylidene)-4-piperidone (EF31): anti-inflammatory and anti-cancer properties. Int Immunopharmacol 12(2):368–377
24. Brietzke E, Mansur RB, Zugman A, Carvalho AF, Macêdo DS, Cha DS et al (2013) Is there a role for curcumin in the treatment of bipolar disorder? Med Hypotheses 80(5):606–612
25. Abdollahi E, Momtazi AA, Johnston TP, Sahebkar A (2018) Therapeutic effects of curcumin in inflammatory and immune-mediated diseases: a nature-made jack-of-all-trades? J Cell Physiol 233(2):830–848
26. Hasanzadeh S, Read MI, Bland AR, Majeed M, Jamialahmadi T, Sahebkar A (2020) Curcumin: an inflammasome silencer. Pharmacol Res 159:104921. https://doi.org/10.1016/j.phrs.2020.104921. Epub 2020 May 25. PMID: 32464325.
27. Teymouri M, Pirro M, Johnston TP, Sahebkar A (2017) Curcumin as a multifaceted compound against human papilloma virus infection and cervical cancers: A review of chemistry, cellular, molecular, and preclinical features. BioFactors 43(3): 331–346.
28. Aggarwal BB, Sundaram C, Malani N, Ichikawa H (2007) Curcumin: the Indian solid gold. In: The molecular targets and therapeutic uses of curcumin in health and disease. Springer, pp 1–75
29. Mollazadeh H, Cicero AFG, Blesso CN, Pirro M, Majeed M, Sahebkar A (2019) Immune modulation by curcumin: the role of interleukin-10. Crit Rev Food Sci Nutr 59(1):89–101
30. Momtazi AA, Sahebkar A (2016) Difluorinated curcumin: a promising curcumin analogue with improved anti-tumor activity and pharmacokinetic profile. Curr Pharm Des 22(28):4386–4397
31. Sahebkar, Amirhossein & Serban, Maria-Corina & Urosniu, Sorin & Banach, Maciej (2015) Effect of curcuminoids on oxidative stress: A systematic review and meta-analysis of randomized controlled trials. Journal of Functional Foods 241. https://doi.org/10.1016/j.jff.2015.01.005
32. Sahebkar A (2013) Why it is necessary to translate curcumin into clinical practice for the prevention and treatment of metabolic syndrome? Biofactors 39(2):197–208
33. Srivastava R, Dikshit M, Srimal R, Dhawan B (1985) Anti-thrombotic effect of curcumin. Thromb Res 40(3):413–417
34. Simental-Mendía LE, Pirro M, Gotto AM, Jr., Banach M, Atkin SL, Majeed M, Sahebkar A (2019) Lipid-modifying activity of curcuminoids: A systematic review and meta-analysis of randomized controlled trials. Critical Reviews in Food Science and Nutrition 59(7):1178–1187.
35. Chuengsamarn S, Rattanamongkolgul S, Luechapudiporn R, Phisalaphong C, Jirawatnotai S (2012) Curcumin extract for prevention of type 2 diabetes. Diabetes Care 35(11):2121–2127
36. Forouzanfar F, Read MI, Barreto GE, Sahebkar A (2020) Neuroprotective effects of curcumin through autophagy modulation. IUBMB Life 72(4):652–664. https://doi.org/10.1002/iub.2209. Epub 2019 Dec 5. PMID: 31804772.
37. Panahi Y, Badeli R, Karami GR, Sahebkar A (2015) Investigation of the efficacy of adjunctive therapy with bioavailability-boosted curcuminoids in major depressive disorder. Phytother Res 29(1):17–21
38. Lao CD, Ruffin MTT, Normolle D, Heath DD, Murray SI, Bailey JM et al (2006) Dose escalation of a curcuminoid formulation. BMC Complement Altern Med 6:10. https://doi.org/10.1186/1472-6882-6-10
39. Lopresti AL (2017) Curcumin for neuropsychiatric disorders: a review of in vitro, animal and human studies. J Psychopharmacol 31(3):287–302

40. Lopresti AL, Hood SD, Drummond PD (2012) Multiple antidepressant potential modes of action of curcumin: a review of its anti-inflammatory, monoaminergic, antioxidant, immune-modulating and neuroprotective effects. J Psychopharmacol 26(12):1512–1524
41. Ji HF, Shen L (2014) The multiple pharmaceutical potential of curcumin in Parkinson's disease. CNS & Neurological Disorders-Drug Targets (Formerly Current Drug Targets-CNS & Neurological Disorders) 13(2):369–373
42. Miodownik C, Lerner V, Kudkaeva N, Lerner PP, Pashinian A, Bersudsky Y et al (2019) Curcumin as add-on to antipsychotic treatment in patients with chronic schizophrenia: a randomized, double-blind, placebo-controlled study. Clin Neuropharmacol 42(4):117–122
43. Baum L, Lam CWK, Cheung SK-K, Kwok T, Lui V, Tsoh J et al (2008) Six-month randomized, placebo-controlled, double-blind, pilot clinical trial of curcumin in patients with Alzheimer disease. J Clin Psychopharmacol 28(1):110–113
44. Pandaran Sudheeran S, Jacob D, Natinga Mulakal J, Gopinathan Nair G, Maliakel A, Maliakel B et al (2016) Safety, tolerance, and enhanced efficacy of a bioavailable formulation of curcumin with Fenugreek dietary fiber on occupational stress: a randomized, double-blind, placebo-controlled pilot study. J Clin Psychopharmacol 36(3):236–243
45. Zhang WY, Guo YJ, Han WX, Yang MQ, Wen LP, Wang KY et al (2019) Curcumin relieves depressive-like behaviors via inhibition of the NLRP3 inflammasome and kynurenine pathway in rats suffering from chronic unpredictable mild stress. Int Immunopharmacol 67:138–144
46. Yi Y, Li J, Chen W (2019) Curcumin evokes antidepressant-like effects in mice by regulating miR-124/brain derived neurotrophic factor. Trop J Pharm Res 18(3):603–609
47. Aubry AV, Khandaker H, Ravenelle R, Grunfeld IS, Bonnefil V, Chan KL et al (2019) A diet enriched with curcumin promotes resilience to chronic social defeat stress. Neuropsychopharmacology 44(4):733–742
48. Kanchanatawan B, Tangwongchai S, Sughondhabhirom A, Suppapitiporn S, Hemrunrojn S, Carvalho AF et al (2018) Add-on treatment with curcumin has antidepressive effects in thai patients with major depression: results of a randomized double-blind placebo-controlled study. Neurotoxicity Res 33(3):621–633
49. Lopresti AL, Drummond PD (2017) Efficacy of curcumin, and a saffron/curcumin combination for the treatment of major depression: a randomised, double-blind, placebo-controlled study. J Affect Disord 207:188–196
50. Lopresti AL, Maes M, Meddens MJ, Maker GL, Arnoldussen E, Drummond PD (2015) Curcumin and major depression: a randomised, double-blind, placebo-controlled trial investigating the potential of peripheral biomarkers to predict treatment response and antidepressant mechanisms of change. Eur Neuropsychopharmacol 25(1):38–50
51. Yu JJ, Pei LB, Zhang Y, Wen ZY, Yang JL (2015) Chronic supplementation of curcumin enhances the efficacy of antidepressants in major depressive disorder: a randomized, double-blind, placebo-controlled pilot study. J Clin Psychopharmacol 35(4):406–410
52. Lopresti AL, Maes M, Maker GL, Hood SD, Drummond PD (2014) Curcumin for the treatment of major depression: a randomised, double-blind, placebo controlled study. J Affect Disord 167:368–375
53. Bergman J, Miodownik C, Bersudsky Y, Sokolik S, Lerner PP, Kreinin A et al (2013) Curcumin as an add-on to antidepressive treatment: a randomized, double-blind, placebo-controlled, pilot clinical study. Clin Neuropharmacol 36(3):73–77
54. Asadi S, Gholami MS, Siassi F, Qorbani M, Sotoudeh G (2019) Beneficial effects of nano-curcumin supplement on depression and anxiety in diabetic patients with peripheral neuropathy: a randomized, double-blind, placebo-controlled clinical trial. Phytother Res 34(4):896–903
55. Saberi-Karimian M, Parizadeh SMR, Ghayour-Mobarhan M, Salahshooh MM, Dizaji BF, Safarian H et al (2018) Evaluation of the effects of curcumin in patients with metabolic syndrome. Comp Clin Path 27(3):555–563
56. Setiawati MCN, Ikawati Z, Nyoman K (2017) Effect of Curcuma Xanthorrhiza Roxb. Extract on TNF α concentration and depression's score in patient with systemic lupus erythematosus. Indones J Pharm 28(3):185–190
57. Esmaily H, Sahebkar A, Iranshahi M, Ganjali S, Mohammadi A, Ferns G et al (2015) An investigation of the effects of curcumin on anxiety and depression in obese individuals: a randomized controlled trial. Chin J Integr Med 21(5):332–338
58. Slavich GM, Giletta M, Helms SW, Hastings PD, Rudolph KD, Nock MK et al (2020) Interpersonal life stress, inflammation, and depression in adolescence: testing social signal transduction theory of depression. Depress Anxiety 37(2):179–193
59. Strimpakos AS, Sharma RA (2008) Curcumin: preventive and therapeutic properties in laboratory studies and clinical trials. Antioxid Redox Signal 10(3):511–546
60. Zhou H, Beevers CS, Huang S (2011) The targets of curcumin. Curr Drug Targets 12(3):332–347
61. Xu Y, Li S, Vernon MM, Pan J, Chen L, Barish PA et al (2011) Curcumin prevents corticosterone-induced neurotoxicity and abnormalities of neuroplasticity via 5-HT receptor pathway. J Neurochem 118(5):784–795
62. Xu Y, Ku B, Cui L, Li X, Barish PA, Foster TC et al (2007) Curcumin reverses impaired hippocampal neurogenesis and increases serotonin receptor 1A mRNA and brain-derived neurotrophic factor expression in chronically stressed rats. Brain Res 1162:9–18

63. Kulkarni SK, Bhutani MK, Bishnoi M (2008) Antidepressant activity of curcumin: involvement of serotonin and dopamine system. Psychopharmacology (Berl) 201(3):435–442
64. Holmes A, Heilig M, Rupniak NM, Steckler T, Griebel G (2003) Neuropeptide systems as novel therapeutic targets for depression and anxiety disorders. Trends Pharmacol Sci 24(11):580–588
65. Alldredge B (2010) Pathogenic involvement of neuropeptides in anxiety and depression. Neuropeptides 44(3):215–224
66. Ebner K, Singewald N (2006) The role of substance P in stress and anxiety responses. Amino Acids 31(3):251–272
67. Panahi Y, Sahebkar A, Amiri M, Davoudi SM, Beiraghdar F, Hoseininejad SL et al (2012) Improvement of sulphur mustard-induced chronic pruritus, quality of life and antioxidant status by curcumin: results of a randomised, double-blind, placebo-controlled trial. Br J Nutr 108(7):1272–1279
68. Sellers MM, Stallone JN (2008) Sympathy for the devil: the role of thromboxane in the regulation of vascular tone and blood pressure. Am J Physiol Heart Circ Physiol 294(5):H1978–H1986
69. Piccirillo G, Fimognari FL, Infantino V, Monteleone G, Fimognari GB, Falletti D et al (1994) High plasma concentrations of cortisol and thromboxane B2 in patients with depression. Am J Med Sci 307(3):228–232
70. Shah BH, Nawaz Z, Pertani SA, Roomi A, Mahmood H, Saeed SA et al (1999) Inhibitory effect of curcumin, a food spice from turmeric, on platelet-activating factor-and arachidonic acid-mediated platelet aggregation through inhibition of thromboxane formation and Ca^{2+} signaling. Biochem Pharmacol 58(7):1167–1172
71. Holsboer F, Gerken A, Stalla GK, Muller OA (1987) Blunted aldosterone and ACTH release after human CRH administration in depressed patients. Am J Psychiatry 144(2):229–231
72. Mokhtari M, Arfken C, Boutros N (2013) The DEX/CRH test for major depression: a potentially useful diagnostic test. Psychiatry Res 208(2):131–139
73. Wagner JA, Tennen H, Finan PH, White WB, Burg MM, Ghuman N (2012) Lifetime history of depression, type 2 diabetes, and endothelial reactivity to acute stress in postmenopausal women. Int J Behav Med 19(4):503–511
74. Burg MM, Martens EJ, Collins D, Soufer R (2011) Depression predicts elevated endothelin-1 in patients with coronary artery disease. Psychosom Med 73(1):2–6. https://doi.org/10.1097/PSY.0b013e3181fdfb25
75. Zupancic ML, Mahajan A (2011) Leptin as a neuroactive agent. Psychosom Med 73(5):407–414
76. Lu X-Y (2007) The leptin hypothesis of depression: a potential link between mood disorders and obesity? Curr Opin Pharmacol 7(6):648–652
77. Lee HI, McGregor RA, Choi MS, Seo KI, Jung UJ, Yeo J et al (2013) Low doses of curcumin protect alcohol-induced liver damage by modulation of the alcohol metabolic pathway, CYP2E1 and AMPK. Life Sci 93(18–19):693–699
78. Chuengsamarn S, Rattanamongkolgul S, Phonrat B, Tungtrongchitr R, Jirawatnotai S (2014) Reduction of atherogenic risk in patients with type 2 diabetes by curcuminoid extract: a randomized controlled trial. J Nutr Biochem 25(2):144–150
79. Freitas-Ferrari MC, Hallak JE, Trzesniak C, Filho AS, Machado-de-Sousa JP, Chagas MH et al (2010) Neuroimaging in social anxiety disorder: a systematic review of the literature. Prog Neuropsychopharmacol Biol Psychiatry 34(4):565–580
80. Chimakurthy J, Murthy TE (2010) Effect of curcumin on quinpirole induced compulsive checking: an approach to determine the predictive and construct validity of the model. N Am J Med Sci 2(2):81–86
81. Kumar A, Singh A (2008) Possible nitric oxide modulation in protective effect of (Curcuma longa, Zingiberaceae) against sleep deprivation-induced behavioral alterations and oxidative damage in mice. Phytomedicine 15(8):577–586
82. Gilhotra N, Dhingra D (2010) GABAergic and nitriergic modulation by curcumin for its antianxiety-like activity in mice. Brain Res 1352:167–175
83. Akimova E, Lanzenberger R, Kasper S (2009) The serotonin-1A receptor in anxiety disorders. Biol Psychiatry 66(7):627–635
84. Nabiuni M, Nazari Z, Abdolhamid Angaji S, Safayi Nejad Z (2011) Neuroprotective effects of Curcumin. Aust J Basic Appl Sci 5(9):2224–2240
85. Sadeghian M, Rahmani S, Jamialahmadi T, Johnston TP, Sahebkar A (2020) The effect of oral curcumin supplementation on healthrelated quality of life: A systematic review and meta-analysis of randomized controlled trials. J Affect Disord 29;278:627–636. https://doi.org/10.1016/j.jad.2020.09.091. Epub ahead of print. PMID: 33038707
86. Fusar-Poli L, Vozza L, Gabbiadini A, Vanella A, Concas I, Tinacci S et al (2019) Curcumin for depression: a meta-analysis. Crit Rev Food Sci Nutr:1–11. https://doi.org/10.1080/10408398.2019.1653260

18 A Systematic Review of the Clinical Use of Curcumin for the Management of Gastrointestinal Diseases

Masoumeh Atefi, Mina Darand, Mohammad Hassan Entezari, Tannaz Jamialahmadi, Mohammad Bagherniya, and Amirhossein Sahebkar

Abstract

Gastrointestinal (GI) diseases are highly prevalent worldwide, with considerable morbidity and mortality. Curcumin has been used for many years as a plant-derived product for the management of various conditions such as abdominal pain and poor digestion. This systematic review was undertaken with the aim of investigating the effect of curcumin or turmeric supplementation on GI diseases. A comprehensive systematic search was conducted in PubMed, Scopus, Web of Science and Google Scholar up to March 2020 to identify clinical trials assessing the effect of curcumin/turmeric alone or in combination with other herbs or nutrients on GI diseases. Twenty-one studies comprising 1478 GI patients were included in the study. Four out of seven studies showed a beneficial effect of curcumin/turmeric supplementation on irritable bowel syndrome (IBS) and six out of seven showed positive effects of these herbs on ulcerative colitis. Two out of four studies highlighted the potential role of curcumin/turmeric in eradication of *H. pylori* infection. Both studies conducted on peptic ulcer disease and two out of four studies performed on Crohn's disease demonstrated positive effects of curcumin/turmeric supplementation. One study showed curcumin supplementation had

M. Atefi · M. H. Entezari · M. Bagherniya (✉)
Food Security Research Center, Department of Community Nutrition, School of Nutrition and Food Science, Isfahan University of Medical Sciences, Isfahan, Iran

M. Darand
Department of Community Nutrition, School of Nutrition and Food Science, Food Security Research Center, Isfahan University of Medical Sciences, Isfahan, Iran

T. Jamialahmadi
Department of Food Science and Technology, Quchan Branch, Islamic Azad University, Quchan, Iran

Department of Nutrition, Faculty of Medicine, Mashhad University of Medical Sciences, Mashhad, Iran

A. Sahebkar (✉)
Biotechnology Research Center, Pharmaceutical Technology Institute, Mashhad University of Medical Sciences, Mashhad, Iran

Neurogenic Inflammation Research Center, Mashhad University of Medical Sciences, Mashhad, Iran

Polish Mother's Memorial Hospital Research Institute (PMMHRI), Lodz, Poland

Halal Research Center of IRI, FDA, Tehran, Iran
e-mail: sahebkara@mums.ac.ir

P. C. Guest (ed.), *Studies on Biomarkers and New Targets in Aging Research in Iran*, Advances in Experimental Medicine and Biology 1291, https://doi.org/10.1007/978-3-030-56153-6_18

no effect on familial adenomatous polyposis. However, in another study, curcumin had favorable effects on proctosigmoiditis. Nine studies reported some minor adverse effects. The results of this systematic review suggest a beneficial effect of curcumin/turmeric supplementation on the management of GI diseases. More randomized clinical controlled trials are needed to confirm these results.

Keywords

Gastrointestinal diseases · Curcumin · Turmeric · IBS · Ulcerative colitis

18.1 Introduction

Gastrointestinal (GI) diseases have a high prevalence worldwide with considerable morbidity and mortality. GI diseases lead to around eight million deaths worldwide each year [1]. Herbal medicine has been traditionally used to prevent and manage various GI diseases [2]. Curcumin (diferuloylmethane) is one of the most commonly used medicinal plants for the treatment of conditions such as poor digestion and abdominal pain [3].

Curcumin is a bright yellow phytochemical that is present in turmeric (*Curcuma longa*) and is part of the Ginger family [4, 5]. It is a nutraceutical agent with a good safety and tolerability profile [6] and numerous known pharmacological effects [7–11]. The use of curcumin is increasing worldwide due to its proposed medicinal effects and properties [12–15]. These effects include antioxidant, antibacterial, antifungal, anti-inflammatory, antiangiogenic, and proapoptotic properties [9, 16–19]. Curcumin has been recommended for the treatment of GI diseases such as irritable bowel syndrome (IBS), colitis, Crohn's disease, gallbladder inflammation, and to improve the symptoms of diarrhea [20]. Turmeric has been approved by the German Commission E for the management of dyspeptic symptoms and as an appetite suppressant [21].

Preclinical studies have shown that curcumin has inhibitory effects on *Helicobacter pylori* (*H. pylori*) [9]. It has also been shown to have anti-inflammatory and antioxidant effects on human lymphocytes and gut epithelial cells, and it has been shown to improve murine experimental colitis. Another study showed that taking curcumin for 12 months improved peptic ulcer [22]. Furthermore, curcumin has anticancer and chemopreventive effects and slows the growth of GI cancers [20]. The protective effects of mucosal curcumin have been shown in colitis with potentially beneficial effects on intestinal microbes [5]. The mechanism of curcumin action includes inhibition of transcriptional nuclear factor - κb (NF-KB) activation and CD4 T-cell penetration into the colonic mucosa. Curcumin also reduces the levels of the pro-inflammatory cytokine tumor necrosis factor alpha (TNF-α) via down-regulation of NF-κB transcription factors [12]. Other inflammatory cytokines inhibited by curcumin include interleukin 1(IL-1), IL-2, cyclooxygenase 2 (COX-2), lipoxygenase (LOX), and it induces the anti-inflammatory molecule nitric oxide [3, 23, 24].

Given the economic and social burden of GI diseases on society, and the fact that there are no effective treatments for some GI disorders such as IBS, GI disorder patients use a variety of methods to relieve symptoms, including complementary and alternative medications. Herbal medicinal products and nutritional supplements are the most common adjuvant therapies used for the management of GI disorders [1, 25].

To the best of our knowledge, no systematic review has examined the efficacy of curcumin in patients with various GI disorders such as ulcerative colitis, IBS, and *H. pylori* infection. Here, we have performed a systematic review for assessing the evidence on the use of curcumin for the management of various GI pathologies.

18.2 Material and Methods

This systematic review conformed to the guidelines of the preferred reporting items for systematic reviews and meta-analyses (PRISMA) [26].

18.2.1 Search Strategy

A comprehensive online search was conducted of the medical databases ISI web of science, Scopus, PubMed, and Google Scholar until March 2, 2020. The search did not limit the publication period. The following search terms were used to identify the potential articles of interest in titles and abstracts:

"curcumin" OR "curcuminoids" OR "turmeric".

AND

"peptic ulcer hemorrhage" OR "bleeding peptic ulcer" OR "bleeding duodenal ulcer" OR "bleeding gastric ulcer" OR "peptic ulcer" OR "duodenal ulcer" OR "gastric ulcer" OR "irritable bowel syndrome" OR "Crohn disease" OR "inflammatory bowel disease" OR "colitis" OR "ileitis" OR "ulcerative colitis" OR "regional enteritis".

AND

"Intervention Study" OR "Intervention Studies" OR "Controlled trial" OR "Randomized controlled trial" OR "Randomized clinical trial" OR "Non-Randomized Controlled Trials" OR "Clinical Trial" OR "Non-Randomized Controlled Trials" OR "Cross-Over study" OR "Cross-Over trial" OR "Cross Over trial" OR "Cross Over study" OR "Double-Blind Method" OR "Double-Blind" OR "Double-Blind trial" OR "Double-Blind study".

Whenever possible, Medical Subject Headings (MESH) terms were used. Finally, we performed a direct search of the reference lists of the included original papers and review articles to identify other relevant works for inclusion.

18.2.2 Study Selection

After the initial electronic search and omission of duplicated references, a careful, independent screening of the titles and abstracts was conducted by two independent researchers (MB and MD). If required, full text versions of remaining publications were obtained. Any disagreements were discussed and a solution arrived at. Finally, all human clinical trials that examined the effects of curcumin/turmeric supplementation alone or in combination with the other herbs or nutrients on GI conditions were included.

18.2.3 Inclusion and Exclusion Criteria

Original studies with following criteria were included in our systematic review: (1) having clinical trials design; (2) considering intervention with curcumin or turmeric; and (3) conducted on participants with GI diseases. Studies were excluded if they: (1) reported duplicate data, (2) were non-English language records; and (3) were reviews, letters, editorial articles, study protocols, or case reports.

18.2.4 Data Extraction

After reading the title and abstracts, appropriate articles were selected based on the inclusion and exclusion criteria. The following information was extracted from articles and recorded in Table 18.1: (i) publication information (first author's last name, publication date, and study location); (ii) participant characteristics (total sample size, diagnosis, mean age); (iii) details of the clinical trial (study design, interventions, and trial duration); (iv) main results; (v) side effects; and (vi) Jadad score. For the latter, the SR and MB independently assessed the quality of the studies using the Jadad scale [27]. This covers three domains as follows:

1. Randomization (0-2 points).
2. Blinding (0-2 points).
3. Dropouts and withdrawals (0-1 points).

In this way, the overall quality of each study was calculated as either low (score ≤ 2) or high (score ≥ 3).

Table 18.1 Summary of the studies included in the systematic review (arranged alphabetically by first author's last name)

Author, year, country	Sample size (treatment / control)	Target population	Diagnosis	Age Range (years) or mean	Study design	Intervention	Control	Duration	Main results	Side effects	Jadad Score
Portincasa et al. (2016), Italy [23]	116 (58/58)	IBS (mild-to-moderate symptoms)	Rome III	In:39/Co: 41.5	Parallel	Two capsules of either curcumin and fennel essential oil (curcumin 42 mg and fennel essential oil 25 mg)	Two capsules placebo	30 days	(1) Significant decrease in mean relative IBS-SSS[1] in intervention compared with control group (2) % symptom-free patients significantly higher in intervention compared with control group (3) % symptom-free patients significantly higher in intervention than placebo group (4) IBS-QoL total score significantly greater in intervention than placebo group	None	3 111

Table 18.1 (continued)

Author, year, country	Sample size (treatment / control)	Target population	Diagnosis	Age Range (years) or mean	Study design	Intervention	Control	Duration	Main results	Side effects	Jadad Score
Lior et al. (2019), Israel [24]	22	IBS	Rome III	53.7	Cross-over	Coltect: 500 mg curcumin, 250 mg green tea, and 100 μg selenomethionine (oral tablets)	Placebo	4 weeks, 2 weeks washout	(1) IBS symptom and severity did not significantly change in intervention compared with control group (2) Abdominal pain, bloating, and effects of IBS on daily activity, symptoms, IBS-related QoL and general health quality of life were unchanged	None	5

(continued)

Table 18.1 (Continued)

Author, year, country	Sample size (treatment / control)	Target population	Diagnosis	Age Range (years) or mean	Study design	Intervention	Control	Duration	Main results	Side effects	Jadad Score
Di Ciaulaa et al. (2018), Italy [28]	211	IBS	Rome III	51 ± 17	Trial	42 mg curcumin and 25 mg fennel essential oil (for first 30 days, dose consisted of 2 capsules b.i.d., before meals) (in next 30 days, reduced to 2 capsules q.d. before a meal, for another 30 days)	–	2 months	(1) Significant reduction in severity index and improved QoL in all IBS subgroups (2) IBS-D patients showed worst clinical picture at entry, with highest IBS severity index and the lowest QoL score, compared with IBS-constipation-predominant and IBS-mixed subtypes. IBS-D patients showed the most pronounced response to therapy, considering both scores.	None	0

Table 18.1 (continued)

Author, year, country	Sample size (treatment / control)	Target population	Diagnosis	Age Range (years) or mean	Study design	Intervention	Control	Duration	Main results	Side effects	Jadad Score
Alt et al. (2017), Germany [30]	90 (43/47)	IBS	ROME-III	In:44.0 ± 13.0/ Co:47.5 ± 14.8	Parallel	IQP-CL-101 softgel: 330 mg proprietary blend of curcuminoids and essential oils, 70 mg fish oil, 15 mg peppermint oil and 8 mg caraway oil and 263 μg thiamine, 39 μg folic acid and 625 μg vitamin D3)	Placebo	8 weeks	Significant reduction in IBS-SSS in intervention group compared with placebo	None	5
Brinkhaus et al. (2005), Germany [4]	106 (24/24/58)	IBS	Extensive clinical exam ruling out organic causes	Mean of G1, G2, and G3: 48 ± 12	Parallel	G1: *Curcuma xanthorrhiza* 60 mg daily, G2: *Fumaria officinalis* 1500 mg daily	G3: placebo	18 weeks	IBS-related pain, IBS-related distension, and global assessment of changes in IBS symptoms and psychological stress did not differ significantly among treatment groups	In curcumin group, one patient had constipation and distension	3 111

(continued)

Table 18.1 (continued)

Author, year, country	Sample size (treatment / control)	Target population	Diagnosis	Age Range (years) or mean	Study design	Intervention	Control	Duration	Main results	Side effects	Jadad Score
Lauche et al. (2016), Germany [25]	32	IBS	ROME-III	50.3 ± 11.9	Parallel	5 g Ayurvedic powder mixture (curry (*Murraya koenigii*)), pomegranate (*Punica granatum*) and turmeric (*Curcuma longa* rhizome pulvis) in a 6:3:1 ratio) dissolved in 100 mL warm water	Placebo	4 weeks	No group differences found between groups for IBS symptom intensity	1/3 patients registered at least one minor adverse event	5
Bundy et al. (2004), UK [3]	207	IBS	Rome II criteria	Majority over 50 years old	Partially blinded, randomized	Standardized turmeric extract: First group: 72 mg (1 tablet/day) Second group: 144 mg (2 tablets/ day)	–	8 weeks	(1) IBS prevalence decreased significantly in both groups (53% and 60%) after treatment, in the 1- and 2-tablet groups, respectively (2) Abdominal pain/discomfort score reduced significantly by 22% and 25% in the 1- and 2-tablet group, respectively (3) Significant improvements in all but one IBS-QOL scales of 5–36% in both groups	Minor side effects including flatulence, which affected approx. 1/4 subjects, and dry mouth	3 111

Table 18.1 (continued)

Author, year, country	Sample size (treatment / control)	Target population	Diagnosis	Age Range (years) or mean	Study design	Intervention	Control	Duration	Main results	Side effects	Jadad Score
Khonche et al. (2016), Iran [9]	60 (30/30)	Peptic ulcer Patients	Biopsy	In:35.0 ± 9.3/ Co:35.1 ± 9.0	Parallel	Standard *Helicobacter pylori* (*H. pylori*) therapy + curcumin (500 mg) + piperine (5 mg)	Standard *H. pylori* therapy + placebo (cellulose)	4 weeks	(1) Improvement of dyspepsia symptoms in intervention compared with control group (2) Reduction of subjects with dyspepsia in intervention compared with control group (3) No differences between groups regarding *H. pylori* eradication	No serious adverse event	4 121

(continued)

Table 18.1 (continued)

Author, year, country	Sample size (treatment / control)	Target population	Diagnosis	Age Range (years) or mean	Study design	Intervention	Control	Duration	Main results	Side effects	Jadad Score
Lang et al. (2015), Israel, Hong Kong, and Cyprus [22]	50 (26/24)	Mild-to-moderate ulcerative colitis	$5 \leq$ SCCAI[2]	In:40.4 ± 12.7/ Co:41.4 ± 13.9	Parallel	Mesalamine + curcumin capsules (3 g/day)	Mes-alamine + placebo	1 month	(1) More patients achieved clinical remission in intervention compared with placebo (2) Clinical response (reduction of ≥ 3 points in SCCAI) in more patients in curcumin versus placebo group (3) Higher endoscopic remission (partial Mayo score ≤ 1) observed in curcumin compared with patients (no patients) in the placebo group	Mild nausea, transient increase in stool frequency, and abdominal bloating (no differences between groups)	4 121

Table 18.1 (continued)

Author, year, country	Sample size (treatment / control)	Target population	Diagnosis	Age Range (years) or mean	Study design	Intervention	Control	Duration	Main results	Side effects	Jadad Score
Hanai et al. (2006), Japan [31]	89 (45/44)	Quiescent ulcerative colitis	Endoscopic index	In:45.2 ± 15.8/ Co:39.7 ± 14.2	Parallel	Curcumin (2 capsules; each capsule contained 1 g)	Placebo	6 months	(1) Lower relapse found in curcumin compared with placebo group (2) CAI and EI improved in curcumin compared with control group	Seven patients in curcumin group had abdominal bulging, nausea, transient hypertension, transient increase in number of stools, and elevated Y–guanosine triphosphate level	4 121

(continued)

Table 18.1 (continued)

Author, year, country	Sample size (treatment / control)	Target population	Diagnosis	Age Range (years) or mean	Study design	Intervention	Control	Duration	Main results	Side effects	Jadad Score
Singla et al. (2014), India [32]	45 (23/22)	Mild-to-moderate distal ulcerative colitis-b25 cm involvement	UCDAI[3] score, 20 3–9	In:32.7 ± 8.9/ Co:35.5 ± 13.8	Parallel	NCB-02: extract *Curcuma longa* with 72% curcumin, 18.08% demethoxy curcumin and 9.42% bis-demethoxy curcumin (140 mg)	Placebo	8 weeks	(1) No significant differences found in clinical remission and improvement on endoscopy between groups (2) Per protocol analysis (patients who completed 8 weeks treatment) revealed significantly better outcomes in NCB-02 group, in terms of clinical response (92.9% vs. 50%, $p = 0.01$), clinical remission (71.4% vs. 31.3%, $p = 0.03$), and improvement on endoscopy	No serious side effect	5

Table 18.1 (continued)

Author, year, country	Sample size (treatment / control)	Target population	Diagnosis	Age Range (years) or mean	Study design	Intervention	Control	Duration	Main results	Side effects	Jadad Score
Kedia et al. (2017), India [33]	62 (29/33)	Mild-to-moderate ulcerative colitis	UCDAI	In:36 ± 12/ Co:34 ± 7	Parallel	Curcumin (1 capsule 3/day curcumin, 150 mg) and mesalamine 2.4 g	Placebo and mes-alamine 2.4 g	8 weeks	No significant difference in: (1) rates of clinical remission (2) clinical response (3) mucosal healing (4) treatment failure	None	5
Shapira et al. (2018), Israel [29]	20	Ulcerative colitis	Clinical activity index (CAI) 4–8	>18 years	Trial	Curcumin (500 mg), green tea (250 mg) and selenium (100 μg) two tablets twice daily	–	8 weeks	(1) 14 subjects (70%) improved (2) CAI decreased significantly at 4 and 8 weeks (3) Endoscopic improvement was observed in 11 (69%) patients, and 4 patients (25%) achieved complete remission	None	1 001

(continued)

Table 18.1 (continued)

Author, year, country	Sample size (treatment / control)	Target population	Diagnosis	Age Range (years) or mean	Study design	Intervention	Control	Duration	Main results	Side effects	Jadad Score
Abbas et al. (2017), Iraq [34]	40 (21/19)	Peptic ulcer	Endoscopic	In:44.1 ± 15.4/ Co:41.2 ± 12.3	Parallel	*H. pylori* eradication triple therapy (clarithromycin [500 mg] tablets [Limassol, Cyprus] + amoxicillin [1 g] capsules [Bristol, UK] + esomeprazole [20 mg] tablets (2/ day) + curcumin (500 mg) capsules 3/day	*H. pylori* eradication triple therapy (clarithromycin [500 mg] tablets [Limassol, Cyprus] + amoxicillin [1 g] capsules [Bristol, UK] + esomeprazole [20 mg] tablets (twice daily)	6 weeks	(1) Curcumin as adjuvant therapy produced significant effect (2) Improvement in healing efficacy in curcumin compared with control group (3) Significant reduction in pro-inflammatory IL-1β level in curcumin compared with control group	Not reported	1 100

Table 18.1 (continued)

Author, year, country	Sample size (treatment / control)	Target population	Diagnosis	Age Range (years) or mean	Study design	Intervention	Control	Duration	Main results	Side effects	Jadad Score
Sadeghi et al. (2019), Iran [12]	70 (35/35)	Mild to moderate ulcerative colitis	$5 \leq$ SCCAI ≤ 12 score	In:40.1 ± 13.2/ Co:40.6 ± 12.4	Parallel	Curcumin (1500 mg/day)	Placebo	8 weeks	(1) Higher reduction in SSCI score in curcumin compared to placebo group (2) Scores of Inflammatory Bowel Disease Questionnaire-9 and QoL significantly higher in intervention compared to control group (3) Curcumin supplementation reduced serum hs-CRP and ESR levels significantly compared with control group (4) No significant changes observed in TNF-α levels of both groups	One patient in curcumin group had skin allergy symptoms, dyspepsia, and heartburn	5

(continued)

Table 18.1 (continued)

Author, year, country	Sample size (treatment / control)	Target population	Diagnosis	Age Range (years) or mean	Study design	Intervention	Control	Duration	Main results	Side effects	Jadad Score
Judaki et al. (2017), Iran [35]	50 (24/26)	Chronic gastritis associated with H. pylori	Biopsy	In:40.4 ± 12.7/ Co:41.4 ± 13.9	Parallel	Triple therapy (omeprazole/ 20 mg, amoxicillin/1 g, and metronidazole/ 800 mg, each given orally 2/day) + turmeric tablet (700 mg orally 3/ day)	Triple therapy	8 weeks	(1) Significantly Malondialdehyde (MDA) and peroxides with increased total antioxidant capacity in intervention compared with control group (2) Oxidative damage to DNA significantly decreased in triple therapy with curcumin compared to triple therapy group (3) Triple therapy group in combination with curcumin significantly decreased all active, chronic, and endoscopic inflammation scores of patients compared with triple therapy group (4) Eradication rate by triple therapy + curcumin significantly increased compared to triple therapy alone	Not reported	1 100

Table 18.1 (continued)

Author, year, country	Sample size (treatment / control)	Target population	Diagnosis	Age Range (years) or mean	Study design	Intervention	Control	Duration	Main results	Side effects	Jadad Score
Koosirirat et al. (2010), Thailand [37]	36 (17/19)	H. pylori infected patients	Positive to H. pylori in 2/3 tests; rapid urease test	In: 48.1 ± 10.0/ Co:46.7 ± 11.0	Trial	Curcumin 120 mg/ day (turmeric: 700 mg 3/day, each turmeric tablet contained 40 mg curcumin)	OAM	OAM: 1 week Turmeric: 4 weeks	(1) Eradication rate of *H. pylori* in patients that received OAM treatment significantly higher than patients who received curcumin (78.9% vs. 5.9%) (2) Levels of IL-8 mRNA expression in OAM group significantly decreased after treatment, but no change observed in curcumin group	Not reported	1 100

(continued)

Table 18.1 (continued)

Author, year, country	Sample size (treatment / control)	Target population	Diagnosis	Age Range (years) or mean	Study design	Intervention	Control	Duration	Main results	Side effects	Jadad Score
Holt et al. (2005), USA [38]	5	Proctosig-moiditis	Proctitis symptoms for 1–32 years	28–54	Trial	550 mg curcumin 2/day for 1 month and 550 mg 3/day for another month	–	2 months	(1) All 5 subjects with proctitis improved at the end of the study, judged by a global score (2) Major changes found were number and quality of stools (3) Two subjects eliminated their prestudy 5ASA medications, one reduced their medications (including termination of prednisone therapy in 1 subject), and one continued taking 5ASA suppositories (4) In these patients with limited ulcerative colitis, serologic indexes of inflammation, sedimentation rate, and CRP returned to normal limits	Not reported	0

Table 18.1 (continued)

Author, year, country	Sample size (treatment / control)	Target population	Diagnosis	Age Range (years) or mean	Study design	Intervention	Control	Duration	Main results	Side effects	Jadad Score
Holt et al. (2005), USA [38]	5	Crohn's disease	Patients with established diagnosis of Crohn's disease	28–54	Trial	The subjects treated with curcumin, 360 mg (1 capsule) 3/day for 1 month and then 360 mg (4 capsules) 4/day for remaining 2 months	–	3 months	(1) CDAI scores for all completed subjects decreased, with mean reduction of 55 points; CRP and ESR also reduced. (2) Four completed subjects reported improvement in clinical symptoms (more formed stools, less frequent bowel movements, and less abdominal pain and cramping)	Not reported	0
Masoodi et al. (2018), Iran [36]	56 (28/28)	Ulcerative colitis	SCCAI	In:38.2 ± 16.4/ Co: 36.0 ± 11.8	Parallel	Curcuminoid nanomicelles (80 mg, 3/day, orally) plus mesalamine (3 g/24 h, orally)	Placebo plus mes-alamine (3 g/24 h, orally)	4 weeks	(1) Scores for urgency of defecation reduced significantly more in case compared with control group (2) Patients in case group experienced better general condition than controls (3) Mean SCCAI score significantly lower in intervention compared with control group ($p = 0.05$)	Flatulence, dyspepsia, headache, increased appetite, nausea, and yellow stool were rarely seen (no differences between groups)	4 211

(continued)

Table 18.1 (continued)

Author, year, country	Sample size (treatment / control)	Target population	Diagnosis	Age Range (years) or mean	Study design	Intervention	Control	Duration	Main results	Side effects	Jadad Score
Cruz-Correa et al. (2018), USA [39]	44 (21/23)	Familial adenom-atous polyposis	5 or more intestinal adenomatous polyps	In:44.5 ± 15.4/ Co:38.7 ± 15.0	Parallel	100% curcumin in three 500 mg capsules (1500 mg orally, 2/day)	Placebo capsules	12 months	(1) No significant difference in mean number of polyps in placebo versus curcumin group (2) No significant difference between mean polyp size in curcumin versus placebo group	Adverse events were few with no significant differences between groups	5

Table 18.1 (continued)

Author, year, country	Sample size (treatment / control)	Target population	Diagnosis	Age Range (years) or mean	Study design	Intervention	Control	Duration	Main results	Side effects	Jadad Score
Bomm-elaer et al. (2019), France [40]	62 (31/31)	Post-operative recurrence of Crohn's disease	Ileo-colonoscopy	In: 37.6 ± 13.8/ Co:35.0 ± 10.5	Parallel	Azathioprine (2.5 mg/kg) + oral curcumin (3 g/day)	Azathioprine (2.5 mg/kg) + placebo	6 months	(1) No differences in postoperative recurrence (Rutgeerts' index score ≥ i2) between groups (2) Significantly higher rate of severe recurrence of CD (Rutgeerts' index score ≥ i3) in intervention compared with control group (3) No differences between groups regarding clinical recurrence of CD (CD activity index score > 150) (4) QoL scores at month 6 did not differ significantly between groups	Severe adverse events in 6% of patients receiving placebo and 16% of patients receiving curcumin (not significant between groups)	5

SSCI: simple clinical colitis activity index, 3 ulcerative colitis disease activity index, *IBS* irritable bowel syndrome, *IBS-D* IBS-diarrhea-predominant, *IBS-SSS* IBS symptom severity score (parameters: abdominal distension, dissatisfaction with bowel habit, interference with quality of life, number of days over the last 10 days with abdominal pain), *OAM* omeprazole, amoxicillin, and metronidazole, *QoL* quality of life, *CAI* clinical activity index, *CDAI* Crohn's disease activity index, *EI* endoscopic index, *Hs-CRP* high sensitivity C-reactive protein, *ESR* erythrocyte sedimentation rate, *TNF-α* tumor necrosis factor alpha, *MDA* malondialdehyde, *GPX* glutathione peroxidase

18.3 Results

18.3.1 Search Results and Study Selection

In the initial search of the database, 1846 articles were identified, of which 82 were duplicates (Fig. 18.1). After reading the titles and abstracts, 1731 papers were deleted, leaving 33 articles. The full texts of these 33 articles were reviewed, and according to the inclusion and exclusion criteria, 12 articles were removed due to the following reasons:

1. Contained duplicated data ($n = 4$).
2. Involved other diseases ($n = 1$).
3. Congress abstract ($n = 3$).
4. Molecular study ($n = 1$).
5. Case report ($n = 1$).
6. Not in English ($n = 1$).
7. No access to full text ($n = 1$).

This left 21 trials, which were analyzed in this systematic review.

18.3.2 Characteristics of the Included Studies

All 21 clinical trial studies were carried out between 2004 and 2019 and consisted of 16 randomized clinical trials (15 parallel and 1 cross-

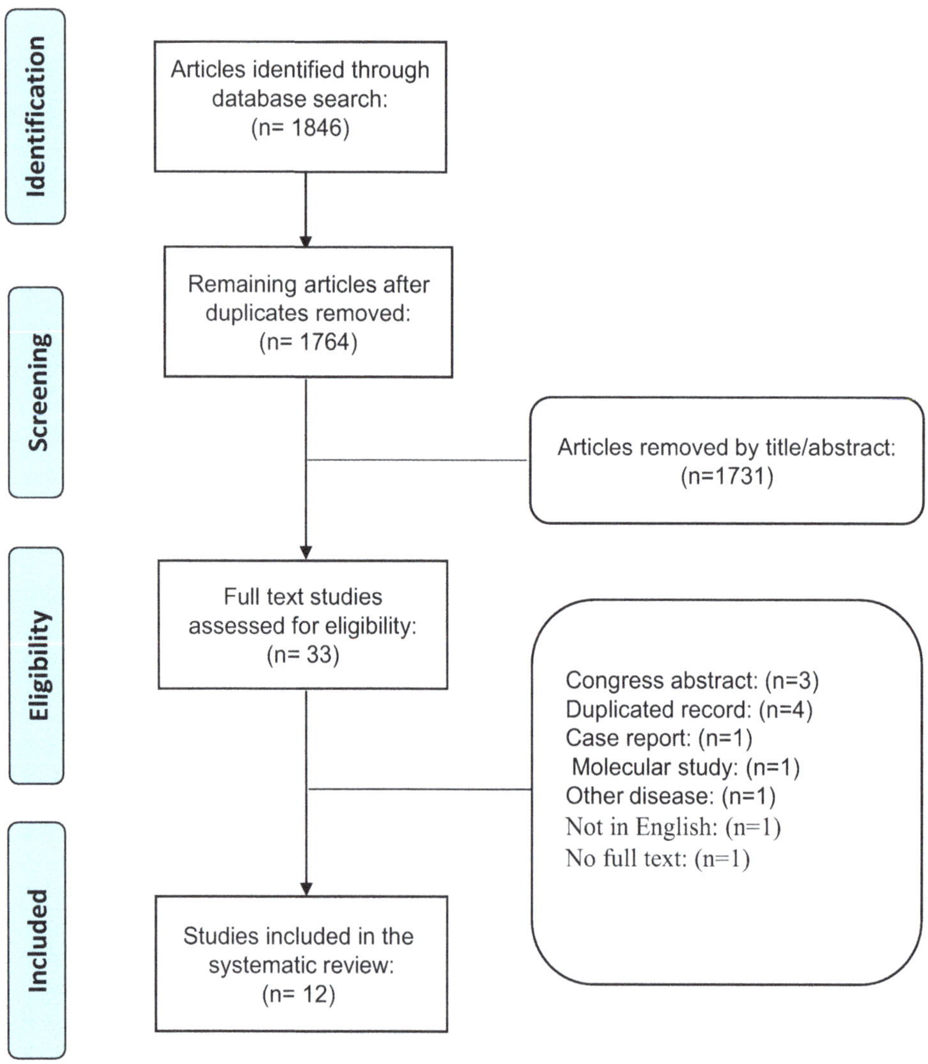

Fig. 18.1 Schematic diagram showing the study selection protocol

over study), 4 trial studies, and 1 partially blinded randomized study. All examine the effect of curcumin on various GI conditions. These studies were conducted in Italy [23, 28], Israel [24, 29], Germany [4, 25, 30], the UK [3], Japan [31], India [32, 33], Iraq [34], Iran [9, 12, 35, 36], Thailand [37], USA [38, 39] and France [40]. Another study was conducted at multiple sites in Israel, Hong Kong and Cyprus [22]. The main characteristics of these studies are summarized in Table 18.1.

The total number of participants in these studies was 1478, of which the smallest sample size was $n = 5$ [4] and the largest was $n = 211$ [28]. Parallel randomized clinical trials included a total of 976 participants, including 503 controls and 473 cases. From the total studies, seven were performed on patients with IBS [3, 4, 23–25, 28, 30], seven on patients with ulcerative colitis [12, 22, 29, 31–33, 36], 2 on those with peptic ulcers [9, 34], four on people infected with *H. pylori* [9, 34, 35, 37], two involved patients with Crohn's disease [38, 40], one study was on familial adenomatous polyposis [39], and one studied proctosigmoiditis [38]. The duration of the interventions in these studies varied from 4 weeks [9, 25, 36] to 12 months [39]. The dosage of curcumin ranged from 42 mg [23, 28] to 3 g [40] per day.

The types of interventions were different in separate studies. Two studies [tested turmeric alone 3, 35], six studies used curcumin alone [4, 12, 31, 32, 38, 39] and one study used a combination of curcumin with turmeric [37]. The other studies used curcumin along with other foods, supplements or medications [9, 22–25, 28–30, 33, 34, 36, 40].

18.3.3 Studies on the Effect of Curcumin Consumption on IBS

Seven studies from 2004 to 2019 tested the effects of curcumin on 784 people with IBS [3, 4, 23–25, 28, 30]. Rome III (ROM3) criteria [23–25, 28, 30] were used to diagnose IBS in five studies, one study used ROM2 criteria [24], and another study diagnosed IBS by extensive clinical examination [4]. These studies were conducted in Italy [23, 28], Israel [24], Germany [4, 25, 30], and the UK [3]. The duration of interventions in these studies ranged from 4 to 18 weeks.

The design of four studies was parallel [4, 23, 25, 30], one was a crossover study [15], one study used questionnaires [28] and the other study was a partially blinded randomized format [3]. Two of these studies did not have a control group [3, 28]. Of these seven studies, one study used curcumin alone in the intervention group [4], in one study, it was administered with turmeric [3], and in other studies, it was given in combination with other substances or in other forms [23–25, 28, 30]. Side effects of taking this supplement were observed in three studies [3, 4, 25].

In studies on the effect of curcumin consumption in patients with IBS, no effect on disease recovery was observed in three studies [4, 24, 25]. However, in three studies in which curcumin was given in combination with other supplements, and in one study using turmeric, positive effects on disease symptoms were observed. In 2016, a study by Portincasa et al. investigated the effect of curcumin and fennel on symptoms and quality of life on 116 patients with IBS (intervention, $n = 58$; control, $n = 58$) for 30 days [23]. Individuals in the intervention group took two daily doses of curcumin capsules at a dose of 42 mg and essential fennel oil at a dose of 25 mg, daily. The results showed an improvement in symptoms of the disease and quality of life, without side effects. Another study by Di Ciaulaa et al. in 2018 investigated the efficacy of biooptimized extracts of turmeric and essential fennel oil on the quality of life in patients with IBS [28]. This study had no control group and involved supplementation of 211 patients with 42 mg of curcumin and 25 mg of fennel essential oil daily for 2 months. This resulted in a significant reduction in the severity and symptoms of the disease and a significant increase in the quality of life. In addition, this protocol resulted in no side effects. In 2017, a study by Alt et al. evaluated the benefit and tolerability of IBP-CL-101 (Xanthofen) in improving the symptoms in patients with IBS [30]. IBP-CL-101 contained a 330 mg proprietary blend of curcuminoids and essential oils, 70 mg fish oil, 15 mg peppermint oil and 8 mg caraway oil, as well as 263 μg thiamine, 39 μg folic acid and 625 μg vitamin D3. This was given daily to 43 people as an intervention group over a period of 8 weeks. The results showed a significant reduction in the severity of symptoms of IBS without any significant side effects in the intervention group. In 2004, a study by Bundy et al. was conducted on 207 people with IBS for 8 weeks to investigate the effect of turmeric extract [3]. One group received 72 mg (1 tablet/day) and the second group received 144 mg (2 tablets/day) of the turmeric supplement. The study had no control group. The results showed a reduction in the prevalence and symptoms of the disease and an increase in the quality of life in these patients.

However, other studies found no favorable effects of curcumin in improving symptoms of IBS. In 2019, Lior et al. conducted a crossover study with the aim of investigating the combination of curcumin, green tea, and selenium methionine on improving symptoms of 22 people with IBS [24]. The supplement contained 500 mg curcumin, 250 mg green tea, and 100 μg selenomethionine, and was given to patients for 4 weeks followed by a 2-week washout. The study did not show any significant improvement of symptoms and severity of IBS or quality of life, although there were no side effects. Another study by Brinkhaus et al. was conducted on 106 patients with IBS with the aim of investigating herbal medicines using either *Curcuma xanthorrhiza* or *Fumaria officinalis* as potential treatments for IBS [4]. Supplements containing 60 mg *Curcuma* or 1500 mg *Fumaria* were given to patients for 18 weeks. After the intervention, side effects of constipation and distension were seen in one patient in the curcumin group and the stress and symptoms associated with IBS did not differ significantly between groups. In 2016, a study by Lauche et al. evaluated the efficacy and safety of Ayurvedic herbs in 32 individuals with diarrhea-predominant IBS over an 8-week treatment period [25]. The supplement contained 5 g Ayurvedic curry powder mixture (*Murraya koenigii*), pomegranate (*Punica granatum*), and turmeric (*Curcuma longa* rhizome pulvis) in a 6:3:1 ratio, dissolved in 100 mL warm water. There were no significant differences in the severity of symptoms between the intervention and control groups and one-third of the patients reported at least one minor adverse event that might be related to the study interventions.

18.3.4 Studies on the Effect of Curcumin Consumption in Ulcerative Colitis

A total of seven studies evaluated the effect of curcumin consumption on 392 people with ulcerative colitis between 2006 and 2019 with an intervention duration of 4 weeks to 6 months [12, 22, 29, 31–33, 36]. These studies were conducted in India [32, 33], Iran [12, 36], Japan [31], Israel [29], and one was a multi-site study (Israel, Hong Kong and Cyprus) [22]. The methods of diagnosis in these studies used the Simple Clinical Colitis Activity Index (SCCAI), the endoscopic index (IE), the UC Disease Activity Index (UCDAI), and the Clinical Activity Index (CAI) [12, 22, 29, 31–33, 36]. In these studies, six studies were of parallel design and included 372 individuals (190 control group participants and 182 intervention group participants) [12, 22, 31–33, 36], and one study had 20 individuals in the intervention group without a control group [29]. In five of the studies, curcumin supplementation was combined with other substances [22, 29, 32, 33, 36] and in two studies, curcumin supplementation alone was used [12, 31].

Out of these seven studies, six studies showed a positive effect of the curcumin supplementation on ulcerative colitis. A 2006 study by Hanai et al. on 89 individuals was aimed at investigating the effect of curcumin on ulcerative colitis using two capsules containing 1 g curcumin given daily to patients for 6 months [31]. The results of this study showed a lower recurrence rate in the curcumin compared to placebo group. The CAI and EI in the curcumin group significantly improved compared to the control group. However, seven patients in the curcumin group had abdominal bloating, nausea, transient hypertension, a transient increase in the number of stools and elevated guanosine triphosphate levels. In 2014, another study by Singla et al. examined the effect of NCB-02 (curcumin) enema in 45 people with mild-to-moderate distal ulcerative colitis for 8 weeks [32]. The supplements contained 140 mg extract with a composition of 72% curcumin, 18.1% demethoxy curcumin and 9.4% bis-demethoxy curcumin. There was a significant difference in clinical repair and endoscopic improvement between groups. The intervention group showed better clinical results and no side effects.

In 2018, a study by Shapira et al. evaluated the effects of a new dietary supplement (curcumin, green tea, and selenium) on 20 people with ulcerative colitis for 8 weeks [29]. This study did not have a control group and the supplement contained curcumin (500 mg), green tea (250 mg), and selenium (100 µg). The study showed that 70% of the participants recovered. There were no significant side effects and there was a significant reduction in CAI after 4 and 8 weeks of intervention. There was also an endoscopic improvement of ulcerative colitis in this patient group. In 2019, a study by Sadeghi et al. evaluated the effect of curcumin supplementation on clinical outcomes and inflammatory symptoms in 70 people with ulcerative colitis [12]. Curcumin supplementation at a dose of 1500 mg/day was given to patients for 8 weeks. There was a significant reduction in C-reactive protein (CRP), SCCAI score, and a significant increase in the quality of life in the intervention group. Side effects were seen in one person in the intervention group, including symptoms of skin allergy, dyspepsia, and heartburn.

In 2018, a study by Masoodi et al. evaluated the effectiveness of curcuminoids on 56 patients with ulcerative colitis for 4 weeks [36]. The supplementation contained curcuminoid nanomicelles (80 mg, three times daily, orally) plus the standard ulcerative colitis drug, mesalamine. The results showed that SCCAI score in the intervention group was significantly lower than that in the control group. Side effects of flatulence, dyspepsia, headache, increased appetite, nausea, and yellow stool were seen in both intervention and control groups. In 2015, a study by Lang et al. evaluated the effectiveness of curcuminoids on 50 patients with ulcerative colitis for 1 month [22]. The supplementation contained curcumin (3 g daily, oral) plus mesalamine. This led to an improvement in clinical symptoms and EI. Mild side effects such as nausea, abdominal distension, and transient increase in stool frequency were seen in both intervention and control groups.

However, one trial showed that administration of curcumin was not effective in improving ulcerative colitis. The study by Kedia et al. examined the effect of low-dose oral curcumin for 8 weeks on mild-to-moderate ulcerative colitis in 62 people [33]. The supplement contained 150 mg curcumin and 2.4 g mesalamine. No significant difference was observed in clinical response of the patients and there were no significant side effects observed in the study.

18.3.5 Studies on the Effect of Curcumin Supplementation on *H. pylori*

The effect of curcumin supplementation on 186 individuals (intervention, $n = 96$ group; control, $n = 90$) was performed in four parallel randomized clinical trials from 2010 to 2017 on patients with *H. pylori* infection [9, 34, 35, 37]. The results regarding the eradication of *H. pylori* were positive in two of the studies.

In 2017, a study by Abbas et al. investigated the effect for 6 weeks treatment with curcumin on the eradication of *H. pylori* infection in 40 individuals with peptic ulcer [34]. The results of this study did not show any side effects and the curcumin treatment resulted in a significant reduction in IL-1β levels in the intervention group. In 2017, a study by Judaki et al. evaluated the effect of a triple therapy with curcumin supplementation for 8 weeks on oxidative stress and histopathologic changes in chronic *H. pylori* gastritis in 16 patients [35]. The results of the study showed a significant reduction in malondialdehyde (MDA), glutathione peroxide (GPX) and inflammation in the intervention group compared to the control group, with no side effects.

In 2010, Koosirirat et al. studied the anti-inflammatory effects of curcumin in 36 patients with *H. pylori* infection [37]. OAM (oxetacaine, aluminum hydroxide, and magnesium hydroxide) supplementation was administered for 4 weeks followed by 120 mg of curcumin supplementation for 4 weeks. There were no significant differences seen and no side effects occurred. In 2016, a study by Khonche et al. investigated the effect of curcumin in 60 people with peptic ulcer [9]. The intervention included a standard treatment for *H. pylori* with 500 mg of curcumin and 5 mg of piperine daily for 4 weeks. The results of the study did not show any significant differences but there were no significant side effects.

18.3.6 Studies on the Effect of Curcumin on Peptic Ulcer Disease

In 2005 and 2019, two studies on 67 people with peptic ulcers examined the effect of curcumin as a potential treatment. In both studies, the use of curcumin supplementation with other supplements showed beneficial effects. In 2016, a study by Khonche et al. examined the effect of curcumin in 60 people with peptic ulcer [9]. The intervention consisted of administering 500 mg curcumin supplement plus 5 mg piperine daily for 4 weeks. The results of the study showed an improvement of dyspepsia symptoms in the intervention group compared to the control group with no side effects. In 2017, a study conducted by Abbas et al. investigated the potential benefit of adjuvant therapy with curcumin to the standard *H. pylori* eradication therapy in patients [34]. The results of the study showed that the use of curcumin supplementation improved inflammatory indices in the intervention group compared to the control group without any side effects.

18.3.7 Studies on the Effect of Curcumin on Crohn's Disease

Two studies were performed in 2005 and 2019 on the effects of curcumin on 62 people with Crohn's disease [38, 40]. Only one of these studies showed a beneficial effect. In 2005, a study by Holt et al. looked at the effects of curcumin on inflammatory bowel disease in 5 people [38]. The intervention included taking a 360 mg curcumin capsule 3 times a day for 2 months and a 360 mg curcumin capsule 4 times a day for 3 months. This led to a significant reduction in Crohn's Disease Activity Index scores, CRP levels, and an improvement in clinical symptoms in the intervention group without side effects. In 2019, a study by Bommela et al. on 62 individuals assessed the effect of curcumin on prevention and recurrence of Crohn's disease after surgery [40]. The intervention included 2.5 mg of azathioprine/kg body weight per day and 3 g curcumin per day for 6 months. There was no significant difference between the curcumin and placebo groups in quality of life, symptoms, or disease severity indicators.

18.3.8 Studies of the Effect of Curcumin on Familial Adenomatous Polyposis

In 2018, a study by Cruz-Correa et al. investigated the efficacy and safety of curcumin in the treatment of intestinal adenomas in 44 patients with familial adenomatous polyposis [39]. The capsule contained 1500 mg pure curcumin and this was taken twice per day for 12 months. The results of the study did not show any difference in number or size of polyps between the intervention and placebo groups.

18.3.9 Studies of the Effect of Curcumin on Proctosigmoiditis

A study by Holt et al. examined the effect of curcumin on 5 patients with proctosigmoiditis. Capsules containing 550 mg of curcumin were given to patients twice daily for 1ne month and three times daily for 1 month [38]. This resulted in an overall improvement of symptoms in all participants without any side effects.

18.3.10 Quality Assessment

Overall, 15 studies out of 21 were categorized as high quality [3, 4, 9, 12, 22–25, 30–33, 36, 39, 40], and 6 were low quality [28, 29, 34, 35, 37, 38], using the Jadad scale.

18.4 Discussion

This systematic review shows that curcumin administration has a positive effect on many GI-related diseases such as IBS, ulcerative colitis, *H. pylori* infection, peptic ulcer disease, Crohn's disease, and familial adenomatous polyposis. In these 21 studies, 16 showed positive effects on the improvement of GI diseases. Given the burden that GI diseases can cause to individuals, society and the healthcare systems, the current findings of some positive effects of curcumin treatment are promising.

18.4.1 The Effect of Curcumin on IBS

In this review, we examined seven articles from 2004 to 2019 on the effects of curcumin consumption in 784 people with IBS [3, 4, 23–25, 28, 30]. Four studies showed positive effects [3, 23, 28, 30]. The beneficial effects of curcumin on IBS could be due to its antioxidant and anti-inflammatory properties. Curcumin can lower pro-inflammatory cytokine levels in the blood and regulates key mediators, the inflammation process [5]. Recent studies of IBS patients have shown that they have a pro-inflammatory phenotype associated with increased levels of IL-6 and mucosal inflammation of the GI tract [41]. Curcumin has been shown to have protective effects on intestinal mucosa in animal models [42] and its efficacy has been reported in inflammatory bowel disease [38].

A human IBS study showed that curcumin could exert its beneficial effects through the brain-gut axis [43]. Curcumin increases serotonin (5-HT) and brain-derived neutrophilic factor (BDNF), as well as the expression of cAMP phosphorus response element-binding protein (PCREB) in the hippocampal region of the brain as well as in the colon [43]. This is important as psychosocial stress factors can negatively impact the gut-brain axis [44]. This may be due to the fact that stress and other psychiatric conditions are linked with increased inflammatory responses [45].

Laboratory studies have shown that curcumin treatment can ameliorate some of the effects of stress such as the reduced BDNF levels and overstimulation of the hypothalamic pituitary adrenal (HPA) axis [46]. Curcumin has poor bioavailability [47] and most of the molecule is excreted in the stool unmetabolized [48]. However, it still appears to have beneficial effects on the gut microbiota [49], and studies in animal models have shown that it can have positive effects on intestinal permeability [50].

In the study by Brinkhaus et al., curcumin was not effective [4]. This may be due to the fact that the patients in this study had not been diagnosed using the Rome criteria and they were resistant to standard IBS treatments. The other studies which showed no effect of curcumin on IBS may have been limited by statistical power. In the trial by Lior et al., 25 patients were needed to achieve a power of 80%, although only 18 patients completed the study [24]. In the study by Lauche et al., the sample size was also limited as 11 of the 32 participants were excluded and the blinding was not successful [25].

18.4.2 The Effect of Curcumin on Ulcerative Colitis

We analyzed seven studies which investigated the effects of curcumin consumption on ulcerative colitis from 2006 to 2019, involving a total of 392 patients [12, 22, 29, 31–33, 36]. Six of these studies showed positive effects [12, 22, 29, 31, 32, 36], while one study found that the curcumin intervention had no effect [33]. In the latter study, there was no significant difference in clinical improvement or mucosal healing after 8 weeks of treatment between curcumin and placebo, although this may have been due to the low dose of the supplement used [33].

As above, the anti-inflammatory properties of curcumin make it a natural treatment option for inflammatory diseases such as ulcerative colitis [51]. Experimental models have shown that curcumin prevents colitis by inducing tri-nitrobenzene sulfonic acid and dextran sodium sulfate. Its mechanism of action is through suppressing the nuclear factor of K-light chain in B-lymphocytes during T helper cell (Th1) and (Th2) expression [41]. As curcuminoids are poorly and variably digested in the GI tract [52], it is still not certain if the observed therapeutic effects depend on bioavailability.

18.4.3 The Effect of Curcumin Supplementation on *H. pylori* Infection

H. pylori can cause infection of the gastric mucosa resulting in mucosal ulcer and gastric cancer [35]. In this study, we examined the effect of curcumin supplementation on *H. pylori* in four parallel randomized clinical trials conducted on 186 individuals (90 cases and 96 control individuals) from 2010 to 2017 [9, 34, 35, 37]. Two of the studies had positive effects on *H. pylori* eradication [34, 35], while the other two studies had no effect [9, 37]. One reason for these differences could be due to combined use of curcumin and triple eradication therapy in the former studies [34, 35]. In addition, the latter studies had a shorter duration of curcumin treatment [9, 37].

In the Judaki et al. study above, curcumin in combination with triple therapy was found to significantly decrease all active, chronic, and endoscopic inflammation scores in gastritis-associated *H. pylori* infected patients compared to the triple therapy treatment alone [35]. In addition, a study on mice showed that it had beneficial effects on the eradication of *H. pylori* [53]. Animal studies have shown that the mechanism of curcumin supplementation may involve inhibition of the metalloproteinase matrix protein (MMP)-3 and MMP-9 enzymes, which are involved in the pathogenesis of *H. pylori* infection [54]. Another study on mice showed that curcumin could inhibit expression of the inflammatory cytokines and chemokines involved in *H. pylori* pathophysiology [55].

The antioxidant properties of curcumin are associated with the phenolic, β-diketone and methyl groups on the molecule, which have free radical-destroying effects. This is associated with increased expression of antioxidant molecules such as superoxide dismutase (SOD), glutathione peroxidase (GPX), glutathione reductase, glutathione transferase, and catalase [35]. In addition, curcumin suppresses IL8 expression via deactivation of NF-κB signaling, which plays a key role in *H. pylori* pathogenesis [56].

18.4.4 The Effect of Curcumin on Peptic Ulcer Disease

Peptic ulcer disease is a disease affecting the lining of the stomach and duodenum, often caused by *H. pylori* infection in about half of the patients, which induces inflammation in gastric epithelial cells. We examined the effect of curcumin on peptic ulcer disease in two clinical trials with 67 patients carried out in 2005 and 2019 [9, 34]. In both studies, patients were given curcumin supplements along with other supplements and this led to improvements in the disease. The anti-ulcer activity of curcumin could be due to reduction of various ulcerative factors such as gastric acid hypersecretion, and increased myeloperoxidase activity, apoptosis, peroxides, and pepsin activity [34].

18.4.5 The Effect of Curcumin on Crohn's Disease

Crohn's disease is a chronic disorder that can lead to intestinal damage. The European Crohn's and Colitis Organization's guideline states that the drugs recommended for prevention of recurrence have an anti-TNF-α mechanism of action. As curcumin is known to reduce TNF-a levels and other other pro-inflammatory factors, we examined its effects on Crohn's disease in two studies performed on 62 people in 2005 and 2019 [38, 40].

The study by Holt et al. found a positive effect of curcumin on improving the symptoms of Crohn's disease [38], although the study by Bommelaer et al. did not show any beneficial effects [40]. This difference could be due to the lower dosage of curcumin used in the latter study [40]. The efficacy of curcumin in the former study is likely to be due to its anti-inflammatory and antioxidant effects in the layers of gut epithelial cells, as found in in vitro studies [57, 58]. In addition, the ability of curcumin to prevent and improve murine experimental colitis has been demonstrated [59, 60].

18.4.6 The Effect of Curcumin on Familial Adenomatous Polyposis

Familial adenomatous polyposis is an autosomal recessive disease characterized by colorectal adenomas in adolescents and young adults [39, 61]. Almost all of these people are at risk for colorectal cancer [61]. Anti-inflammatory medications have been shown to cause regression of intestinal adenomas in these patients [39]. A study by Cruz-Correa et al. on 44 people with familial adenomatous polyposis showed that there were no significant effects after curcumin supplementation for 12 months [39]. The small sample size and low doses of curcumin used may have led to this result. One potential mechanism of curcumin in chemoprevention is by changing the colonic microbiome. In a study on mice, curcumin increased native lactobacilli bacteria and reduced coriobacteria [62]. Curcumin also increased bifidobacteria causing aberrant crypt foci in murine colorectal carcinogenesis models [62]. Curcumin may also affect colonic bile acid compounds, which can alter the secretion and deconjugation of bile acids through interplay with microbiomes [63].

There are some limitations to the present review, arising mainly from the nature and the characteristics of the included studies. The small number of participants and the heterogenous nature of the included studies with regard to the intervention approaches made it difficult to draw definitive conclusions for the use of curcumin in the treatment of GI diseases. Nevertheless, some positive results were obtained supporting the case for further work in this field, focusing on the use of carefully characterized patient groups, sufficient statistical power and standardized protocols. This will allow more accurate comparisons to be drawn both within and across studies.

18.5 Conclusions

Within the limitations of this systematic review, the results show a beneficial effect of curcumin/turmeric supplementation in various GI diseases. Curcumin/turmeric supplementation resulted in an improvement of clinical outcomes of patients with IBS, ulcerative colitis, *H. pylori*, peptic ulcer, Crohn's disease, and proctosigmoiditis. However, some studies had conflicting results and some showed the presence of some side effects of curcumin-related treatments. Therefore, additional randomized clinical controlled trials are needed using well-characterized patient groups and standardized protocols to clarify these results.

References

1. Chan JSH, Chao ACW, Cheung VCH, Wong SSK, Tang W, Wu JCY et al (2019) Gastrointestinal disease burden and mortality: a public hospital-based study from 2005 to 2014. J Gastroenterol Hepatol 34(1):124–131
2. Langmead L, Rampton D (2001) Herbal treatment in gastrointestinal and liver disease—benefits and dangers. Aliment Pharmacol Ther 15(9):1239–1252
3. Bundy R, Walker AF, Middleton RW, Booth J (2004) Turmeric extract may improve irritable bowel syndrome symptomology in otherwise healthy adults: a pilot study. J Altern Complement Med 10(6):1015–1018
4. Brinkhaus B, Hentschel C, Keudell CV, Schindler G, Lindner M, Stützer H et al (2005) Herbal medicine with curcuma and fumitory in the treatment of irritable bowel syndrome: a randomized, placebo-controlled, double-blind clinical trial. Scand J Gastroenterol 40(8):936–943
5. Ng QX, Soh AYS, Loke W, Venkatanarayanan N, Lim DY, Yeo WS (2018) A meta-analysis of the clinical use of curcumin for irritable bowel syndrome (IBS).

J Clin Med 7(10):E298. https://doi.org/10.3390/jcm7100298
6. Soleimani V, Sahebkar A, Hosseinzadeh H (2018) Turmeric (Curcuma longa) and its major constituent (curcumin) as nontoxic and safe substances: Review. Phytother Res 32(6):985–995. https://doi.org/10.1002/ptr.6054. Epub 2018 Feb 26. PMID: 29480523
7. Abdollahi E, Momtazi AA, Johnston TP, Sahebkar A (2018) Therapeutic effects of curcumin in inflammatory and immune-mediated diseases: a nature-made jack-of-all-trades? J Cell Physiol 233(2):830–848
8. Iranshahi M, Sahebkar A, Takasaki M, Konoshima T, Tokuda H (2009) Cancer chemopreventive activity of the prenylated coumarin, umbelliprenin, in vivo. Eur J Cancer Prev 18(5):412–415
9. Khonche A, Biglarian O, Panahi Y, Valizadegan G, Soflaei S, Ghamarchehreh M et al (2016) Adjunctive therapy with curcumin for peptic ulcer: a randomized controlled trial. Drug Res 66(8):444–448
10. Mollazadeh H, Cicero AFG, Blesso CN, Pirro M, Majeed M, Sahebkar A (2019) Immune modulation by curcumin: the role of interleukin-10. Crit Rev Food Sci Nutr 59(1):89–101
11. Hasanzadeh S, Read MI, Bland AR, Majeed M, Jamialahmadi T, Sahebkar A (2020) Curcumin: an inflammasome silencer. Pharmacol Res 159:104921. https://doi.org/10.1016/j.phrs.2020.104921. Epub 2020 May 25. PMID: 32464325.
12. Sadeghi N, Mansoori A, Shayesteh A, Hashemi SJ (2020) The effect of curcumin supplementation on clinical outcomes and inflammatory markers in patients with ulcerative colitis. Phytother Res 34(5):1123–1133
13. Momtazi AA, Sahebkar A (2016) Difluorinated curcumin: a promising curcumin analogue with improved anti-tumor activity and pharmacokinetic profile. Curr Pharm Des 22(28):4386–4397
14. Momtazi AA, Derosa G, Maffioli P, Banach M, Sahebkar A (2016) Role of microRNAs in the therapeutic effects of curcumin in non-cancer diseases. Mol Diagn Ther 20(4):335–345
15. Teymouri M, Pirro M, Johnston TP, Sahebkar A (2017) Curcumin as a multifaceted compound against human papilloma virus infection and cervical cancers: A review of chemistry, cellular, molecular, and preclinical features. BioFactors 43(3):331–346.
16. Talebi S, Bagherniya M, Atkin SL, Askari G, Orafai HM et al (2020) The beneficial effects of nutraceuticals and natural products on small dense LDL levels, LDL particle number and LDL particle size: a clinical review. Lipids Health Dis 19:1–21
17. Bagherniya M, Nobili V, Blesso CN, Sahebkar A (2018) Medicinal plants and bioactive natural compounds in the treatment of non-alcoholic fatty liver disease: a clinical review. Pharmacol Res 130: 213–240
18. Bagherniya M, Johnston T, Sahebkar A (2020) Regulation of apolipoprotein B by natural products and nutraceuticals: a comprehensive review. Curr Med Chem. https://doi.org/10.2174/0929867327666200427092114. [Epub ahead of print]
19. Rezaee R, Momtazi AA, Monemi A, Sahebkar A (2017) Curcumin: a potentially powerful tool to reverse cisplatin-induced toxicity. Pharmacol Res 117:218–227
20. Rajasekaran SA (2011) Therapeutic potential of curcumin in gastrointestinal diseases. World J Gastrointest Pathophysiol 2(1):1
21. Blumenthal M, Busse WR, Goldberg A, Gruenwald J, Hall T, Riggins CW et al (eds) (1998) The complete German commission E monographs. Therapeutic guide to herbal medicines. American Botanical Council, Austin, TX. ISBN-10: 096555550X
22. Lang A, Salomon N, Wu JC, Kopylov U, Lahat A, Har-Noy O et al (2015) Curcumin in combination with mesalamine induces remission in patients with mild-to-moderate ulcerative colitis in a randomized controlled trial. Clin Gastroenterol Hepatol 13(8):1444–1449. e1441
23. Portincasa P, Bonfrate L, Scribano ML, Kohn A, Caporaso N, Festi D et al (2016) Curcumin and fennel essential oil improve symptoms and quality of life in patients with irritable bowel syndrome. J Gastrointestin Liver Dis 25(2):151–157
24. Lior O, Sklerovsy-Benjaminov F, Lish I, Konokoff F, Naftali T (2019) Treatment of irritable bowel syndrome with a combination of curcumin, Green Tea and Selenomethionine has a positive effect on satisfaction with Bowel habits. J Biosci Med 7(5):170–179
25. Lauche R, Kumar S, Kreppenhofer J, Lüdtke R, Rampp T, Dobos G et al (2014) Efficacy and safety of ayurvedic herbs in diarrhea-predominant irritable Bowel syndrome—a randomized controlled crossover trial. J Altern Complement Med 20(5):A81–A81
26. Moher D, Liberati A, Tetzlaff J, Altman DG (2009) Preferred reporting items for systematic reviews and meta-analyses: the PRISMA statement. Ann Intern Med 151(4):264–269
27. Moher D, Cook D, Jadad A, Tugwell P, Moher M, Jones A et al. (1999) Assessing the quality of reports of randomised trials: implications for the conduct of meta-analyses. Health Technol Assess 3(12):i–iv., 1–98
28. Di Ciaula A, Portincasa P, Maes N, Albert A (2018) Efficacy of bio-optimized extracts of turmeric and essential fennel oil on the quality of life in patients with irritable Bowel syndrome. Ann Gastroenterol 31(6):685–691
29. Shapira S, Leshno A, Katz D, Maharshak N, Hevroni G, Jean-David M et al (2018) Of mice and men: a novel dietary supplement for the treatment of ulcerative colitis. Ther Adv Gastroenterol 11:1756283X17741864. https://doi.org/10.1177/1756283X17741864
30. Alt F, Chong PW, Teng E, Uebelhack R (2017) Evaluation of benefit and tolerability of IQP-CL-101 (xanthofen) in the symptomatic improvement of irritable bowel syndrome: a double-blinded, randomised, placebo-controlled clinical trial. Phytother Res 31(7):1056–1062

31. Hanai H, Iida T, Takeuchi K, Watanabe F, Maruyama Y, Andoh A et al (2006) Curcumin maintenance therapy for ulcerative colitis: randomized, multicenter, double-blind, placebo-controlled trial. Clin Gastroenterol Hepatol 4(12):1502–1506
32. Singla V, Pratap Mouli V, Garg SK, Rai T, Choudhury BN, Verma P et al (2014) Induction with NCB-02 (curcumin) enema for mild-to-moderate distal ulcerative colitis—a randomized, placebo-controlled, pilot study. J Crohns Colitis 8(3):208–214
33. Kedia S, Bhatia V, Thareja S, Garg S, Mouli VP, Bopanna S et al (2017) Low dose oral curcumin is not effective in induction of remission in mild to moderate ulcerative colitis: results from a randomized double blind placebo controlled trial. World J Gastrointest Pharmacol Ther 8(2):147–154
34. Abbas SH, Abdulridha MK, Najeb AA (2017) Potential benefit of curcumin adjuvant therapy to the standard Helicobacter pylori eradication therapy in patients with peptic ulcer disease. Asian J Pharm Clin Res 10(5):313–317
35. Judaki A, Rahmani A, Feizi J, Asadollahi K, Hafezi Ahmadi MR (2017) Curcumin in combination with triple therapy regimes ameliorates oxidative stress and histopathologic changes in chronic gastritis-associated helicobacter pylori infection. Arq Gastroenterol 54(3):177–182
36. Masoodi M, Mahdiabadi MA, Mokhtare M, Agah S, Kashani AH, Rezadoost AM et al (2018) The efficacy of curcuminoids in improvement of ulcerative colitis symptoms and patients' self-reported well-being: a randomized double-blind controlled trial. J Cell Biochem 119(11):9552–9559
37. Koosirirat C, Linpisarn S, Changsom D, Chawansuntati K, Wipasa J (2010) Investigation of the anti-inflammatory effect of Curcuma longa in Helicobacter pylori-infected patients. Int Immunopharmacol 10(7):815–818
38. Holt PR, Katz S, Kirshoff R (2005) Curcumin therapy in inflammatory bowel disease: a pilot study. Dig Dis Sci 50(11):2191–2193
39. Cruz-Correa M, Hylind LM, Marrero JH, Zahurak ML, Murray-Stewart T, Casero RA Jr et al (2018) Efficacy and safety of curcumin in treatment of intestinal adenomas in patients with familial adenomatous polyposis. Gastroenterology 155(3):668–673
40. Bommelaer G, Laharie D, Nancey S, Hebuterne X, Roblin X, Nachury M et al (2019) Oral curcumin no more effective than placebo in preventing recurrence of Crohn's disease after surgery in a randomized controlled trial. Clin Gastroenterol Hepatol. https://doi.org/10.1016/j.cgh.2019.08.041. pii: S1542-3565(19)30916-4 [Epub ahead of print]
41. Chandan S, Mohan BP, Chandan OC, Ahmad R, Challa A, Tummala H et al (2020) Curcumin use in ulcerative colitis: is it ready for prime time? A systematic review and meta-analysis of clinical trials. Ann Gastroenterol 33(1):53–58
42. Lubbad A, Oriowo M, Khan I (2009) Curcumin attenuates inflammation through inhibition of TLR-4 receptor in experimental colitis. Mol Cell Biochem 322(1–2):127–135
43. Yu Y, Wu S, Li J, Wang R, Xie X, Yu X et al (2015) The effect of curcumin on the brain-gut axis in rat model of irritable bowel syndrome: involvement of 5-HT-dependent signaling. Metab Brain Dis 30(1):47–55
44. Ng QX, Soh AYS, Loke W, Venkatanarayanan N, Lim DY, Yeo WS (2019) Systematic review with meta-analysis: The association between post-traumatic stress disorder and irritable bowel syndrome. J Gastroenterol Hepatol 34(1):68–73
45. Pace TW, Mletzko TC, Alagbe O, Musselman DL, Nemeroff CB, Miller AH et al (2006) Increased stress-induced inflammatory responses in male patients with major depression and increased early life stress. Am J Psychiatry 163(9):1630–1633
46. Xu Y, Ku B, Tie L, Yao H, Jiang W, Ma X et al (2006) Curcumin reverses the effects of chronic stress on behavior, the HPA axis, BDNF expression and phosphorylation of CREB. Brain Res 1122(1):56–64
47. Ammon HP, Wahl MA (1991) Pharmacology of Curcuma longa. Planta Med 57(1):1–7
48. Sharma RA, McLelland HR, Hill KA, Ireson CR, Euden SA, Manson MM et al (2001) Pharmacodynamic and pharmacokinetic study of oral Curcuma extract in patients with colorectal cancer. Clin Cancer Res 7(7):1894–1900
49. Feng W, Wang H, Zhang P, Gao C, Tao J, Ge Z et al (2017) Modulation of gut microbiota contributes to curcumin-mediated attenuation of hepatic steatosis in rats. Biochim Biophys Acta 1861(7):1801–1812
50. Wang J, Ghosh SS, Ghosh S (2017) Curcumin improves intestinal barrier function: modulation of intracellular signaling, and organization of tight junctions. Am J Physiol Cell Physiol 312(4):C438–C445
51. Sahebkar A, Cicero AF, Simental-Mendía LE, Aggarwal BB, Gupta SC (2016) Curcumin downregulates human tumor necrosis factor-α levels: A systematic review and meta-analysis of randomized controlled trials. Pharmacol Res 107:234–242
52. Shakeri A, Panahi Y, Johnston TP, Sahebkar A (2019) Biological properties of metal complexes of curcumin. BioFactors 45(3):304–317
53. De R, Kundu P, Swarnakar S, Ramamurthy T, Chowdhury A, Nair GB et al (2009) Antimicrobial activity of curcumin against Helicobacter pylori isolates from India and during infections in mice. Antimicrob Agents Chemother 53(4):1592–1597
54. Kundu P, De R, Pal I, Mukhopadhyay AK, Saha DR, Swarnakar S (2011) Curcumin alleviates matrix metalloproteinase-3 and-9 activities during eradication of Helicobacter pylori infection in cultured cells and mice. PLoS One 6(1):e16306. https://doi.org/10.1371/journal.pone.0016306
55. Santos AM, Lopes T, Oleastro M, Gato IV, Floch P, Benejat L et al (2015) Curcumin inhibits gastric inflammation induced by Helicobacter pylori infection in a mouse model. Nutrients 7(1):306–320

56. Münzenmaier A, Lange C, Glocker E, Covacci A, Moran A, Bereswill S et al (1997) A secreted/shed product of Helicobacter pylori activates transcription factor nuclear factor-kappa B. J Immunol 159(12):6140–6147
57. Midura-Kiela MT, Radhakrishnan VM, Larmonier CB, Laubitz D, Ghishan FK, Kiela PR (2 012) Curcumin inhibits interferon-γ signaling in colonic epithelial cells. Am J Physiol Gastrointest Liver Physiol 302(1):G85–G96
58. Larmonier CB, Midura-Kiela M, Ramalingam R, Laubitz D, Janikashvili N, Larmonier N et al (2011) Modulation of neutrophil motility by curcumin: implications for inflammatory bowel disease. Inflamm Bowel Dis 17(2):503–515
59. Sugimoto K, Hanai H, Tozawa K, Aoshi T, Uchijima M, Nagata T et al (2002) Curcumin prevents and ameliorates trinitrobenzene sulfonic acid–induced colitis in mice. Gastroenterology 123(6):1912–1922
60. Cooney JM, Barnett MP, Dommels YE, Brewster D, Butts CA, McNabb WC et al (2016) A combined omics approach to evaluate the effects of dietary curcumin on colon inflammation in the Mdr1a−/− mouse model of inflammatory bowel disease. J Nutr Biochem 27:181–192
61. McConnell R (1976) Familial polyposis Coli. family studies, histopathology, differential diagnosis, and results of treatment. J Med Genet 13(5):415–416
62. McFadden R-MT, Larmonier CB, Shehab KW, Midura-Kiela M, Ramalingam R, Harrison CA et al (2015) The role of curcumin in modulating colonic microbiota during colitis and colon cancer prevention. Inflamm Bowel Dis 21(11):2483–2494
63. Dey N, Wagner VE, Blanton LV, Cheng J, Fontana L, Haque R et al (2015) Regulators of gut motility revealed by a gnotobiotic model of diet-microbiome interactions related to travel. Cell 163(1):95–107

19 The Use of Curcumin for the Treatment of Renal Disorders: A Systematic Review of Randomized Controlled Trials

Mohammad Bagherniya, Davood Soleimani, Mohammad Hossein Rouhani, Gholamreza Askari, Thozhukat Sathyapalan, and Amirhossein Sahebkar

Abstract

Chronic kidney disease (CKD) is one of the significant causes of morbidity and mortality worldwide, which could develop and progress to end-stage renal disease. Increased inflammation and reduced antioxidant capacity commonly occur in CKD and hemodialysis patients. Curcumin is a natural bioactive compound with antioxidant and anti-inflammatory properties. This systematic review was undertaken with the main aim of assessing the effects of curcumin/turmeric supplementation on renal diseases based on clinical trials. A comprehensive search was performed in PubMed/MEDLINE, Scopus, ISI Web of Science, and Google Scholar from inception up to April 6, 2020 to identify clinical trials assessing the effects of curcumin or turmeric alone, or in combination with other herbs or nutrients on renal diseases. Twelve studies met the eligibility criteria. These randomized controlled trials (RCTs) comprised 631 patients with either chronic kidney diseases (CKD), hemodialysis, diabetic proteinuria and nephropathy, and lupus nephritis. Curcumin/turmeric supplementation had favorable effects on renal diseases, particularly in terms of inflammation and oxidative stress. However, with the exception for proteinuria, their impact on clinical parameters, such as blood urea nitrogen, creatinine, glomerular filtration rate (GFR), and serum albumin, was weak and not significant. No serious adverse effects were reported following curcumin/turmeric supplementation. Within the limitations of this review, it can be concluded

M. Bagherniya · M. H. Rouhani · G. Askari
Food Security Research Center, Department of Community Nutrition, School of Nutrition and Food Science, Isfahan University of Medical Sciences, Isfahan, Iran

D. Soleimani
School of Nutrition Sciences and Food Technology, Kermanshah University of Medical Sciences, Kermanshah, Iran

T. Sathyapalan
Department of Academic Diabetes, Endocrinology and Metabolism, Hull York Medical School, University of Hull, Hull, UK

A. Sahebkar (✉)
Biotechnology Research Center, Pharmaceutical Technology Institute, Mashhad University of Medical Sciences, Mashhad, Iran

Neurogenic Inflammation Research Center, Mashhad University of Medical Sciences, Mashhad, Iran

Halal Research Center of IRI, FDA, Tehran, Iran

Polish Mother's Memorial Hospital Research Institute (PMMHRI), Lodz, Poland
e-mail: sahebkara@mums.ac.ir

P. C. Guest (ed.), *Studies on Biomarkers and New Targets in Aging Research in Iran*, Advances in Experimental Medicine and Biology 1291, https://doi.org/10.1007/978-3-030-56153-6_19

that curcumin/turmeric supplementation might have some beneficial effects on inflammatory and oxidative stress parameters of patients but no considerable positive impact on clinical outcomes of kidney diseases, apart from proteinuria.

Keywords

Curcumin · Turmeric · Kidney · Renal diseases · Proteinuria · Diabetic nephropathy

19.1 Introduction

Chronic kidney disease (CKD) is a significant cause of morbidity and mortality worldwide, which imposes a high economic cost to the healthcare systems [1]. CKD has a direct correlation with increased risks of cardiovascular morbidity, premature death, hospitalization, cognitive dysfunction, and decreased quality of life [1–4]. A recent meta-analysis indicated a high global prevalence of CKD (11–13% with the majority in stage 3) [1]. End-stage renal disease (ESRD) occurs when the glomerular filtration rate (GFR) is substantially decreased and accompanied by signs and symptoms of kidney failure that necessitate replacement therapy [5].

Today, diabetes and hypertension are the primary cause of ESRD [6]. The ever-increasing high prevalence of type 2 diabetes has led to a significant rise in the number of patients with CKD and the number requiring end-stage renal failure management, particularly with the need for dialysis [6].

It has been demonstrated that inflammation plays a central role in the etiology of chronic diseases such as CKD, cardiovascular disease, diabetes, metabolic syndrome, and hypertension [7–11]. Increased inflammation and reduced antioxidant capacity is a common complication among CKD and hemodialysis patients [11, 12]. One of the significant factors involved in the development and progression of nondiabetic or diabetic proteinuric CKD and its related complications is increased oxidative stress among patients with renal disorders [13, 14]. Controlling proteinuria and delaying the progression of CKD in adults are the main targets of the current treatment of kidney diseases [15, 16]. Likewise, although dialysis and transplantation are an effective way of managing end-stage renal failure, these approaches are not optimal therapies for kidney failure since they are associated with several adverse effects [5]. In CKD and hemodialysis patients, it has been shown that high levels of inflammatory and hyperlipidemic factors are associated with the increased risk of cardiovascular disease [17–20].

Finding new, safe, and practical therapeutic approaches to treat kidney diseases as one of the strategies or complementary therapy, with an emphasis on reducing inflammation, has recently attracted significant attention. In this regard, medicinal plants and herbal bioactive compounds with antidiabetic, antilipidemic, antihypertension, and anti-inflammatory properties could be potential candidates to consider as an alternative or complementary medicine in the treatment of renal diseases [21–25].

Curcumin is the main active ingredient in turmeric with several proven health benefits [26–31]. It has been shown that this natural bioactive compound has several unique properties such as anti-tumor, anti-inflammatory, antioxidant, antithrombotic, chemosensitizing and chemopreventive, neuroprotective and cardioprotective, lipid-modifying, analgesic, and antirheumatic activities [22, 32–36]. Recently, the favorable effects of curcumin on kidney diseases were shown in preclinical studies [37–39]. In addition, the results of these medicinal herbs on kidney diseases have been assessed in clinical trials [40–42].

Although the effects of curcumin/turmeric on renal diseases were assessed in some clinical trials, to our knowledge, there has been no review of the main findings of these studies. Thus, the main aim of this systematic review was to assess the effects of curcumin or turmeric supplementation on kidney diseases in clinical trials.

19.2 Methods

This systematic review was designed and reported based on the guidelines of the preferred reporting items for systematic reviews and meta-analyses (PRISMA) [43].

19.2.1 Search Strategy

ISI web of science, Scopus, PubMed, and Google Scholar databases systematically searched up to April sixth, 2020 without any limitation for language and date using the following keywords: ("curcumin" OR "curcuminoids" OR "turmeric") AND ("hemodialysis" OR "Chronic Kidney Disease" OR "Acute Kidney Disease" OR "nephritis" OR "renal diseases" OR "kidney diseases" OR "nephro" OR "Nephrotic syndrome") AND ("Intervention Study" OR "Intervention Studies" OR "Controlled trial" OR "Randomized controlled trial" OR "Randomised clinical trial" OR "Non-Randomized Controlled Trials" OR "Clinical Trial" OR "Non-Randomized Controlled Trials" OR "Cross-Over study" OR "Cross-Over trial" OR "Cross Over trial" OR "Cross Over study" OR "Double-Blind Method" OR "Double-Blind" OR "Double-Blind trial" OR "Double-Blind study"). Whenever possible, Medical Subject Headings (MESH) terms were also used. Finally, a direct search was performed of the reference lists of the included original papers and review articles to identify other relevant works for inclusion in the current review.

19.2.2 Study Selection

Two independent authors (MB and GA) reviewed the title and abstract of all papers found in the first stages of the search. Articles were excluded if they did not meet the inclusion criteria using a screen form with a hierarchical approach based on study design, population, exposure, and outcome, and reference lists of relevant review articles were reviewed to explore additional studies. The full texts of eligible citations were considered. Any disagreements were discussed and a final agreement made.

19.2.3 Inclusion and Exclusion Criteria

The main aims of the search were to identify articles investigating the effects of curcumin/turmeric supplementation alone or in combination with other herbs or nutrients on kidney or renal diseases. Original articles were included in the present systematic review which followed these criteria: (1) use of a clinical trial design; (2) curcumin or turmeric supplementation was applied; and (3) the study was conducted on patients with kidney or renal diseases as a primary disease or secondary condition.

Studies were excluded if they (1) were uncontrolled; (2) reported duplicate data; or (3) were reviews, letters, editorial articles, study protocol, or case reports.

19.2.4 Data Extraction

Relevant articles were selected after screening records from the initial search. The following information was extracted from eligible studies and reported in Table 19.1: publication information including the first author's last name, publication date, study location, details of the clinical trial including sample size, patients, mean age (years), study design, intervention (treatment), control, duration of the study, findings of the studies including main results, side effects, and study quality using Jadad Scores.

19.2.5 Quality Assessment

Two independent reviewers (MB and GA) evaluated the quality of the included studies using the Jadad scale, which contained three domains as follows: (1) randomization (0-2 points); (2) blinding (0-2 points); and (3) dropouts and withdrawals (0-1 points) [44]. According to the mentioned domains, the overall quality of each study was considered as low or high quality if the scores were less than or equal to 2 or greater than or equal to 3, respectively [45].

Table 19.1 Summary of studies included in the systematic review

Author, year, country	Sample size (male/female)	Patients	Age Range (years) or mean	Study design	Intervention	Control	Duration	Main results	Side effects	Jadad score
Afshar (2020), Iran [42]	57 (34/20)	Hemodialysis	57.2 ± 10.7	RCT	Nano-curcumin (3 × 40 mg: 120 mg/day)	Placebo soft gel (paraffin)	12 weeks	(1) hs-CRP level, and adhesion molecules (ICAM-1, VCAM-1) significantly decreased in the intervention group compared with placebo (2) FBS and lipid profile was not change in intervention compared with control group	No Side effects	4
Alvarenga (2020), Brazil [41]	28 (14/14)	Hemodialysis	53.5 ± 13.3	RCT	100 mL of orange juice with 12 g of carrot and 2.5 g of turmeric after each dialysis session per week	Same juice without curcumin	3 months	(1) Turmeric supplementation resulted in significant decrease in NF-kB mRNA expression and hs-CRP levels (2) mRNA expression of Nrf2, NLRP3 inflammasome, and IL-1β in peripheral blood mononuclear cells did not change (3) BMI, waist circumferences, lipid profile, HbA1c, glucose, albumin, urea, hemoglobin, potassium, and phosphorus did not change in both groups	No Side effects	5

Author, year, country	Sample size (male/female)	Patients	Age Range (years) or mean	Study design	Intervention	Control	Duration	Main results	Side effects	Jadad score
Samadian (2017), Iran [53]	71 (?/?)	Hemodialysis	49.6 ± 16.8	RCT	1500 mg of turmeric (3 caps/day each capsule contained 500 mg of turmeric or 22.1 mg of curcumin) (1500 mg of turmeric or 66.3 mg of curcumin/day)	Placebo capsule (sorbitol)	12 weeks	(1) Although significant reduction in hs-CRP, IL-6, and TNF-α levels in turmeric group, there was no statistical difference between intervention and control groups. (2) Albumin level significantly increased in turmeric group but not significant compared to placebo group (3) Potassium and liver function tests did not change in both groups.	No side effects	5
Hami (2019), Iran [40]	60 (35/25)	Moderate-to-severe CKD (serum creatinine >1.2 mg/dL or GFR between 15 and 60 mL/min (stage 3–4) who were candidates for cardiac angiography or angioplasty)	63.8 ± 11.5	RCT	Curcumin (3 × 500 mg: 1500 mg/day)	Placebo	Two days before coronary angiography or angioplasty to three days after	(1) CIN occurred in 5 (16.7%) patients in curcumin group and 7 (23.3%) in placebo group (no difference between groups) (CIN was defined by increased serum creatinine ≥0.3 mg/dL or increase to ≥1.5 times of the baseline within 48 h after procedure) (2) Serum creatinine increased after 72 h intervention in both groups (no difference between groups)	No side effects	4

(continued)

Table 19.1 (continued)

Author, year, country	Sample size (male/female)	Patients	Age Range (years) or mean	Study design	Intervention	Control	Duration	Main results	Side effects	Jadad score
Jim!enez-Osorio (2016), Mexico, [46]	101 (61/40)	Nondiabetic or diabetic proteinuric CKD	48.2 ± 7.6	RCT	Curcumin (107 × 3: 320 mg/day	Placebo (starch)	8 weeks	(1) Curcumin treatment did not improve proteinuria, estimated GFR, or lipid profile (2) Curcumin attenuated lipid peroxidation in individuals with nondiabetic proteinuric CKD (3) Curcumin enhanced antioxidant capacity in subjects with diabetic proteinuric CKD (4) Curcumin had no effect on antioxidant enzyme activities or Nrf2 activation	Not reported	2
Moreillon (2013), United States [49]	16 (?/?)	Non-dialysis, mild and moderate CKD	56 ± 16	RCT	Curcumin and Boswellia serrata (each capsule contained 824 mg of purified turmeric extract, 95% curcuminoids, and 516 mg of Boswellia serrata extract, 10% 3-acetyl-11-keto-β--boswellic acid) (2 caps/day)	Placebo (roasted rice powder)	8 weeks	No significant changes observed in IL-6, CRP, TNFα, creatinine, BUN, albumin, total protein, ALT, AST, glucose levels	Only minor side effects reported	4

Author, year, country	Sample size (male/ female)	Patients	Age Range (years) or mean	Study design	Intervention	Control	Duration	Main results	Side effects	Jadad score
Ortiz (2019), Mexico [50]	40 (26/14)	CKD and iron overload undergoing hemodialysis	37.0 ± 11.5	RCT	Resveratrol + curcumin (oral dose of 500 mg of resveratrol and 500 mg of curcumin/day)	Placebo	12 weeks	(1) Triglycerides, VLDL, cholesterol significantly deceased in intervention compared with the placebo group (2) FBS and uric acid did not significantly change between groups (3) BMI, muscle mass (kg), and bone (kg) significantly increased in intervention group, while fat% and subjective global evaluation significantly decreased (data of control group not reported) (4) Ferritin significantly decreased in intervention compared with control group (5) TBARS and carbonyl values did not change in the study groups	Not reported	2

(continued)

Table 19.1 (continued)

Author, year, country	Sample size (male/female)	Patients	Age Range (years) or mean	Study design	Intervention	Control	Duration	Main results	Side effects	Jadad score
Pakfetrat (2014), Iran [52]	100 (60/40)	Hemodialysis—suffering from pruritus	53.3 ± 15.8	RCT	1500 mg of turmeric (3 caps/day each capsule contained 500 mg of turmeric or 22.1 mg of curcumin) (1500 mg of turmeric or 66.3 mg of curcumin/day)	Placebo (starch)	8 weeks	(1) Turmeric treatment significantly reduced hs-CRP compared to placebo group (2) Reduction in pruritus scores significantly greater in turmeric than in placebo group (3) BUN and creatinine did not significantly change in intervention compared with control group	No side effect	5
Pakfetrat, 2015, Iran [51]	48 (29/19)	Hemodialysis	49.4 ± 14.7	RCT	1500 mg of turmeric(3 caps/day each capsule contained 500 mg of turmeric or 22.1 mg of curcumin) (1500 mg of turmeric or 66.3 mg of curcumin/day)	Placebo (starch)	8 weeks	(1) Serum albumin significantly increased in turmeric than in placebo group (2) Plasma MDA significantly decreased and RBC CAT activity increased in turmeric group compared with placebo (3) GPX and GR increased in both groups, with no significant differences between groups (4) Hemoglobin, lipid profile, and ALT, AST did not significantly change in the turmeric group compared with placebo	No side effect	4

Author, year, country	Sample size (male/female)	Patients	Age Range (years) or mean	Study design	Intervention	Control	Duration	Main results	Side effects	Jadad score
Khajehdehi (2011), Iran [47]	40 (22/18)	Overt type 2 diabetic nephropathy	52.8 ± 9.3	RCT	1500 mg of turmeric(3 caps/day each capsule contained 500 mg of turmeric or 22.1 mg of curcumin) (1500 mg of turmeric or 66.3 mg of curcumin/day)	Placebo (starch)	8 weeks	(1) Systolic and diastolic blood pressure, FBS, BUN, creatinine and lipid profile did not significant between groups (2) Proteinuria and serum TGF-β significantly reduced in turmeric compared with placebo group (3) Serum TNF-α and IL-8 levels were not significantly reduced in the turmeric group compared with placebo	No side effects	3
Khajehdehi (2012), Iran [48]	24 (12/12)	Relapsing or refractory biopsy-proven lupus nephritis	33.6 ± 10.8	RCT	1500 mg of turmeric (3 caps/day each capsule contained 500 mg of turmeric or 22.1 mg of curcumin) (1500 mg of turmeric or 66.3 mg of curcumin/day)	Placebo (starch)	3 months	(1) Systolic blood pressure and proteinuria significantly decreased in trial group compared with baseline. After 2 months, reduction in proteinuria was significant compared with placebo, but in 3 months no significant difference was observed (2) GFR, BUN, creatinine, albumin, and diastolic blood pressure did not significantly change in both groups	No side effects	3

(continued)

Table 19.1 (continued)

Author, year, country	Sample size (male/female)	Patients	Age Range (years) or mean	Study design	Intervention	Control	Duration	Main results	Side effects	Jadad score
Vanaie (2019), Iran [54]	46 (27/19)	T2DM, overt albuminuria ≥300 mg/24 h, and estimated glomerular filtration rate (eGFR) ≥30 mL/min/1.73 m^2	59.8 ± 8.4	RCT	Curcumin (3 × 500 mg: 1500 mg/day)	Placebo	16 weeks	(1) Albuminuria decreased significantly in curcumin compared to placebo group (2) Compared with placebo, curcumin treatment had no effects on serum BUN, creatinine, FBS, 2-h pp. BS, HbA1C, lipid profile, and albumin	Epigastric pain in one case	4

RCT randomized controlled trial, *hs-CRP* high-sensitivity C-reactive protein, *FBS* fasting blood sugar, *NF-kB* nuclear factor Kappa B, *Nrf2* nuclear factor erythroid factor 2-related, *IL* interleukin, *NLRP3* Nod-like receptor pyrin domain containing 3, *HbA1c* hemoglobin A1C, *BMI* body mass index, *CKD* chronic kidney diseases, *CIN* contrast-induced nephropathy, *BUN* blood urea nitrogen, *VLDL* very low density lipoprotein, *TBARS* thiobarbituric acid reactive substances, *MDA* malondialdehyde, *RBC* red blood cell, *GPX* glutathione peroxidase, *GR* glutathione reductase, *CAT* catalase, *AST* aspartate aminotransferase, *ALT* alanine aminotransferase, *TGF-β* transforming growth factor, *TNF-α* tumor necrosis factor-α, *GFR* glomerular filtration rate, *T2DM* Type 2 diabetes mellitus, *2-h pp BS* 2-h postprandial blood sugar

19.3 Results

The online search yielded in 679 non-duplicated records, of which, after reading the title and abstracts, 660 irrelevant records were omitted. Nineteen articles were selected for a full assessment, of which 12 records were appropriate (Fig. 19.1). Thus, data extraction was performed on 12 articles [40–42, 46–54]. The characteristics of each selected paper are shown in Table 19.1. The studies were published between 2011 and 2020 and comprised a total of 631 patients. Eight studies were conducted in Iran [40, 42, 47, 48, 51–54], two in Mexico [46, 50], one in Brazil [41], and one in the United States of America [49]. Six studies worked on hemodialysis patients [41, 42, 50–53], three studies were conducted on CKD patients (with or without diabetes) [40, 46, 49], and one of these articles worked on contrast-induced nephropathy patients [40]. One study was conducted on patients with overt type 2 diabetic nephropathy [47], one in patients with type 2 diabetes mellitus with overt albuminuria [54], and one in patients with relapsing or refractory biopsy-

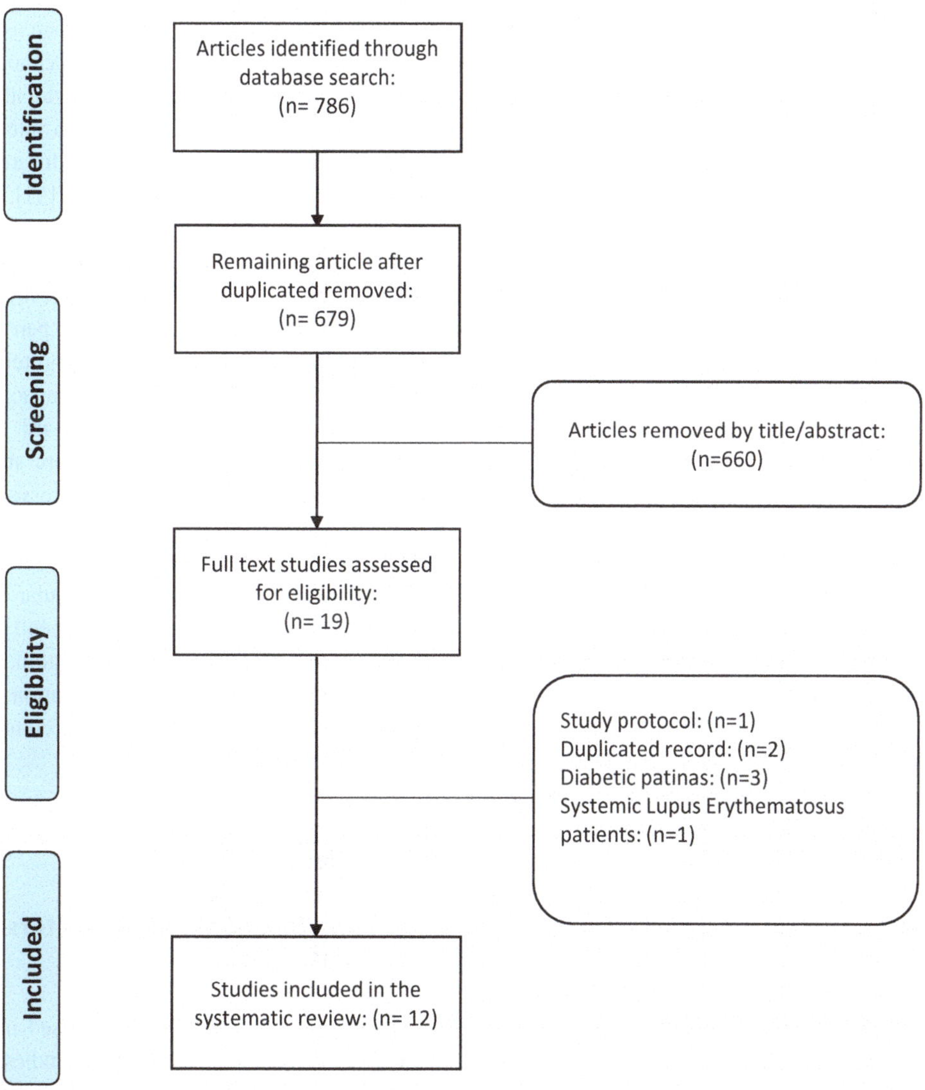

Fig. 19.1 Flowchart of the process of study selection

proven lupus nephritis [48]. All of the studies were conducted on both male and females, except for two studies which did not mention the gender of the participants [49, 53]. The mean age of the participants varied from 33.6 to 63.8 years. All of the included studies were conducted in the context of randomized controlled trials, all applied randomization, and all had a placebo group. One study used nano-curcumin (120 mg/day) [42], six studies used turmeric (500 mg–2.5 g daily) [41, 47, 48, 51–53], three studies used curcumin (320–1500 mg daily) [40, 46, 54], one applied curcumin and Boswellia serrata (each capsule contained 824 mg of purified turmeric extract, 95% curcuminoids, and 516 mg of Boswellia serrata extract; 2 capsules/day) [49], and one study used curcumin (500 mg/daily) in combination with resveratrol (500 mg/day) [50]. The duration of the studies was in the range of 8–16 weeks [41, 42, 46–54]. One study used curcumin two days before coronary angiography or angioplasty to three days afterwards in CKD patients [40].

19.3.1 Quality Assessment

Overall, ten studies out of 12 were categorized as high quality [40–42, 47–49, 51–54], and two had a low-quality methodological approach [46, 50].

19.3.2 Effects of Curcumin on the Inflammatory and Stress Oxidative Status of Patients

In three studies, curcumin/turmeric consumption significantly reduced high-sensitivity C-reactive protein (hs-CRP) compared with placebo [41, 42, 52]. In one study, hs-CRP levels were significantly reduced after turmeric consumption, although no differences were found compared with the placebo group [53]. In one study, curcumin and Boswellia serrata did not affect CRP levels [49]. Four studies assessed the effects of curcumin/turmeric on serum interleukins (ILs), and one of these found that IL-6 levels were significantly reduced in response to treatment [49]. However, no significant difference was found than in placebo groups. In other studies, curcumin/turmeric had no significant effects on IL-6 [53], IL-8 [47], and IL-1β [41] compared with the placebo group. Likewise, in comparison to the placebo, curcumin/turmeric had no significant effects on TNF-α levels [47, 49, 53]. In one study, turmeric treatment significantly decreased nuclear factor kappa B (NF-κb) mRNA expression. However, mRNA expression of nuclear factor erythroid factor 2-related (Nrf2), Nod-like receptor pyrin domain containing 3 (NLRP3) inflammasome did not significantly change [41]. In one study, plasma malondialdehyde (MDA) was significantly decreased, and red blood cell catalase (CAT) activity was increased in the turmeric group compared with placebo, and glutathione peroxidase (GPX) and glutathione reductase (GR) were increased in both intervention and control groups. However, differences between groups were not significant [51]. One study showed that curcumin and resveratrol did not affect thiobarbituric acid reactive substance (TBARS) levels [50]. Another investigation showed that curcumin attenuated lipid peroxidation in individuals with nondiabetic proteinuric CKD and enhanced the antioxidant capacity in subjects with diabetic proteinuric CKD [46]. In the study of Ortiz et al., ferritin levels were significantly decreased in the curcumin and resveratrol group compared with placebo in patients with CKD and iron overload undergoing hemodialysis [50]. The adhesion molecules, intercellular adhesion molecule 1 (ICAM-1), and vascular cell adhesion protein 1 (VCAM-1), were significantly decreased after 12 weeks of intervention with nano-curcumin in hemodialysis patients compared with the placebo [42].

19.3.3 Effects of Curcumin on the Clinical and Biochemical Parameters of the Kidney

Blood urea nitrogen (BUN) was assessed in five studies [47–49, 52, 54]. In all of these studies, curcumin/turmeric had no significant effect on this parameter compared with placebo. Likewise, creatinine was assessed in six studies, and curcumin/turmeric did not affect this in comparison to the control group [40, 47–49, 52, 54]. In the study conducted by Hami et al., serum creatinine was increased after 72-h intervention in both groups, but there was no difference between the groups [40]. In another study, curcumin treatment did not affect proteinuria in nondiabetic or diabetic proteinuric CKD [46]. However, turmeric treatment remarkably reduced proteinuria in relapsing or refractory lupus nephritis and overt type 2 diabetic nephropathy patients compared with placebo [47, 48]. Glomerular filtration rate (GFR) did not improve after curcumin/turmeric supplementation in nondiabetic or diabetic proteinuric CKD [46] and relapsing or refractory lupus nephritis patients [48]. After turmeric supplementation, serum albumin was significantly increased compared with placebo in a study conducted in hemodialysis patients [51]. However, there was no significant effect in five studies conducted on hemodialysis patients [41, 53], CKD patients [49], patients with relapsing or refractory lupus nephritis [48], and diabetic patients with overt albuminuria [54]. In the study of Vanaie et al., curcumin supplementation for 16 weeks resulted in a substantial reduction in albuminuria among diabetic patients with overt albuminuria in comparison to the placebo [54]. Other biochemical factors, such as serum uric acid [50], urea [41], phosphorus [41], potassium [41, 53], and hemoglobin [41, 51], were evaluated in a few studies, which found no significant effects. Moreover, compared with placebo, turmeric treatment had no effects on systolic and diastolic blood pressure, as assessed in two studies in patients with overt type 2 diabetic nephropathy and relapsing or refractory lupus nephritis [47, 48].

19.3.4 Effects of Curcumin on Lipid Profile, Glycemic Indices, and Liver Functional Tests

Curcumin/turmeric supplementation had no significant effects on lipid profile [41, 42, 46, 47, 51, 54], except for one study in which 12 weeks' treatment with curcumin and resveratrol led to decreased triglycerides, very low density lipoprotein (VLDL), and cholesterol compared with placebo [50]. Fasting blood sugar (FBS) [42, 47, 50, 54] and hemoglobin A1C (HbA1c) [41, 54] levels did not change in response to curcumin/turmeric treatment. Likewise, after curcumin/turmeric supplementation, liver function tests such as measurement of alanine aminotransferase (ALT) and aspartate aminotransferase (AST) did not change in the study groups [49, 53].

19.3.5 Effects of Curcumin/Turmeric on Other Outcomes and Adverse Effects

In their study, Ortiz et al. showed that 12 weeks' intervention with curcumin and resveratrol significantly increased body mass index (BMI), muscle mass (kg), and bone mass (kg). At the same time, the percentage of fat and subjective global evaluation were significantly decreased (data of the control group were not reported) in patients with CKD and iron overload undergoing hemodialysis [50]. In the study conducted by Alvarenga et al., turmeric did not affect BMI or waist circumference [41]. Overall supplementation with curcumin/turmeric had no adverse side effects in renal diseases, although two studies stated that a few patients reported some adverse effects [49, 54] (Table 19.1).

19.4 Discussion

To the best of our knowledge, this is the first review which systematically assessed the effects of curcumin/turmeric supplementation on renal diseases. The main findings were that curcumin/turmeric supplementation had some favorable effects on renal diseases, particularly in terms of inflammation and stress oxidative, although the effects on clinical parameters such as BUN, creatinine, GFR, and serum albumin were weak and insignificant.

Atherosclerotic cardiovascular disease is the common and important cause of morbidity and mortality among patients with CKD. This appears to be a consequence of abnormalities of lipid metabolism, and a systemic increase in inflammation and oxidative stress in CKD patients [55]. Clinical evidence has documented that consumption of curcumin and turmeric can improve endothelial function and serum lipid profiles among patients with cardiovascular disease by modulating several pathways involved in lipid metabolism, including lipoprotein lipase (LPL), cholesteryl ester transfer protein (CETP), peroxisome proliferator-activated receptor gamma (PPAR-γ), and PPAR-α [56, 57]. In addition, it has well-established anti-inflammatory and antioxidative properties [29–32].

Both preclinical and clinical evidence suggests that curcumin can interfere with multiple signaling pathways linked to tissue injury, including the NF-κB signaling pathway, arachidonic acid (AA) metabolism, and the damaging effects of reactive oxygen species (ROS) [58]. NF-κB is the most critical transcriptional activator protein in the inflammatory cascade. Clinical evidence on hemodialysis patients showed that turmeric supplementation resulted in significant reduction in NF-kB mRNA expression [41]. In vivo and in vitro studies have shown that curcumin downregulates NF-κB expression through inhibition of the c-Jun N-terminal kinase (JNK), p38 mitogen-activated protein kinases (p38 MAPK), and extracellular signal-regulated kinase (ERK) pathways in response to pro-inflammatory stimuli and redox status [59–62]. Also, curcumin appears to inhibit nuclear NF-κB translocation through IκB kinase (IKK) activation and prevent NF-κB binding to the promoter region of inflammation-relevant genes [63].

Some effects of curcumin on inflammation can also be attributed to modulation of leukocyte recruitment into injured tissue, thereby altering AA metabolism and down-regulating NF-κB target genes, such as CXCL1, CXCL2, ICAM-1, and VCAM-1 [64, 65]. In line with this, a nanoformulation of curcumin has been found to reduce the levels of ICAM-1 and VCAM-1 adhesion molecules in hemodialysis patients after 12 weeks' intervention [42]. AA metabolites, such as leukotrienes, have been implicated in the inflammatory cascade through increasing endothelial membrane permeability and leukocyte recruitment. Curcumin switches AA metabolism through inactivation of phospholipase A2 (*PLA2G2A*) and arachadonate 5-lipoxygenase (*ALOX5*) and down-regulation of cyclooxygenase-2 (*PTGS2*) gene expression via the NF-κB pathway [66]. This is consistent with findings which have shown that curcumin could act as a selective COX-2 inhibitor [67].

Patients with renal diseases have consistently exhibited an imbalance between oxidants and antioxidants status. Elevated free radical levels not only cause damage to DNA and increase lipid peroxidation, but also stimulate signaling pathways that switch on inflammation-relevant gene expression. Curcumin as the main phenolic compound in turmeric directly scavenges free radicals. In addition, increasing evidence supports its roles in enhancing both endogenous enzymatic and non-enzymatic antioxidants, perhaps due to induction of the Nrf2 transcription factor [68–70].

One of the major drawbacks of curcumin use is its poor absorption and bioavailability, which has severely limited its applications [71]. To overcome this limitation, one of the studies reviewed here used nano-curcumin [42]. The use of new formulations of curcumin such as curcumin-piperine or phospholipid curcumin with greater bioavailability may lead to better results in treatment outcomes for kidney disease patients [72].

Although this systematic review comprehensively assessed the effects of curcumin and turmeric supplementation on renal diseases based on randomized clinical trials, several limitations should be acknowledged. First, the small number of included studies and corresponding low numbers of participants made it difficult to draw robust evidence-based conclusions on its efficacy. Second, it was difficult to make comparisons across the studies due to their heterogeneous nature, regarding use of different forms of curcumin, different dosages, and treatment durations, as well as dissimilar outcome measures. Finally, more than a half of the studies (eight studies) were conducted in Iran, which restricts the ability to extend the findings to other populations.

19.5 Conclusions

Within the limitations of this review, it can be concluded that curcumin or turmeric supplementation might have some beneficial effects on inflammatory and stress oxidative parameters of patients with renal diseases. However, curcumin/turmeric had no considerable positive impact on clinical outcomes of kidney diseases apart from alleviation of proteinuria. The patients reported no serious adverse effects following curcumin/turmeric supplementation, supporting the case that these natural products are relatively safe. Thus, more randomized controlled trials using larger patient numbers and standardized protocols are needed for more accurate testing of the effects of curcumin supplementation on renal diseases.

References

1. Hill NR, Fatoba ST, Oke JL, Hirst JA, O'Callaghan CA, Lasserson DS et al (2016) Global prevalence of chronic kidney disease–a systematic review and meta-analysis. PLoS One 11(7):e0158765. https://doi.org/10.1371/journal.pone.0158765
2. Etgen T, Chonchol M, Forstl H, Sander D (2012) Chronic kidney disease and cognitive impairment: a systematic review and meta-analysis. Am J Nephrol 35(5):474–482
3. Perlman RL, Finkelstein FO, Liu L, Roys E, Kiser M, Eisele G et al (2005) Quality of life in chronic kidney disease (CKD): a cross-sectional analysis in the Renal Research Institute-CKD study. Am J Kidney Dis 45(4):658–666
4. Gansevoort RT, Correa-Rotter R, Hemmelgarn BR, Jafar TH, Heerspink HJ, Mann JF et al (2013) Chronic kidney disease and cardiovascular risk: epidemiology, mechanisms, and prevention. Lancet 382(9889):339–352
5. Alebiosu C, Ayodele O (2005) The global burden of chronic kidney disease and the way forward. Ethn Dis 15(3):418
6. Atkins RC (2005) The epidemiology of chronic kidney disease. Kidney Int 67:S14–S18
7. Dandona P, Aljada A, Bandyopadhyay A (2004) Inflammation: the link between insulin resistance, obesity and diabetes. Trends Immunol 25(1):4–7
8. Haffner SM (2006) The metabolic syndrome: inflammation, diabetes mellitus, and cardiovascular disease. Am J Cardiol 97(2):3–11
9. Manabe I (2011) Chronic inflammation links cardiovascular, metabolic and renal diseases. Circ J 75(12):2739–2748
10. Pawelec G, Goldeck D, Derhovanessian E (2014) Inflammation, ageing and chronic disease. Curr Opin Immunol 29:23–28
11. Akchurin M, Kaskel F (2015) Update on inflammation in chronic kidney disease. Blood Purif 39(1–3):84–92
12. Yao Q, Axelsson J, Stenvinkel P, Lindholm B (2004) Chronic systemic inflammation in dialysis patients: an update on causes and consequences. ASAIO J 50(6):lii–lvii. https://doi.org/10.1097/01.mat.0000147958.87989.eb
13. Forbes JM, Coughlan MT, Cooper ME (2008) Oxidative stress as a major culprit in kidney disease in diabetes. Diabetes 57(6):1446–1454
14. Small DM, Coombes JS, Bennett N, Johnson DW, Gobe GC (2012) Oxidative stress, anti-oxidant therapies and chronic kidney disease. Nephrology 17(4):311–321
15. Chandar J, Abitbol C, Montane B, Zilleruelo G (2007) Angiotensin blockade as sole treatment for proteinuric kidney disease in children. Nephrol Dial Transpl 22(5):1332–1337
16. Alicic RZ, Tuttle KR (2014) Novel therapies for diabetic kidney disease. Adv Chronic Kidney Dis 21(2):121–133
17. Pandya V, Rao A, Chaudhary K (2015) Lipid abnormalities in kidney disease and management strategies. World J Nephrol 4(1):83
18. Keane WF, Tomassini JE, Neff DR (2012) Lipid abnormalities in patients with chronic kidney disease: implications for the pathophysiology of atherosclerosis. J Atheroscler Thromb 20(2):123–133
19. Wanner C, Zimmermann J, Schwedler S, Metzger T (2002) Inflammation and cardiovascular risk in dialysis patients. Kidney Int 61:S99–S102
20. Qureshi AR, Alvestrand A, Divino-Filho JC, Gutierrez A, Heimbürger O, Lindholm B et al (2002) Inflammation, malnutrition, and cardiac disease as predictors of mortality in hemodialysis patients. J Am Soc Nephrol 13(Suppl 1):S28–S36
21. Talebi S, Bagherniya M, Atkin SL, Askari G, Orafai HM, Sahebkar A (2020) The beneficial effects of nutraceuticals and natural products on small dense LDL levels, LDL particle number and LDL particle size: a clinical review. Lipids Health Dis 19(1):66. https://doi.org/10.1186/s12944-020-01250-6
22. Bagherniya M, Nobili V, Blesso CN, Sahebkar A (2018) Medicinal plants and bioactive natural compounds in the treatment of non-alcoholic fatty liver disease: a clinical review. Pharmacol Res 130:213–240
23. Bagherniya M, Johnston TP, Sahebkar A (2020) Regulation of apolipoprotein B by natural products and nutraceuticals: a comprehensive review. Curr Med Chem. https://doi.org/10.2174/0929867327666200427092114. Online ahead of print

24. Musabayane C (2012) The effects of medicinal plants on renal function and blood pressure in diabetes mellitus. Cardiovasc J Afr 23(8):462–468
25. Iranshahi M, Sahebkar A, Hosseini ST, Takasaki M, Konoshima T, Tokuda H (2010) Cancer chemopreventive activity of diversin from Ferula diversivittata in vitro and in vivo. Phytomedicine 17(3–4):269–273
26. Kocaadam B, Şanlier N (2017) Curcumin, an active component of turmeric (Curcuma longa), and its effects on health. Crit Rev Food Sci Nutr 57(13):2889–2895
27. Ringman JM, Frautschy SA, Cole GM, Masterman DL, Cummings JL (2005) A potential role of the curry spice curcumin in Alzheimer's disease. Curr Alzheimer Res 2(2):131–136
28. Wongcharoen W, Phrommintikul A (2009) The protective role of curcumin in cardiovascular diseases. Int J Cardiol 133(2):145–151
29. Mollazadeh H, Cicero AFG, Blesso CN, Pirro M, Majeed M, Sahebkar A (2019) Immune modulation by curcumin: the role of interleukin-10. Crit Rev Food Sci Nutr 59(1):89–101
30. Hasanzadeh S, Read MI, Bland AR, Majeed M, Jamialahmadi T, Sahebkar A (2020) Curcumin: an inflammasome silencer. Pharmacol Res 159:104921. https://doi.org/10.1016/j.phrs.2020.104921. Epub 2020 May 25. PMID: 32464325
31. Teymouri M, Pirro M, Johnston TP, Sahebkar A (2017) Curcumin as a multifaceted compound against human papilloma virus infection and cervical cancers: A review of chemistry, cellular, molecular, and preclinical features. BioFactors 43(3):331–346
32. Mohajeri M, Bianconi V, Ávila-Rodriguez MF, Barreto GE, Jamialahmadi T, Pirro M, Sahebkar A (2020) Curcumin: a phytochemical modulator of estrogens and androgens in tumors of the reproductive system. Pharmacol Res 156:104765. https://doi.org/10.1016/j.phrs.2020.104765. Epub 2020 Mar 23. PMID: 32217147
33. Sahebkar A (2013) Why it is necessary to translate curcumin into clinical practice for the prevention and treatment of metabolic syndrome? BioFactors 39(2):197–208
34. Momtazi AA, Derosa G, Maffioli P, Banach M, Sahebkar A (2016) Role of microRNAs in the therapeutic effects of curcumin in non-cancer diseases. Mol Diagn Ther 20(4):335–345
35. Momtazi AA, Sahebkar A (2016) Difluorinated curcumin: a promising curcumin analogue with improved anti-tumor activity and pharmacokinetic profile. Curr Pharm Des 22(28):4386–4397
36. Bagheri H, Ghasemi F, Barreto GE, Rafiee R, Sathyapalan T, Sahebkar A (2020) Effects of curcumin on mitochondria in neurodegenerative diseases. Biofactors. 46(1):5–20. https://doi.org/10.1002/biof.1566. Epub 2019 Oct 3. PMID: 31580521
37. Huang SJ, Huang J, Yan YB, Qiu J, Tan RQ, Liu Y et al (2020) The renoprotective effect of curcumin against cisplatin-induced acute kidney injury in mice: involvement of miR-181a/PTEN axis. Ren Fail 42(1):350–357
38. Damiano S, Andretta E, Longobardi C, Prisco F, Paciello O, Squillacioti C et al (2020) Effects of curcumin on the renal toxicity induced by Ochratoxin A in rats. Antioxidants (Basel) 9(4):E332. https://doi.org/10.3390/antiox9040332
39. Akomolafe SF, Olasehinde TA, Adewale OO, Ajayi OB (2020) Curcumin improves biomolecules associated with renal function and attenuates oxidative injury and histopathological changes in potassium-induced toxicity in rats. Kidney Biol Trace Elem. https://doi.org/10.1007/s12011-020-02113-y. Online ahead of print
40. Hami M, Bigdeli A, Rajabi O, Salehi M (2019) The effect of curcumin in prevention of contrast nephropathy following coronary angiography or angioplasty in CKD patients. Iran J Kidney Dis 13(5):304–309
41. Alvarenga L, Salarolli R, Cardozo LFMF, Santos RS, de Brito JS, Kemp JA et al (2020) Impact of curcumin supplementation on expression of inflammatory transcription factors in hemodialysis patients: a pilot randomized, double-blind, controlled study. Clin Nutr S0261-5614(20):30110–30112. https://doi.org/10.1016/j.clnu.2020.03.007. Online ahead of print
42. Afshar GV, Rasmi Y, Yagmaye P, Khadem-Ansari MH, Makhdomi K, Rasooli J (2020) The effects of nano-curcumin supplementation on serum level of hs-CRP, adhesion molecules, and lipid profiles in hemodialysis patients, a randomized controlled clinical trial. Iran J Kidney Dis 14(1):52–61
43. Moher D, Liberati A, Tetzlaff J, Altman DG (2009) Preferred reporting items for systematic reviews and meta-analyses: the PRISMA statement. Ann Intern Med 151(4):264–269
44. Jadad AR, Moore RA, Carroll D, Jenkinson C, Reynolds DJM, Gavaghan DJ et al (1996) Assessing the quality of reports of randomized clinical trials: is blinding necessary? Control Clin Trials 17(1):1–12
45. Moher D, Cook D, Jadad A, Tugwell P, Moher M, Jones A et al (1999) Assessing the quality of reports of randomised trials: implications for the conduct of meta-analyses. Health Technol Assess 3(12):i–iv., 1–98
46. Jiménez-Osorio AS, García-Niño WR, González-Reyes S, Álvarez-Mejía AE, Guerra-León S, Salazar-Segovia J et al (2016) The effect of dietary supplementation with curcumin on redox status and Nrf2 activation in patients with nondiabetic or diabetic proteinuric chronic kidney disease: a pilot study. J Ren Nutr 26(4):237–244
47. Khajehdehi P, Pakfetrat M, Javidnia K, Azad F, Malekmakan L, Nasab MH et al (2011) Oral supplementation of turmeric attenuates proteinuria, transforming growth factor-β and interleukin-8 levels in patients with overt type 2 diabetic nephropathy: a randomized, double-blind and placebo-controlled study. Scand J Urol Nephrol 45(5):365–370
48. Khajehdehi P, Zanjaninejad B, Aflaki E, Nazarinia M, Azad F, Malekmakan L et al (2012) Oral supplementation of turmeric decreases proteinuria, hematuria, and systolic blood pressure in patients

suffering from relapsing or refractory lupus nephritis: a randomized and placebo-controlled Study. J Ren Nutr 22(1):50–57

49. Moreillon JJ, Bowden RG, Deike E, Griggs J, Wilson R, Shelmadine B et al (2013) The use of an anti-inflammatory supplement in patients with chronic kidney disease. J Complement Integr Med 10(1):1–10
50. Ortiz BOM, Preciado ARF, Emiliano JR, Garza SM, Rodríguez ER, de Alba Macías LA (2019) Recovery of bone and muscle mass in patients with chronic kidney disease and iron overload on hemodialysis and taking combined supplementation with curcumin and resveratrol. Clin Interv Aging 14:2055–2062
51. Pakfetrat M, Akmali M, Malekmakan L, Dabaghimanesh M, Khorsand M (2015) Role of turmeric in oxidative modulation in end-stage renal disease patients. Hemodial Int 19(1):124–131
52. Pakfetrat M, Basiri F, Malekmakan L, Roozbeh J (2014) Effects of turmeric on uremic pruritus in end stage renal disease patients: a double-blind randomized clinical trial. J Nephrol 27(2):203–207
53. Samadian F, Dalili N, Gholi FPR, Fattah M, Malih N, Nafar M et al (2017) Evaluation of Curcumin's effect on inflammation in hemodialysis patients. Clin Nutr ESPEN 22:19–23
54. Vanaie A, Shahidi S, Iraj B, Siadat ZD, Kabirzade M, Shakiba F et al (2019) Curcumin as a major active component of turmeric attenuates proteinuria in patients with overt diabetic nephropathy. J Res Med Sci 24:77. https://doi.org/10.4103/jrms.JRMS_1055_18
55. Cachofeiro V, Goicochea M, De Vinuesa SG, Oubiña P, Lahera V, Luño J (2008) Oxidative stress and inflammation, a link between chronic kidney disease and cardiovascular disease: New strategies to prevent cardiovascular risk in chronic kidney disease. Kidney Int 74:S4–S9
56. Hallajzadeh J, Milajerdi A, Kolahdooz F, Amirani E, Mirzaei H, Asemi Z (2019) The effects of curcumin supplementation on endothelial function: a systematic review and meta-analysis of randomized controlled trials. Phytother Res 33(11):2989–2995
57. Qin S, Huang L, Gong J, Shen S, Huang J, Ren H et al (2017) Efficacy and safety of turmeric and curcumin in lowering blood lipid levels in patients with cardiovascular risk factors: a meta-analysis of randomized controlled trials. J Nutr 16(1):68. https://doi.org/10.1186/s12937-017-0293-y
58. Jurenka JS (2009) Anti-inflammatory properties of curcumin, a major constituent of Curcuma longa: a review of preclinical and clinical research. Altern Med Rev 14(2):141–153
59. Camacho-Barquero L, Villegas I, Sánchez-Calvo JM, Talero E, Sánchez-Fidalgo S, Motilva V et al (2007) Curcumin, a Curcuma longa constituent, acts on MAPK p38 pathway modulating COX-2 and iNOS expression in chronic experimental colitis. Int Immunopharmacol 7(3):333–342
60. Cho JW, Lee KS, Kim CW (2007) Curcumin attenuates the expression of IL-1β, IL-6, and TNF-α as well as cyclin E in TNF-α-treated HaCaT cells; NF-κB and MAPKs as potential upstream targets. Int J Mol Med 19(3):469–474
61. Goel A, Kunnumakkara AB, Aggarwal BB (2008) Curcumin as "Curecumin": from kitchen to clinic. Biochem Pharmacol 75(4):787–809
62. Ilbey YO, Ozbek E, Cekmen M, Simsek A, Otunctemur A, Somay A (2009) Protective effect of curcumin in cisplatin-induced oxidative injury in rat testis: mitogen-activated protein kinase and nuclear factor-kappa B signaling pathways. Hum Reprod 24(7):1717–1725
63. Marín YE, Wall BA, Wang S, Namkoong J, Martino JJ, Suh J et al (2007) Curcumin downregulates the constitutive activity of NF-κB and induces apoptosis in novel mouse melanoma cells. Melanoma Res 17(5):274–283
64. Bachmeier BE, Mohrenz IV, Mirisola V, Schleicher E, Romeo F, Höhneke C et al (2008) Curcumin downregulates the inflammatory cytokines CXCL1 and-2 in breast cancer cells via NFκB. Carcinogenesis 29(4):779–789
65. Pan Y, Zhang X, Wang Y, Cai L, Ren L, Tang L et al (2013) Targeting JNK by a new curcumin analog to inhibit NF-kB-mediated expression of cell adhesion molecules attenuates renal macrophage infiltration and injury in diabetic mice. PLoS One 8(11):e79084. https://doi.org/10.1371/journal.pone.0079084. eCollection 2013
66. Hong J, Bose M, Ju J, Ryu J-H, Chen X, Sang S et al (2004) Modulation of arachidonic acid metabolism by curcumin and related β-diketone derivatives: effects on cytosolic phospholipase A 2, cyclooxygenases and 5-lipoxygenase. Carcinogenesis 25(9):1671–1679
67. Jiang H, Deng C-S, Zhang M, Xia J (2006) Curcumin-attenuated trinitrobenzene sulphonic acid induces chronic colitis by inhibiting expression of cyclooxygenase-2. World J Gastroenterol 12(24):3848–3853
68. Yang C, Zhang X, Fan H, Liu Y (2009) Curcumin upregulates transcription factor Nrf2, HO-1 expression and protects rat brains against focal ischemia. Brain Res 1282:133–141
69. Balogun E, Hoque M, Gong P, Killeen E, Green CJ, Foresti R et al (2003) Curcumin activates the haem oxygenase-1 gene via regulation of Nrf2 and the antioxidant-responsive element. Biochem J 371(3):887–895
70. Pizzo P, Scapin C, Vitadello M, Florean C, Gorza L (2010) Grp94 acts as a mediator of curcumin-induced antioxidant defence in myogenic cells. J Cell Mol Med 14(4):970–981
71. Lopresti AL (2018) The problem of curcumin and its bioavailability: could its gastrointestinal influence contribute to its overall health-enhancing effects? Adv Nutr 9(1):41–50
72. Anand P, Kunnumakkara AB, Newman RA, Aggarwal BB (2007) Bioavailability of curcumin: problems and promises. Mol Pharm 4(6):807–818

Curcumin for the Treatment of Prostate Diseases: A Systematic Review of Controlled Clinical Trials

20

Mohammad Bagherniya, Gholamreza Askari, Babak Alikiaii, Saeed Abbasi, Davood Soleimani, Thozhukat Sathyapalan, Tannaz Jamialahmadi, and Amirhossein Sahebkar

Abstract

Prostate cancer is one of the significant causes of morbidity and mortality worldwide. Benign prostatic hyperplasia is another condition of the prostate which, like prostate cancer, is more common among ageing men and is linked to inflammation. In this study, a systematic review was undertaken to estimate the effect of turmeric or curcumin supplementation on prostate diseases. A comprehensive search was conducted in PubMed, Scopus, ISI Web of Science and Google Scholar up to 15 April 2020 to identify clinical trials assessing the effects of curcumin/turmeric alone or in combination with other herbs on prostate diseases. This led to the identification of 11 records comprising 745 patients who met the eligibility criteria. Eight studies were conducted on patients with prostate cancer, and three were on other diseases of the prostate. Although outcomes across the studies were heterogeneous, in some studies curcumin/turmeric supplementation had some favourable effects. This included beneficial effects on the levels of prostate-specific antigen (PSA) (2/6 studies), quality of life (1/2 studies), as well as on oxidative stress markers, feelings of incom-

M. Bagherniya · G. Askari
Food Security Research Center, Department of Community Nutrition, School of Nutrition and Food Science, Isfahan University of Medical Sciences, Isfahan, Iran

B. Alikiaii · S. Abbasi
Anesthesiology and Critical Care Research Center, Faculty of Medicine, Isfahan University of Medical Sciences, Isfahan, Iran

D. Soleimani
School of Nutrition Sciences and Food Technology, Kermanshah University of Medical Sciences, Kermanshah, Iran

T. Sathyapalan
Department of Academic Diabetes, Endocrinology and Metabolism, Hull York Medical School, University of Hull, Hull, UK

T. Jamialahmadi
Department of Food Science and Technology, Quchan Branch, Islamic Azad University, Quchan, Iran

Department of Nutrition, Faculty of Medicine, Mashhad University of Medical Sciences, Mashhad, Iran

A. Sahebkar (✉)
Biotechnology Research Center, Pharmaceutical Technology Institute, Mashhad University of Medical Sciences, Mashhad, Iran

Neurogenic Inflammation Research Center, Mashhad University of Medical Sciences, Mashhad, Iran

Halal Research Center of IRI, FDA, Tehran, Iran

Polish Mother's Memorial Hospital Research Institute (PMMHRI), Lodz, Poland
e-mail: sahebkara@mums.ac.ir

P. C. Guest (ed.), *Studies on Biomarkers and New Targets in Aging Research in Iran*, Advances in Experimental Medicine and Biology 1291, https://doi.org/10.1007/978-3-030-56153-6_20

plete bladder emptying, urination frequency, intermittency, urgency, weak stream, straining and nocturia. Curcumin/turmeric supplementation had no significant adverse effects among patients. This study demonstrated that turmeric or curcumin supplementation might have beneficial effects on some parameters related to prostate diseases, but it should be noted that some studies showed no effect. Therefore, further studies using curcumin-related compounds, particularly in highly bioavailable forms, are needed to assess the impact of curcumin on prostate conditions.

Keywords

Curcumin · Turmeric · Prostate · Prostate-specific antigen · Prostate cancer

20.1 Introduction

Prostate cancer is a substantial cause of morbidity and mortality worldwide [1]. It is the second most common cause of cancer and the sixth leading cause of cancer-related death among men [2]. Another prostate condition which, like prostate cancer, is more common among ageing men is benign prostatic hyperplasia (BPH). BPH and lower urinary tract symptoms (LUTS) are prevalent in about 50–75% of men aged 50 years and older and about 80% among men 70 years of age and older [3]. BPH can negatively affect several health outcomes, resulting in decreased quality of life and could be associated with the development of prostate cancer [3, 4].

Several studies have shown that in prostate diseases, age, race, genetic background, unhealthy lifestyle (particularly diet and low physical activity), metabolic diseases such as obesity and inflammation are major risk factors [5–8]. To prevent this group of diseases, following a healthy lifestyle considering a diet consisting of high amounts of fruits and vegetables (rich in flavonoids, fibres, vitamin C and carotenoids) and being physically active are strongly recommended [5, 6].

For the treatment of prostatic diseases, pharmacotherapy, radiotherapy and surgical options are the usual approaches. However, all of these approaches are associated with adverse effects. For example, alpha-blockers and 5-alpha reductase inhibitors, which are commonly used to manage BPH, are associated with erectile dysfunction, decreased libido, dizziness and hypotension [9]. Although the success rate of transurethral resection of the prostate is around 75% of BPH patients, it could potentially lead to blood loss, urinary incontinence, infection and sexual dysfunction [10]. Likewise, several side effects have been attributed to radiotherapy and other surgical approaches [11, 12]. Therefore, recently, alternative and complementary approaches such as herbal medicine have attracted significant attention to prevent and treat prostate diseases.

Curcumin is a primary component of turmeric with several proven health benefits and is considered as a safe natural agent for both prevention and treatment of several diseases [13–19]. It has been shown that this active compound of turmeric has several unique properties such as antitumour, anti-inflammatory, antioxidant, antithrombotic, chemosensitising and chemopreventive, anti-atherosclerotic and cardioprotective, lipid-modifying, antibacterial, antifungal, antiviral, analgesic, antidepressant and antirheumatic activities [20–24]. Evidence suggests that curcumin has anticancer activity [25–29] and, based on preclinical studies, it might be used to downregulate gene expression in prostate cancer cells [30–32]. Also, due to the anti-inflammatory effects of curcumin [33], it has been suggested that this bioactive compound might have a beneficial impact on BPH and other inflammatory prostate conditions [34, 35].

Although the effects of curcumin or turmeric alone or in combination with other herbs and nutrients on human prostate diseases were assessed in some clinical trials, to our knowledge, there is no systematic review that has evaluated and summarised the effects of curcumin/turmeric on various prostate conditions. Thus, this systematic review was conducted with the main aim of assessing the impact of curcumin/turmeric on prostate conditions based on controlled clinical trials.

20.2 Methods

This systematic review was designed and reported using the guidelines of the preferred reporting items for systematic reviews and meta-analyses (PRISMA) [36].

20.2.1 Search Strategy

We performed a conclusive systematic search on medical databases including ISI web of science, Scopus, PubMed and Google Scholar databases from inception up to 15 April 2020 without any limitation for language or date using the following keywords: ("curcumin" OR "curcuminoids" OR "turmeric") AND ("prostate cancer" OR "prostatic intraepithelial neoplasia" OR "prostatic malignancy" OR "prostate carcinoma" OR "prostate") AND ("Intervention Study" OR "Intervention Studies" OR "Controlled trial" OR "Randomized controlled trial" OR "Randomised clinical trial" OR "Non-Randomized Controlled Trials" OR "Clinical Trial" OR "Non-Randomized Controlled Trials" OR "Cross-Over study" OR "Cross-Over trial" OR "Cross Over trial" OR "Cross Over study" OR "Double-Blind Method" OR "Double-Blind" OR "Double-Blind trial" OR "Double-Blind study"). Whenever possible, Medical Subject Headings (MESH) terms were used.

20.2.2 Study Selection

The title and abstract of all papers, which were found in early search, were independently reviewed by two authors (MB and GA). Articles which did not meet the inclusion criteria were excluded using a screening form with a hierarchical approach incorporating study design, population, exposure and outcome. To explore additional studies, the reference lists of relevant review articles were reviewed. The full text of eligible citations was considered. Any disagreements were discussed and a final agreement made.

20.2.3 Inclusion and Exclusion Criteria

The search was conducted with the aims of identifying articles examining the effects of curcumin/turmeric supplementation alone or in combination with the other herb or herbal bioactive compounds on prostate diseases including prostate cancer or benign prostate hyperplasia. In the present systematic review, only original articles which met the following criteria were included:

1. Use of a clinical trials design.
2. Application of curcumin or turmeric supplementation.
3. Use of a control group.
4. Conducted on patients with prostate disease as a primary disease.

Studies were excluded if they:

1. Had no control.
2. Reported duplicate data.
3. Were reviews, letters, editorial articles, study protocol or case reports.

20.2.4 Data Extraction

Relevant articles were selected after screening records that were found in the initial search. The following information was extracted from eligible and included articles and reported in Table 20.1:

1. Publication information including first author's last name, publication date and study location details of the clinical trial including sample size, patients and mean age (years).
2. Study design.
3. Intervention (treatment).
4. Control.
5. Duration.
6. Main results.
7. Side effects.
8. Study quality (see below).

Table 20.1 Summary of the studies included in the systematic review

Author, year, country	Sample size	Patients	Age range (years) or mean	Study design	Intervention	Control	Duration	Main results	Side effects	Jadad score
Choi, (2019), Korea [48]	82	Patients with prostate cancer who received intermittent androgen deprivation	72.23 ± 7.56	RCT	Curcumin 1440 mg/day (6 caps × 240 mg)	Placebo	6 months	(1) Median off-treatment duration was 16.3 months and 18.5 months in curcumin and placebo groups, respectively (2) Proportion of patients with PSA progression during active curcumin treatment period (6 months) was significantly lower in the curcumin than in the placebo group (3) No significant difference found in treatment duration between the two groups (4) Change of PSA, testosterone levels during 6 months, and HRQOL scores at 6 months were not different between curcumin and placebo groups	No adverse effects	4

Author, year, country	Sample size	Patients	Age range (years) or mean	Study design	Intervention	Control	Duration	Main results	Side effects	Jadad score
Hejazi, (2013), Iran [44]	40	Prostate cancer patients undergoing external beam RT	70.71 ± 8.18	RCT	Curcumin 3 g/day (6 × 500 mg capsules)	Placebo	Since 1 week before onset of radiotherapy until completion of RT	(1) Patients in the curcumin group experienced milder urinary symptoms compared to the placebo group (2) No group differences in bowel symptoms, treatment-related symptoms, or sexual activity	Not reported	4
Hejazi, (2016), Iran [43]	40	Prostate cancer patients undergoing external beam RT	70.71 ± 8.18	RCT	Curcumin 3 g/day (6 × 500 mg capsules)	Placebo	Since 1 week before onset of radiotherapy until completion of RT	(1) Compared with the placebo group, TAC significantly increased and SOD significantly decreased in the curcumin group 3 months after completion of RT (2) Catalase and GPx did not significantly change in the study groups (3) After RT, PSA decreased in both groups, but no difference between groups (the treatment was successful in both groups) (4) No significant differences regarding MRI/MRS images between study groups	No side effects	4

(continued)

Table 20.1 (Continued)

Author, year, country	Sample size	Patients	Age range (years) or mean	Study design	Intervention	Control	Duration	Main results	Side effects	Jadad score
Saadipoor (2019), Iran [42]	64	Patients with histologically confirmed adenocarcinoma of the prostate and the performance status of 0–2 who were candidates for 3D conformal or intensity-modulated RT (IMRT)	70.21 ± 8.03	RCT	Nanocurcumin (120 mg/day)	Placebo	3 days before and during RT course	(1) Radiation-induced proctitis noted in 18/31 (58.1%) of placebo-treated patients versus 15/33 (45.5%) of nanocurcumin-treated patients ($p = 0.313$) (2) Differences between groups regarding radiation-induced cystitis, duration of radiation toxicities, haematological nadirs, and tumour response were not significant	No serious adverse effects	5
Morgia, (2017), Italy [41]	60	Chronic prostatitis/chronic pelvic pain syndrome type III	Median age of 32 years	RCT	Curcumin extract 350 mg (95%) and Calendula extract 80 mg (1 suppository/die for 1 month)	Placebo	3 months	(1) After 3 months of therapy, compared with the placebo, in the intervention group, a significant improvement of NIH-CPSI, IIEF, PEDT, peak flow and VAS was observed (2) PVR did not change in the study groups	A rate of adverse events of 5% in the treatment group, including diarrhoea (80%) and rectal pruritus (20%)	4

Author, year, country	Sample size	Patients	Age range (years) or mean	Study design	Intervention	Control	Duration	Main results	Side effects	Jadad score
Ledda, (2012), Italy [34]	61	Patients with benign prostatic hyperplasia (BPH)	58.51 ± 4.49	Trial	Best standard management (BSM) + Meriva® 2 tablets/day (2 × 500 mg of Meriva®/day, corresponding to 2 × 100 mg curcumin/day)	BSM alone	24 weeks	(1) All International IPSS readings, with exception of the stream weakness score in the BSM group, were improved in both groups (2) Overall results in Meriva® group were significantly better than in the BSM-only group (3) Quality of life improved in both groups, but was significantly better in Meriva® group (4) Significantly more important decrease found in clinical and subclinical episodes of urinary infections and urinary block in Meriva® group (5) Feeling of incomplete bladder emptying, urination frequency (urinate again <2 h), intermittency (flow stop/start), urgency (difficult to postpone), weak stream, straining (push or strain) and nocturia were significantly improved in Meriva® compared with the control group	No side effects	0

(continued)

Table 20.1 (continued)

Author, year, country	Sample size	Patients	Age range (years) or mean	Study design	Intervention	Control	Duration	Main results	Side effects	Jadad score
Noguchi (2017), Japan [40]	36	Postoperative prostate cancer patients who elevated PSA	Median age 67 and 68 years in each group	RCT	High-absorbance curcumin (180 mg/day)	Placebo	6 months	No significant difference in PSA levels between the two groups. However, in patients with preoperative hormonal therapy, high-dose curcumin significantly inhibited elevation of PSA levels	No adverse effects	No full text
Ledda (2017), Italy [39]	61	Patients with genitourinary cancers (prostate or bladder malignancies)	58–74 years	Trial	Standard management (SM) plus daily supplementation, 2 tablets multi-component formulation (Oncotris™): curcumin Enterosoma® [curcuma containing *Curcuma longa L.* extract standardised to 95% curcuminoids, N-acetylcysteine, grapefruit seeds, chitosan from shellfish, resveratrol, black pepper, emulsifier (polysorbate 80)]; astaxanthin (derived from *Haematococcus pluvialis*); cordyceps (*Cordyceps sinensis*)	SM	6 weeks	(1) Intensity of signs and symptoms evaluated significantly decreased in the treatment group. However, no significant changes observed in the control group (2) Supplementation with Oncotris™ associated with significant improvement in haematological parameters, particularly in haemoglobin level, number of red blood cells and white blood cells (3) Significant decrease observed in the level of PSA and free radicals in the treatment group, no significant change observed in the control group	No adverse effects	1

Author, year, country	Sample size	Patients	Age range (years) or mean	Study design	Intervention	Control	Duration	Main results	Side effects	Jadad score
van Die (2017), Australia [46]	22	Patients with recurrent prostate cancer	69 years in active group, 75 in placebo	RCT	Two tablets 2/day of: (1) turmeric rhizome extract [TUMEP25] 25:1, standardised for curcumin 95%, assay 95–105% total curcuminoids; curcumin 100 mg (400 mg/day); (2) resveratrol from *Polygonum cuspidatum* extract dry concentrate [POLEP100], standardised, 100:1, containing not less than 50% resveratrol, 30 mg (120 mg/day); (3) Green tea leaf dry concentrate [GRTEPE 50%], standardised, 25:1, containing not less than 50% polyphenols; catechins 100 mg (400 mg/day [details of the analytic methods for standardisation available on request] and 2 capsules 2/day of: broccoli sprout concentrate [BROEPE] 20:1, equivalent to fresh sprouts 2000 mg (8 g/day)	Placebo	12 weeks	(1) No statistical difference between groups on clinical outcomes (2) There was between-subject variation in the PSA post treatment, but on average the active treatment group experienced a non-significant increase in log-slope of PSA (pre-treatment doubling time = 10.2 months, post-treatment doubling time = 5.5 months). The placebo group experienced no change in log-slope of PSA (pre-treatment doubling time = 10.8 months, post-treatment doubling time = 10.9 months) (3) There was non-significant reduction in depression scores in the active treatment arm, and a non-significant reduction in IPSS scores in the placebo group	Some adverse effects in both groups	4

(continued)

Table 20.1 (continued)

Author, year, country	Sample size	Patients	Age range (years) or mean	Study design	Intervention	Control	Duration	Main results	Side effects	Jadad score
Thomas (2014), United kingdom [47]	199	Patients with localised prostate cancer	74	RCT	Tablet 3/day containing: blend of pomegranate, green tea, broccoli and turmeric: broccoli powder 100 mg, turmeric powder 100 mg, pomegranate whole fruit powder 100 mg, green tea 5:1 extract 20 mg equivalent to 100 mg of green tea and bulking agent (di-calcium phosphate), anti-caking agents (modified maize-based starch, maltodextrin and magnesium stearate) removed post trial	Placebo	6 months	(1) Median rise in PSA in the treatment group was 14.7% as opposed to 78.5% in the placebo group ($P = 0.0008$) (2) No significant differences within the predetermined subgroups for age, Gleason grade, treatment category or body mass index (3) No differences in cholesterol, blood pressure, blood sugar, C-reactive protein or adverse events	Some gastrointestinal adverse effects in the intervention group, however, no significant differences were found between groups	5

Author, year, country	Sample size	Patients	Age range (years) or mean	Study design	Intervention	Control	Duration	Main results	Side effects	Jadad score
Cosentino (2016), Italy [45]	80	Patients with transurethral resection of prostate (TURP) and transurethral resection of bladder (TURB), and subjects with BPH	50–85 years	Trial	Killox® (curcuminoids 190 mg, resveratrol 20 mg, NAC 100 mg, zinc 6 mg). Killox® tablets administered 2/day to patients since discharge from hospital (treatment duration: 20 days for TURP and 10 days for TURB). Killox® tablets were administered once a day to BPH patients of the case study group for 60 days	NSAID (200 mg per day, for 1 week)	20 days for TUR, 10 days for TURB, 60 days for BPH	(1) All individuals treated with Killox®, except one in TURP group, completely recovered in 14 days after surgery, whereas among controls 21 (52.5%) out of 40TURP subjects and 4 (40%) out of 10 TURB subjects were found with symptoms of inflammation and urinary burning. Effect of Killox® was statistically significant in both trials (for TURP and TURB) in comparison with controls (2) In BPH patients, Killox® was effective, shortening duration of irritation symptoms, in the treatment group, and IPSS significantly reduced compared with the control group	No adverse effects	0

IAD intermittent androgen deprivation, *RCT* randomised controlled trial, *HRQOL* health-related quality of life, *PSA* prostate-specific antigen, *QLQ-PR25* prostate cancer-specific quality of life questionnaire module, *TAC* total antioxidant capacity, *SOD* activity of superoxide dismutase, *GPx* glutathione peroxidase, *RT* radiotherapy, *NIH-CPSI* National Institutes of Health Chronic Prostatitis Symptom Index, *IIEF* International Index of Erectile Function questionnaire, *PEDT* premature ejaculation diagnostic tool, *VAS* visual analogue score, *PVR* post-void residual, *IPSS* International Prostate Symptom Score, *BSM* best standard management, *BPH* benign prostatic hyperplasia, *NSAID* non-steroidal anti-inflammatory drug

20.2.5 Quality Assessment

The quality of the included studies was assessed by two independent reviewers (MB and GA) using the Jadad scale. This scale includes three domains as follows [37]:

1. Randomisation (0-2 points).
2. Blinding (0-2 points).
3. Dropouts and withdrawals (0-1 points).

According to the mentioned domains, the overall quality of each study was considered as low quality if the score was ≤2 and high quality for scores ≥3 [38].

20.3 Results

20.3.1 Search Results and Study Selection

A total of 1176 studies were acknowledged during the initial search, of which 51 records were duplicates. After reading the title and abstracts, 1107 irrelevant records were excluded and the full text of the 18 remaining articles was reviewed. According to our inclusion and exclusion criteria, 8 articles were omitted due to the following reasons: duplicated data ($n = 1$); no control group ($n = 4$); or conference report ($n = 2$). Thus, data extraction was done on 11 articles [34, 39–48]. The results of the search are shown in Fig. 20.1. The characteristics of each selected paper are shown in Table 20.1.

20.3.2 Characteristics of the Included Studies

Studies were published between 2012 and 2019, and comprised a total of 745 patients. Four studies were conducted in Italy [34, 39, 41, 45], three in Iran [42–44], one in Korea [48], one in Australia [46], one in the United Kingdom [47] and one in Japan [40]. Eight studies were conducted on patients with prostate cancer [40, 42–44, 46–48], and three studies were conducted on other diseases of the prostate such as chronic prostatitis, chronic pelvic pain syndrome type III, benign prostatic hyperplasia (BPH), and patients with transurethral resection of the prostate and transurethral resection of the bladder [34, 41, 45]. The mean or median age of the participants varied between 32 and 74 years. All of the included studies were controlled clinical trials. Six studies used curcumin alone (in different forms such as curcumin, nanocurcumin, high absorption curcumin) to treat patients [34, 40, 42–44, 48]. One study used nanocurcumin (120 mg/day) [42], three studies used native curcumin (1440 to 3000 mg/day) [43, 44, 48], one study used high-absorbance curcumin (180 mg/day) [40] and another study used Meriva®, containing 200 mg curcumin/day [34]. The other studies used curcumin or turmeric in combination with other herbs or nutrients [39, 41, 45–47]. The duration of the studies varied from 3 days before and during radiotherapy to 6 months.

20.3.3 Quality Assessment

Overall, 7 out of 11 studies had high-quality scores using the Jadad scale [41–44, 46–48], three studies were categorised as low-quality studies in terms of methodological approach [34, 39, 45]. We did not access the full text of one study [40].

20.3.4 Outcome Assessments

In six studies, the effect of curcumin (alone or in combination with other herbs) supplementation on prostate-specific antigen (PSA) was assessed. In four of these studies, curcumin did not have a significant effect on PSA in comparison to the control group among prostate cancer patients [40, 43, 46, 48]. Nevertheless, in one study, turmeric was used in combination with other herbs on patients with prostate cancer, and this found a median rise in PSA in the treatment group of 14.7%, compared to 78.5% in the placebo group ($p < 0.05$) [47]. In another study, curcumin treatment, in combination with the other herbs, sig-

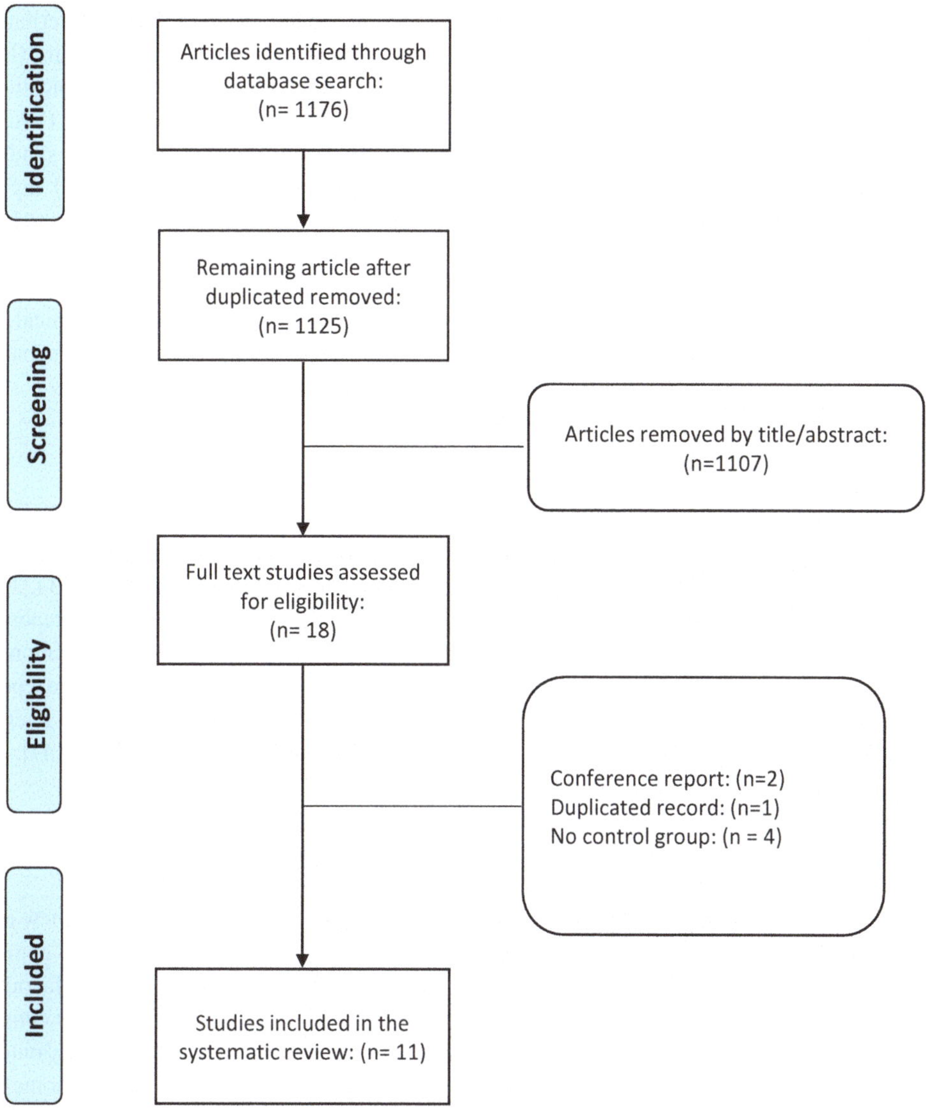

Fig. 20.1 Flow chart of the study selection process

nificantly decreased PSA in patients with genitourinary cancers (prostate or bladder malignancies) and no significant change was observed in the control group [39]. Four studies reported the effects of curcumin on signs and symptoms of the patients. In one study, patients in the curcumin group experienced milder urinary symptoms in comparison to the placebo group. However, no group differences were observed in bowel symptoms, treatment-related symptoms or sexual activity of patients with prostate cancer [44]. In another study, conducted on BPH patients, all components of the international prostate symptom score (IPSS) were improved in both curcumin and control groups, except the stream weakness score was not altered in the curcumin group [34]. In another study conducted on genitourinary cancers, the intensity of signs and symptoms was significantly decreased in the treatment group, with no significant change in the control group. In addition, supplementation with Oncotris™ (curcumin in combination with other herbs) resulted in a significant improvement in haematological parameters, particularly in haemoglobin levels, and numbers of red and white blood cells [39]. In one study, a significantly higher number of patients recovered completely in the Killox® treatment (contained curcuminoids 190 mg, resveratrol 20 mg, NAC 100 mg, zinc 6 mg) compared to the control group [45]. This included patients with transurethral resection of the prostate (TURP) and transurethral resection of the bladder (TURB) who had symptoms of inflammation. In addition, the Killox® treatment was effective in shortening the duration of irritation symptoms in BPH patients. Finally, there was also a significant reduction in IPSS scores in the Killox® treatment compared with the control group. However, another study found no significant differences in treatment duration, testosterone levels, and quality of life between curcumin and placebo groups in prostate cancer patients [48]. A study conducted among BPH patients showed that the quality of life of patients was increased in the curcumin group compared with the control group [34]. Besides, there was a significant reduction in the clinical and subclinical episodes of urinary infections and urinary blockage in the curcumin group than in the control group. Moreover, feelings of incomplete bladder emptying, urination frequency (repeat urination <2 h), intermittency (flow stop/start), urgency (difficult to postpone), weak stream, straining and nocturia (need for urination at night time) were significantly improved in the Meriva® (curcumin) group compared with the control group. In one study, the effect of curcumin on oxidative stress was assessed in which supplementation with 3 g/day curcumin led to a significant increase in total antioxidant capacity and decreased activity of superoxide dismutase (SOD) 3 months after completion of radiotherapy [43]. However, catalase and glutathione peroxidase (GPx) did not significantly change in the study groups. In addition, this study found no significant differences regarding magnetic resonance imaging/magnetic resonance spectroscopy analyses between the study groups [43]. In the intervention group, a significant improvement was observed regarding National Institutes of Health Chronic Prostatitis Symptom Index (NIH-CPSI), International Index of Erectile Function questionnaire (IIEF), peak flow, premature ejaculation diagnostic tool (PEDT) and visual analogue score (VAS) in chronic prostatitis/chronic pelvic pain syndrome type III patients who consumed 350 mg curcumin and 80 mg calendula extracts [41]. As shown in Table 20.1, supplementation with curcumin alone or in combination with other herbs had no significant effects on the clinical, biochemical and psychological factors in patients with prostate conditions.

20.3.5 Adverse Effects

In seven studies, curcumin/turmeric supplementation had no adverse effects on patients [34, 39, 40, 42, 43, 45, 48]. One study did not report any adverse effects of curcumin supplementation [44]. In three studies, some adverse effects were reported, particularly gastrointestinal side effects after supplementation with natural herbs [41, 46, 47].

20.4 Discussion

The main finding of this systematic review is that curcumin supplementation alone or in combination with the other herbs had some favourable effects on prostate diseases without any considerable adverse effects. However, the study outcome measures were heterogeneous and, in some studies, curcumin turmeric treatment had no significant beneficial effects. To the best of our knowledge, this is the first review which systematically assessed the effects of curcumin or turmeric supplementation on prostate diseases.

Several preclinical studies have revealed anticancer effects of curcumin, including breast cancer and prostate cancer [49–51]. Although the exact mechanisms regarding the effects on prostate cancer are unclear, it has been shown that curcumin might exert effects on androgen receptor (AR) signalling and numerous downstream targets such as vascular endothelial growth factor (VEGF), phosphatase and tensin homologue (PTEN) and nuclear factor kappa-light-chain-enhancer of activated B cells (NF-κB) [52–54]. It has also been shown that curcumin inhibits AR binding to the androgen response element of the PSA gene (*KLK3*) as well as the PSA expression in androgen-sensitive human prostate adenocarcinoma LNCaP cells [55, 56].

Out of the eight studies conducted on prostate cancer patients, some favourable effects of curcumin treatment were achieved as seen by improvement in clinical parameters. However, the heterogeneous nature of measured outcomes and the use of different forms of curcumin alone or in combination with the other herbs made it difficult to draw definitive conclusions. Nevertheless, the results showed that curcumin has some promise as a safe and natural herbal bioactive compound to improve the health situation of patients with prostate cancer. One of the most common approaches to treat many cancers such as prostate cancer is radiotherapy, which can have well known adverse side effects. For example, active oxygen radicals such as hydroxyl radicals can be produced by the ionising radiation, actually increasing the oxidative stress burden in radiotherapy patients [57, 58]. In experimental studies, it has been shown that inflammation plays a crucial role in both tumour resistance and the incidence of various side effects following both chemotherapy and radiotherapy. In addition, the development of second cancer during the years after this treatment could develop as a result of the associated chronic inflammation [59]. Therefore, reducing the oxidative stress levels of cancer patients using safe and natural agents may offer a means of minimising or preventing such effects [60, 61].

The antioxidant and anti-inflammatory properties of curcumin have been demonstrated by several preclinical and clinical studies [20, 33, 62, 63]. In addition, curcumin was shown to have a radioprotective effect reported by in vivo and clinical trial studies [64–66]. In our systematic review, only one study assessed the effects of curcumin on oxidative stress status of prostate cancer patients, and this confirmed that curcumin had positive effects on this outcome [43]. This stresses the need for further studies on the specific effects of curcumin on inflammatory factors and oxidative stress levels in prostate cancer and other malignancies.

Likewise, several studies have shown that inflammation plays a central role in developing BPH [67, 68]. Furthermore, in enlarged prostate tissues, higher levels of inflammatory prostanoids such as prostaglandins and leukotrienes have been observed as well as chronic activation of T-lymphocytes, B-lymphocytes and macrophages [34, 69, 70]. Anti-hormonal agents including 5α-reductase inhibitors and α-adrenergic blockers are used to manage BPH, although these have no significant anti-inflammatory actions. However, non-steroidal anti-inflammatory drugs (NSAIDs) cannot be prescribed to treat BPH as these drugs could induce prostate damage [34]. As elevated IL-8 has been strongly linked to inflammatory prostate conditions, such as chronic prostatitis/chronic pelvic pain syndrome (CP/CPPS) and BPH [71], the anti-inflammatory properties of curcumin could offer a useful treatment option, particularly since it is known to reduce IL-8 levels [72, 73]. Therefore, using curcumin to treat BPH or the other prostate inflammatory conditions has a strong rational and our findings of favourable effects of curcumin supplementation on BPH signs, symptoms and clinical outcomes support this.

Several limitations of this review should be addressed. First, due to the small number of studies and the small number of included participants, drawing a robust evidence-based conclusion is not possible. Second, the heterogeneous nature of the included studies in terms of using different forms and dosages of curcumin and various combinations of curcumin with other herbs makes it difficult to pin down the effects to any specific procedure. One possible difference may be related to the poor absorption and bioavailability of curcumin in its natural form [74]. To overcome this limitation, one of the studies reviewed here used nanocurcumin [42], and one used a high bioavailability form of curcumin [40]. We suggest that new formulations of curcumin with high bioavailability should be applied in future studies, such as curcumin-piperine or phospholipid curcumin [75].

20.5 Conclusion

Within the limitations of this review, it can be concluded that curcumin supplementation has beneficial effects on some parameters related to prostate diseases. In particular, curcumin appeared to have anti-inflammatory effects which could offer some therapeutic effects and improve patient outcomes. However, in some studies, these natural products had no favourable effects. Further studies using curcumin, particularly high bioavailability forms of curcumin, are needed to assess the impact of curcumin on prostate diseases. Such studies should include higher numbers of participants and standardised procedures.

References

1. Scher HI, Solo K, Valant J, Todd MB, Mehra M (2015) Prevalence of prostate cancer clinical states and mortality in the United States: estimates using a dynamic progression model. PLoS One 10(10):e0139440. https://doi.org/10.1371/journal.pone.0139440
2. Jain S, Saxena S, Kumar A (2014) Epidemiology of prostate cancer in India. Meta Gene 2:596–605
3. Egan KB (2016) The epidemiology of benign prostatic hyperplasia associated with lower urinary tract symptoms: prevalence and incident rates. Urol Clin North Am 43(3):289–297
4. Ilic D, Misso M (2012) Lycopene for the prevention and treatment of benign prostatic hyperplasia and prostate cancer: a systematic review. Maturitas 72(4):269–276
5. Daniyal M, Siddiqui ZA, Akram M, Asif H, Sultana S, Khan A (2014) Epidemiology, etiology, diagnosis and treatment of prostate cancer. Asian Pac J Cancer Prev 15(22):9575–9578
6. Lim KB (2017) Epidemiology of clinical benign prostatic hyperplasia. Asian J Urol 4(3):148–151
7. Bashir MN (2015) Epidemiology of prostate cancer. Asian Pac J Cancer Prev 16(13):5137–5141
8. Allott EH, Masko EM, Freedland SJ (2013) Obesity and prostate cancer: weighing the evidence. Eur Urol 6(5):800–809
9. Tacklind J, Fink HA, MacDonald R, Rutks I, Wilt TJ (2010) Finasteride for benign prostatic hyperplasia. Cochrane Database Syst Rev (10):CD006015. https://doi.org/10.1002/14651858.CD006015.pub3
10. Hoffman RM, MacDonald R, Wilt T (2000) Laser prostatectomy for benign prostatic obstruction. Cochrane Database Syst Rev (1):CD001987. https://doi.org/10.1002/14651858.CD001987.pub2
11. Michaelson MD, Cotter SE, Gargollo PC, Zietman AL, Dahl DM, Smith MR (2008) Management of complications of prostate cancer treatment. CA Cancer J Clin 58(4):196–213
12. Darwish-Yassine M, Berenji M, Wing D, Copeland G, Demers RY, Garlinghouse C et al (2014) Evaluating long-term patient-centered outcomes following prostate cancer treatment: findings from the Michigan Prostate Cancer Survivor study. J Cancer Surviv 8(1):121–130
13. Bagherniya M, Johnston TP, Sahebkar A (2020) Regulation of apolipoprotein B by natural products and nutraceuticals: a comprehensive review. Curr Med Chem. https://doi.org/10.2174/0929867327666200427092114. Online ahead of print
14. Talebi S, Bagherniya M, Atkin SL, Askari G, Orafai HM, Sahebkar A (2020) The beneficial effects of nutraceuticals and natural products on small dense LDL levels, LDL particle number and LDL particle size: a clinical review. Lipids Health Dis 19:1–21
15. Kocaadam B, Şanlier N (2017) Curcumin, an active component of turmeric (*Curcuma longa*), and its effects on health. Crit Rev Food Sci Nutr 57(13):2889–2895
16. Ringman JM, Frautschy SA, Cole GM, Masterman DL, Cummings JL (2005) A potential role of the curry spice curcumin in Alzheimer's disease. Curr Alzheimer Res 2(2):131–136
17. Wongcharoen W, Phrommintikul A (2009) The protective role of curcumin in cardiovascular diseases. Int J Cardiol 133(2):145–151
18. Iranshahi M, Sahebkar A, Hosseini ST, Takasaki M, Konoshima T, Tokuda H (2010) Cancer chemopreventive activity of diversin from Ferula diversivittata in vitro and in vivo. Phytomedicine 17(3-4):269–273
19. Soleimani V, Sahebkar A, Hosseinzadeh H (2018) Turmeric (Curcuma longa) and its major constituent (curcumin) as nontoxic and safe substances: Review. Phytother Res 32(6):985–995. https://doi.org/10.1002/ptr.6054. Epub 2018 Feb 26. PMID: 29480523
20. Bagherniya M, Nobili V, Blesso CN, Sahebkar A (2018) Medicinal plants and bioactive natural compounds in the treatment of non-alcoholic fatty liver disease: a clinical review. Pharmacol Res 130:213–240
21. Zorofchian Moghadamtousi S, Abdul Kadir H, Hassandarvish P, Tajik H, Abubakar S, Zandi K (2014) A review on antibacterial, antiviral, and antifungal activity of curcumin. Biomed Res Int 2014:186864. https://doi.org/10.1155/2014/186864
22. Mollazadeh H, Cicero AFG, Blesso CN, Pirro M, Majeed M, Sahebkar A (2019) Immune modulation by curcumin: The role of interleukin-10. Crit Rev Food Sci Nutr 59(1):89–101
23. Momtazi AA, Derosa G, Maffioli P, Banach M, Sahebkar A (2016) Role of microRNAs in the thera-

peutic effects of curcumin in non-cancer diseases. Mol Diagn Ther 20(4):335–345
24. Hasanzadeh S, Read MI, Bland AR, Majeed M, Jamialahmadi T, Sahebkar A (2020) Curcumin: an inflammasome silencer. Pharmacol Res 159:104921. https://doi.org/10.1016/j.phrs.2020.104921. Epub 2020 May 25. PMID: 32464325
25. Agrawal DK, Mishra PK (2010) Curcumin and its analogues: potential anticancer agents. Med Res Rev 30(5):818–860
26. Aggarwal BB, Kumar A, Bharti AC (2003) Anticancer potential of curcumin: preclinical and clinical studies. Anticancer Res 23(1/A):363–398
27. Momtazi AA, Sahebkar A (2016) Difluorinated curcumin: A promising curcumin analogue with improved anti-tumor activity and pharmacokinetic profile. Curr Pharm Des 22(28):4386–4397. https://doi.org/10.2174/1381612822666160527113501
28. Teymouri M, Pirro M, Johnston TP, Sahebkar A (2017) Curcumin as a multifaceted compound against human papilloma virus infection and cervical cancers: A review of chemistry, cellular, molecular, and preclinical features. BioFactors 43(3):331–346
29. Mohajeri M, Bianconi V, Ávila-Rodriguez MF, Barreto GE, Jamialahmadi T, Pirro M, Sahebkar A (2020) Curcumin: a phytochemical modulator of estrogens and androgens in tumors of the reproductive system. Pharmacol Res 156:104765. https://doi.org/10.1016/j.phrs.2020.104765. Epub 2020 Mar 23. PMID: 32217147
30. Chendil D, Ranga RS, Meigooni D, Sathishkumar S, Ahmed MM (2004) Curcumin confers radiosensitizing effect in prostate cancer cell line PC-3. Oncogene 23(8):1599–1607
31. Mukhopadhyay A, Bueso-Ramos C, Chatterjee D, Pantazis P, Aggarwal BB (2001) Curcumin downregulates cell survival mechanisms in human prostate cancer cell lines. Oncogene 20(52):7597–7609
32. Nakamura K, Yasunaga Y, Segawa T, Ko D, Moul JW, Srivastava S et al (2002) Curcumin down-regulates AR gene expression and activation in prostate cancer cell lines. Int J Oncol 21(4):825–830
33. Menon VP, Sudheer AR (2007) Antioxidant and anti-inflammatory properties of curcumin. In: Aggarwal BB, Surh YJ, Shishodia S (eds) The molecular targets and therapeutic uses of curcumin in health and disease. Springer, Boston, MA, pp 105–125. ISBN-13: 9780387464008
34. Ledda A, Belcaro G, Dugall M, Luzzi R, Scoccianti M, Togni S et al (2012) Meriva(R), a lecithinized curcumin delivery system, in the control of benign prostatic hyperplasia: a pilot, product evaluation registry study. Panminerva Med 54(1 Suppl 4):17–22
35. Cosentino V, Fratter A, Cosentino M (2016) Anti-inflammatory effects exerted by Killox®, an innovative formulation of food supplement with curcumin, in urology. Eur Rev Med Pharmacol Sci 20(7):1390–1398
36. Moher D, Liberati A, Tetzlaff J, Altman DG (2009) Preferred reporting items for systematic reviews and meta-analyses: the PRISMA statement. Ann Intern Med 151(4):264–269
37. Jadad AR, Moore RA, Carroll D, Jenkinson C, Reynolds DJM, Gavaghan DJ et al (1996) Assessing the quality of reports of randomized clinical trials: is blinding necessary? Control Clin Trials 17(1):1–12
38. Moher D, Cook D, Jadad A, Tugwell P, Moher M, Jones A et al (1999) Assessing the quality of reports of randomised trials: implications for the conduct of meta-analyses. Health Technol Assess 3(12):i–iv., 1–98
39. Ledda A, Belcaro G, Dugall M, Luzzi R, Hosoi M, Feragalli B et al (2017) A natural pharma standard supplement formulation to control treatment-related toxicity and oxidative stress in genitourinary cancer: a preliminary study. Eur Rev Med Pharmacol Sci 21(18):4196–4202
40. Noguchi T (2017) The effects of curcumin to the PSA recurrence of prostate cancer after surgical treatment. Teikyo Med J 40(6):263–272
41. Morgia G, Russo GI, Urzi D, Privitera S, Castelli T, Favilla V et al (2017) A phase II, randomized, single-blinded, placebo-controlled clinical trial on the efficacy of Curcumina and Calendula suppositories for the treatment of patients with chronic prostatitis/chronic pelvic pain syndrome type III. Arch Ital Urol Androl 89(2):110–113
42. Saadipoor A, Razzaghdoust A, Simforoosh N, Mahdavi A, Bakhshandeh M, Moghadam M et al (2019) Randomized, double-blind, placebo-controlled phase II trial of nanocurcumin in prostate cancer patients undergoing radiotherapy. Phytother Res 33(2):370–378
43. Hejazi J, Rastmanesh R, Taleban FA, Molana SH, Hejazi E, Ehtejab G et al (2016) Effect of curcumin supplementation during radiotherapy on oxidative status of patients with prostate cancer: a double blinded, randomized, placebo-controlled study. Nutr Cancer 68(1):77–85
44. Hejazi J, Rastmanesh R, Taleban FA, Molana SH, Ehtejab G (2013) A pilot clinical trial of radioprotective effects of curcumin supplementation in patients with prostate cancer. J Cancer Sci Ther 5(10):320–324
45. Cosentino V, Fratter A, Cosentino M (2016) Anti-inflammatory effects exerted by Killox (R), an innovative formulation of food supplement with curcumin, in urology. Eur Rev Med Pharmacol Sci 20(7):1390–1398
46. van Die MD, Williams SG, Emery J, Bone KM, Taylor JM, Lusk E et al (2017) A placebo-controlled double-blinded randomized pilot study of combination phytotherapy in biochemically recurrent prostate cancer. Prostate 77(7):765–775
47. Thomas R, Williams M, Sharma H, Chaudry A, Bellamy P (2014) A double-blind, placebo-controlled randomised trial evaluating the effect of a polyphenol-rich whole food supplement on PSA progression in men with prostate cancer-the UK NCRN

Pomi-T study. Prostate Cancer Prostatic Dis 17(2): 180–186
48. Choi YH, Han DH, Kim SW, Kim MJ, Sung HH, Jeon HG et al (2019) A randomized, double-blind, placebo-controlled trial to evaluate the role of curcumin in prostate cancer patients with intermittent androgen deprivation. Prostate 79(6):614–621
49. Yallapu MM, Jaggi M, Chauhan SC (2010) β-Cyclodextrin-curcumin self-assembly enhances curcumin delivery in prostate cancer cells. Colloids Surf B Biointerfaces 79(1):113–125
50. Kakarala M, Brenner DE, Korkaya H, Cheng C, Tazi K, Ginestier C et al (2010) Targeting breast stem cells with the cancer preventive compounds curcumin and piperine. Breast Cancer Res Treat 122(3):777–785
51. Aggarwal BB, Shishodia S, Takada Y, Banerjee S, Newman RA, Bueso-Ramos CE et al (2005) Curcumin suppresses the paclitaxel-induced nuclear factor-κB pathway in breast cancer cells and inhibits lung metastasis of human breast cancer in nude mice. Clin Cancer Res 11(20):7490–7498
52. Jordan BC, Mock CD, Thilagavathi R, Selvam C (2016) Molecular mechanisms of curcumin and its semisynthetic analogues in prostate cancer prevention and treatment. Life Sci 152:135–144
53. Du Y, Long Q, Zhang L, Shi Y, Liu X, Li X et al (2015) Curcumin inhibits cancer-associated fibroblast-driven prostate cancer invasion through MAOA/mTOR/HIF-1α signaling. Int J Oncol 47(6):2064–2072
54. Wei X, Du Z-Y, Cui X-X, Verano M, Mo RQ, Tang ZK et al (2012) Effects of cyclohexanone analogues of curcumin on growth, apoptosis and NF-κB activity in PC-3 human prostate cancer cells. Oncol Lett 4(2):279–284
55. Hong JH, Lee G, Choi HY (2015) Effect of curcumin on the interaction between androgen receptor and Wnt/β-catenin in LNCaP xenografts. Korean J Urol 56(9):656–665
56. Teiten M-H, Gaascht F, Eifes S, Dicato M, Diederich M (2010) Chemopreventive potential of curcumin in prostate cancer. Genes Nutr 5(1):61–74
57. Anscher MS, Vujaskovic Z (2005) Mechanisms and potential targets for prevention and treatment of normal tissue injury after radiation therapy. Semin Oncol 32(2 Suppl 3):S86–S91
58. Huang X, Song C, Zhong C, Wang F (2012) Research progress in the radioprotective effect of superoxide dismutase. Drug Discov Ther 6(4):169–177
59. Farhood B, Mortezaee K, Goradel NH, Khanlarkhani N, Salehi E, Nashtaei MS et al (2019) Curcumin as an anti-inflammatory agent: Implications to radiotherapy and chemotherapy. J Cell Physiol 234(5):5728–5740
60. Moss RW (2007) Do antioxidants interfere with radiation therapy for cancer? Integr Cancer Ther 6(3):281–292
61. Simone CB II, Simone NL, Simone V, Simone CB (2007) Antioxidants and other nutrients do not interfere with chemotherapy or radiation therapy and can increase kill and increase survival, Part 2. Altern Ther Health Med 13(1):22–29
62. Abdollahi E, Momtazi AA, Johnston TP, Sahebkar A (2018) Therapeutic effects of curcumin in inflammatory and immune-mediated diseases: a nature-made jack-of-all-trades? J Cell Physiol 233(2):830–848
63. Ghandadi M, Sahebkar A (2017) Curcumin: An effective inhibitor of interleukin-6. Curr Pharm Des 23(6):921–931
64. Ryan JL, Heckler CE, Ling M, Katz A, Williams JP, Pentland AP et al (2013) Curcumin for radiation dermatitis: a randomized, double-blind, placebo-controlled clinical trial of thirty breast cancer patients. Radiat Res 180(1):34–43
65. Akpolat M, Kanter M, Uzal MC (2009) Protective effects of curcumin against gamma radiation-induced ileal mucosal damage. Arch Toxicol 83(6):609–617
66. Inano H, Onoda M (2002) Radioprotective action of curcumin extracted from *Curcuma longa* LINN: inhibitory effect on formation of urinary 8-hydroxy-2′-deoxyguanosine, tumorigenesis, but not mortality, induced by γ-ray irradiation. Int J Radiat Oncol Biol Phys 53(3):735–743
67. Nickel JC (2008) Inflammation and benign prostatic hyperplasia. Urol Clin North Am 35(1):109–115
68. Kramer G, Mitteregger D, Marberger M (2007) Is benign prostatic hyperplasia (BPH) an immune inflammatory disease? Eur Urol 51(5):1202–1216
69. Altavilla D, Minutoli L, Polito F, Irrera N, Arena S, Magno C et al (2012) Effects of flavocoxid, a dual inhibitor of COX and 5-lipoxygenase enzymes, on benign prostatic hyperplasia. Br J Pharmacol 167(1):95–108
70. Notara M, Ahmed A (2012) Benign prostate hyperplasia and stem cells: a new therapeutic opportunity. Cell Biol Toxicol 28(6):435–442
71. Penna G, Mondaini N, Amuchastegui S, Degli Innocenti S, Carini M, Giubilei G et al (2007) Seminal plasma cytokines and chemokines in prostate inflammation: interleukin 8 as a predictive biomarker in chronic prostatitis/chronic pelvic pain syndrome and benign prostatic hyperplasia. Eur Urol 51(2):524–533
72. Cho JW, Lee KS, Kim CW (2007) Curcumin attenuates the expression of IL-1β, IL-6, and TNF-α as well as cyclin E in TNF-α-treated HaCaT cells; NF-κB and MAPKs as potential upstream targets. Int J Mol Med 19(3):469–474
73. Jain SK, Rains J, Croad J, Larson B, Jones K (2009) Curcumin supplementation lowers TNF-α, IL-6, IL-8, and MCP-1 secretion in high glucose-treated cultured monocytes and blood levels of TNF-α, IL-6, MCP-1, glucose, and glycosylated hemoglobin in diabetic rats. Antioxid Redox Signal 11(2):241–249
74. Lopresti AL (2018) The problem of curcumin and its bioavailability: could its gastrointestinal influence contribute to its overall health-enhancing effects? Adv Nutr 9(1):41–50
75. Anand P, Kunnumakkara AB, Newman RA, Aggarwal BB (2007) Bioavailability of curcumin: problems and promises. Mol Pharm 4(6):807–818

Curcumin: A Review of Its Effects on Epilepsy

21

Fatemeh Forouzanfar, Muhammed Majeed, Tannaz Jamialahmadi, and Amirhosein Sahebkar

Abstract

Central nervous system (CNS) diseases pose an enormous healthcare burden, at both an individual and a societal level. Epilepsy has now become one of the most prevalent CNS disorders. Pharmaceutical drugs prescribed for epilepsy often have serious side effects and, for this reason, attention has turned to the use of medicinal plants. Curcumin (diferuloylmethane) is a major component of *Curcuma longa* and exhibits various pharmacological effects, including anti-inflammatory, antioxidant, and immunoregulatory properties. Here, we have reviewed the literature relating specifically to the antiepileptic effects of curcumin. The evidence suggests a protective effect of curcumin in the control of epileptic seizures, together with a protective effect on the relief of memory impairment, which may stem from its influence on monoamine levels in the brain.

F. Forouzanfar
Neuroscience Research Center, Mashhad University of Medical Sciences, Mashhad, Iran

Department of Neuroscience, Faculty of Medicine, Mashhad University of Medical Sciences, Mashhad, Iran

M. Majeed
Sabinsa Corporation, East Windsor, NJ, USA

T. Jamialahmadi
Department of Food Science and Technology, Quchan Branch, Islamic Azad University, Quchan, Iran

Department of Nutrition, Faculty of Medicine, Mashhad University of Medical Sciences, Mashhad, Iran

A. Sahebkar (✉)
Biotechnology Research Center, Pharmaceutical Technology Institute, Mashhad University of Medical Sciences, Mashhad, Iran

Neurogenic Inflammation Research Center, Mashhad University of Medical Sciences, Mashhad, Iran

School of Pharmacy, Mashhad University of Medical Sciences, Mashhad, Iran

Polish Mother's Memorial Hospital Research Institute (PMMHRI), Lodz, Poland
e-mail: sahebkara@mums.ac.ir

Keywords

Epilepsy · Curcumin · Herbal medicine · Seizure

21.1 Introduction

Epilepsy is one of the most common neurological disorders and carries an associated stigma, psychiatric comorbidity, and high economic cost. According to the World Health Organization (WHO) 2010 Global Burden of Disease study, epi-

P. C. Guest (ed.), *Studies on Biomarkers and New Targets in Aging Research in Iran*, Advances in Experimental Medicine and Biology 1291, https://doi.org/10.1007/978-3-030-56153-6_21

lepsy is the second most burdensome neurologic disease throughout the world in terms of disability-adjusted life years [1, 2]. It is characterized by repeated unprovoked seizures, which could be generalized or partial. Experimental and clinical studies in genetic and lesional epilepsies suggest the existence of numerous common pathogenic mechanisms, which include alterations in the intrinsic properties of neurons and an imbalance between glutamate and gamma-aminobutyric acid (GABA) neurotransmitter signals for excitation and inhibition, respectively [3, 4]. Voltage-gated Na^+ channels are important for action potentials, while voltage-gated K^+ channels are important in repolarization and hyperpolarization. Genetic mutations or acquired functional alterations in these channels have been identified in relevant animal models and in patients with various forms of epilepsy. These genetic mutations or acquired functional alterations lead to loss of function of GABAergic neurotransmission (alterations in GABA synthesis, release, or GABA receptor signaling) or gain of function of glutamatergic neurotransmission (alterations in glutamate receptors or reuptake). These findings are in accord with the known mechanisms of action of the currently available antiepileptic drugs [4]. Ion channels represent a particularly attractive target for many antiepileptic drugs, promoting inhibition of neuronal excitability. One class of antiepileptic drugs, which include phenytoin, carbamazepine, valproic acid, felbamate, rufinamide, lamotrigine, lacosamide, topiramate, zonisamide, and oxcarbazepine, blocks neuronal voltage-gated Na^+ channels. By contrast, retigabine enhances activation of K^+ channels, and ethosuximide and valproic acid obstruct T-type Ca^{2+} channels [5]. Other antiepileptic drugs have been shown to decrease the efficiency of the excitatory synapse. These targets and drugs include the synaptic vesicle glycoprotein 2A (for levetiracetam), the $\alpha 2\delta$ subunit of the voltage-gated Ca^{2+} channel (for gabapentin and pregabalin), α-amino-3-hydroxy-5-methyl-4-isoxazole propionic acid (for perampanel), and N-methyl-D-aspartate receptors (for felbamate). Some antiepileptic drugs exert their main actions via an increase in synaptic inhibition and enhance their efficiency [5]. These include tiagabine, which inhibits the GABA transporter GAT1, leading to decreased GABA uptake into presynaptic terminals and astrocytes, and vigabatrin, which irreversibly inhibits the catabolic enzyme GABA transaminase, leading to accumulation of GABA in presynaptic terminals and astrocytes [5]. The benzodiazepines and the barbiturates have been found to enhance inhibitory neurotransmission by allosterically modulating $GABA_A$ receptor-mediated Cl^- currents. It has been reported that many antiepileptic drugs (e.g., valproic acid, lamotrigine, phenobarbital, gabapentin, felbamate, and topiramate) can act via multiple complementary mechanisms [5]. However, it has been noted that the current drugs fail to suppress the occurrence of seizure in almost 30% of epileptic patients [6].

21.2 Herbal Medicine for Epilepsy

Herbal therapy and the use of natural products have been used by humans throughout the ages [7, 8]. Medicinal plants have been widely used for their beneficial pharmacological effects, which include antifungal, anticancer, antiviral, analgesic, and anti-inflammatory properties [9–12]. The use of herbal medicinal products and phytonutrients has surged over the past three decades [13, 14]. Of note, the high cost and side effects of many modern manufactured drugs have encouraged the use of traditional medicines, which are also much more affordable [11, 13, 15].

Curcumin (diferuloylmethane) is a polyphenol derived from the rhizomes of *Curcuma longa* L. (turmeric). This plant is native to tropical South Asia, but is now widely cultivated in other tropical and subtropical areas throughout the world [16]. Extensive studies on *Curcuma longa* have shown that this medicinal herb possesses significant health benefits. Curcumin is extensively used in Asian cuisine as well as for various medicinal purposes. The discovery of curcumin dates back approximately two centuries to two laboratory scientists from Harvard College, Vogel and Pelletier [17–19]. Subsequently, Milobedzka and colleagues identified the chemical structure of curcumin to bediferuloylmethane or 1,6-hepta diene-3,5-dione-1,7-bis(4-hydroxy-3-methoxyphenyl)-(1E,6E) [20].

Curcumin possesses a wide range of pharmacological effects, including antioxidant [21, 22], anti-inflammatory [22-24], antimicrobial [25], immunoregulatory [26], lipid-modifying [27], epigenetic-modifying [28], anti-tumor [29-31], hypouricemic [32], analgesic [33, 34], hepatoprotective [24]and neuroprotective [35, 36] activities.

In the current chapter, we have reviewed the studies that specifically focus on the antiepileptic effect of curcumin. The efficacy of curcumin in different animal models of epilepsy is summarized in Table 21.1.

21.3 Curcumin in Animal Models of Epilepsy

21.3.1 Pentylenetetrazol (PTZ) Kindling Model

Kindling is a chronic animal model of epilepsy that has been extensively studied in order to understand the process and to discover novel compounds for controlling the disease [52]. A subconvulsive dose of PTZ (30 mg/kg) induced kindling in rats. Curcumin showed a dose-dependent antiseizure effect. Administration of curcumin (300 mg/kg) significantly increased the latency to myoclonic jerks, clonic seizures as well as generalized tonic–clonic seizures, improved seizure scores, and reduced the number of myoclonic jerks. PTZ kindling induced significant oxidative stress including an increase in malondialdehyde (MDA) and a reduction in glutathione which was reversed by pretreatment with curcumin in a dose-dependent manner. Curcumin treatment also showed a protective effect against cognitive impairment in a dose-dependent manner [53]. In one study, a liposomal formulation of curcumin (25 and 50 mg/kg) increased the seizure threshold current and the latency to myoclonic and generalized seizures in a current electroshock seizure (ICES) test and PTZ-induced seizures, respectively. Similarly, in a status epilepticus (SE) model in mice, a liposomal formulation of curcumin increased the latency to onset and reduced the duration of seizures in comparison to vehicle- and liposome (placebo)-treated mice [54].

Akula and Kulkarni showed that pretreatment with curcumin (80 mg/kg, p.o.) raised the seizure threshold for the onset of tonic extension by PTZ. Moreover, the results showed that the anticonvulsant effect of curcumin against the PTZ seizure threshold occurred through a direct or indirect involvement of activation of adenosine A1, but not A2A, receptors in mice [55]. The effect of curcumin-loaded chitosan-alginate-sodium tripolyphosphate (STPP) nanoparticles was evaluated in a PTZ-induced kindling model. Administration of curcumin nanoparticles showed anticonvulsant activity and prevented cognitive impairment in fully kindled animals. The level of cell death and glial activation was lower in animals treated with curcumin nanoparticles than in those that received free curcumin [56]. Choudhary et al. showed that curcumin ameliorated seizure severity, depression-like behavior, and learning and memory impairment in PTZ-kindled mice, possibly by central monoaminergic modulation and an inhibitory effect on nitrosative stress and acetylcholinesterase activity [57]. Another study in the kindled model of epilepsy reported that curcumin treatment ameliorated the cognitive deficits with no significant effect on seizure score. Curcumin was able to restore the activity of mitochondrial complexes as well as cause a reduction in reactive oxygen species (ROS) generation, lipid peroxidation and protein carbonyls in PTZ animals, and administration of curcumin restored glutathione levels in PTZ rats. Curcumin protected mitochondria from seizure-induced structural alterations. Moreover, curcumin-supplemented PTZ rats showed a reversal of the morphological deficits and reduced cell loss [58]. In a study conducted by Kaur et al. [59], PTZ-treated animals showed severe generalized tonic–clonic seizures and longer escape latency, suggesting impaired cognitive function. Further, PTZ-treated animals had increased expression of astrocyte [glial fibrillary acid protein (GFAP)] and microglial [ionized calcium-binding adaptor molecule 1 (Iba-1)] activation markers at the mRNA and protein levels. PTZ-treated animals also showed an increase in

Table 21.1 The efficacy of curcumin in animal models of epilepsy

Animal model	Dose	Route	Observation	Reference
ICES test, PTZ model	25 and 50 mg/kg	Oral	– Increase in seizure threshold current and latency to myoclonic and generalized seizures in ICES test and PTZ-induced seizures, respectively – Increase in the latency to onset and reduced duration of seizures in a status epilepticus model in mice	[54]
Li-Pc model	50, 100, and 200 mg/kg	Oral	Ameliorated SE-induced cognitive dysfunction and oxidative injuries in the hippocampus and striatum regions of the brain	[66]
PTZ model	20–120 mg/kg	Oral	Increase in seizure threshold for the onset of tonic extension by PTZ	[55]
ICES test	Acute doses 50, 100, and 200 mg/kg, chronic dose 100 mg/kg	Oral	Increased the seizure threshold with both acute and chronic administration – Anxiogenic effect of curcumin disappeared on chronic administration	[76]
PTZ model	50, 100, and 200 mg/kg	Ip	– Ameliorated seizure severity, depression-like behavior, learning, and memory impairments	[57]
Pilocarpine model	80 mg/kg	Oral	– No seizure manifestations were observed after the treatment of epileptic animals with curcumin – Reduced MDA levels – Decreased the levels of NO – Increased GSH levels – Increased catalase activity – Increased Na^+, K^+- ATPase activity	[62]
PTZ model	12.5 or 25 mg/kg	Ip	– Prevention of cognitive impairment – Reduction of cell death and glial activation in brain tissue	[56]
KA model	100 mg/kg	Ip	– Did not prevent the appearance of spontaneous recurrent seizures – Decrease in the severity of spontaneous recurrent seizures – Improvement in cognitive impairment	[68]
Iron model	500 ppm, 1500 ppm	Curcumin-supplemented standard food pellets	– Prevented the development of higher-grade seizures – Prevented oxidative modification of the PKC regulatory subunit, lipid peroxidation, protein oxidation, Na–K ATPase activity and DPH anisotropy in the cortex	[73]
PTZ model	100 mg/kg	Oral	– Restored the activity of mitochondrial complexes – Reduction in ROS generation, lipid peroxidation, and protein carbonyls – Restored the glutathione levels – Protected mitochondria – Reversed the morphological deficits and reduced cell loss	[58]

(continued)

Table 21.1 (continued)

Animal model	Dose	Route	Observation	Reference
PTZ model	100 mg/kg	Ip	– Did not prevent or delay PTZ-induced seizures – Prevented cognitive deficits – Decreased pro-inflammatory cytokines and a chemokine – Decline of GFAP and Iba-1 expression	[59]
Pilocarpine model	50 mg/kg	Ip	– Activities of AchE and Na+/K + -ATPase were decreased – Prevented oxidative stress – Improved the elevated level of caspase-3 in the hippocampus and cortex – Reversed the activation of AchE – Failed to return Na+/K + -ATPase activity to its normal value	[63]
PTZ model	300 mg/kg	Oral	– Increased the latency to myoclonic jerks, clonic seizures as well as generalized tonic–clonic seizures – Improved the seizure score – Reduced the number of myoclonic jerks	[53]
Pilocarpine model	80 mg/kg	Oral	– Improved pathological abnormalities – Ameliorated most of the changes in amino acid concentrations	[64]
PTZ model	300 mg/kg	Oral	– Enhanced the action of subtherapeutic doses of valproate, phenytoin, phenobarbitone, and carbamazepine	[60]
KA model	50 mg/kg	Ip	– Delayed the onset of KA-induced seizure without affecting the seizure score – Reduced the c-Fos expression in DG, CA1, and CA3 hippocampal regions – Reduced the expression of c-jun, COX-2, BDNF, and iNOS mRNA	[71]

ICES increasing current electroshock seizures, *PTZ* pentylenetetrazole, *Li-Pc* lithium-pilocarpine, *SE* status epilepticus, *AchE* acetylcholinesterase, *KA* kainic acid, *ip* intraperitoneal

mRNA and protein levels of pro-inflammatory cytokines [interleukin (IL)-1β, IL-6, tumor necrosis factor (TNF)-α] and the chemokine monocyte chemoattractant protein (MCP)-1 in the hippocampus and cortex. Immunoreactivity to GFAP and Iba-1 antibodies was also enhanced in the hippocampus and cortex, suggesting gliosis in PTZ-treated animals. Curcumin supplementation (100 mg/kg) administered to PTZ-treated animals did not prevent or delay the PTZ-induced seizures, although treatment with curcumin at the same dosage did prevent cognitive deficits in kindled animals. Moreover, the expression of pro-inflammatory cytokines and the chemokine MCP-1 were significantly decreased in the hippocampus and cortex of PTZ-treated rats supplemented with curcumin. Furthermore, a decrease in GFAP and Iba-1 expression in PTZ animals treated with curcumin was seen. Curcumin also reduced the number of activated glial cells in PTZ-treated animals [59].

The interaction profile of curcumin (300 mg/kg, p.o.) with four antiepileptic drugs (phenytoin, phenobarbitone, valproate, and carbamazepine) was assessed by Reeta et al. [60]. Therapeutic and subtherapeutic doses of antiepileptic drugs were used and curcumin was co-administered. The results showed that curcumin administration along with a subtherapeutic dose of valproate produced 100% protection against PTZ-induced seizures, while valproate alone (150 mg/kg) produced only 50% protection. When curcumin was administered along with subtherapeutic doses of phenytoin (20 mg/kg) and phenobarbitone (20 mg/kg), 83.3% protection was observed in the maximal-electroshock (MES)-induced seizure model, while only 33.3% protection was seen when the drugs were administered alone at subtherapeutic doses. Protection against seizures using a combination of curcumin and carbamazepine (10 mg/kg) was 66.6%. These data suggest that curcumin enhances the action of subtherapeutic doses of valproate, phenytoin, phenobarbitone, and carbamazepine [60].

21.3.2 Pilocarpine Model

Pilocarpine is a cholinergic agonist used as a model to induce epilepsy [61]. Pilocarpine injection led to the induction of oxidative stress in the hippocampus of rats, as indicated by the increased nitric oxide levels and the decrease in glutathione levels and catalase activity. In addition, a decrease in Na^+/K^+-ATPase activity and an increase in acetylcholinesterase activity were observed in the hippocampus in the pilocarpine-treated rats. Curcumin, *Nigella sativa* oil, or valproate administration ameliorated oxidative stress induced by the pilocarpine treatment and restored the Na^+/K^+-ATPase activity in the hippocampus to control values [62].

An increase in lipid peroxidation and a decrease in reduced glutathione and nitric oxide were seen in a rat model of SE induced by pilocarpine. Further, increments in TNF-α and caspase-3 levels were detected in the cortex and hippocampus regions of a rat model of SE. The activities of acetylcholinesterase (AchE) and Na+/K + -ATPase were significantly reduced in the cortex and hippocampus of the rat model of SE. Treatment with curcumin nanoparticles prevented oxidative stress and ameliorated the elevated level of caspase-3 in the hippocampus and cortex. Curcumin showed an anti-inflammatory activity by restoring hippocampal TNF-α to normal levels. Curcumin nanoparticles reversed the activation of AchE, although they failed to return Na+/K + -ATPase activity to its normal value [63].

The effects of curcumin and *Nigella sativa* oil on the pilocarpine-induced alterations in amino acid neurotransmitters and morphometric changes in the hippocampus and cortex of rats were assessed by Noor et al. [64]. A significant increase in aspartate accompanied by a significant decline in glycine and taurine levels were observed in the hippocampus of the pilocarpine-treated animals. Pilocarpine also induced a significant increase in cortical levels of aspartate, glutamate, GABA, glycine, and taurine. Treatment with curcumin and valproate ameliorated the increase in amino acid concentrations and also improved the pathological abnormalities induced by pilocarpine, while *Nigella sativa* oil was not effective in improving the pilocarpine-induced abnormalities [64]. The antiseizure effect of curcumin was also studied by Du et al. in pilocarpine-induced seizures in rats [65]. The authors observed a potent antioxidant activity and significant anticonvulsant activity at the doses of 100 and 300 mg/kg of curcumin.

Ahmad [66] demonstrated that administration of curcumin in a dose-dependent manner showed antiepileptic activity against lithium–pilocarpine (Li–Pc) induced SE in young rats, and also decreased the intensity and frequency of seizure, peripheral cholinergic signs (PCS), and stereotyped movement (STM) episodes. Analysis of the probe test in the Morris water-maze showed that rats that had undergone Li–Pc treatment spent less time in the target quadrant than in the control group. Pretreatment of Li–Pc treatment rats with curcumin had an attenuating effect and the rats spent more time in the target quadrant. Furthermore, curcumin-treated animals showed a significant and dose-dependent attenuation of the Li–Pc induced increase in thiobarbituric reactive substances (TBARS) in the hippocampus and striatum when compared with the Li–Pc SE group. Determination of reduced glutathione (GSH) as another marker of oxidative stress in hippocampal and striatal tissue revealed that rats treated with Li–Pc showed a significant depletion of GSH, and pretreatment with curcumin significantly, and in a dose-dependent manner, attenuated this effect in comparison to the Li–Pc (SE) group [66].

21.3.3 Kainic Acid Model

The kainic acid model of temporal lobe epilepsy is widely used to understand the molecular, cellular, and pharmacological mechanisms of epilepsy in animal models [67]. A study was conducted to assess the protective effect of continued administration of curcumin during the latent period in adult rats in a post-SE model of temporal lobe epilepsy induced by kainic acid. The results showed that, while the administration of curcumin during the latent period does not prevent the occurrence of spontaneous recurrent seizures after SE, it can be effective in decreasing the severity of spontaneous recurrent seizures and improving cognitive impairment. Moreover, the administration of curcumin after status epilepticus significantly decreased the level of IL-1β and TNF-α, attenuated the activation of astrocytes in the rat hippocampus, and led to less neuronal loss in both the dentate hilus (DH) and Cornu Ammonis (CA)-3 regions [68].

Kainic acid induces long-lasting seizures in rodents that mimic the SE condition in humans, which is a seizure lasting for more than 5 min and poses an emergency situation of brain damage and cognitive impairment. Gupta et al. studied the effect of curcumin administration 30 min before kainic acid challenge in rats. They observed a protective effect of curcumin at doses of 100 and 200 mg/kg [intraperitoneal (i.p.)] in reducing the incidence of convulsions and raising the latency to onset of seizures. Oxidative stress induced by long-lasting seizures could be reversed by curcumin administration at higher doses. The 50 mg/kg dose was not effective in this model of epilepsy induced by kainic acid [69]. In another study, the effect of manganese complexes of curcumin (Cp-Mn) and diacetylcurcumin (DiAc-Cp-Mn) was investigated with regard to the kainic acid (10 mg/kg, i.p.) and L-arginine (L-Arg) (1000 mg/kg, i.p.)-induced enhanced nitric oxide levels in rats. The manganese complexes of curcumin (Cp-Mn) and diacetylcurcumin (DiAc-Cp-Mn) (50 mg/kg, i.p.) significantly suppressed the effects of kainic acid and L-Arg without affecting the basal nitric oxide concentration. Kainic acid-induced seizures resulted in severe neuronal cell damage in the CA1 and CA3 subfields of the hippocampus. Both Cp-Mn and DiAc-Cp-Mn prevented KA-induced neuronal cell damage in both CA1 and CA3 regions of the rat hippocampus when compared with the vehicle control. The Cp-Mn and DiAc-Cp-Mn showed greater neuroprotective effect than their parent compounds, curcumin and diacetylcurcumin [70].

In a similar study, the mechanism underlying the protective effects of Cp-Mn and DiAc-Cp-Mn was evaluated on kainic acid- induced excitotoxicity (at 10 mg/kg i.p.) in the rat hippocampus. Kainic acid injection induced seizures and increased the expression of neurotoxic markers, including c-jun, cyclooxygenase 2 (COX-2), brain-derived neurotrophic factor (BDNF), heat shock protein 70 (hsp70), and inducible nitric oxide synthase (iNOS), measured at 6 and 72 h after injection. Pretreatment with Cp-Mn (50 mg/kg, i.p.) and DiAc-Cp-Mn (50 mg/kg, i.p.) delayed the onset of kainic acid-induced seizure without affecting the seizure score. The expression of c-Fos c-jun, COX-2, BDNF, and iNOS mRNA was suppressed by Cp-Mn and DiAc-Cp-Mn, whereas curcumin attenuated only iNOS mRNA expression. None of the compounds tested had an effect on kainic acid-induced HSP70 expression [71].

21.3.4 Iron Model

Iron-induced epilepsy is a well-known model of human posttraumatic epilepsy (PTE), first proposed by Willmore et al. [72]. Curcumin treatment at high dose (1500 ppm) significantly inhibited the onset of grade III and IV seizures in rats with iron-induced epilepsy. Although treatment with a low dose of curcumin (500 ppm) was ineffective in inhibiting grade III seizures, it did delay the onset and progression of generalized seizures. Furthermore, curcumin treatment prevented oxidative modification of the protein kinase C (PKC) regulatory subunit, lipid peroxidation, protein oxidation, Na^+/K^+ ATPase activity, and diphenylhexatriene (DPH) anisotropy in the cortex of epileptic rats [73].

21.3.5 Other Studies

Phenytoin (75 mg/kg, i.p.) produces significant deficits in learning and memory, as indicated by an increase in retention transfer latency in the elevated plus maze test and a significant reduction in retention latency in the passive avoidance paradigm. In the brain, chronic administration of phenytoin leads to a significant elevation in MDA and a reduction in GSH levels. Administration of curcumin to phenytoin-treated rats was found to reverse cognitive impairment and oxidative stress in a dose-dependent manner [74].

The effect of curcumin on cognitive impairment in addition to oxidative stress induced by phenobarbitone and carbamazepine was also studied [75]. The administration of phenobarbitone and carbamazepine for 21 days resulted in a marked impairment of learning and memory, as well as changes in oxidative stress markers including enhanced MDA and decreased GSH levels. Concomitant curcumin administration was effective in preventing the cognitive impairment and reduced the increased oxidative stress markers induced by phenobarbitone and carbamazepine. Curcumin co-administration did not cause any significant alteration in the serum concentrations of these antiepileptic drugs [75].

Curcumin treatment of animals significantly increased the seizure threshold in the ICES test, using both acute and chronic administration. In the chronic study, curcumin-treated groups showed significant protection against post-ictal mortality [76]. Furthermore, acute administration produced an anxiogenic effect in the elevated plus maze and actophotometer tests. However, with long-term treatment, this anxiogenic effect disappeared.

In a study designed to investigate whether mammalian target of rapamycin (mTOR) inhibition has antiepileptogenic and/or antiseizure effects in the electrogenic post-SE rat model, the results showed that rapamycin did not prevent epilepsy but did have seizure-suppressing properties as long as the rapamycin blood levels were sufficiently high. As an alternative mTOR inhibitor, curcumin had no effect on chronic seizures, possibly because brain curcumin was not present at adequate levels. This result also suggests that the reported beneficial effects of oral curcumin on seizure activity are model-dependent and might not apply to chronic epilepsy models [77].

21.4 Conclusions

Epilepsy is a common central nervous system (CNS) disease that is caused by an imbalance in the excitation and inhibition signals in the brain. There are numerous drugs available to treat epilepsy, although the adverse effects of these drugs limit their usefulness and acceptance. The use of medicinal plants has gained popularity in recent years for the prevention and treatment of numerous diseases. Curcumin is one such medicinal plant that has enormous potential in the treatment of a wide range of diseases such as epilepsy. The results from the present review article suggest that many pathways are involved in the antiepileptic effects of curcumin, including antioxidant and anti-inflammatory mechanisms. Therefore, further biochemical and molecular studies focusing on the mechanism of its antiepileptic effects are warranted. Additionally, there is a need for well-designed clinical trials to evaluate the potential of curcumin for both the prevention and therapy of epilepsy.

References

1. Murray CJ, Vos T, Lozano R, Naghavi M, Flaxman AD, Michaud C et al (2012) Disability-adjusted life years (DALYs) for 291 diseases and injuries in 21 regions, 1990–2010: a systematic analysis for the Global Burden of Disease Study 2010. Lancet 380(9859):2197–2223
2. Fiest KM, Sauro KM, Wiebe S, Patten SB, Kwon CS, Dykeman J et al (2017) Prevalence and incidence of epilepsy: a systematic review and meta-analysis of international studies. Neurology 88(3):296–303
3. McNamara JO (1999) Emerging insights into the genesis of epilepsy. Nature 399(6738 Suppl): A15–A22
4. Bialer M, White HS (2010) Key factors in the discovery and development of new antiepileptic drugs. Nat Rev Drug Discov 9(1):68–82
5. Simonato M (2018) Epilepsy an update on disease mechanisms: the potential role of microRNAs. Front Neurol 9:176. https://doi.org/10.3389/fneur.2018.00176
6. Löscher W, Klitgaard H, Twyman RE, Schmidt D (2013) New avenues for anti-epileptic drug discovery and development. Nat Rev Drug Discov 12(10):757–776

7. Li JW-H, Vederas JC (2009) Drug discovery and natural products: end of an era or an endless frontier? Science 325(5937):161–165
8. Forouzanfar F, Bazzaz BSF, Hosseinzadeh H (2014) Black cumin (Nigella sativa) and its constituent (thymoquinone): a review on antimicrobial effects. Iran J Basic Med Sci 17(12):929–938
9. Forouzanfar F, Afkhami Goli A, Asadpour E, Ghorbani A, Sadeghnia HR (2013) Protective effect of Punica granatum L. against serum/glucose deprivation-induced PC12 cells injury. Evid Based Complement Alternat Med 2013:716730. https://doi.org/10.1155/2013/716730
10. Hosseini A, Forouzanfar F, Rakhshandeh H (2016) Hypnotic effect of nepeta glomerulosa on pentobarbital-induced sleep in mice. Jundishapur J Nat Pharm Prod 11(1). https://doi.org/10.17795/jjnpp-25063
11. Wang X, Zhang A, Zhou X, Liu Q, Nan Y, Guan Y et al (2016) An integrated chinmedomics strategy for discovery of effective constituents from traditional herbal medicine. Sci Rep 6:18997. https://doi.org/10.1038/srep18997
12. Sadeghnia HR, Shaterzadeh H, Forouzanfar F, Hosseinzadeh H (2017) Neuroprotective effect of safranal, an active ingredient of Crocus sativus, in a rat model of transient cerebral ischemia. Folia Neuropathol 55(3):206–213
13. Ekor M (2014) The growing use of herbal medicines: issues relating to adverse reactions and challenges in monitoring safety. Front Pharmacol 4:177. https://doi.org/10.3389/fphar.2013.00177
14. Forouzanfar F, Hosseinzadeh H, Ebrahimzadeh Bideskan A, Sadeghnia HR (2016) Aqueous and ethanolic extracts of Boswellia serrata protect against focal cerebral ischemia and reperfusion injury in rats. Phytother Res 30(12):1954–1967
15. Forouzanfar F, Hosseinzadeh H (2018) Medicinal herbs in the treatment of neuropathic pain: a review. Iran J Basic Med Sci 21(4):347–358
16. Li S, Yuan W, Deng G, Wang P, Yang P, Aggarwal B (2011) Chemical composition and product quality control of turmeric (Curcuma longa L.). Pharm Crops 2:28–54
17. Gupta SC, Patchva S, Aggarwal BB (2013) Therapeutic roles of curcumin: lessons learned from clinical trials. AAPS J 15(1):195–218
18. Prasad S, Tyagi AK, Aggarwal BB (2014) Recent developments in delivery, bioavailability, absorption and metabolism of curcumin: the golden pigment from golden spice. Cancer Res Treat 46(1):2–18
19. Gupta SC, Patchva S, Koh W, Aggarwal BB (2012) Discovery of curcumin, a component of golden spice, and its miraculous biological activities. Clin Exp Pharmacol Physiol 39(3):283–299
20. Prasad S, Gupta SC, Tyagi AK, Aggarwal BB (2014) Curcumin, a component of golden spice: from bedside to bench and back. Biotechnol Adv 32(6):1053–1064
21. Sahebkar A, Serban MC, Ursoniu S, Banach M (2015) Effect of curcuminoids on oxidative stress: a systematic review and meta-analysis of randomized controlled trials. J Funct Foods 18:898–909
22. Abdollahi E, Momtazi AA, Johnston TP, Sahebkar A (2018) Therapeutic effects of curcumin in inflammatory and immune-mediated diseases: a nature-made jack-of-all-trades? J Cell Physiol 233(2):830–848
23. Panahi Y, Hosseini MS, Khalili N, Naimi E, Majeed M, Sahebkar A (2015) Antioxidant and anti-inflammatory effects of curcuminoid-piperine combination in subjects with metabolic syndrome: a randomized controlled trial and an updated meta-analysis. Clin Nutr 34(6):1101–1108
24. Saberi-Karimian M, Keshvari M, Ghayour-Mobarhan M, Salehizadeh L, Rahmani S, Behnam B, Jamialahmadi T, Asgary S, Sahebkar A. Effects of curcuminoids on inflammatory status in patients with non-alcoholic fatty liver disease: A randomized controlled trial. Complement Ther Med. 2020 Mar;49:102322. https://doi.org/10.1016/j.ctim.2020.102322. Epub 2020 Jan 21. PMID: 32147075.
25. Hasanzadeh S, Read MI, Bland AR, Majeed M, Jamialahmadi T, Sahebkar A. Curcumin: an inflammasome silencer. Pharmacol Res. 2020 Sep;159:104921. doi: 10.1016/j.phrs.2020.104921. Epub 2020 May 25. PMID: 32464325.
26. Abdollahi E, Momtazi AA, Johnston TP, Sahebkar A (2018) Therapeutic effects of curcumin in inflammatory and immunemediated diseases: a nature-made jack-of-all-trades? J Cell Physiol 233(2):830–848
27. Luis E. Simental-Mendía, Matteo Pirro, Antonio M. Gotto Jr., Maciej Banach, Stephen L. Atkin, Muhammed Majeed & Amirhossein Sahebkar (2019) Lipid-modifying activity of curcuminoids: A systematic review and meta-analysis of randomized controlled trials, Critical Reviews in Food Science and Nutrition, 59:7, 1178-1187, https://doi.org/10.1080/10408398.2017.1396201
28. Sahebkar A (2014) Are curcuminoids effective C-reactive protein-lowering agents in clinical practice? Evidence from a meta-analysis. Phytother Res 28(5):633–642
29. De R, Kundu P, Swarnakar S, Ramamurthy T, Chowdhury A, Nair GB, Mukhopadhyay AK (2009) Antimicrobial activity of curcumin against Helicobacter pylori isolates from India and during infections in mice. Antimicrob Agents Chemother 53(4):1592–1597
30. Karimian MS, Pirro M, Majeed M, Sahebkar A (2017) Curcumin as a natural regulator of monocyte chemoattractant protein-1. Cytokine Growth Factor Rev 33:55–63
31. Teymouri M, Pirro M, Johnston TP, Sahebkar A (2017) Curcumin as a multifaceted compound against human papilloma virus infection and cervical cancers: a review of chemistry, cellular, molecular, and preclinical features. BioFactors 43(3):331–346
32. Panahi Y, Kianpour P, Mohtashami R, Jafari R, Simental-Mendía LE, Sahebkar A (2016) Curcumin lowers serum lipids and uric acid in subjects with non-alcoholic fatty liver disease: a randomized controlled trial. J Cardiovasc Pharmacol 68(3):223–229

33. Panahi Y, Hosseini MS, Khalili N, Naimi E, Simental-Mendía LE, Majeed M et al (2016) Effects of curcumin on serum cytokine concentrations in subjects with metabolic syndrome: a post-hoc analysis of a randomized controlled trial. Biomed Pharmacother 82:578–582
34. Sahebkar A, Cicero AF, Simental-Mendía LE, Aggarwal BB, Gupta SC (2016) Curcumin downregulates human tumor necrosis factor-α levels: a systematic review and meta-analysis of randomized controlled trials. Pharmacol Res 107:234–242
35. Forouzanfar F, Read MI, Barreto GE, Sahebkar A. Neuroprotective effects of curcumin through autophagy modulation. IUBMB Life. 2020 Apr;72(4):652-664. doi: 10.1002/iub.2209. Epub 2019 Dec 5. PMID: 31804772.
36. Bagheri, H., Ghasemi, F., Barreto, G.E., Rafiee, R., Sathyapalan, T., Sahebkar, A. Effects of curcumin on mitochondria in neurodegenerative diseases (2020) BioFactors, 46 (1), pp. 5-20.
37. Cicero AF, Colletti A, Bajraktari G, Descamps O, Djuric DM, Ezhov M et al (2017) Lipid-lowering nutraceuticals in clinical practice: position paper from an International Lipid Expert Panel. Nutr Rev 75(9):731–767
38. Momtazi AA, Banach M, Pirro M, Katsiki N, Sahebkar A (2017) Regulation of PCSK9 by nutraceuticals. Pharmacol Res 120:157–169
39. Panahi Y, Khalili N, Sahebi E, Namazi S, Reiner Ž, Majeed M et al (2017) Curcuminoids modify lipid profile in type 2 diabetes mellitus: a randomized controlled trial. Complement Ther Med 33:1–5
40. Mirzaei H, Masoudifar A, Sahebkar A, Zare N, Nahand JS, Rashidi B et al (2017) MicroRNA: a novel target of curcumin in cancer therapy. J Cell Physiol 233(4):3004–3015
41. Momtazi, A.A., Derosa, G., Maffioli, P., Banach, M., Sahebkar, A. Role of microRNAs in the Therapeutic Effects of Curcumin in Non-Cancer Diseases (2016) Molecular Diagnosis and Therapy, 20 (4), pp. 335-345.
42. Mohajeri M, Bianconi V, Ávila-Rodriguez MF, Barreto GE, Jamialahmadi T, Pirro M, Sahebkar A. Curcumin: a phytochemical modulator of estrogens and androgens in tumors of the reproductive system. Pharmacol Res. 2020 Jun;156:104765. doi: 10.1016/j.phrs.2020.104765. Epub 2020 Mar 23. PMID: 32217147.
43. Abbas Momtazi A, Sahebkar A (2016) Difluorinated curcumin: a promising curcumin analogue with improved anti-tumor activity and pharmacokinetic profile. Curr Pharm Des 22(28):4386–4397
44. Rezaee R, Momtazi AA, Monemi A, Sahebkar A (2017) Curcumin: a potentially powerful tool to reverse cisplatin-induced toxicity. Pharmacol Res 117:218–227
45. Mohajeri M, Sahebkar A (2018) Protective effects of curcumin against doxorubicin-induced toxicity and resistance: a review. Crit Rev Oncol Hematol 122:30–51
46. Sahebkar A, Henrotin Y (2015) Analgesic efficacy and safety of curcuminoids in clinical practice: a systematic review and meta-analysis of randomized controlled trials. Pain Med 17(6):1192–1202
47. Sahebkar A, Saboni N, Pirro M, Banach M (2017) Curcumin: an effective adjunct in patients with statin-associated muscle symptoms? J Cachexia Sarcopenia Muscle 8(1):19–24
48. Keihanian F, Saeidinia A, Bagheri RK, Johnston TP, Sahebkar A (2017) Curcumin, hemostasis, thrombosis and coagulation. J Cell Physiol 233(6):4497–4511
49. Sahebkar A (2010) Molecular mechanisms for curcumin benefits against ischemic injury. Fertil Steril 94(5):e75–ee6
50. Teymouri M, Pirro M, Johnston TP, Sahebkar A (2017) Curcumin as a multifaceted compound against human papilloma virus infection and cervical cancers: a review of chemistry, cellular, molecular, and preclinical features. BioFactors 43(3):331–346
51. Iranshahi M, Sahebkar A, Takasaki M, Konoshima T, Tokuda H. Cancer chemopreventive activity of the prenylated coumarin, umbelliprenin, in vivo. Eur J Cancer Prev. 2009 Sep;18(5):412-5. doi: 10.1097/CEJ.0b013e32832c389e. PMID: 19531956.
52. Mazzolani F, Togni S (2013) Oral administration of a curcumin-phospholipid delivery system for the treatment of central serous chorioretinopathy: a 12-month follow-up study. Clin Opthalmol 7:939–945
53. Allegri P, Mastromarino A, Neri P (2010) Management of chronic anterior uveitis relapses: efficacy of oral phospholipidic curcumin treatment. Long-term follow-up. Clin Ophthalmol 4:1201–1206
54. Lelli D, Sahebkar A, Johnston TP, Pedone C (2017) Curcumin use in pulmonary diseases: state of the art and future perspectives. Pharmacol Res 115:133–148
55. Esmaily H, Sahebkar A, Iranshahi M, Ganjali S, Mohammadi A, Ferns G et al (2015) An investigation of the effects of curcumin on anxiety and depression in obese individuals: a randomized controlled trial. Chin J Integr Med 21(5):332–338
56. Panahi Y, Badeli R, Karami GR, Sahebkar A (2015) Investigation of the efficacy of adjunctive therapy with bioavailability-boosted curcuminoids in major depressive disorder. Phytother Res 29(1):17–21
57. Rahmani S, Asgary S, Askari G, Keshvari M, Hatamipour M, Feizi A et al (2016) Treatment of non-alcoholic fatty liver disease with curcumin: a randomized placebo-controlled trial. Phytother Res 30(9):1540–1548
58. Amel Zabihi N, Pirro MP, Johnston T, Sahebkar A (2017) Is there a role for curcumin supplementation in the treatment of non-alcoholic fatty liver disease? The data suggest yes. Curr Pharm Des 23(7):969–982
59. Dhir A (2012) Pentylenetetrazol (PTZ) kindling model of epilepsy. Curr Protoc Neurosci Unit 9:37. https://doi.org/10.1002/0471142301.ns0937s58
60. Mehla J, Reeta K, Gupta P, Gupta YK (2010) Protective effect of curcumin against seizures and

cognitive impairment in a pentylenetetrazole-kindled epileptic rat model. Life Sci 87(19–22):596–603
61. Agarwal NB, Jain S, Nagpal D, Agarwal NK, Mediratta PK, Sharma KK (2013) Liposomal formulation of curcumin attenuates seizures in different experimental models of epilepsy in mice. Fundam Clin Pharmacol 27(2):169–172
62. Akula KK, Kulkarni S (2014) Effect of curcumin against pentylenetetrazol-induced seizure threshold in mice: possible involvement of adenosine A1 receptors. Phytother Res 28(5):714–721
63. Hashemian M, Anissian D, Ghasemi-Kasman M, Akbari A, Khalili-Fomeshi M, Ghasemi S et al (2017) Curcumin-loaded chitosan-alginate-STPP nanoparticles ameliorate memory deficits and reduce glial activation in pentylenetetrazol-induced kindling model of epilepsy. Prog Neuro-Psychopharmacol Biol Psychiatry 79(Pt B):462–471
64. Choudhary KM, Mishra A, Poroikov VV, Goel RK (2013) Ameliorative effect of Curcumin on seizure severity, depression like behavior, learning and memory deficit in post-pentylenetetrazole-kindled mice. Eur J Pharmacol 704(1–3):33–40
65. Kaur H, Bal A, Sandhir R (2014) Curcumin supplementation improves mitochondrial and behavioral deficits in experimental model of chronic epilepsy. Pharmacol Biochem Behav 125:55–64
66. Kaur H, Patro I, Tikoo K, Sandhir R (2015) Curcumin attenuates inflammatory response and cognitive deficits in experimental model of chronic epilepsy. Neurochem Int 89:40–50
67. Reeta K, Mehla J, Pahuja M, Gupta YK (2011) Pharmacokinetic and pharmacodynamic interactions of valproate, phenytoin, phenobarbitone and carbamazepine with curcumin in experimental models of epilepsy in rats. Pharmacol Biochem Behav 99(3):399–407
68. Cavalheiro EA (1995) The pilocarpine model of epilepsy. Ital J Neurol Sci 16(1–2):33–37
69. Ezz HSA, Khadrawy YA, Noor NA (2011) The neuroprotective effect of curcumin and Nigella sativa oil against oxidative stress in the pilocarpine model of epilepsy: a comparison with valproate. Neurochem Res 36(11):2195–2204
70. Khadrawy YA, Sawie HG, Hosny EN (2018) Neuroprotective effect of curcumin nanoparticles against rat model of status epilepticus induced by pilocarpine. J Complement Integr Med 15(3). https://doi.org/10.1515/jcim-2017-0117
71. Noor NA, Ezz HSA, Faraag AR, Khadrawy YA (2012) Evaluation of the antiepileptic effect of curcumin and Nigella sativa oil in the pilocarpine model of epilepsy in comparison with valproate. Epilepsy Behav 24(2):199–206
72. Du P, Tang HY, Li X, Lin HJ, Peng WF, Ma Y et al (2012) Anticonvulsive and antioxidant effects of curcumin on pilocarpine-induced seizures in rats. Chin Med J 125(11):1975–1979
73. Ahmad M (2013) Protective effects of curcumin against lithium–pilocarpine induced status epilepticus, cognitive dysfunction and oxidative stress in young rats. Saudi J Biol Sci 20(2):155–162
74. Lévesque M, Avoli M (2013) The kainic acid model of temporal lobe epilepsy. Neurosci Biobehav Rev 37(10 Pt 2):2887–2899
75. Jiang Z, Guo M, Shi C, Wang H, Yao L, Liu L et al (2015) Protection against cognitive impairment and modification of epileptogenesis with curcumin in a post-status epilepticus model of temporal lobe epilepsy. Neuroscience 310:362–371
76. Gupta YK, Briyal S, Sharma M (2009) Protective effect of curcumin against kainic acid induced seizures and oxidative stress in rats. Indian J Physiol Pharmacol 53(1):39–46
77. Sumanont Y, Murakami Y, Tohda M, Vajragupta O, Watanabe H, Matsumoto K (2006) Prevention of kainic acid-induced changes in nitric oxide level and neuronal cell damage in the rat hippocampus by manganese complexes of curcumin and diacetylcurcumin. Life Sci 78(16):1884–1891
78. Sumanont Y, Murakami Y, Tohda M, Vajragupta O, Watanabe H, Matsumoto K (2007) Effects of manganese complexes of curcumin and diacetylcurcumin on kainic acid-induced neurotoxic responses in the rat hippocampus. Biol Pharm Bull 30(9):1732–1739
79. Willmore LJ, Sypert GW, Munson JB (1978) Recurrent seizures induced by cortical iron injection: a model of posttraumatic epilepsy. Ann Neurol 4(4):329–336
80. Jyoti A, Sethi P, Sharma D (2009) Curcumin protects against electrobehavioral progression of seizures in the iron-induced experimental model of epileptogenesis. Epilepsy Behav 14(2):300–308
81. Reeta K, Mehla J, Gupta YK (2009) Curcumin is protective against phenytoin-induced cognitive impairment and oxidative stress in rats. Brain Res 1301:52–60
82. Reeta K, Mehla J, Gupta YK (2010) Curcumin ameliorates cognitive dysfunction and oxidative damage in phenobarbitone and carbamazepine administered rats. Eur J Pharmacol 644(1–3):106–112
83. Bharal N, Sahaya K, Jain S, Mediratta PK, Sharma KK (2008) Curcumin has anticonvulsant activity on increasing current electroshock seizures in mice. Phytother Res 22(12):1660–1664
84. Drion CM, Borm LE, Kooijman L, Aronica E, Wadman WJ, Hartog AF et al (2016) Effects of rapamycin and curcumin treatment on the development of epilepsy after electrically induced status epilepticus in rats. Epilepsia 57(5):688–697

Index

P. C. Guest (ed.), *Studies on Biomarkers and New Targets in Aging Research in Iran*, Advances in Experimental Medicine and Biology 1291, https://doi.org/10.1007/978-3-030-56153-6

D

The manufacturer's authorised representative in the EU is Springer Nature Customer Service Centre GmbH, Europaplatz 3, 69115 Heidelberg, Germany. If you have any concerns regarding our products, please contact ProductSafety@springernature.com

Printed and bound by CPI Group (UK) Ltd, Croydon, CR0 4YY
30/09/2026
02225694-0020